Electroanalysis at the Nanoscale

Durham University, United Kingdom
1–3 July 2013

FARADAY DISCUSSIONS
Volume 164, 2013

RSC Publishing

The Faraday Division of the Royal Society of Chemistry, previously the Faraday Society, founded in 1903 to promote the study of sciences lying between Chemistry, Physics and Biology.

EDITORIAL STAFF

Editor
Philip Earis

Deputy editor
Heather Montgomery

Development editor
Rowan Frame

Senior publishing editor
Susan Weatherby

Publishing editors
Sarah Farley, Michael Spencelayh

Publishing assistants
Victoria Bache, Sian Gordon, Ruba Miah

Publisher
Niamh O'Connor

Faraday Discussions (Print ISSN 1359-6640, Electronic ISSN 1364-5498) is published 8 times a year by the Royal Society of Chemistry, Thomas Graham House, Science Park, Milton Road, Cambridge, UK CB4 0WF. Volume 163 ISBN-13: 978-1-84973-691-6

2013 annual subscription price: print+electronic £765, US $1428; electronic only £727, US $1356. Customers in Canada will be subject to a surcharge to cover GST. Customers in the EU subscribing to the electronic version only will be charged VAT. All orders, with cheques made payable to the Royal Society of Chemistry, should be sent to RSC Order Department, Royal Society of Chemistry, Thomas Graham House, Science Park, Milton Road, Cambridge, CB4 0WF, UK. Tel +44 (0) 1223 432398; E-mail orders@rsc.org

If you take an institutional subscription to any RSC journal you are entitled to free, site-wide web access to that journal. You can arrange access *via* Internet Protocol (IP) address at www.rsc.org/ip. Customers should make payments by cheque in sterling payable on a UK clearing bank or in US dollars payable on a US clearing bank.

PRINTED IN THE UK

Faraday Discussions documents a long-established series of *Faraday Discussion* meetings which provide a unique international forum for the exchange of views and newly acquired results in developing areas of physical chemistry, biophysical chemistry and chemical physics.

SCIENTIFIC COMMITTEE, Volume 164

Chair
Professor Richard Compton (University of Oxford, UK)

Professor Damien Arrigan (Curtin University, Perth, Australia)
Professor Craig Banks (Manchester Metropolitan University, UK)
Professor Peter Fielden (University of Manchester, UK)
Dr Ritu Kataky (Durham University, UK)
Dr Jay Wadhawan (University of Hull, UK)

FARADAY STANDING COMMITTEE ON CONFERENCES

Chair
A Mount (Edinburgh, UK)

W A Brown (UCL, UK)
I Hamley (Reading, UK)
J Hirst (Nottingham, UK)
G Hutchings (Cardiff, UK)
C Percival (Manchester, UK)

© The Royal Society of Chemistry 2013. Apart from fair dealing for the purposes of research or private study, or criticism or review, as permitted under the Copyright, Designs and Patents Act 1988 and Related Rights Regulations 2003, this publication may only be reproduced, stored or transmitted, in any form or by any means, with the prior permission in writing of the Publishers or in the case of reprographic reproduction in accordance with the terms of licences issued by the Copyright Licensing Agency in the UK. US copyright law applicable to users in the USA. The Royal Society of Chemistry takes reasonable care in the preparation of this publication but does not accept liability for the consequences of any errors or omissions.

Royal Society of Chemistry:
Registered Charity No. 207890.

⊗The paper used in this publication meets the requirements of ANSI/NISO Z39.48-1992 (Permanence of Paper).

Electroanalysis at the Nanoscale

Faraday Discussions

www.rsc.org/faraday_d

A General Discussion on Electroanalysis at the Nanoscale was held in Durham, UK on the 1st, 2nd and 3rd of July 2013.

RSC Publishing is a not-for-profit publisher and a division of the Royal Society of Chemistry. Any surplus made is used to support charitable activities aimed at advancing the chemical sciences. Full details are available from www.rsc.org

CONTENTS

ISSN 1359-6640; ISBN 978-1-84973-691-6

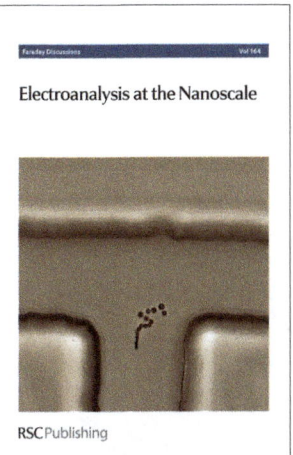

Cover
See Wang,
Faraday Discuss., 2013, **164**, 9.

Electrochemically-prepared antibody-functionalized catalytic micromotors for capturing and transporting target biomolecules in lab-on-a-chip devices.

Image reproduced by permission of Professor Joseph Wang from *Faraday Discuss.*, 2013, **164**, 9.

INTRODUCTORY LECTURE

9   **Template electrodeposition of catalytic nanomotors**
Joseph Wang

PAPERS AND DISCUSSIONS

19   **Microelectrochemical visualization of oxygen consumption of single living cells**
Michaela Nebel, Stefanie Grützke, Nizam Diab, Albert Schulte and Wolfgang Schuhmann

33   **Vesicular release of neurotransmitters: converting amperometric measurements into size, dynamics and energetics of initial fusion pores**
Alexander Oleinick, Frédéric Lemaître, Manon Guille Collignon, Irina Svir and Christian Amatore

57   **Potential-dependent single molecule blinking dynamics for flavin adenine dinucleotide covalently immobilized in zero-mode waveguide array of working electrodes**
Jing Zhao, Lawrence P. Zaino III and Paul W. Bohn

CO-SPONSORS

IJ CAMBRIA SCIENTIFIC

71	**Networks of DNA-templated palladium nanowires: structural and electrical characterisation and their use as hydrogen gas sensors**
Mariam N. Al-Hinai, Reda Hassanien, Nicholas G. Wright, Alton B. Horsfall, Andrew Houlton and Benjamin R. Horrocks
93	**General discussion**
107	**Anodic TiO_2 nanotubes: double walled *vs.* single walled**
Ning Liu, Hamed Mirabolghasemi, Kiyoung Lee, Sergiu P. Albu, Alexei Tighineanu, Marco Altomare and Patrik Schmuki
117	**The simplest model of charge storage in single file metallic nanopores**
Alexei A. Kornyshev
135	**Carbon nanotube based electrochemical sensor for the sensitive detection of valacyclovir**
Badal Shah, Todd Lafleur and Aicheng Chen
147	**Electroanalysis using modified hierarchical nanoporous carbon materials**
Rusbel Coneo Rodriguez, Angelica Baena Moncada, Diego F. Acevedo, Gabriel A. Planes, Maria C. Miras and Cesar A. Barbero
175	**Pd@Au core–shell nanocrystals with concave cubic shapes: kinetically controlled synthesis and electrocatalytic properties**
Ling Zhang, Wenxin Niu, Jianming Zhao, Shuyun Zhu, Yali Yuan, Tao Yuan, Lianzhe Hu and Guobao Xu
189	**Electrochemical mechanical micromachining based on confined etchant layer technique**
Ye Yuan, Lianhuan Han, Jie Zhang, Jingchun Jia, Xuesen Zhao, Yongzhi Cao, Zhenjiang Hu, Yongda Yan, Shen Dong, Zhong-Qun Tian, Zhao-Wu Tian and Dongping Zhan
199	**Decoration of active sites to create bimetallic surfaces and its implication for electrochemical processes**
Blake J. Plowman, Ilija Najdovski, Andrew Pearson and Anthony P. O'Mullane
219	**General discussion**
241	**Mapping fluxes of radicals from the combination of electrochemical activation and optical microscopy**
Sorin Munteanu, Jean Paul Roger, Yasmina Fedala, Fabien Amiot, Catherine Combellas, Gilles Tessier and Frédéric Kanoufi
259	**Electrochemically assisted self-assembly of ordered and functionalized mesoporous silica films: impact of the electrode geometry and size on film formation and properties**
Grégoire Herzog, Emilie Sibottier, Mathieu Etienne and Alain Walcarius
275	**Metallic impurities availability in reduced graphene is greatly enhanced by its ultrasonication**
Rou Jun Toh and Martin Pumera
283	**Highly sensitive detection of nitroaromatic explosives at discrete nanowire arrays**
Sean Barry, Karen Dawson, Elon Correa, Royston Goodacre and Alan O'Riordan
295	**A systematic study of the influence of nanoelectrode dimensions on electrode performance and the implications for electroanalysis and sensing**
Ilka Schmueser, Anthony J. Walton, Jonathan G. Terry, Helena L. Woodvine, Neville J. Freeman and Andrew R. Mount

315 **General discussion**

339 **Double layer effects at nanosized electrodes**[†]
Andreas Bund and Clemens Kubeil

349 **Pulse electroanalysis at gold–gold micro-trench electrodes: Chemical signal filtering**
Sara E. C. Dale and Frank Marken

361 **Effects of adsorption and confinement on nanoporous electrochemistry**
Je Hyun Bae, Ji-Hyung Han, Donghyeop Han and Taek Dong Chung

377 **Gold nanowire electrodes in array: simulation study and experiments**
Amélie Wahl, Karen Dawson, John MacHale, Seán Barry, Aidan J. Quinn and Alan O'Riordan

391 **Nanoscale control of interfacial processes for latent fingerprint enhancement**
Rachel M. Sapstead (nee Brown), Karl S. Ryder, Claire Fullarton, Maximilian Skoda, Robert M. Dalgliesh, Erik B. Watkins, Charlotte Beebee, Robert Barker, Andrew Glidle and A. Robert Hillman

411 **General discussion**

CONCLUDING REMARKS

437 **Closing remarks: looking back and ahead at `nano' electroanalytical chemistry**
David E. Williams

ADDITIONAL INFORMATION

441 **Poster titles**
443 **List of participants**
445 **Index of contributors**

Template electrodeposition of catalytic nanomotors

Joseph Wang*

Received 30th July 2013, Accepted 31st July 2013
DOI: 10.1039/c3fd00105a

The combination of nanomaterials with electrode materials has opened new horizons in electroanalytical chemistry, and in electrochemistry in general. Over the past two decades we have witnessed an enormous activity aimed at designing new electrochemical devices based on nanoparticles, nanotubes or nanowires, and towards the use of electrochemical routes – particularly template-assisted electrodeposition – for preparing nanostructured materials. The power of template-assisted electrochemical synthesis is demonstrated in this article towards the preparation and the realization of self-propelled catalytic nanomotors, ranging from Pt–Au nanowire motors to polymer/Pt microtube engines. Design considerations affecting the propulsion behavior of such catalytic nanomotors are discussed along with recent bioanalytical and environmental applications. Despite recent major advances, artificial nanomotors have a low efficiency compared to their natural counterparts. Hopefully, the present *Faraday Discussion* will stimulate other electrochemistry teams to contribute to the fascinating area of artificial nanomachines.

1 Introduction

The first Royal Society Discussion Meeting held under the name "Faraday Discussions", was held in 1947, and was devoted to the topic of "Electrode Processes".[1] This Introduction is intended to provide some background to the topic of the present *Faraday Discussion* 164 on Electroanalysis at the Nanoscale, starting with a short history of the subject. The trend to miniaturization of electrodes started in the mid 1970s, prior to the nanotechnology boom. Leading electrochemists have recognized for a long time that the miniaturization of working electrodes has obvious practical advantages, and opens up fundamentally new possibilities.[2–4] Ultramicroelectrodes have been widely used for over three decades for local electrochemical investigations, and particularly for *in vivo* monitoring of neurotransmitters.[2,4] The emergence of nanotechnology, and the introduction of nanoparticles, nanotubes or nanowires and in the 1980s and 1990s, has opened new horizons for designing new electrochemical devices and electroanalytical strategies. Over the same periods we have witnessed

Department of NanoEngineering, University of California, San Diego La Jolla, CA 92093, USA. E-mail: josephwang@ucsd.edu; Fax: +1 (858) 534 9553; Tel: +1 (858) 246 0128

considerable efforts in the use of electrochemical routes for preparing nanostructured materials. The field of nanoscale electrochemistry has thus been growing very fast.

2 Use of nanomaterials in electroanalysis and electrochemistry

Major efforts, starting in the 1980s, have greatly improved the way of making, assembling, positioning and imaging nanomaterials of different compositions with controlled sizes, shapes and functionality. The unique properties of nanoscale materials have paved the way to new and improved electrochemical sensing devices.[5,6] The range of nanomaterials used in electrochemistry is wide and diverse. Nanomaterials such as carbon nanotubes, gold nanoparticles, or silicon nanowires have thus made a major impact on the field of electrochemical biosensors, ranging from glucose enzyme electrodes to DNA hybridization sensors.[5-7] For example, various nanoparticles have been used towards effective electrical communication between the redox proteins and the electrodes or as amplification tags for ultrasensitive electrochemical bioaffinity assays.[8,9] Nanomaterial–biomolecule hybrid systems have also led to the development of new and improved biofuel cells. Electrocatalytic sensing and energy-conversion applications have also benefited from the ability to vary the size, composition and shape of nanomaterials, and hence tailoring their electrochemical reactivity. The newest nanoscale carbon material, graphene, has already been shown to be extremely useful for enhancing greatly a wide range of electrochemical sensing and energy storage applications.[10] Considerable attention has been given recently to understanding the various factors and parameters that affect the electrochemical reactivity of nanoscale carbon materials, owing to uncertainties associated with the source or quality of these nanomaterials.[11]

3 Use of electrochemical methods for preparing nanostructured materials

Electrochemical deposition has been shown to be extremely useful for preparing nanomaterials. In particular, the membrane-template electrosynthesis method, introduced by Charles Martin in the mid 1990s, has been one of the most widely used electrochemical routes for preparing nanostructured materials.[12] Such template-assisted electrochemical growth of different nanowires involves electrodeposition into the cylindrical void nanopores of a host porous membrane template, followed by dissolution of the template.[13,14] Nanoporous membranes (e.g., track-etched polycarbonate, anodized alumina), with a wide range of pore diameters (0.03–10 μm) and pore densities (10^5–10^9 pores cm^{-2}), are available commercially from companies such as Millipore or Whatman, and are the most widely used sacrificial template.

The template electrodeposition method is very general as it entails synthesis of the desired material within the cylindrical pores. The method has thus been extremely useful for preparing nanowires with broad range of chemical compositions, including metallic, polymeric and semiconductor nanowires. This is accomplished by depositing first a thin film of metal on one side of the template to create the working electrode and electrical contact. The membrane is then assembled in an electrochemical cell with the open pores facing a plating bath to allow deposition of the wire segments. Applying a potential to this metal film contact in the presence of an electrolyte containing the metal ions or monomer to be deposited thus results in controlled bottom-up growth of nanowires in the pores of the template. Nanowires of tailor-made lengths can be obtained by controlling the charge passed during synthesis. Microstructures of different shapes, including cylindrical microtubes, conical, double-conical or bilayer microtubes or core–shell microwires, have also been realized by the template-assisted electrochemical route. Multi-segment nanowires have been prepared through sequential electrodeposition of several materials, with different pre-determined lengths, into the pores of the template (Fig. 1).[15] Such multisegment nanowires have found important applications as product barcodes[16] or tags for multiplexed biodetection.[17,18]

The power and versatility of template-assisted electrochemical syntheses is demonstrated in the following sections towards the realization of self-propelled catalytic nanomotors of different designs and propulsion mechanisms.

4 Template synthesis of self-propelled catalytic micromotors

Template electrodeposition has been shown to be extremely attractive for preparing catalytic nanowire and microtube motors.[15,19,20] Locomotion of small scale objects through fluid environments is one of the most exciting areas of nanotechnology.[19,21,22] Man-made nano/microscale catalytic motors represent a major step towards the development of practical nanomachines. Such synthetic motors have received considerable recent attention over the past decade owing to their great potential for diverse applications ranging from directed drug delivery to environmental remediation.[23–25] Particular attention has been given to the design of efficient synthetic micro/nanoscale motors that convert chemical energy into autonomous motion, based on the decomposition of hydrogen peroxide fuel.

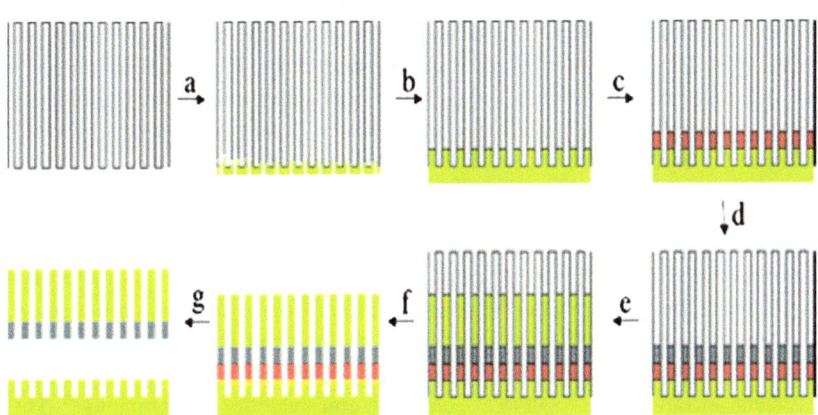

Fig. 1 Template-assisted electrodeposition of multi-segment nanowires: (a) Gold sputtering onto the alumina membrane template. (b) Electrodeposition of gold plugs. (c) Electrodeposition of a sacrificial layer of copper. (d) Electrodeposition of nickel segment. (e) Electrodeposition of gold segment. (f) Dissolution of alumina. (g) Dissolution of copper. (Reproduced from ref. 15.)

4.1 Template prepared catalytic nanowires

Initial efforts in this direction, pioneeered by Sen and Mallouk at Pen State University[26] and Ozin's group in Toronto,[15] have led to the introduction of catalytic nanomotors based on bisegment (Pt–Au) nanowires. The sequential deposition of the platinum and gold segments thus leads to asymmetric nanowires with spatially defined catalytic zones, with the oxidation and reduction of the hydrogen-peroxide fuel occurring preferentially at the Pt anode and Au cathode. Such asymmetry is essential for generating a directional force and movement through a self-electrophoretic propulsion mechanism. The speed of the bimetallic nanowire motors is proportional to the mixed potential difference (ΔE) of the fuel at the corresponding metal segments. Tafel plots of the anodic and cathodic reactions on the corresponding electrode materials can be used for obtaining the ΔE.

We demonstrated the ability to increase the velocity and force of synthetic bimetallic nanowire motors by exploring new motor compositions. For example, we illustrated that the incorporation of carbon nanotubes (CNT) into the platinum segment of catalytic nanowire motors leads to a dramatically enhanced speed and power.[27] The resulting nanomotors are capable of moving autonomously at speeds approaching 100 body lengths per second, reflecting the greatly increased fuel decomposition rate at the anodic Pt–CNT segment. We also demonstrated a dramatic increase of the speed of nanowire motors to over 100 $\mu m\ s^{-1}$ by replacing their cathodic gold segment with a Ag–Au alloy segment.[28] Such behavior is attributed to the substantially enhanced electrochemical reactivity of Ag–Au alloys, compared to silver or gold alone. The speed of these alloy nanowire motors is strongly affected by the composition of the Ag–Au segment, being nearly linearly proportional the silver level in the growth mixture solution from 0 to 75% (v/v). Such dependence is attributed to the influence of the Ag–Au alloy composition upon the reduction process of the peroxide fuel.

Guiding the movement of catalytic nanowire motors and regulating their speed are essential for diverse future applications of catalytic nanomotors. Most commonly such motion control is accomplished by adding a nickel segment[29] to allow guided movement along preselected paths as well collective motion of multiple nanomotors through alignment with an external magnetic field. Speed control of catalytic nanowire motors has been realized through a temperature control. Balasubramanian et al.[30] demonstrated the use of heat pulses for modulating rapidly the speed of catalytic nanowire motors. Such temperature control reflects the controlled reaction kinetics, analogous to that observed in thermoelectrochemistry at heated electrodes.

4.2 Template prepared catalytic microtube engines

Bubble-propelled microtube engines ('microrockets') have been developed to address the limitation of catalytic nanowire motors to low ionic-strength media.[31,32] As desired for diverse practical applications, the oxygen-bubble propulsion mechanism of microtube engines (in the presence of hydrogen peroxide fuel) leads to efficient locomotion in salt-rich solutions and real-life environments.[33] These tubular microengines thrust forward in discrete increments, reflecting the release of the individual oxygen microbubbles (generated at their inner catalytic surface) through the larger opening of the microcone. The speed of the tubular microengine corresponds to the product of the bubble radius and expelling frequency.

Tubular microengines were prepared initially by Mei and Schmidt using a rolled-up lithographic fabrication route.[31,32] Our team described a simplified template-membrane based electrodeposition synthesis of highly efficient and smaller (8-μm long) polymer/Pt bilayer conical microtube engines, illustrated in the SEM images of Fig. 2.[20,34] These peroxide-driven bubble-propelled micro-rockets have been electrosynthesized using the conical polycarbonate membrane template. Such a template directed route involves sequential deposition of the polymeric and catalytic metal tubular layers within the conical-shaped micropores of the membrane. The nucleation and growth of conducting polymer microtubes within such conical micropores involve electrostatic and solvophobic interactions between the polymers and pore wall.[35] An inner catalytic platinum tubular layer is subsequently plated inside the resulting polymeric microtube. The influence of the composition and electropolymerization conditions upon the propulsion of new template-prepared polymer-based bilayer microtubular microbots has been investigated. Such investigations examined the effect of different electropolymerized outer layers, including polyaniline (PANI), polypyrrole (PPy), or poly(3,4-ethylenedioxythiophene) (PEDOT) (Fig. 2), and of various inner catalytic metal surfaces (Ag, Pt, Au, Ni–Pt alloy), upon the movement of such bilayer microtubes.[34] Such polymeric layers were selected owing to their inherent chemical stability. Electropolymerization conditions, such as the monomer concentration and medium (*e.g.*, surfactant, electrolyte) or plating parameters (*e.g.*, charge) have a profound effect upon the exact morphology of the resulting polymer/metal bilayer microtubes and hence upon their propulsion behavior. The most efficient propulsion was observed using PEDOT/Pt microengines that offer a remarkable speed of over 1400 body lengths s^{-1}, which is the fastest relative speed reported to date for man-made micro/nanomotors. Such ultrafast nanomotor speed corresponds to a large force and power essential to execute different cargo-towing and delivery tasks. Magnetic guidance of such template-prepared microengines is achieved through the deposition of an intermediate Ni layer. In addition to the polymer–Pt microtube bilayer, it is possible to achieve bubble propulsion using template-assisted deposition of bimetallic Cu–Pt microjects.[36]

Inner biocatalytic layers can serve as attractive alternatives to electrocatalytic metals for propelling peroxide-driven nano/microscale motors.[34,37] Template electrodeposition of the polymer–Au microtube can lead to a very rough gold surface suitable for immobilizing large amounts of the enzyme catalase through a mixed alkanethiol monolayer and EDC coupling.[34] The resulting catalase-powered microengines propel favorably in the presence of a low peroxide level (0.5% H_2O_2).

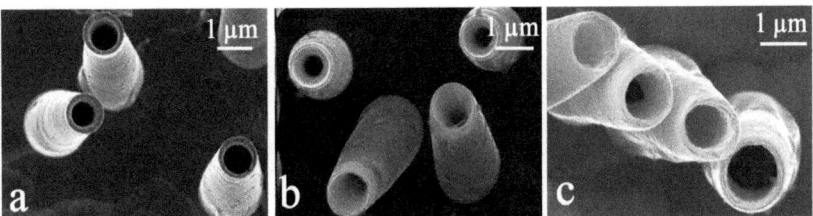

Fig. 2 SEM images of three template-prepared polymer/Pt bilayer microtube engines based on different outer polymeric layers: (a) PANI/Pt; (b) PPy/Pt; (c) PEDOT/Pt. (Reproduced from ref. 34.)

The template electrodeposition can be used for preparing microtube engines based on other fuels. For example, Gao et al.[38] described the efficient acid-driven propulsion of polymer/zinc bilayer microrockets. The effective propulsion of the zinc-based microtubes in acidic media reflects the continuous thrust of hydrogen bubbles generated by the spontaneous redox reaction occurring at the inner Zn surface. The exploration of new alternative fuels, through studies of new motor materials and reactions, should expand the scope of operation and environments of catalytic nanomotors.

4.2.1 Recent applications of template-prepared microtube engines.
Various bioanalytical ('Capture–Transport') separation applications of catalytic microtube engines have been accomplished by functionalizing their outer surface with an immobilized bioreceptor.[39] For example, we have also demonstrated the rational functionalization of tubular microengines with different bioreceptors, such as antibodies or oligonucleotide probes (Fig. 3). This has been accomplished *via* a mixed self-assembled monolayer (SAM) chemistry on a sputtered gold surface (Fig. 3) or through carboxy groups on the outer (PEDOT–carboxy) polymer layer. Such surface modification procedures ensure efficient binding processes while minimizing nonspecific binding, and have a minor effect upon the propulsion efficiency in biological media. The resulting functionalized microoenegines have been shown to be extremely useful for capturing, transporting, isolating and detecting a wide range of target bioanalytes. The 'on-the-fly' isolation of cancer cells,[40] proteins,[41] or DNA targets[42] from complex raw biological matrices has thus been accomplished. The unique features of these motion-driven bio-isolation protocols make them an extremely attractive alternative for current sample processing protocols in connection to microfabricated lab-on-chip microsystems.

Recently, we introduced new microtube engines possessing a 'built-in' recognition capability within their outer polymeric layer itself, hence eliminating the need for an additional receptor functionalization step.[43,44] These included nanomotors for recognizing monosaccharides based on a poly-3-aminophenylboronic acid (PAPBA) outer layer,[43] and nanomotors based on a molecularly imprinted polymer (MIP) outer layer that contained artificial receptor cavities.[44] The latter involves introduction of the imprinted recognition sites for the target protein (FITC-labeled avidin) template, by adsorption on the walls of the micropores, during the electropolymerization of the outer PEDOT layer (Fig. 4). Direct extraction and isolation of Av-FITC from raw serum and saliva samples has thus

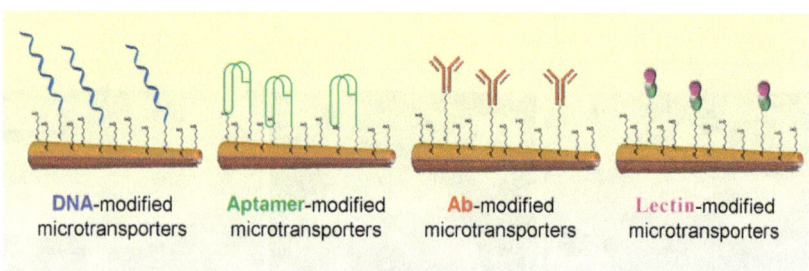

Fig. 3 Microtube engines functionalized with ss-DNA (a), aptamer (b), antibody (Ab), and lectin (d) receptors, for the 'on-the-fly' isolation of nucleic acids, proteins, cancer cells and bacteria, respectively, from unprocessed biological media (Reproduced from ref. 39.)

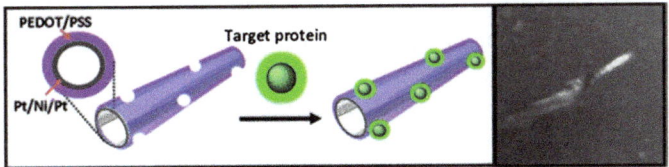

Fig. 4 Scheme illustrating the preparation and characterization of the MIP-based microtube engine, containing artificial receptor cavities within its outer layer, along with the strategy for capture and transport of the target protein. Also shown (right) is an image of the MIP-based microengine after movement in a solution containing the fluorescent-tagged analyte. (Reproduced with permission from ref. 44.)

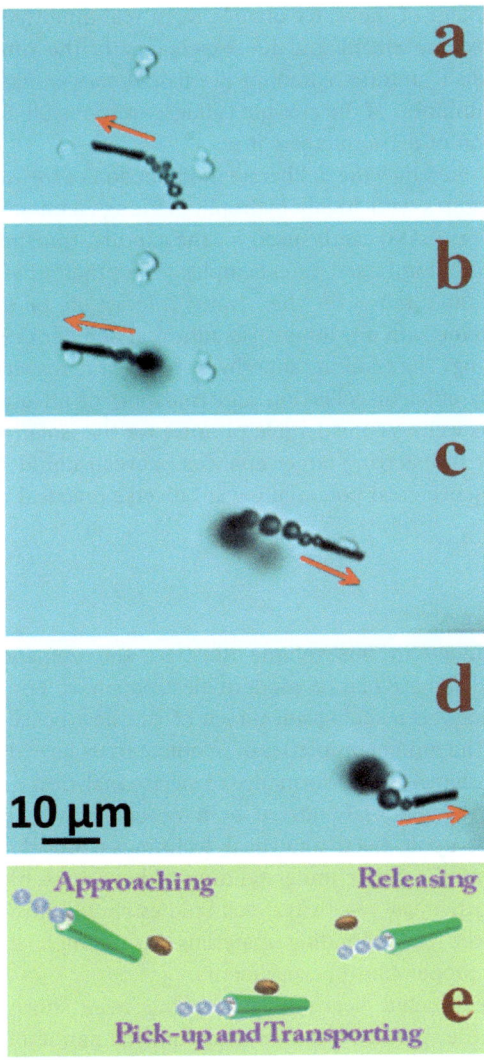

Fig. 5 'Capture–Transport–Release' of yeast cells using a PAPBA/Ni/Pt tubular microengine approaching (a), capturing (b), transporting (c) and releasing (d) the yeast cell. (e) Schematic representation of the 'Capture–Transport–Release' process. (Reproduced with permission from ref. 43.)

been accomplished following short navigation times. The images of Fig. 5 illustrate the use of the PAPBA/Pt microtube engine 'on-the-fly' binding and transport of yeast cells (containing sugar residues on their wall), along with the fructose-triggered release of the cells. Such use of the recognition polymeric layer does not hinder the efficient propulsion of the microengine in aqueous and physiological media.

Environmental applications. Catalase-powered microtube motors have been used for rapid water-quality testing based on changes in their propulsion behavior and lifetime in the presence of common pollutants[45] Enzyme-powered biocompatible polymeric (PEDOT)/Au-catalase tubular microengines have thus offered highly-sensitive direct optical visualization of changes in the swimming behavior in the presence of common contaminants (toxins) and hence a direct real-time assessment of the water quality. Such real-time tracking of the toxin-induced hindered movement and life expectancy of the tubular microengine ('artificial microfish') mimics common live-fish water testing and relies on the toxin-induced inhibition of the enzyme catalase, responsible for the biocatalytic bubble propulsion of these microengines.

We also illustrated how the deliberate modification of the rough outer surface of microtube engines with highly hydrophobic long-chain self-assembled alkanethiol monolayers (SAM) can be used for the capture, transport and removal of oil droplets from contaminated water samples.[25] This motor-based environmental remediation method relies on the strong interaction of the SAM-modified microtubular engine with oily liquids *via* adhesion and permeation onto its long alkanethiol coating. The resulting superhydrophobic microswimmers thus offer a rapid and highly efficient collection and transport of oil droplets in aqueous environments, as well as considerable promise for the isolation of hydrophobic molecules or for transferring target analytes between liquid–liquid immiscible interfaces, and hence great potential toward diverse practical applications.

Conclusions

The papers included in this volume illustrate the tremendous impact that nanoscale materials have had on modern electroanalysis. For over two decades, electroanalysis has been taking advantage of the new possibilities that nanomaterials offer. The unique properties of nanomaterials have thus been exploited successfully to enhance the performance of electroanalytical devices. There is no doubt that the new materials offered by nanoscience will continue to have a profound impact upon electroanalytical techniques. Considerable efforts are required for gaining a better understanding of how these nanoscale materials affect the electrochemical reactivity. Such studies should involve a careful characterization of the corresponding materials (using high resolution imaging techniques) and proper control experiments.

The template-directed electrodeposition has been shown to be extremely useful and versatile approach for preparing catalytic nanomotors, ranging from Pt–Au nanowires to polymer/Pt microtube engines. The template-assisted electrochemical growth can also be used for preparing externally-powered (fuel-free) nanowire motors, which are particularly suitable for *in vivo* biomedical applications.[46,47] A wide range of other useful nanoobjects can be prepared by the template electrodeposition technique, and by electrochemical methods in general.

The use of synthetic nanomotors to power nanomachines is one of the most exciting challenges facing nanotechnology. Microscale catalytic motors, converting chemical energy into autonomous motion, have been redesigned over the past decade for faster movement, larger force, and for enhanced functionality. Despite of these major recent advances, both catalytic nanowires and microtube motors still have a low efficiency with which they convert the free energy stored in chemical fuel to mechanical energy.[48] Accordingly, I hope that this article will stimulate other electrochemistry groups to bring their expertise and talent to address the important challenge of powering artificial nanomotors and to accelerate the pace of developing functional nanomachines for meeting future societal needs.

Acknowledgements

This work was supported by the Defense Threat Reduction Agency-Joint Science and Technology Office for Chemical and Biological Defense (Grant no. HDTRA1-13-1-0002).

References

1 D. D. Eley, *Discuss. Faraday Soc.*, 1947, **1**, 129.
2 R. M. Wightman, *Anal. Chem.*, 1981, **53**, 1125A.
3 M. Fleischmann and S. Pons, *Anal. Chem.*, 1987, **59**, 1391A.
4 A. M. Bond, *Analyst*, 1994, **119**, 1R.
5 J. Wang, *Analyst*, 2005, **130**, 421.
6 A. Merkoci, *Electroanalysis*, 2007, **19**, 739.
7 J. Wang, *Anal. Chim. Acta*, 2003, **500**, 247.
8 Y. Xiao, F. Patolsky, E. Katz, J. F. Hainfeld and I. Willner, *Science*, 2003, **299**, 1877.
9 J. Wang, *Small*, 2005, **1**, 1036.
10 Y. Shao, J. Wang, H. Wu, J. Liu, I. A. Aksay and Y. Lin, *Electroanalysis*, 2010, **22**, 1027.
11 X. Ji, R. O. Kadara, J. Krussma, Q. Chen and C. E. Banks, *Electroanalysis*, 2010, **22**, 7.
12 C. R. Martin, *Chem. Mater.*, 1996, **8**, 1739.
13 C. Schönenberger, B. M. I. Van der Zande, L. G. J. Fokkink, M. Henny, C. Schmid, H. Birk and U. Staufer, *J. Phys. Chem B*, 1977, **101**, 5505.
14 A. Walcarius, *Anal. Bioanal. Chem.*, 2009, **396**, 261.
15 S. Fournier-Bidoz, A. C. Arsenault, I. Manners and G. A. Ozin, *Chem. Commun.*, 2005, **441**.
16 S. Nicewarner-Pena, R. G. Freeman, B. D. Reiss, L. He, D. J. Pena, I. D. Walton, R. Remy Cromer, C. D. Keating and M. J. Natan, *Science*, 2001, **294**, 137.
17 J. B. H. Tok, F. Chuang, M. C. Kao, K. A. Rose, S. S. Pannu, M. Y. Sha, G. Chakarova, S. G. Penn and G. M. Dougherty, *Angew. Chem., Int. Ed.*, 2006, **45**, 6900.
18 J. Wang, G. Liu and G. Rivas, *Anal. Chem.*, 2003, **75**, 4667.
19 J. Wang, *Nanomachines: Fundamentals and Applications*, Wiley-VCH, Weinheim, 2013, (ISBN 978-3-527-33120-8).
20 W. Gao, S. Sattayasamitsathit, J. Orozco and J. Wang, *J. Am. Chem. Soc.*, 2011, **133**, 11862.
21 A. Goel and V. Vogel, *Nat. Nanotechnol.*, 2008, **3**, 465.
22 M. G. L. van den Heuvel and C. Dekker, *Science*, 2007, **317**, 333.
23 J. Wang and W. Gao, *ACS Nano*, 2012, **6**, 5745.
24 S. Sengupta, M. E. Ibele and A. Se, *Angew. Chem., Int. Ed.*, 2012, **51**, 8434.
25 M. Guix, J. Orozco, M. García, W. Gao, S. Sattayasamitsathit, A. Merkoçi, A. Escarpa and J. Wang, *ACS Nano*, 2012, **6**, 4445.
26 W. F. Paxton, K. C. Kistler, C. C. Olmeda, A. Sen, S. K. St. Angelo, Y. Cao, T. E. Mallouk, P. E. Lammert and V. H. Crespi, *J. Am. Chem. Soc.*, 2004, **126**, 13424.
27 R. Laocharoensuk, J. Burdick and J. Wang, *ACS Nano*, 2008, **2**, 1069.
28 U. Demirok, R. Laocharoensuk, M. Manesh and J. Wang, *Angew. Chem., Int. Ed.*, 2008, **47**, 9349.
29 T. R. Kline, W. F. Paxton, T. E. Mallouk and A. Sen, *Angew. Chem., Int. Ed.*, 2005, **44**, 744.

30 S. Balasubramanian, D. Kagan, K. Manesh, P. Calvo-Marzal, G. U. Flechsig and J. Wang, *Small*, 2009, **5**, 1569.
31 Y. Mei, A. A. Solovev, S. Sanchez and O. G. Schmidt, *Chem. Soc. Rev.*, 2011, **40**, 2109.
32 G. Huang, J. Wang and Y. Mei, *J. Mater. Chem.*, 2012, **22**, 6519.
33 K. M. Manesh, M. Cadona, R. Yuan, D. Kagan, S. Balasubramanian and J. Wang, *ACS Nano*, 2010, **4**, 1799.
34 W. Gao, S. Sattayasamitsathit, A. Uygun, A. Pei, A. Ponedal and J. Wang, *Nanoscale*, 2012, **4**, 2447.
35 S. I. Cho and S. B. Lee, *Acc. Chem. Res.*, 2008, **41**, 699.
36 G. Zhao and M. Pumera, *RSC Adv.*, 2013, **3**, 3963.
37 S. Sanchez, S. A. Solovev, Y. F. Mei and O. G. Schmidt, *J. Am. Chem. Soc.*, 2010, **132**, 13144.
38 W. Gao, A. Uygun and J. Wang, *J. Am. Chem. Soc.*, 2012, **134**, 897.
39 S. Campuzano, D. Kagan, J. Orozco and J. Wang, *Analyst*, 2011, **136**, 4621.
40 S. Balasubramanian, D. Kagan, C. M. Hu, S. Campuzano, M. J. Lobo-Castañon, N. Lim, D. Y. Kang, M. Zimmerman, L. Zhang and J. Wang, *Angew. Chem., Int. Ed.*, 2011, **50**, 4161.
41 M. Garcia, J. Orozco, M. Guix, W. Gao, S. Sattayasamitsathit, A. Escarpa, A. Merkoci and J. Wang, *Nanoscale*, 2013, **5**, 1325.
42 D. Kagan, S. Campuzano, S. Balasubramanian, F. Kuralay, G. Flechsig and J. Wang, *Nano Lett.*, 2011, **11**, 2083.
43 F. Kuralay, S. Sattayasamitsathit, W. Gao, A. Uygun, A. Katzenberg and J. Wang, *J. Am. Chem. Soc.*, 2012, **134**, 15217.
44 J. Orozco, A. Cortes, G. Cheng, S. Sattayasamitsathit, W. Gao, X. Feng, Y. Shen and J. Wang, *J. Am. Chem. Soc.*, 2013, **135**, 5336.
45 J. Orozco, W. Gao, V. Garcia, M. D'Agostino, A. Cortes and J. Wang, *ACS Nano*, 2013, **7**, 818.
46 P. Calvo-Marzal, S. Sattayasamitsathit, S. Balasubramanian, J. R. Windmiller, C. Dao and J. Wang, *Chem. Commun.*, 2010, **46**, 1623.
47 W. Wang, L. A. Castro, M. Hoyos and T. E. Mallouk, *ACS Nano*, 2012, **6**, 6122.
48 W. Wang, T. Y. Chiang, D. Velegol and T. E. Mallouk, *J. Am. Chem. Soc.*, 2013, **135**, 10557.

> # Faraday Discussions

PAPER

Microelectrochemical visualization of oxygen consumption of single living cells

Michaela Nebel,[a] Stefanie Grützke,[a] Nizam Diab,[b] Albert Schulte[c] and Wolfgang Schuhmann[*a]

Received 10th February 2013, Accepted 18th March 2013
DOI: 10.1039/c3fd00011g

The detection of cellular respiration activity is important for the assessment of the status of a biological cell. Due to its non-invasive character and high spatial resolution scanning electrochemical microscopy (SECM) is a powerful tool for single cell measurements. Common limitations of respiration studies performed by SECM are discussed and strategies provided to further adapt SECM detection schemes to the specific requirements for the investigation of single cell respiration. In particular the combination of a potential pulse technique in the redox competition mode of SECM with a shearforce-based constant-distance positioning of the SECM tip is proposed for characterising the impact of the tip reaction during SECM imaging. The adjustment of the driving force of the tip reaction and the selection of the time for data acquisition after applying the potential pulse allowed a successful visualization of cell respiration activity.

Introduction

Single cells, as the smallest independent unit of a living organism, facilitate the study of complex biochemical processes in a simplified manner. According to the principle of reductionism this simplification enables elucidating implications of the role of individual cells inside the whole organism, a better understanding of selected reaction pathways and the species involved therein.[1] A specific advantage of electrochemical methods in biological studies is that some of the most important species involved in biological processes like NO, catecholamine neurotransmitters, oxygen, and reactive nitrogen (RNS) or reactive oxygen species (ROS) are directly detectable under physiological and non-invasive conditions. Furthermore, microelectrodes facilitate local electrochemical investigations in the vicinity of an inspected cell and a first application was already described in

[a]Lehrstuhl für Analytische Chemie – Elektroanalytik & Sensorik, Ruhr-Universität Bochum, Universitätsstr. 150, 44780 Bochum, Germany. E-mail: Wolfgang.Schuhmann@rub.de; Fax: (+49)234-32-14783; Tel: (+49) 2343226200
[b]Chemistry Department, Faculty of Arts and Sciences, The Arab American University, P.O. Box 240, Jenin, Palestine
[c]Biochemistry-Electrochemistry Research Unit, School of Chemistry and Biochemistry, Institute of Science, Suranaree University of Technology, Nakhon Ratchasima 30000, Thailand

1976.[2] The flexibility of electrochemical detection schemes in combination with the broad range of traceable species led to an increasing number of different applications of stationary microelectrodes as reviewed *e.g.* in ref. 3. Scanning electrochemical microscopy (SECM)[4] as a spatially resolving microelectrochemical technique enables accurate positioning of a microelectrode accompanied by visualization of local electrochemical activity distributions. Beside the investigation of fluxes through the cellular membrane during exocytosis events[5,6] and the mapping of intracellular activity,[7–9] the detection of the respiration activity of individual cells is of high importance. The respiratory chain is directly connected with the metabolism of a cell and therefore the consumed oxygen is an essential measure of the cell status.

Due to a comparatively simple experimental setup and a fast detection procedure, a large number of single cell SECM investigations have been performed in the constant-height mode in which the SECM tip approaches the sample surface at x,y-coordinates far away from the specimen under investigation and scanned laterally keeping the z-position fixed.[10,11] In order to prove that the obtained constant-height SECM images were caused by the respiration activity of the cell, several studies utilized KCN to block mitochondrial respiration by inhibition of cytochrome c oxidase in complex IV of the respiratory chain.[10,12] A subsequent diminution or absence of a current decrease above the cell body is used as evidence for a successful visualization of cell respiration. These experiments are based on the assumption that the cell morphology is not changed due to respiration inhibition. However, it is known that morphology fluctuations occur during cell death, like *e.g.* swelling or shrinking. These variations are not considered but might influence the tip current.

Further refinements of the SECM detection scheme for single cell analysis were dedicated to strategies minimizing limitations of constant-height mode experiments. Beside the use of specially designed double barrel electrodes for the simultaneous detection of the tip-to-sample distance and the biological activity,[13,14] the embedding of cells into cavities,[15] the evaluation of SECM approach curves on top and at the side of the investigated cell,[16,17] and time-lapse SECM[18] have been reported. However, studies utilizing different redox mediators are based on the assumption that oxygen and the added mediator behave in a comparable way, *e.g.* showing similar diffusion coefficients. Furthermore, these strategies lack a true topography-free investigation of the structure–activity-relationships necessary for an in-depth understanding of cellular processes. Therefore, constant-distance mode (cd-mode) SECM experiments in which the tip follows the contour of the investigated cell in a continuously controlled and constant working distance are considered to be superior to constant-height mode investigations.

The easiest constant-distance mode is the constant-current mode in which the tip current is used for maintaining the working distance during the experiment and the stored z-position of the tip generates a visualization of the cell topography.[19,20] However, a basic requirement for a precise constant-current mode experiment is a uniform electrochemical activity of the investigated surface. Since the examination of variations in reaction rates at the sample surface is the intrinsic aim of SECM investigations, the constant-current mode is strongly limited in that respect. Alternatively, current-independent distance control mechanisms include the adjustment of the tip-to-sample distance by impedance-based signals,[21] through shearforce interactions,[22] and *via* merging other scanning probe techniques like atomic force microscopy (AFM)[23] or scanning ion conductance microscopy (SICM).[24] Recently, these strategies have been successfully applied for cell investigations.[5,6,20,25–27]

An additional aspect often neglected in SECM respiration studies is the high permeability of the cell membrane for certain molecules and the consequent possibility of diffusional exchange across the lipid bilayer due to the formed concentration gradients. Therefore, we focus in this contribution on the elucidation of the complex interaction of different reaction rates involved in the SECM detection scheme for the investigation of the respiration activity of single living cells. Particularly, the influence of the driving force of the tip reaction on the overall result of the SECM experiment is investigated. In order to further study and control the role of the scanned tip, a potential pulse profile applied at the SECM tip in combination with a time dependent data acquisition is utilized.

Experimental

Cell cultures and solutions

Retzius cells were from medicinal leeches and cultured according to a procedure described in ref. 28. PC12 cells were grown on poly(ornithine)-covered glass slides as described in ref. 29, and transformed human umbilical vein endothelial cells (HUVEC cells) were cultured as reported previously.[13] Samples of PC12 and HUVEC were kindly provided by Dr A. Blöchl and Prof. Dr I. D. Dietzel-Meyer (Lehrstuhl für Molekulare Neurobiochemie, Ruhr-Universität Bochum, Germany). Human embryonic kidney cells (HEK293) were cultured as described in ref. 30 and were kindly supplied by Dr C. H. Wetzel (Lehrstuhl für Zellphysiologie, Ruhr-Universität Bochum, Germany). Measuring solutions contained 125 mM NaCl, 5 mM KCl, 1.2 mM NaH_2PO_4, 1 mM $CaCl_2$, 1.2 mM $MgCl_2$, 10 mM glucose and 25 mM 4-(2-hydroxyethyl)-1-piperazine-1-ethanesulfonic acid (HEPES), pH 7.4 for PC12 and for HUVEC. An extracellular solution with 140 mM NaCl, 5 mM KCl, 2 mM $CaCl_2$, 1 mM $MgCl_2$ and 10 mM HEPES, pH 7.3 were used for HEK293 cells. $K_4[Fe(CN)_6]$ for the determination of the tip-to-sample separation in the feedback mode was from Merck (Darmstadt, Germany) and used as received.

SECM experiments

SECM measurements have been performed with a Bio-SECM setup based on the instrument described in.[6] The positioning system of the SECM is mounted on an inverted microscope (Axiovert 25 C, Carl Zeiss, Jena, Germany with an integrated digital CCD camera from IDS Imaging Development Systems, Obersulm, Germany) allowing easy positioning of the tip next to the desired cells and continuous optical inspection of the cell status. The SECM positioning system consists of three orthogonal stepper motors (SPI Robot Systems, Oppenheim, Germany) with a nominal resolution of 10 nm per microstep and an additional piezo element (NanoCube P-611.ZS with an E-665 controller from Physik Instrumente, Waldbronn, Germany) for the z-direction. Furthermore, the SECM includes components necessary for the optical shearforce-based constant-distance mode with a laser (CDM 14S/S70/1, Atos, Pfungstadt, Germany), a split photodiode (Spot 4D, Laser2000, Wessling, Germany), a lock-in amplifier (model 5210, Ametek, Meerbusch, Germany), and an agitation piezo element (PSt 500/5/15,

Piezomechanik Pickelmann, München, Germany) for vibrating the tip in its characteristic resonance frequency. For constant-distance mode SECM experiments the laser is positioned at the end of the vibratory tip electrode. The resulting Fresnel diffraction pattern impinges on the split photodiode and the difference signal of the photodiode is used for the readout of the tip vibration. This signal is phase-sensitively amplified with respect to the sinus shaped agitation signal applied at the agitation piezo element *via* a lock-in amplifier and used for the current-independent distance control procedure. For electrochemical measurements a potentiostat model VA-10 (Npi Electronics, Tamm, Germany) was used together with an ADDA board (CIO-DAS 802/16, Plug-In Electronics, Eichenau, Germany) for potential control and data acquisition. All potentials were measured *versus* a miniaturized Ag/AgCl (3 M KCl) reference electrode in a two electrode configuration. The whole setup is controlled by software programmed in Visual Basic and placed on a vibration damped base inside a faraday cage.

SECM tips were vibratory 10 μm Pt electrodes fabricated according to ref. 31 for the PC12 cell experiments or otherwise highly flexible polymer-insulated and platinized carbon fibre electrodes with diameters of 7 to 9 μm fabricated following a procedure described in ref. 5, 6 & 32. Either a cathodic electrodeposition paint (ClearClad HSR401 from LVH coatings, UK) or alternatively an anodic paint (Canguard, BASF Coatings, Münster, Germany) were used for the insulation of the carbon fibres (type Grafil E/XA-S, Courtaulds Limited Carbon Fibres Division, Coventry, UK). The precipitation of the soluble polymer was repeated twice by means of a two-step procedure of subsequent precipitation invoked by an electrochemically induced pH shift and heat curing at 180 °C for 20 min in order to crosslink the precipitated polymer. After insulation and cutting to expose a fresh carbon surface, the carbon fibre electrodes were electrochemically characterised by means of cyclic voltammetry in a solution containing 5 mM $[Ru(NH_3)_6]Cl_3$ (ABCR, Karlsruhe, Germany) and 100 mM KCl (VWR International, Darmstadt, Germany). To enable oxygen detection, a thin layer of platinum was deposited on the fibre electrodes using a solution of 2 mM H_2PtCl_6 (Merck, Darmstadt, Germany) and potentiodynamic cycling (300 mV to −500 mV, scanrate: 100 mV s^{-1}, 3 cycles). The resonance frequency of the vibratory SECM tips was typically in the range of 0.5–5 kHz. A set point of 2–5% change from the shearforce signal in bulk was defined as the condition of the distance control procedure and used for the adjustment of the tip-to-sample separation. The corresponding magnitude value was kept constant during continuous shearforce-based constant-distance mode. For this purpose the *z*-position of the tip was adjusted after each lateral displacement *via* the *z*-piezo element until this set point was reached again. Alternatively, for 4D SF/CD-SECM[33] a magnitude change of 5% was defined as the stop criterion of the shearforce-based tip approach. At each grid point the tip was moved towards the sample surface under shearforce control until this change in the tip vibration was reached. After current detection at the point of closest approach, a retraction of the tip was accomplished in defined *z*-increments and the corresponding signal was measured at each of the different but constant tip-to-sample separations. Performing this sequence of shearforce-based positioning and subsequent retraction steps leads to a 4D dataset that includes complete current images in various constant working distances towards the topography contour of the sample surface.

Results and discussion

Limitations of constant-height mode SECM for single cell experiments

The most frequently used experimental approach for studying the respiration activity by means of SECM is a variation of the substrate-generation/tip-collection (SG/TC) mode. In this competitive SG/TC arrangement the tip is continuously polarized to the oxygen reduction potential in order to detect the amount of oxygen present in the gap between the tip and sample. Due to the competition between the tip and sample for the same amount of oxygen, a decrease of the reduction current of the scanned tip represents a locally higher respiratory activity of the investigated cell. An exemplary constant-height mode SECM experiment of adherently growing Retzius cells using this competition type SG/TC mode is displayed in Fig. 1.

The image shows clearly the expected current decrease above the pair of individual Retzius cells that correspond in size and shape with the optical appearance of the investigated cells. Although this image shows a clear contrast and good signal-to-noise ratio, the interpretation of this type of experiments might be more difficult than it initially seems to be. The delicate situation for the interpretation of oxygen measurements at living cells in constant-height mode SECM is explained in more detail in Fig. 2. A breathing cell decreases the oxygen concentration in its proximity and hence a smaller tip current is expected. However, the tip current is distance dependent. A decreased tip-to-sample distance leads likewise to a reduced tip current due to the hindered diffusion of oxygen from the bulk solution into the gap between the tip and sample. Due to the comparatively large dimensions of a cell body, which is comparable with the positioning distance of a conventional 10 μm diameter SECM tip, the topography contribution is not negligible. In order to demonstrate this superposition of both effects, a control experiment was performed, adding a freely diffusing redox mediator to determine the tip-to-sample separation (scheme in Fig. 2a) similar to an approach described by Gonsalves *et al.*[17] for array scan experiments. $[Fe(CN)_6]^{4-}$ was chosen as a hydrophilic mediator that does not cross the cell membrane and is therefore suitable to investigate the topography in the negative feedback mode.[7,8] The current signal of the competition arrangement in Fig. 2b shows a current decrease at the positions of the cells. Taking the tip current for the additionally added redox mediator into account (grey line in Fig. 2b) it

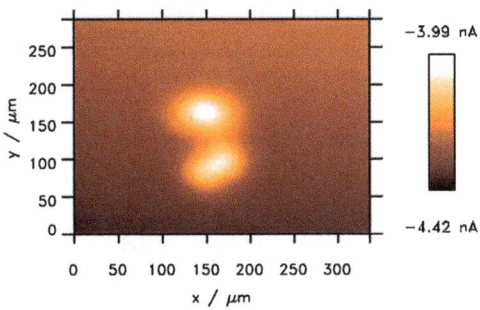

Fig. 1 Constant-height SG/TC mode SECM image of a pair of adherently growing Retzius cells from the medicinal leech in a competition mode arrangement.

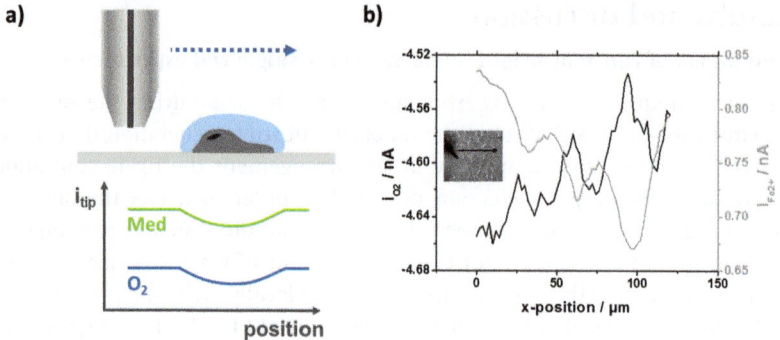

Fig. 2 (a) Scheme of the current signal in a constant-height mode SECM line scan across an adherently growing single cell. At each x,y-grid point the tip potential was changed between oxygen detection in a competition arrangement and the feedback mode detection of an additionally added free diffusing redox mediator for the investigation of the tip-to-sample distance. (b) SECM line scan across a HEK293 cell in a solution containing 0.5 mM [Fe(CN)$_6$]$^{4-}$. Tip: platinized carbon fibre electrode (diameter 7 μm), $E_{tip,O2}$ = −600 mV, $E_{tip,Fe2+}$ = +500 mV. (black line: O$_2$ reduction; grey line: [Fe(CN)$_6$]$^{4-}$ oxidation). Insert: Optical micrograph taken with the camera of the inverted microscope of the SECM setup showing the direction and position of the line scan.

becomes obvious that the topography expectedly affects the tip signal. Consequently, the particular challenge for respiration studies in constant-height mode at single cells is in fact seen; that a current decrease is caused on the one hand by the oxygen consumed by the living cell and on the other hand by a smaller tip-to-sample distance. As a matter of fact, a deconvolution of both contributions is difficult. This is even more important because oxygen consumption due to respiration is slow while the current decrease due to the negative feedback effect is substantial.

Beside this superposition of topographic and activity effects, the steep elevated structure of the cells leads to a high risk of contact between tip and cell body causing a mechanical stimulation or a destruction of the fragile cell membrane. Furthermore, the utilization of an additional redox mediator might influence the metabolism of the investigated cell. On the basis of these results it is obvious that a current-independent distance control mechanism is required for a reliable investigation of living cells.

Distance controlled SECM investigation in the competitive SG/TC mode

In order to overcome the above-mentioned drawbacks, shearforce-based constant-distance mode SECM experiments have been applied in a further step for SECM respiration studies. A line scan in the shearforce-based constant-distance mode utilizing the competitive SG/TC mode across a PC12 cell is shown in Fig. 3. The topographic line scan is represented by the z-position of the tip when the set point of the shearforce-based distance control is reached. The tip followed the contour of the PC12 cell in a working distance of about 100 to 200 nm, starting at a position above the Petri dish and moving in the direction of the PC12 cell body. The topography scan shows clearly the elevated structure of the cell body with a height of about 11 μm. Contrary to the expected current decrease, a current increase was detected above the cell body at the tip.

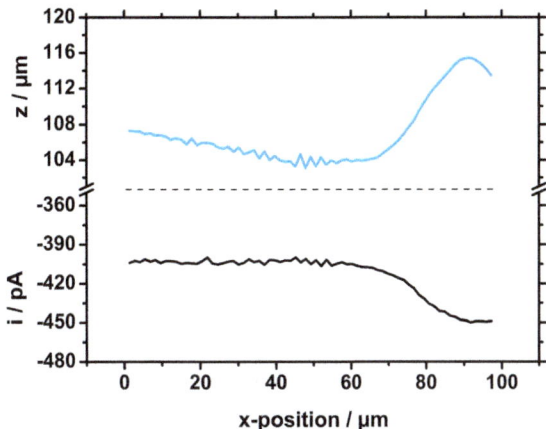

Fig. 3 Shearforce-based constant-distance mode SECM line scan above a PC12 cell. Investigation of the respiration activity in the competitive SG/TC mode beginning from a position above the Petri dish in the direction of the cell body.

Due to the semipermeability of the cell membrane, single cells exhibit similarities to a liquid/liquid interface in a SECM experiment.[34] At high oxygen reduction rates at the tip, a depletion of oxygen in the vicinity of the cell occurs. Oxygen is known to easily cross the cellular membrane.[8] As a result, a diffusional flux from the intracellular medium occurs along the concentration gradient, unintentionally leading to a situation similar to the previously described SECM-induced transfer (SECM-IT) mode.[35] Evidently, the living cell has to be considered like a sponge filled with oxygen and hence as an additional source for oxygen. The tip reaction actively intervenes and dominantly determines the oxygen concentration inside the gap between tip and cell which is even more pronounced due to the extremely short working distance established by shearforce positioning. As a matter of fact, this additional flux of oxygen interferes with the detection of respiration activity. Thus, the tip cannot be considered any longer as a passive spectator, which implies that the properties of the tip itself will largely alter the obtained SECM image. It is clear that under these conditions an error-free detection of cell respiration is impossible. Although this effect has been described in literature[8,17,26,27,36] it was not considered in cell respiration studies by means of SECM. Most studies try to circumvent this effect by using smaller electrodes; however, in order to achieve measurable currents even small electrodes may compete substantially with the slow impact of cell respiration on the oxygen concentration in the gap between tip and cell.

To verify this observation in more detail and to prove that the current increase is caused by the proposed transmembrane flux of oxygen into the gap, a control experiment was performed at another PC12 cell (Fig. 4). In contrast to the first experiment, $[Fe(CN)_6]^{4-}$ was added and used as redox mediator for an additional determination of the tip-to-sample distance in a negative feedback mode configuration. In a reliable constant-distance mode experiment no change in the oxidation current of $[Fe(CN)_6]^{4-}$ is expected.

In order to enable a simultaneous evaluation of two different redox species, the potential at the tip was changed at each point of the x,y-grid between −600 mV and

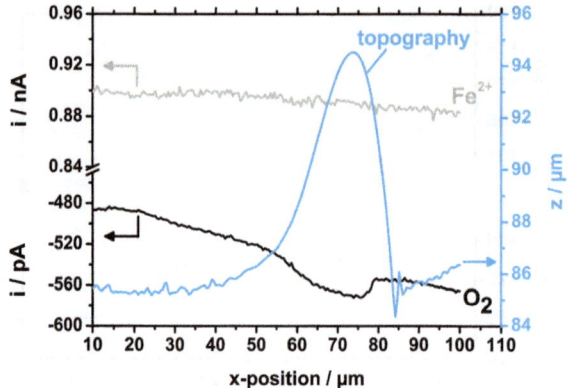

Fig. 4 Constant-distance mode SECM line scan above a PC12 cell in a cell buffer solution containing 30 µM $K_4[Fe(CN)_6]$ for the independent determination of the working distance in the feedback mode. The potential was changed between −600 mV for oxygen detection and +500 mV for the diffusion-limited oxidation of the $[Fe(CN)_6]^{4-}$ species. A current increase during oxygen detection was observed above the investigated cell. The oxidation current of $[Fe(CN)_6]^{4-}$ is not changed by passing the cell body, proving the accuracy of the shearforce distance control.

+500 mV, respectively. Again, the topography of the investigated cell body is clearly visualized using the shearforce-based z-positioning and a height of about 9 µm can be derived for the specific cell. As in the previous experiment an increase of the oxygen current was detected above the PC12 cell. However, no current change was observed for $[Fe(CN)_6]^{4-}$ oxidation proving the reliability of the shearforce-based constant-distance positioning. Hence, any possible impact of an error in the distance control can be excluded as the source for the unexpected increase in the oxygen reduction current above the cells. Furthermore, the observed behaviour is clearly due to the chosen experimental design and not an effect detected at a specific cell type that is demonstrated by Fig. 5 with the results of a 4D SF/CD-SECM experiment at a HEK293 cell. The detection procedure of the 4D SF/CD mode facilitates constant-distance mode experiments in various tip-to-sample separations. Therefore, this mode is able to visualize complete diffusion fields in front of a sample surface.[33] The topography image (Fig. 5a) clearly reflects the orientation and shape of the investigated cell agglomeration as shown in the optical micrograph (Fig. 5b). The tip current response is displayed in Fig. 5c. A current increase was detected above the investigated cells and confirms the results achieved with the PC12 cells.

In order to analyse further the distance dependence of the tip current, an electrochemical tomogram (Fig. 5d) was extracted from the 4D dataset at the line marked in Fig. 5c. The observed current increase reaches far into the solution and it is detectable even at a distance of 15 µm above the point of shearforce contact. Control experiments at glass beads and polymer spheres with comparable sizes to the cell bodies as model systems were performed to further evaluate the extraction of oxygen out of the cell body. As expected, no increase of the oxygen reduction current was observed during similar experiments to those shown in Fig. 5 using glass beads with a 10 µm diameter instead of the cells. However, if polystyrene beads of similar size were used to replace the cells, a small current increase was at least detected, most likely due to the soft outer layer of the beads which is soaked with electrolyte solution (results not shown).

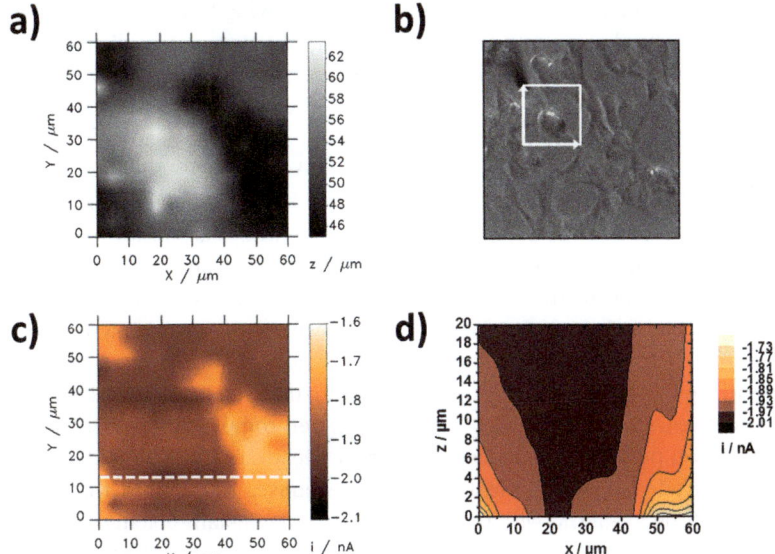

Fig. 5 Investigation of a HEK293 cell sample by means of 4D SF/CD-SECM in the competition mode arrangement. (a) Topography image displayed as the z-position of the tip after reacting the pre-set stop criterion (the point of closest approach) during the shearforce-based z-approach curve, (b) optical micrograph of the cell sample marking the investigated area and the scan directions. (c) Tip current in the plane of closest approach at $E_{tip} = -600$ mV. A significant current increase is observed above the inspected cell. (d) Electrochemical tomogram (x,z,i-image) extracted at the line marked in (c). The position of shearforce contact is defined as 0 μm.

Based on these observations three sources of oxygen can be supposed affecting the tip current: (1) Oxygen consumed by respiration activity of the investigated cell ($O_{2,R}$). The amount of oxygen consumed by a single cell during the timescale of a SECM experiment is supposed to be small compared to the overall oxygen dissolved in the measuring solution. The consumption of oxygen by a breathing cell causes only small current variations as compared with the high background current. (2) Oxygen that is transported from the bulk solution to the tip by diffusion ($O_{2,D}$). The diffusion of oxygen inside the gap depends on the working distance and on the geometrical size of the tip, thus a similar contribution is assumed for each tip-to-sample separation of the constant-distance mode experiment. (3) Oxygen reaching the tip due to permeation through the cell membrane ($O_{2,P}$). This flux is induced by a tip that consumes a large amount of oxygen in the surroundings of the inspected cell and evokes therefore a concentration gradient between the intracellular and extracellular medium. This contribution is not related to the cell activity and superimposes the current signal significantly.

Characterisation and adjustment of the influence of the tip reaction onto the overall imaging result

In order to visualize respiration activity successfully, a variation of the SECM detection scheme is required. The central parameter initiating a disturbance in the established detection procedure is the driving force of the oxygen reduction at the tip electrode. A simple method to reduce the driving force is a decrease of the applied potential. However, even at a weak reduction potential of −200 mV visualization of cell respiration was not successful (data not shown). Further decreasing the electrode size or limiting the diffusion field in front of the tip through fast scan cyclic voltammetry (FSCV) or potential pulses are additional strategies to diminish the oxygen reduction rate. While FSCV enables the polarisation of the tip for short timescales and a detection of different redox species,[37] a time-resolved current detection during short timeframes is possible *via* the redox competition mode of scanning electrochemical microscopy (RC-SECM).[38] Originally designed for the investigation of heterogeneous oxygen reduction catalysts, the RC-SECM mode includes the competition between tip and catalyst sample for the oxygen inside the gap. Furthermore, oxygen depletion is avoided by a potential pulse profile applied at the tip comprising injection of oxygen by means of oxidative water splitting and a following oxygen competition pulse. During the competition pulse a time dependent current decay curve is recorded enabling a time-resolved analysis of the tip current response. Combining the detection scheme of the RC-SECM with the shearforce-based constant-distance mode leads to a SECM mode (SF/CD-RC-SECM, Fig. 6) that is able to limit the tip reaction by short potential pulses in a constant working distance.

The application of the SF/CD-RC-SECM to HUVEC cells is shown in Fig. 7. A crucial experimental detail is the choice of the potential applied during electrode movement. In order to minimize the overall amount of oxygen consumed by the tip, a potential of +500 mV was applied during the z-positioning and lateral motion of the tip to the next *x,y*-grid point. At this potential no oxygen reduction occurs in a buffer solution of pH 7.4. Immediately before the competition experiment, oxygen is injected by a short potential pulse to 1.4 V. The topography image of the investigated cell is shown in Fig. 7a and the corresponding optical micrograph of the cells with an indication of the scan area and scan direction is displayed in Fig. 7b. Due to the time-resolved data acquisition of the RC-SECM detection scheme, a dataset of 100 current images is achieved during one single array scan. A selection of current images at times of

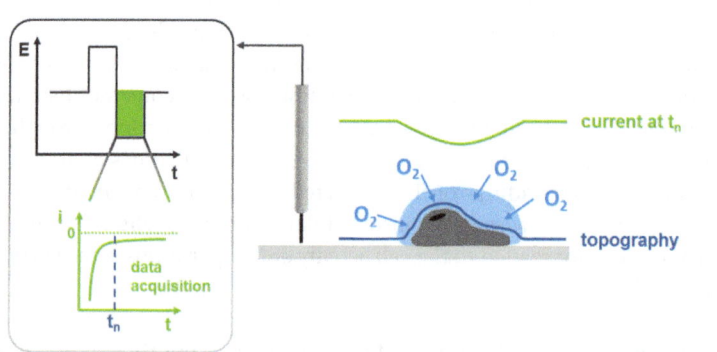

Fig. 6 Scheme of the redox competition mode (RC SECM) implemented in the shearforce-based constant-distance SECM. A flexible potential pulse profile is applied at the tip at each point of the *x,y*-grid. In a first pulse oxygen is injected in the gap through oxidative water splitting. During the following competition pulse, the tip and cell compete for the oxygen present inside the gap and a time-dependent data acquisition is realized leading to a transient detection scheme.

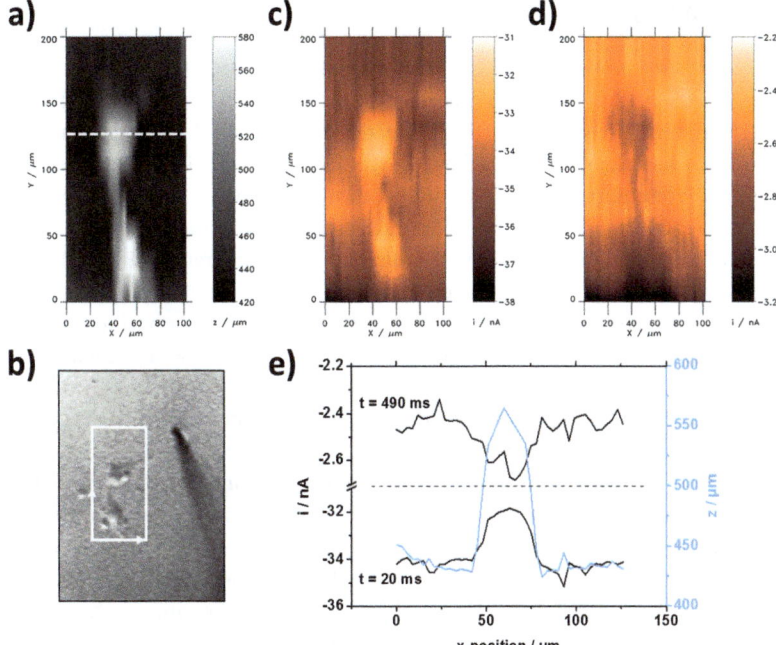

Fig. 7 Constant-distance mode SECM experiment in the RC mode at HUVEC cells. (a) Topography image of the investigated cells displayed as the tip z-position after establishing a constant working distance. (b) Optical micrograph of the investigated cells with a mark of the scanned area and the scan directions, (c) current signal at $t = 20$ ms and d) current image at $t = 490$ ms after the competition pulse was applied at the SECM tip. In both current images a lighter colour represents a current decrease while a current increase appears in a darker colour. (e) Line scans extracted from the array scans at the position marked in (a). Potential pulse profile: base potential: 0.5 V, pulse 1 (oxygen injection pulse): 0.2 s at 1.4 V, and pulse 2 (competition pulse): 0.5 s at −0.65 V.

20 ms and 490 ms after the competition pulse was applied are shown in Fig. 7c and figure 7d, respectively. For a direct comparison of both tip responses at the chosen acquisition times, line scans (Fig. 7e) have been extracted at the position marked in Fig. 7a.

At short pulse times a current decrease was observed directly above the inspected cell bodies. Obviously, the influence of the positioned tip was successfully diminished enabling a visualization of the locally reduced amount of oxygen due to cell respiration. However, during the course of the potential pulse a change of the tip signal was detected. At 490 ms after applying the competition pulse at the tip a current increase was measured at the x,y-positions where the topography image confirmed the presence of the cell body. At this time the expansion of the diffusion field in front of the tip electrode was sufficient to cause a depletion of oxygen inside the gap, thus disturbing the respiration detection. By means of the SF/CD-RC-SECM a visualization of the transition from a situation in which the tip acts as a passive observer to a tip that actively sucks out oxygen from the cell body was achieved. Furthermore, the time-dependent data acquisition of this detection scheme provides a tool to adapt the tip reaction to the unique requirements of the performed investigation of the cell activity.

Conclusions

The respiration activity of single cultured cells was investigated by means of constant-distance mode SECM and existing limitations of established detection procedures were presented. Furthermore, the influence of the reaction rate at the scanned tip was identified as the key element for a successful respiration study. In order to overcome restrictions arising from a high driving force of the electrochemical reaction at the SECM tip, a flexible potential pulse profile in combination with a time-dependent data acquisition was integrated in the detection scheme of the shearforce-based constant-distance mode. The proposed SF/CD-RC-SECM enabled a detailed study of the interaction of the involved reaction rates through the time-resolved detection procedure. At short pulse times a successful visualization of the respiration activity was achieved due to the minimized interference from the tip reaction. Additionally, the expanding diffusion field in front of the tip electrode within the duration of the applied competition pulse caused the formation of a concentration gradient that was compensated by a transmembrane diffusion of oxygen. At the end of the competition pulse the flux of oxygen from the intracellular medium to the outer side of the cell wall was superimposed over the contribution of the cell respiration. The SF/CD-RC-SECM detection scheme enabled the direct observation of the change from a passive observer that is able to detect the tiny changes in the local oxygen concentration caused by the breathing cell to an electrode that provokes actively a depletion of oxygen inside the gap between the tip and cell body and thus disturbing the respiration detection. Due to the detection of a series of individual current images at different pulse lengths during one single SF/CD-RC-SECM experiment, a detailed knowledge of the optimal time for the data acquisition is not required and the image with the best contrast can be chosen after the experiment was completed. For future work an additional reduction of the tip electrode size is proposed as a supplementary strategy to further decrease the amount of oxygen consumed by the scanned tip concomitantly increasing the lateral resolution.

Acknowledgements

The authors are thankful to all colleagues who contributed to this project over many years of attempts to finally detect oxygen consumption of living cells in a reliable way, particularly Dr Andreas Hengstenberg, Dr Sonnur Isik-Uppenkamp and Dr Kathrin Eckhard. Moreover, valuable discussions with Prof. Dr Tomokazu Matsue are acknowledged. Dr A. Blöchl and Prof. Dr I. D. Dietzel-Meyer (Lehrstuhl für Molekulare Neurobiochemie, Ruhr-Universität Bochum, Germany) as well as Dr D. Hollatz, S. Zielke and Dr C. H. Wetzel (Lehrstuhl für Zellphysiologie, Ruhr-Universität Bochum, Germany) are acknowledged for providing cell samples. This work was supported by the EU and the state NRW in the frame work of the HighTech.NRW program and the Center for Electrochemical Sciences – CES.

References

1 J. V. Sweedler and E. A. Arriaga, *Anal. Bioanal. Chem.*, 2006, **387**, 1–2.
2 R. N. Adams, *Anal. Chem.*, 1976, **48**, 1128A–1138A.
3 (*a*) C. Amatore, S. Arbault, M. Guille and F. Lemaitre, *Chem. Rev.*, 2008, **108**, 2585–2621; (*b*) A. Schulte and W. Schuhmann, *Angew. Chem., Int. Ed.*, 2007, **46**, 8760–8777; (*c*)

S. Borgmann, *Anal. Bioanal. Chem.*, 2009, **394**, 95–105; (d) A.-S. Cans and A. G. Ewing, *J. Solid State Electrochem.*, 2011, **15**, 1437–1450.
4 (a) A. J. Bard, F. R. F. Fan, J. Kwak and O. Lev, *Anal. Chem.*, 1989, **61**, 132–138; (b) R. C. Engstrom, M. Weber, D. J. Wunder, R. Burgess and S. Winquist, *Anal. Chem.*, 1986, **58**, 844–848.
5 A. Hengstenberg, A. Blöchl, I. D. Dietzel and W. Schuhmann, *Angew. Chem., Int. Ed.*, 2001, **40**, 905–908.
6 L. P. Bauermann, W. Schuhmann and A. Schulte, *Phys. Chem. Chem. Phys.*, 2004, **6**, 4003–4008.
7 B. Liu, S. Rotenberg and M. Mirkin, *Proc. Natl. Acad. Sci. U. S. A.*, 2000, **97**, 9855–9860.
8 B. Liu, W. Cheng, S. A. Rotenberg and M. V. Mirkin, *J. Electroanal. Chem.*, 2001, **500**, 590–597.
9 X. Li and A. J. Bard, *J. Electroanal. Chem.*, 2009, **628**, 35–42.
10 T. Yasukawa, Y. Kondo, I. Uchida and T. Matsue, *Chem. Lett.*, 1998, 767–768.
11 (a) T. Yasukawa, T. Kaya and T. Matsue, *Chem. Lett.*, 1999, 975–976; (b) M. Nishizawa, K. Takoh and T. Matsue, *Langmuir*, 2002, **18**, 3645–3649; (c) P. M. Diakowski and Z. F. Ding, *Phys. Chem. Chem. Phys.*, 2007, **9**, 5966–5974.
12 (a) T. Kaya, Y. S. Torisawa, D. Oyamatsu, M. Nishizawa and T. Matsue, *Biosens. Bioelectron.*, 2003, **18**, 1379–1383; (b) L. L. Zhu, N. Gao, X. L. Zhang and W. R. Jin, *Talanta*, 2008, **77**, 804–808.
13 S. Isik, M. Etienne, J. Oni, A. Blochl, S. Reiter and W. Schuhmann, *Anal. Chem.*, 2004, **76**, 6389–6394.
14 (a) S. Isik, J. Castillo, A. Blochl, E. Csoregi and W. Schuhmann, *Bioelectrochemistry*, 2007, **70**, 173–179; (b) T. Yasukawa, T. Kaya and T. Matsue, *Anal. Chem.*, 1999, **71**, 4637–4641.
15 (a) H. Shiku, T. Shiraishi, H. Ohya, T. Matsue, H. Abe, H. Hoshi and M. Kobayashi, *Anal. Chem.*, 2001, **73**, 3751–3758; (b) H. Shiku, T. Shiraishi, S. Aoyagi, Y. Utsumi, M. Matsudaira, H. Abe, H. Hoshi, S. Kasai, H. Ohya and T. Matsue, *Anal. Chim. Acta*, 2004, **522**, 51–58; (c) Y. S. Torisawa, T. Kaya, Y. Takii, D. Oyamatsu, M. Nishizawa and T. Matsue, *Anal. Chem.*, 2003, **75**, 2154–2158; (d) Y. Torisawa, H. Shiku, T. Yasukawa, M. Nishizawa and T. Matsue, *Sens. Actuators, B*, 2005, **108**, 654–659; (e) R. Obregon, Y. Horiguchi, T. Arai, S. Abe, Y. Zhou, R. Takahashi, A. Hisada, K. Ino, H. Shiku and T. Matsue, *Talanta*, 2012, **94**, 30–35.
16 M. Tsionsky, Z. G. Cardon, A. J. Bard and R. B. Jackson, *Plant Physiol.*, 1997, **113**, 895–901.
17 M. Gonsalves, A. L. Barker, J. V. Macpherson, P. R. Unwin, D. O'Hare and C. P. Winlove, *Biophys. J.*, 2000, **78**, 1578–1588.
18 (a) Y. Hirano, Y. Nishimiya, K. Kowata, F. Mizutani, S. Tsuda and Y. Komatsu, *Anal. Chem.*, 2008, **80**, 9349–9354; (b) M. M. N. Zhang, Y.-T. Long and Z. Ding, *J. Inorg. Biochem.*, 2012, **108**, 115–122.
19 (a) J. M. Liebetrau, H. M. Miller, J. E. Baur, S. A. Takacs, V. Anupunpisit, P. A. Garris and D. O. Wipf, *Anal. Chem.*, 2003, **75**, 563–571; (b) W. Wang, Y. Xiong, F.-Y. Du, W.-H. Huang, W.-Z. Wu, Z.-L. Wang, J.-K. Cheng and Y.-F. Yang, *Analyst*, 2007, **132**, 515–518; (c) P. Sun, F. O. Laforge, T. P. Abeyweera, S. A. Rotenberg, J. Carpino and M. V. Mirkin, *Proc. Natl. Acad. Sci. U. S. A.*, 2008, **105**, 443–448; (d) Y. Takahashi, A. I. Shevchuk, P. Novak, B. Babakinejad, J. Macpherson, P. R. Unwin, H. Shiku, J. Gorelik, D. Klenerman, Y. E. Korchev and T. Matsue, *Proc. Natl. Acad. Sci. U. S. A.*, 2012, **109**, 11540–11545.
20 R. T. Kurulugama, D. O. Wipf, S. A. Takacs, S. Pongmayteegul, P. A. Garris and J. E. Baur, *Anal. Chem.*, 2005, **77**, 1111–1117.
21 M. A. Alpuche-Aviles and D. O. Wipf, *Anal. Chem.*, 2001, **73**, 4873–4881.
22 (a) M. Ludwig, C. Kranz, W. Schuhmann and H. E. Gaub, *Rev. Sci. Instrum.*, 1995, **66**, 2857–2860; (b) P. I. James, L. F. Garfias-Mesias, P. J. Moyer and W. H. Smyrl, *J. Electrochem. Soc.*, 1998, **145**, L64–L66; (c) B. Ballesteros Katemann, A. Schulte and W. Schuhmann, *Chem.-Eur. J.*, 2003, **9**, 2025–2033.
23 J. V. Macpherson, P. R. Unwin, A. C. Hillier and A. J. Bard, *J. Am. Chem. Soc.*, 1996, **118**, 6445–6452.
24 D. J. Comstock, J. W. Elam, M. J. Pellin and M. C. Hersam, *Anal. Chem.*, 2010, **82**, 1270–1276.
25 (a) X. C. Zhao, P. M. Diakowski and Z. F. Ding, *Anal. Chem.*, 2010, **82**, 8371–8373; (b) S. Isik and W. Schuhmann, *Angew. Chem., Int. Ed.*, 2006, **45**, 7451–7454; (c) R. J. Fasching, S. J. Bai, T. Fabian and F. B. Prinz, *Microelectron. Eng.*, 2006, **83**, 1638–1641.
26 Y. Takahashi, Y. Hirano, T. Yasukawa, H. Shiku, H. Yamada and T. Matsue, *Langmuir*, 2006, **22**, 10299–10306.
27 Y. Takahashi, A. I. Shevchuk, P. Novak, Y. Murakami, H. Shiku, Y. E. Korchev and T. Matsue, *J. Am. Chem. Soc.*, 2010, **132**, 10118–10126.
28 I. D. Dietzel, P. Drapeau and J. G. Nicholls, *J. Physiol.*, 1986, **372**, 191–205.
29 L. A. Greene and A. S. Tischler, *Proc. Natl. Acad. Sci. U. S. A.*, 1976, **73**, 2424–2428.

30 K. Klasen, D. Hollatz, S. Zielke, G. Gisselmann, H. Hatt and C. H. Wetzel, *Pfluegers Arch.*, 2012, **463**, 779–797.
31 A. Hengstenberg, C. Kranz and W. Schuhmann, *Chem.-Eur. J.*, 2000, **6**, 1547–1554.
32 A. Schulte and R. H. Chow, *Anal. Chem.*, 1996, **68**, 3054–3058.
33 M. Nebel, K. Eckhard, T. Erichsen, A. Schulte and W. Schuhmann, *Anal. Chem.*, 2010, **82**, 7842–7848.
34 S. Amemiya, J. D. Guo, H. Xiong and D. A. Gross, *Anal. Bioanal. Chem.*, 2006, **386**, 458–471.
35 A. L. Barker, J. V. Macpherson, C. J. Slevin and P. R. Unwin, *J. Phys. Chem. B*, 1998, **102**, 1586–1598.
36 K. Nagamine, Y. Takahashi, K. Ino, H. Shiku and T. Matsue, *Electroanalysis*, 2011, **23**, 1168–1174.
37 (*a*) L. Díaz-Ballote, M. Alpuche-Aviles and D. O. Wipf, *J. Electroanal. Chem.*, 2007, **604**, 17–25; (*b*) D. S. Schrock, D. O. Wipf and J. E. Baur, *Anal. Chem.*, 2007, **79**, 4931–4941; (*c*) D. S. Schrock and J. E. Baur, *Anal. Chem.*, 2007, **79**, 7053–7061; (*d*) K. L. Adams, M. Puchades and A. G. Ewing, *Annu. Rev. Anal. Chem.*, 2008, **1**, 329–355.
38 (*a*) K. Eckhard, X. Chen, F. Turcu and W. Schuhmann, *Phys. Chem. Chem. Phys.*, 2006, **8**, 5359–5365; (*b*) K. Eckhard and W. Schuhmann, *Electrochim. Acta*, 2007, **53**, 1164–1169.

Faraday Discussions

Vesicular release of neurotransmitters: converting amperometric measurements into size, dynamics and energetics of initial fusion pores†

Alexander Oleinick, Frédéric Lemaître, Manon Guille Collignon, Irina Svir and Christian Amatore*

Received 4th March 2013, Accepted 15th March 2013
DOI: 10.1039/c3fd00028a

Amperometric currents displaying a pre-spike feature (PSF) may be treated so as to lead to precise information about initial fusion pores, *viz.*, about the crucial event initiating neurotransmitter vesicular release in neurons and medullary glands. However, amperometric data alone are not self-sufficient, so their full exploitation requires external calibration to solve the inverse problem. For this purpose we resorted to patch-clamp measurements published in the literature on chromaffin cells. Reported pore radii were thus used to evaluate the diffusion rate of neurotransmitter cations in the partially altered matrix located near the fusion pore entrance. This allowed an independent determination of each initial fusion pore radius giving rise to a single PSF event. The statistical distribution of the radii thus obtained provided for the first time an experimental access to the potential energy well governing the thermodynamics of such systems. The shape of the corresponding potential energy well strongly suggested that, after their creation, initial fusion pores are essentially controlled by the usual physicochemical laws describing pores formed in bilayer lipidic biological membranes, *i.e.*, they have an essentially lipidic nature.

Introduction

In exocytotic cells neurotransmitters are transported inside vesicles, in which they are densely packed.[1] Whenever release into the extracellular environment is needed, the gated-entrance of calcium ions into the cytoplasm initiates a process stimulating primed vesicles to fuse with the cytoplasmic cell membrane. This fusion begins through the opening of a nanometric liquid channel connecting the vesicle with the extracellular space so that neurotransmitters may be released towards a target cell (synaptic cells: neurons and nerve terminals) or into

Ecole Normale Supérieure, Département de Chimie, UMR ENS-CNRS-UPMC 8640 "PASTEUR", 24 rue Lhomond, 75005 Paris, France. E-mail: christian.amatore@ens.fr; Tel: +33-(0)1-4432-3788

† Dedicated to the late Professor Louis Nadjo, a lamented friend and colleague but most of all Christian Amatore's first tutor in experimental electrochemistry.

circulating fluids (gland cells).[1] It is commonly agreed that, irrespective of the cell nature (synaptic or medullar gland cell), this central process occurs through several distinct steps initiated by the flow of calcium ions into the cell cytoplasm. Calcium ion entry promotes the assembly of specific proteins (SNAREs) anchored in the cell and vesicle membranes. Progressive tightening of the SNAREs complexes forces the vesicle to enter in close contact with the cell membrane (the "docking" stage, Fig. 1). After docking, a narrow pore of nanometric dimensions, named the initial fusion pore, spontaneously forms across the two membranes so that the vesicle content may diffuse into the synaptic cleft (synaptic cells) or into the circulating fluids (gland cells).

Unless the initial fusion pore closes after some duration (as is believed to occur in neurons, or has been observed during "kiss and run" events for gland cells)[3] it eventually suddenly quickly enlarges through a rapid flow of the vesicle bilayer membrane into that of the cell.[4a] This process, named "full fusion", is completed within a few milliseconds and unmasks a larger fraction of the vesicle matrix surface, thus allowing a massive release compared to the extremely limited one which may occur through the small initial fusion pore (Fig. 1).

Since its first characterization by patch-clamp techniques,[5,6] the initial fusion pore has stimulated numerous studies, echoing its crucial importance in initiating the whole process as well as crucial fundamental demands concerning its size, nature and energetics. It now seems clear that its initial formation requires the essential involvement of proteins such as SNAREs and calcium ions, which permit the two membranes to overcome their repulsion and come into sufficiently close contact to initiate a local merging.[7,8] Many experiments support this view, and in particular, the exocytotic frequency is drastically decreased when SNAREs are disabled, e.g., after botulic treatment.[9,10] However, it seems that other local cellular sub-structures (e.g., the surface cytoskeleton,[11] molecular motors, etc.) may also be involved during this phase, since exocytotic events appear to occur only at a few precisely localized spots on the cell membranes.[4b,12] Therefore there is no doubt that the initial pore formation involves a series of organized bio-structures on

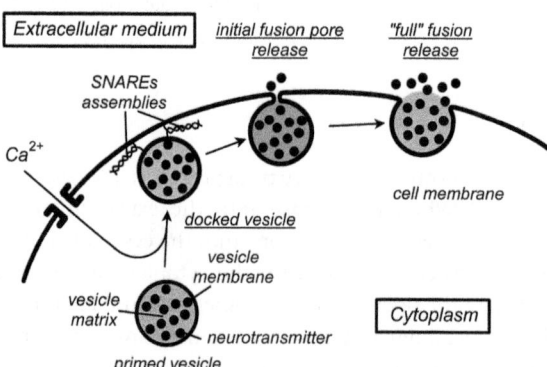

Fig. 1 Schematic representation of the main sequential stages of a vesicular exocytosis event. Neurotransmitter molecules are represented by filled circles initially contained inside the fusing vesicle. Note that, for the sake of simplicity, possible SNAREs assemblies existing in the fusion stages after creation of the initial fusion pore are not shown since this is still a matter of debate; similarly, only a toroid topology is shown for the "full" fusion stage but the vesicle opening may be connected to that in the cell membrane through a lipidic tube (see Fig. 6 in the Appendix).[2]

which specific bio-machineries operate. Nevertheless, there is still no agreement about the very nature of the initial fusion pore after it has been created: does it involve purely lipidic components? proteic ones? any combination thereof? In other words, does it obey purely physicochemical laws or remain controlled biologically, at least in part?[13,14]

The present work aims to introduce a new approach for experimental characterization of the initial fusion pore dimensions and providing statistical elements about its energy since such data are critical for assessing its very nature. To the best of our knowledge, most of the quantitative experimental information about the initial biological fusion pores is topological and stems from two main sources: patch-clamp experiments (topological information reconstructed from dynamic events);[5–7,15–17] or high-power electron microscopy of prepared cell slices (direct topological observation from events "frozen" at different completion stages during the preparation).[18] Both techniques concur to establish that initial fusion pore radii are around 1.2 nm.

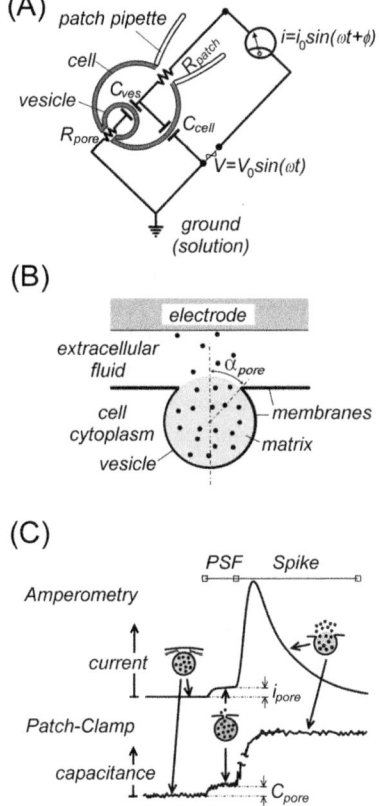

Fig. 2 Schematic principles of patch-clamp (A) and amperometry (B) and comparison of their respective measurements of a vesicular exocytotic event (C). (A) shows a "whole cell" patch-clamp configuration but several other configurations may be used.[5–7] (B) indicates the geometry of amperometric measurements in the "artificial synapse" configuration[19] and defines the opening angle, α_{pore}, characterizing the size of the pore at any fusion stage (this is shown for a toroid connection; see Fig. 6 in the Appendix for a tubular one[2]). In (C) the subscript "pore" stands for "initial fusion pore".

Patch-clamp measurements rely on dynamic measurement of impedance changes induced locally in the cell membrane by pore opening. These data are extracted in terms of their capacitive and resistive components, recorded as functions of time at an adequate frequency, on the basis of models initially developed by E. Neher et al. for the investigation of ion channels (see Fig. 2A).[15–17] Briefly, the vesicle and its opening fusion pore are represented by an equivalent electrical RC circuit in which the resistive component changes are mainly those of the liquid tube present in the pore and in the vesicle, while the capacitive ones mostly reflect the excess of surface area provoked by the pore opening, viz., is related to the pore radius (Fig. 2A).[20] These are converted into topological information through independent calibration.[21] Based on such investigations it is now widely accepted that fusion pores suddenly open (i.e., at submillisecond durations from the closed stage to complete opening).[22] Note that the experimental range of observed pore diameters (viz., 2.4 ± 0.7 nm; 68% confidence)[22] is much wider than the accuracy of patch-clamp measurements. Such a dispersion of measured values then suggests that the initial fusion pores display a rather wide intrinsic distribution of sizes, a fact that strongly mitigates against a purely proteic nature.

Initial fusion pores are also observable by amperometry, viz., through monitoring electrochemically the fluxes of released neurotransmitters through their rapid quantitative collection by their Faradaic oxidation at a carbon fiber ultramicroelectrode surface positioned at ca. 100 nm from the cell membrane so as to create an "artificial synapse" (Fig. 2B).[19] The full fusion release stage (see Fig. 1) is featured by an intense amperometric spike whose rising section characterizes mostly the exponential pore enlargement rate while its smoother descending branch represents mainly the progressive diffusional release of molecules from inside the vesicle (Fig. 2C).[23–25] More precisely, the exponential time decay of the amperometric current during this latest phase establishes that the pore radius has already completed its rapid growth phase and reached a constant value.[26,27] About 30% of such spikes are preceded by comparatively small specific current features (termed a "foot" or "pre-spike-feature", PSF) whose diffusional characteristics point to the fact that they represent diffusion of neurotransmitters through the nanometric initial fusion pore.[28–30] Diffusion laws are well established,[31] so one may think that PSF currents could be precisely deconvoluted from amperometric data so as to provide a fine description of the initial fusion pore topology.[32] Should it be feasible, amperometry could become a method of choice for the statistical investigation of initial fusion pore dimensions, since a single amperometric trace monitored from a single stimulated cell already offers several hundred PSF events with different characteristics (i.e., different durations, plateau intensities and shapes).

To develop such an amperometric approach one may rely on our previous theory of diffusion from spherical bodies impermeable to diffusing molecules over a fraction of their surface.[27] Yet, this would suppose knowing a priori either the rate of diffusion across the inside of the vesicle, $\kappa = D_{ves}/R_{ves}^2$ (where D_{ves} is the diffusion coefficient inside the vesicle and R_{ves} the radius of the vesicle), or the maximum cone angle α_{open}^{max} (see Fig. 2B) limiting the diffusion-permeable surface area when the fusion is complete, or the concentration of neurotransmitter, C_{ves}^{rel}, contained initially inside the vesicle and effectively released at the end of the process.[27] However, if the quantitative and kinetic information contained in the amperometric spike provides a useful relationship between these key parameters (e.g., see below eqn (1)),[27b] it cannot give access to any one of these three crucial values without blunt assumptions based on mean values.[23,24,33] Nevertheless, the large variability of vesicular events (see below) and present doubts about the completeness of full fusion[23,24] (*i.e.*, does $\alpha_{\text{open}}^{\max}$ really approach 180°?) preclude using mean values for the treatment of any particular event (note that hereafter, and in all Figures, angle values are always reported in degrees unless specified otherwise so that, when necessary, their conversion into radians requires a re-scaling by the factor $\pi/180$).

Hence, solving the inverse problem at hand requires an independent entry. We wish to show hereafter that comparing patch-clamp and PSF amperometric data provides such an external entry. This permits a quantitative access to the diffusion rate κ and therefore allows the extraction of topological information from individual PSF currents. Note that such a strategy does not differ conceptually and practically from that used in extracting pore dimensions from impedance measurements. Indeed, this also requires external input, *viz.*, independent calibrations of differential capacitances and resistivities.[21]

The following analysis is developed in the context of dense core vesicles, such as those of chromaffin cells which are investigated amperometrically here, but it may be readily adapted to other situations provided that precise patch-clamp and amperometric data are available.

Results and discussion

Position of the problem

It is clear that $\alpha_{\text{open}}^{\max}$ cannot be known *ex nihilo* from a current trace unless it is hypothesized that full fusion is total, *viz.*, that $\alpha_{\text{pore}}^{\max} = 180°$, as some of us have done previously[23,24] following prevailing views about full fusion at the time when these former works were performed.[4a] However, this is now an increasingly questioned issue.[13,14,34]

Alternatively, one could think of deriving $C_{\text{ves}}^{\text{rel}}$ through the amperometric charge Q determined during a whole release event since $C_{\text{ves}}^{\text{rel}} = Q/[2F(4\pi R_{\text{ves}}^3/3)]$, (where F is the Faraday constant, *i.e.*, 96,485 C mol^{-1}, and the factor 2 stems from the fact that oxidation of adrenaline cation involves a two-electron process[35]). However, application of this procedure would require that the vesicle radius, R_{ves}, is known with adequate precision for the vesicle undergoing the precise event under investigation. R_{ves} could be formally obtained from coupled experiments involving simultaneous TIRFM and amperometry measurements.[36] However, at their present stage, the maximum resolution of such experiments is not yet sufficient for the precision required here. Since R_{ves} exhibits a rather large distribution,[36] this prompted us to rely on external access to κ through calibrating PSF current extraction based on published patch-clamp data on the initial fusion pore radius value (1.2 nm).[22]

Extraction of diffusional rate by comparing PSF and published patch-clamp data

In the following we consider a statistically representative series of 80 amperometric events displaying distinct pre-spike features with different shapes and current intensities covering the whole spectrum of experimental variability observed for these events. These were first analyzed as described below to extract in each case the value of the diffusion rate $\kappa = D_{\text{ves}}/R_{\text{ves}}^2$, so that the amperometric PSF current intensity agreed with the accepted mean initial fusion pore radius value.

For this purpose the theoretical approach developed previously by some of us[27] was specifically adapted to small opening angles. The whole treatment was performed in terms of angles rather than dimensions since this allowed us to by-pass the unknown value of the radius, R_{ves}, of the particular vesicle giving rise to the PSF of concern. This required establishing a conversion between radius and angle scales. This was readily obtained by convoluting mathematically the two normal distributions reported for R_{ves}[37] and R_{pore}.[22] The median opening angle value corresponding to the initial fusion pore opening (see Fig. 2B) could then be evaluated at $(\alpha_{\text{pore}}^{\text{open}})^{\text{median}} \approx 0.44°$ (note that for the range of $(\alpha_{\text{pore}}^{\text{open}})^{\text{median}}$ values considered here, one has $\sin[\alpha_{\text{pore}}^{\text{open}}(\pi/180)]^{\text{median}} \approx (\pi/180)(\alpha_{\text{pore}}^{\text{open}})^{\text{median}}$ with ample accuracy). This value was taken as a normalization gauge used to derive a κ value from each experimental PSF current plateau through the following procedure.

The experimental current is proportional to the unknown $C_{\text{ves}}^{\text{rel}}$ value. However, it may be readily shown (see Appendix) that:

$$Q/i_{\text{pore}} \approx (\pi/3)(R_{\text{ves}}^3/D_{\text{ves}}R_{\text{pore}}) = \pi/[3(\pi/180)\kappa\alpha_{\text{pore}}^{\text{open}}] = 60/(\kappa\alpha_{\text{pore}}^{\text{open}}) \qquad (1)$$

so that equating $\alpha_{\text{pore}}^{\text{open}}$ to its gauge value (viz., 0.44°) provided a first entry for κ which allowed disentangling of the problem from the unknown $C_{\text{ves}}^{\text{rel}}$ value. This was further refined using a simulation program based on a specific conformal mapping allowing a fast and precise treatment of the high release flux near the edge of the fusion pore which enlarged with time[27] after specifically adapting it to very small angles. At each time step, an automatic feedback loop adjusted the time variations of $\alpha_{\text{pore}}(t)$ so that the simulated release flux tracked the experimental one, $\varphi(t) = i(t)/(2F)$ where $i(t)$ is the experimental amperometric current.[35] Such procedure generated a specific $\alpha_{\text{pore}}(t)$ function describing each single PSF event for the κ value used. Each $\alpha_{\text{pore}}(t)$ function displayed at least one plateau, $\alpha_{\text{pore}}^{\text{open}}$. In cases where a series of successive plateaus was obtained (see below and Fig. 4) the one which immediately preceded the exponential growth phase (see the asterisk marks in Fig. 4) was considered as representative of the initial fusion pore. This procedure was automatically iterated through changing κ values up to when an $\alpha_{\text{pore}}^{\text{open}}$ resulted equal to the gauge one, $\alpha_{\text{pore}}^{\text{gauge}} = 0.44°$. The above strategy was applied to a series of 80 representative amperometric events with PSFs of different shapes (i.e., plateaux or ramps) and different magnitudes (i.e., different $i_{\text{pore}}^{\text{plateau}}$ values) extracted from traces monitored on 5 batches of chromaffin cells with 7 μm diameter beveled carbon fiber ultramicroelectrodes according to a previously published procedure[30] (see the Experimental section and Fig. 2B).

Since we used a constant value for the gauge angle, this procedure led to a statistical distribution of 80 diffusion rate constant values (Fig. 3A). As expected for a series of single events, κ obeys closely an exponential distribution,[38a] viz., $P(\kappa) = \exp(-\kappa/\kappa_{\text{mean}})/\kappa_{\text{mean}}$ with $\kappa_{\text{mean}} = 415$ s^{-1} representing the statistically significant mean κ value.[38b] Note that since $R_{\text{ves}}^{\text{mean}} = 156$ nm[37] it follows that the mean diffusion coefficient of neurotransmitter in the vesicle matrix, partially altered[39] near the pore entrance, is $D_{\text{ves}}^{\text{mean}} = 1.0 \times 10^{-7}$ cm^2 s^{-1}, i.e., resulting in a value ca. 100 times smaller than in the extracellular medium. This justified a posteriori the assumption made in deriving eqn (A1)–(A3) in the Appendix, and hence validated our conclusion that the exact shape of the pore (viz., toroid or tubular) was mostly irrelevant.

Each PSF was then re-analyzed imposing $\kappa_{\text{mean}} = 415$ s^{-1} as the common diffusion rate constant.[38b] This produced a distribution of $\alpha_{\text{pore}}^{\text{open}}$ values (Fig. 3B)

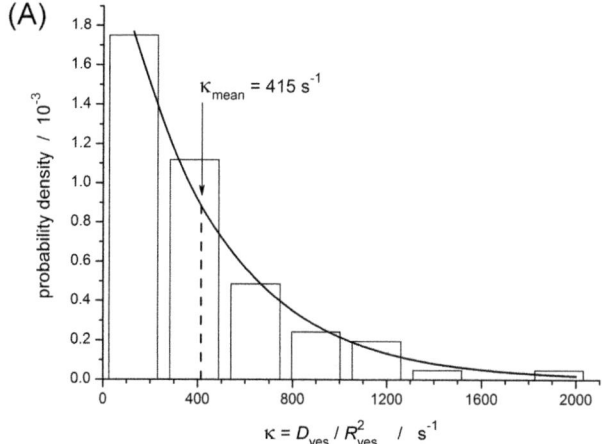
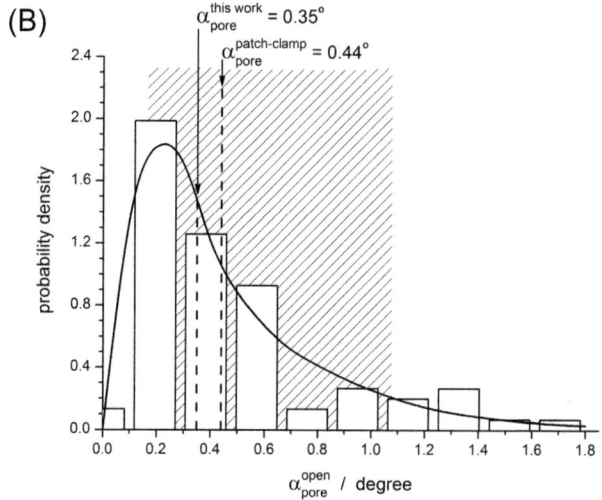
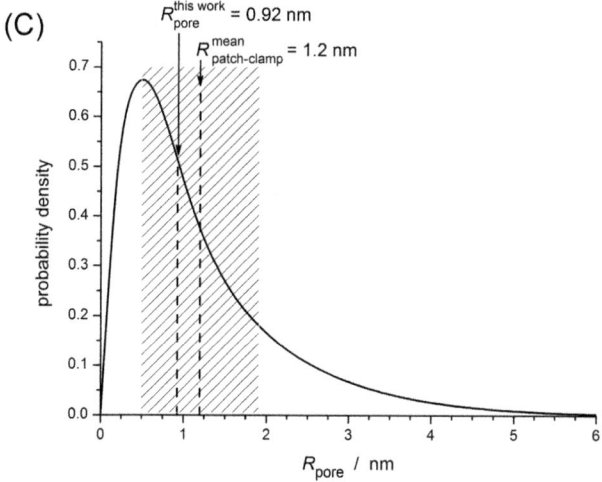

whose mean value was 0.35° instead of 0.44°, which was used for the gauge value. However, this apparent slight discrepancy was statistically insignificant; indeed, the present distribution was in perfect agreement with that predicted based on the analytical convolution between the distributions for patch-clamp (R_{pore}) and R_{ves} (Fig. 3B). Finally, convoluting the α_{pore}^{open} distribution with the reported R_{ves} normal one[37] afforded the distribution of R_{pore} values shown in Fig. 3C. This distribution also overlapped satisfactorily with that reported from patch-clamp measurements though it is more skewed towards high R_{pore} values.

The present analysis allowed the reconstruction of the opening time function of each initial fusion pore. Fig. 4 illustrates a few examples of such reconstructed time functions. Interestingly, some of these data show that the initial fusion pore may pass through two (ca. 10% of PSF) or even three (ca. 4% of PSF) stable successive stages during the pre-spike release, suggesting that the initial fusion pore does not always proceed irrevocably from its stable stage to its exponentially enlarging phase.[23,24,40–42] Such features were already present in the amperometric traces, though less noticeable due to their smearing out by the filtering effect of diffusion.[43] In this respect, the present analysis evidences that there is no fundamental difference between PSF with a plateau or ramp despite their experimental difference in shape.[30] To the best of our knowledge such a staircase transition of initial fusion pores towards their unstable state has never been reported before, except maybe in Fig. 2b of ref. 22, which reports one long patch-clamp trace with features which could be interpreted similarly. This may well represent progressive unlocking of SNARE assemblies[7] or changes of the membrane cytoskeleton[11,12] under the pressure created by the restrained swelling of the matrix near the pore entrance.[23,24,39] Alternatively, this may represent the progressive alteration of the matrix exposed to the extracellular medium (see below). However, the fact that there are no more than three such intermediate plateaus appears consistent with the SNAREs-related interpretation since three SNAREs complexes are supposed to be involved in the docking stage,[7] so that under some circumstances they may untangle sequentially within a few tens of milliseconds delay.

Most patch-clamp measurements indicate that initial fusion pores open suddenly.[4] Conversely, $\alpha_{pore}(t)$ time variations in Fig. 4 display rise times. Yet, this is not at all contradictory. Indeed, patch-clamp measurements involve electrokinetic measurements which track instantly (i.e., are limited only by the experimental apparatus time constants which are exceedingly small)[15–17] any sudden changes in capacitance and resistance. Conversely, amperometry requires that molecules be released by the matrix and diffuse away.[44] Such a process cannot initiate before the matrix volume near the pore entrance has partially swollen,[23,24,39] so as to allow the neurotransmitter to achieve a sufficiently high diffusion coefficient compared to its restricted mobility in the initially highly compacted granule.

Fig. 3 Statistical distributions based on 80 amperometric events displaying well-defined PSF: (A) κ values as obtained through the normalization gauge procedure (see text); (B) α_{pore}^{open} as obtained through using a common diffusion rate constant value $\kappa_{mean} = 415$ s^{-1}; (C) R_{pore} as obtained through the analytical convolution of α_{pore}^{open} distribution in (B) with that of R_{ves}[37] ($P(R_{pore}) = P(\alpha_{pore}^{open}) \otimes P(R_{ves})$). In (B,C) the hatched zones indicate the confidence intervals based on patch-clamp[22] and R_{ves}[37] published measurements (confidence levels: 95%). Values of $\alpha_{pore}^{this\ work}$ and $R_{pore}^{this\ work}$ are the medians of the respective distributions obtained in this work. Solid curves in (A) and (B) represent the statistical regressions determined based on the histograms. That in (C) results from the convolution between that in (B) with the normal distribution of vesicle radii.

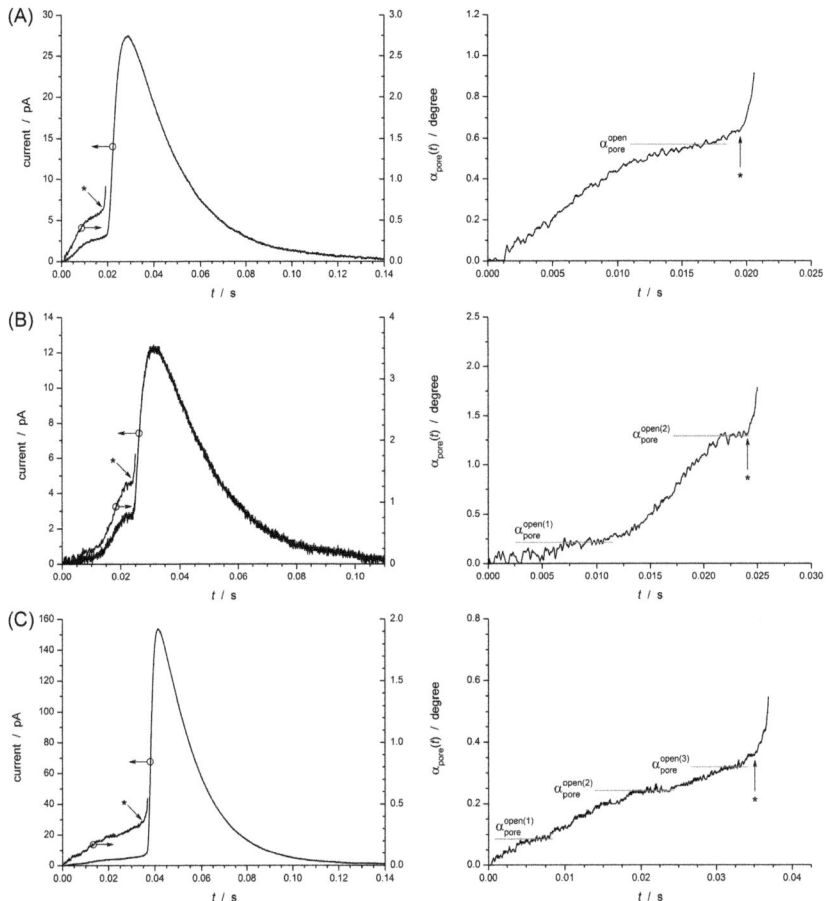

Fig. 4 Representative series of $\alpha_{pore}(t)$ time functions reconstructed using $\kappa = 415\ s^{-1}$ from a typical amperometric current trace exhibiting a well-defined PSF featuring single (A), double (B) and triple-stage (C) initial fusion pore transition towards the exponential growth phase (see insets). In each case the left column shows the actual current trace and the reconstructed $\alpha_{pore}(t)$ function while the right one gives an expanded view of it. For all reconstructed $\alpha_{pore}(t)$ time functions, the beginning of the exponential growth phase of the pores is indicated by an asterisk.

This process was shown to be fast[39a] but it cannot be instantaneous. Such a progressive local matrix distortion must thus appear as an apparent progressive increase in R_{pore} values in our analysis, which assumes a constant (*i.e.*, final) diffusion coefficient value at any time. As such, one may be tempted to conclude that the rise times shown in Fig. 4 feature an intrinsic bias in the present analysis and should then be discarded.[44] However, what matters biologically in the initial fusion pore opening is not really its opening but the flow of neurotransmitter it may release when it does so. Thus, rather than being an artifact, the rise times observed here provide extremely informative data on the biological role of fusion pores that patch-clamp measurements cannot. Although of extreme interest, a proper analysis of such phenomena is beyond the scope of this work and certainly requires an improvement in the time resolution of amperometric measurements (500 molecules per millisecond and a 25 µs sampling time at present).

Finally, and despite its excellent precision, it is noted that this analysis did not evidence initial fusion pore flickering or closing ("kiss-and-run") events).[3,22] The absence of "kiss-and-run" observation in our work follows obviously from the fact that, owing to the necessary use of eqn (1), all PSF which could be treated had to lead to full fusion, *i.e.*, had to be followed by a fully developed spike. However, one would think that flickering could be detected through our analysis. Yet, should a pore flicker, unless this led to a sufficiently long duration of the closed and open phases (*i.e.*, longer than a few milliseconds),[3] diffusional filtering in the matrix (see above) would necessarily average the open/closed phases into a lower and smoother PSF current.[44,45] If so, such events would then lead to smaller apparent values of $\alpha_{\text{pore}}^{\text{open}}$ and R_{pore} for such events, thus biasing the two distributions towards their low-value ranges.

Energy distribution of initial fusion pores

After the two membranes have partially fused so as to create a stable fusion pore, the dimensions of the latter are necessarily imposed by the energetics of the system whatever its initial energy and the constraints (biological or physico-chemical) which apply on it and define its energy potential well. In this respect the initial pore system cannot differ from any other physicochemical one. However, there is one marked difference from usual physicochemical ones in the sense that the pore is created in a slow energy-dissipating environment. Usual quickly-deformable physicochemical systems may rapidly dissipate their excess energy by transfer into configuration changes and vibrations, solvent molecules vibrations, *etc.*, so that they may quickly equilibrate thermodynamically in their potential well. Conversely, irrespective of its toroid or tubular shape, the initial pore is created as a double-fold in a bilayer bilipidic membrane so that its capacity to release energy by usual means is drastically limited. Basically, it must dissipate its energy mostly through viscous transfer implying lateral shear friction of each monolayer against each other, since this is the only possibility for changing the double curvature of the pore. The same membrane displacements are also hampered viscously by the structured electrolyte surrounding it[23,24,40–42] and possibly by the membrane cytoskeleton mesh.[46] In addition, in the present biological case, such viscous dissipation may even be considerably reduced due to the presence of transmembrane proteins (including still locked SNAREs assemblies) which may restrict relative side-slipping of each membrane layer by clipping them locally, or owing to significant folding of the bilayer membrane.[18] As such, the initial fusion pore may maintain its initial energy for unusually long time durations compared to classical physicochemical systems (see the Appendix).

Despite these intrinsic differences, it remains that the statistical probability of observing a given initial pore, *i.e.*, with a given radius, depends on the energetic cost of creating such structure. Hence, the probability $P(R_{\text{pore}})$ of observing a R_{pore} value in the distribution shown in Fig. 3C reflects the Boltzmann probability of achieving an overall energy $W_{\text{pore}}(R_{\text{pore}})$:[47]

$$P(R_{\text{pore}}) \propto \exp[-\Delta W_{\text{pore}}(R_{\text{pore}})/k_B T] \qquad (2)$$

where k_B is the Boltzmann constant and T the temperature. This allows translation of the graph in Fig. 3C into the corresponding $\Delta W_{\text{pore}}(R_{\text{pore}})$ statistical variations shown in Fig. 5A which represents the variations of the initial fusion pore energy with its radius. Note that in eqn (2) and hereafter we prefer using $\Delta W_{\text{pore}}(R_{\text{pore}})$ instead of $W_{\text{pore}}(R_{\text{pore}})$ to recall that this does not provide a proper energy scale but rather a relative one for the population at hand. In this context, let us stress upon one important caveat. Readers less familiar with physical electrochemistry may think that the electrode potential influences the pore energy, *e.g.*, contributing to partial electroporation, due to electrostatic interactions between cell membrane(s) and the electrode placed at sub-micrometric distances from the cell. However, in any electrolyte solution, such as PBS used here, and for electrode potentials up to several tenths of a volt, the electrostatic potential is screened over distances from the electrode surface less than one nanometer by appropriate ions. Hence, since the artificial synaptic cleft width is *ca.* 100 nm, the electrode potential cannot have any influence at all on the exocytotic process nor on the energy of the fusion pore. This electrochemical (Faradaic) situation differs considerably from those pertaining to electrophysiology which rely on so-called "reference electrodes" (*i.e.*, imposing a potential in its surroundings). Henceforth, what we consider in $W_{\text{pore}}(R_{\text{pore}})$ is the energy released into the system and not yet dissipated when the pore opens (see below and the Appendix).

Fig. 5A is constructed using an $\alpha_{\text{pore}}^{\text{open}}$ scale (*viz.*, based on the histogram distribution in Fig. 3B) rather than a R_{pore} one, since our analysis produced primarily $\alpha_{\text{pore}}^{\text{open}}$ values while the R_{pore} distribution in Fig. 3C was reconstructed statistically from the analytical fit of the $\alpha_{\text{pore}}^{\text{open}}$ one. Nonetheless, the plots in Fig. 5A & B provide approximate R_{pore} scales following a straight rescaling of the $\alpha_{\text{pore}}^{\text{open}}$ ones using: $<R_{\text{pore}}> \approx (156\pi/180)\alpha_{\text{pore}}^{\text{open}} \approx 2.72\alpha_{\text{pore}}^{\text{open}}$ where R_{pore} is in nm and $\alpha_{\text{pore}}^{\text{open}}$ in degrees).

Rapid inspection of the plot in Fig. 5A and comparison to those in Fig. 7 in the Appendix supports the view that the initial fusion pores have an essentially lipidic structure. To clarify this conclusion we need to briefly evoke the basic components which define the potential energy of bilayer membranes.[23,24,40–42] First, a pore of radius R_{pore} created in a partially tense lipidic membrane provides a decrease of the membrane surface area by πR_{pore}^2, and hence induces a decrease of its surface tension by $\Delta W_{\text{tension}} = -\sigma \pi R_{\text{pore}}^2$ (σ is the surface tension coefficient).[41] This tends to stabilize the system by enlarging the pore radius. However, by doing so imposes that a larger amount of lipids are transferred from the membrane bulk into unfavorable positions lying on the pore edge, thus increasing their energy proportionally to the pore perimeter, *i.e.*, by $\Delta W_{\text{edge}}^{\text{line}} = \rho(2\pi R_{\text{pore}})$ where ρ is the line tension coefficient.[41] Also, in a biological membrane, such expansion may confront some biological constraints (*e.g.*, still-locked SNAREs assemblies,[7] actin mesh of the cytoskeleton,[12,46] slow diffusing transmembrane proteins,[46] *etc.*). Altogether, this may create an additional positive "edge" energy, $\Delta W_{\text{edge}}^{\text{biol}}$, also opposing the pore opening beyond a given radius (see the Appendix). Finally, when opened, the pore is prevented from closing spontaneously by several factors. First, decreasing the pore radius to extremely small values would necessarily imply a considerable equatorial bending of the membrane (*i.e.*, curvature centered on the symmetry axis of the pore) while this becomes negligible *vs.* the fold-bending at larger R_{pore} values whose effect on the energy was included in the $\Delta W_{\text{edge}}^{\text{line}}$ expression above. Second, facing membranes brought to angstromic distances from each other experience electrostatic or Van der Waals repulsions.[8] Third, closing up a nanometric pore requires squeezing out rather structured solvent molecules and electrolyte ions.[48] Except for the first one,[42] these

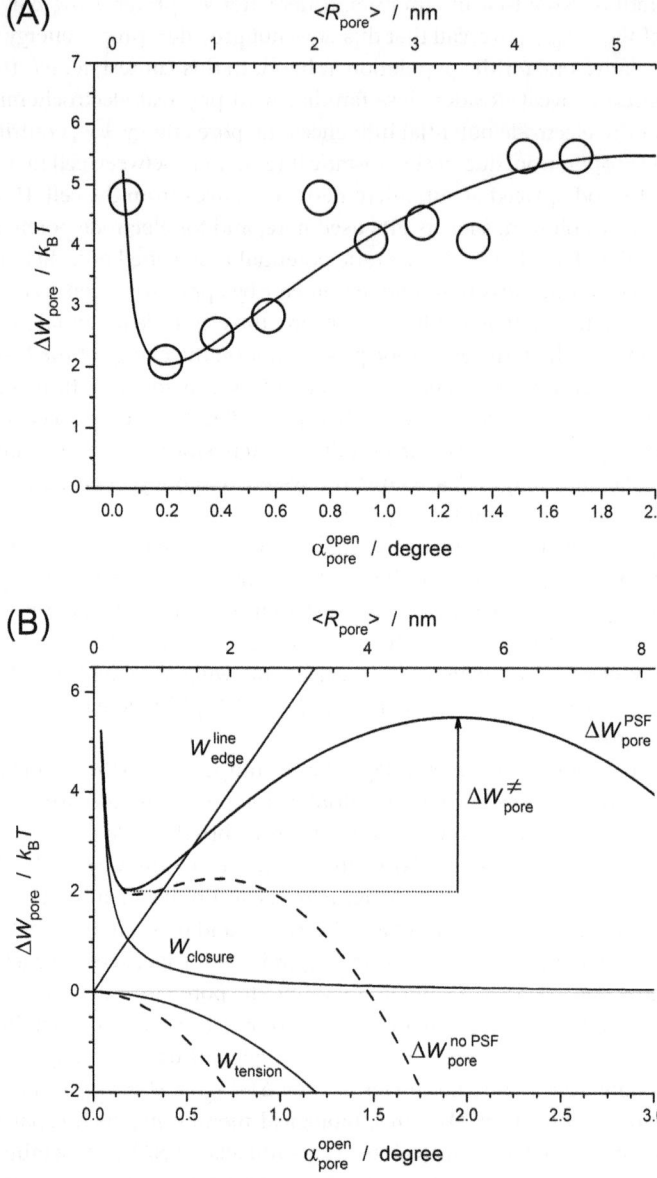

Fig. 5 (A) Experimental variation of the pore potential energy as a function of its opening angle assuming a Boltzmann control (eqn (2)) of the size distribution (open circles; each circle corresponds to one bin of the histogram in Fig. 3B) and its fit according to eqn (3) (solid curve) with: $\gamma_{mean} = 2.2 \times 10^{-30}$ J m and $n = 1$, $\rho_{mean} = 1.3$ pN, $\sigma_{mean} = 0.25$ mN m^{-1}. (B) Individual components of the fitting equation in (A). In (A,B) the top horizontal axis shows a straight conversion of the α_{pore}^{open} bottom axis into the R_{pore} one using $<R_{pore}> = \alpha_{pore}^{open}(\pi/180)R_{ves}^{mean}$, where α_{pore}^{open} is in degrees, R_{pore} in nm and $R_{ves}^{mean} = 156$ nm.[37] In (B) the dashed curves correspond to the same γ_{mean} and ρ_{mean} values as the solid ones but to $\sigma_{mean} = 0.68$ mN m^{-1} (see text).

contributions are difficult to predict quantitatively in the absence of precise information on the pore structure. Yet, all three add collectively into an additional positive "edge" energy, $\Delta W_{edge}^{closure}$, prevailing at angstromic distances but vanishing after a few angstroms radius. In absence of any better knowledge one may thus consider in a first approximation that $\Delta W_{\text{edge}}^{\text{closure}}$ varies as γ/R_{pore}^n where γ and n are two constants depending on the exact operating repulsive forces.

Altogether, these energy components define the potential well, $\Delta W_{\text{pore}}(R_{\text{pore}}) = (\Delta W_{\text{edge}}^{\text{closure}} + \Delta W_{\text{edge}}^{\text{line}} + \Delta W_{\text{edge}}^{\text{biol}} + \Delta W_{\text{tension}})_{R_{\text{pore}}}$, in which the pore lies depending on its initial energy. If rapid energy dissipation could occur, the pore would reach its stable radius value corresponding to the potential energy minimum such as $d\Delta W_{\text{pore}}/dR_{\text{pore}} = 0$ and $d^2\Delta W_{\text{pore}}/dR_{\text{pore}}^2 > 0$.[23,24,40,42] Fig. 7 in the Appendix represents the typical expected shapes of $\Delta W_{\text{pore}}(R_{\text{pore}})$ depending on the relative magnitudes of ρ/σ either in the absence (viz., when $\Delta W_{\text{edge}}^{\text{biol}}$ is negligible vs. the other energies involved) or in the presence of biological edge constraints (viz., when $\Delta W_{\text{edge}}^{\text{biol}}$ is significant). Comparison of the plot in Fig. 5A to those in Fig. 7 strongly suggests that the pore structure is mostly governed by its lipid-related energy components, i.e. that $\Delta W_{\text{edge}}^{\text{biol}}$ may be neglected. This is further supported by the decomposition of the regression energy curve into its three individual components as shown in Fig. 5B (arbitrarily using $n = 1$ in describing $\Delta W_{\text{edge}}^{\text{closure}}$, since there is not enough data in the pertinent range to estimate the n value):

$$\Delta W_{\text{pore}}^{\text{lipid}}(R_{\text{pore}}) = (\gamma/R_{\text{pore}}) + 2\pi\rho R_{\text{pore}} - \pi\sigma R_{\text{pore}}^2 \qquad (3)$$

Analysis of the data in Fig. 5A according to eqn (3) yielded (correlation coefficient = 0.81, 80 PSF) $\gamma_{\text{mean}} = 2.2 \times 10^{-30}$ J m, $\rho_{\text{mean}} = 1.3$ pN and $\sigma_{\text{mean}} = 0.25$ mN m^{-1} for room temperature (i.e., $k_\text{B}T = 4.1 \times 10^{-21}$ J). To the best of our knowledge it is impossible to evaluate the plausibility of the ensuing γ value whose precision is extremely low. However, rationalizing γ as mostly stemming from electrostatics[49] would imply considerably less than one elementary charge being located within the pore area. This result evidences that if any charges or dipoles do exist they are totally screened by the electrolyte.[49] Hence, such γ value seems more coherent with the assumption that $\Delta W_{\text{edge}}^{\text{closure}}$ features the squeezing of electrolyte ions and water molecules out of fusion pores.[48] Conversely, the above values for ρ and σ are determined with a rather good precision and are perfectly consistent with those reported for bilipidic membranes.[40,42]

To conclude this section we wish to discuss a specific aspect of the potential energy well shown in Fig. 5. Indeed, for the above σ and ρ values, a pore with an excess energy exceeding $\Delta W_{\text{pore}}^{\neq} \approx 3.5 k_\text{B}T$ vs. the minimum of the potential well (at ca. $2k_\text{B}T$), i.e., created with an initial radius larger than ca. 5.4 nm, should proceed immediately to its exponential growth phase since then $d\Delta W_{\text{pore}}/dR_{\text{pore}} < 0$ and increases in magnitude almost proportionally to R_{pore}. Conversely, pores created with a radius less than 5.4 nm should be poised in the potential well and be stable. In coherence with this analysis, the application of eqn (2) shows that 97% of the pores created under such conditions should be stable, a proportion which compares satisfactorily with the fact that, by construction, all the release events treated here gave rise to a PSF.

However, in normally tense bilipidic membranes σ may range up to 1 mN m^{-1} (and up to a few mN m^{-1} for tense ones; note however that above 10 mN m^{-1} bilipidic membranes cannot keep their cohesion and destructure by spontaneously creating pores to release their excess surface tension energy).[40,41] Thus, one may reasonably expect that among all release events (i.e., including those which do not display a PSF) the initial pore may be created in membranes displaying a wider spectrum of surface tensions than those which led to the observation of PSF.

Experimentally it is commonly observed that ca. 70% of fusion events in chromaffin cells and related ones do not evidence any detectable PSF.[19,28] Within the framework of the present analysis, they correspond to an unstable initial fusion pore, i.e., their 70% probability features that of the events giving rise to an activation barrier $\Delta W^{\neq}_{\text{pore}}$ being less than $\Delta W^{\neq}_{\text{pore}} < -(ln 0.7)k_B T \approx 0.36 k_B T$. Hence, assuming that the above ρ and γ values are constant features for the systems at hand, using eqn (3) leads to the prediction that whenever $\sigma \geq 0.68$ mN m^{-1} initial fusion pores are created under unstable conditions they proceed to their exponentially expanding phase from their very creation (compare Fig. 5B). Such a σ value is not unrealistic at all, being within the usual range of normally tense membranes, so this may well explain why ca. 70% of release events do not lead to any PSF observation.

Conclusion

The present work evidenced that, provided that external data entry is available, amperometric currents displaying a pre-spike feature (PSF) may be inverted theoretically so as to provide precise topological information about the initial fusion pores, viz., about the crucial initial step which engages the full process of vesicular release. We choose here to rely on patch-clamp literature data on chromaffin cells as such an external input. This allowed evaluation of the diffusion rate, $\kappa = D_{\text{ves}}/R^2_{\text{ves}} = 415$ s^{-1}, of neurotransmitter cations in the partially altered matrix located near the pore entrance. In turn this led to an evaluation of the distribution of initial fusion pore radii and provided for the first time an experimental description of the potential energy well for such systems.

The shape of the corresponding potential energy well thus-reconstructed strongly suggests that after their creation initial fusion pores are essentially controlled by the usual physicochemical laws used in describing pores formed in bilayer lipidic membranes, i.e., they have an essentially lipidic nature. This is an essential piece of information since there is still much debate about the structure of the initial fusion pores. Furthermore, upon assuming that the membranes may be partially tense, the conclusion of the energetic analysis performed here predicts that a rather large proportion of events should not lead to any thermodynamically stable fusion pores. This is fully coherent with the noted absence of any detectable PSF in 70% of amperometric events. Therefore, even if solving the inverse problem requires an external entry (taken here from patch-clamp literature data), the information contained in amperometric currents offer extremely important additional information.

Finally, assuming that the value of $\kappa = D_{\text{ves}}/R^2_{\text{ves}}$ remains constant[38b] for matrixes of chromaffin cells when "full" fusion occurs (i.e., after the exponential growth phase of the fusion pore occurs), the same analysis may be extended so as to characterize the final extent of fusion pores at the end of their exponential growth phase. This would provide a first correct statistical evaluation of the extent of full fusion and afford a definitive answer to the question of how full "full" fusion is. Work is presently in progress in our laboratory in this direction.

Experimental

The 80 amperometric spikes with PSF analyzed here were extracted from amperometric traces of exocytosis monitored at Ba^{2+}-stimulated single bovine chromaffin cells using beveled and polished carbon fiber ultramicroelectrodes (7-μm diameter). Details of electrode preparation and single cell experiments were identical to previously published procedures.[28] Electrodes were held at +0.65 V *versus* a silver/silver chloride reference electrode using a commercially available picoamperometer (model AMU-130, Radiometer Analytical Instruments, Copenhagen, Denmark) with a time response set at 50 μs. The output was digitized at 40 kHz, displayed in real time and stored on a computer (Powerlab-4SP A/D converter and software Chart, ADinstruments, Colorado Springs, CO) with no subsequent digital filtering. Each amperometric trace obtained during cell secretion was visually inspected and signals were marked as exocytotic spikes whenever their maximum current intensities were three times higher than the rms noise (0.4–0.7 pA), determined from the baseline current recorded before each signal (30 ms minimum time trace). Special attention was paid to verify the baseline stability before and after each spike to avoid any bias due to baseline drift. Overlapping spikes were excluded. Generally, 50–200 spikes fulfilling these criteria could be isolated from each amperometric trace, and among them only those displaying a clear PSF were retained. Signals from five different cell batches obtained from five different animals were gathered to produce the set of 80 spikes with PSF analyzed in this work.

Appendix

I. Influence of the shape of the pore: toroid *vs.* tubular

One expects intuitively that the exact shape of the pore, *viz.*, toroid or tubular (see Fig. 6) matters for the extraction of the opening angle from PSF currents, *i.e.*, conditions the validity of the present results. In fact, we show below that this is not true as soon as D_{ves}, the diffusion coefficient of the neurotransmitter in the altered part of the matrix near the pore entrance, is much smaller than its value, D_{cleft}, in the cleft of the artificial synapse (see Fig. 2B in the main text). Indeed, provided that the possible tubule length remains moderate, *viz.*, does not exceed a few tens of nanometers, for dense core vesicles such as those of chromaffin cells the only possible bias is that the D_{ves} value afforded by our toroid-based analysis may be slightly underestimated if a tubular connection was involved.

To establish this point let us evaluate the different diffusional fluxes of the neurotransmitter in all compartments that are visited when the pore is tubular (Fig. 6B). In the vesicle, owing to the relative values of the pore and vesicle radii (*i.e.*, ca. 1 nm vs. 150 nm), from the matrix side, the pore entrance can be viewed as creating a diffusional field akin to that of a disk nanoelectrode in an infinite insulating plane.[50] Indeed, the vesicle membrane curvature may be neglected with respect to the scales considered here. Furthermore, owing to the extremely short dimension scales considered here, steady state diffusion is achieved within microseconds. Thus:[50]

$$\phi_{ves} \cong 4D_{ves}(C_{ves}^{rel} - C_{ves}^{exit})R_{pore} \quad (A1)$$

where C_{ves}^{exit} is the neurotransmitter concentration in the matrix at the entrance of the pore and other variables are defined in the main text.

Considering a tubular pore of length L, the flux within the pore is given by:[31]

$$\phi_{pore} = \pi D_{pore}(C_{pore}^{enter} - C_{pore}^{exit})R_{pore}^2/L \quad (A2)$$

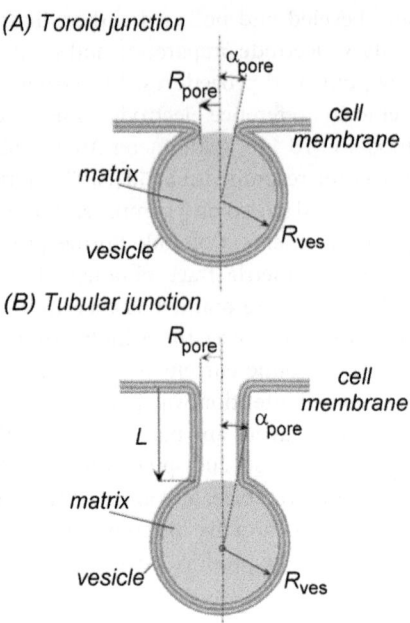

Fig. 6 Limiting geometries of the cell/vesicle membranes junction considered in this work: toroid (A) or tubular (B). Note that for the sake of simplicity and definition of the different variables used in the present model the pores are shown as being purely lipidic. Similarly, for clarity, the thickness of the membranes (ca. 3 nm) and the pore radius are enlarged with respect to the vesicle diameter.

where $C_{\text{pore}}^{\text{enter}} = C_{\text{ves}}^{\text{exit}}/\chi_{\text{matrix}}$ is the neurotransmitter concentration in the electrolyte at the pore entrance and $C_{\text{pore}}^{\text{exit}}$ that at its exit; χ_{matrix} is the partition coefficient of the neurotransmitter cation in the partially altered matrix (i.e., smaller than that in the fully compact matrix which prevails before the pore is opened, but larger than in the fully swollen matrix).[34,39] D_{pore} is the diffusion coefficient of the neurotransmitter in the tubular pore. This may be considered as extremely close to its value, D_{cleft}, in the artificial synaptic cleft as soon as the pore diameter approaches or exceeds 2 nm.[51] Note that the expression for ϕ_{pore} in eqn (A2) considers only straight diffusion. Should an electroosmotic drive be also involved, owing to the direction of the pH gradient (i.e., ca. 5.2 in the vesicle vs. physiological pH in the cleft),[52] a larger ϕ_{pore} would result. Hence, $(C_{\text{pore}}^{\text{enter}} - C_{\text{pore}}^{\text{exit}})$ would be smaller than predicted here considering the diffusional expression in eqn (A2) with the result that the pore length L would appear smaller. The ensuing analysis thus provides a maximum maximorum evaluation of the possible bias produced upon using a toroid junction based analysis instead of considering a tubular one when this applies. In other words, what follows evaluates the maximum distortion possibly incurred upon considering a toroid pore, when it may actually be tubular.

Since (i) the electrode maintains a zero concentration of neurotransmitter at its surface and (ii) the thickness of the cleft (ca. 100 nm) greatly exceeds the pore radius (a few nm at most), the flux inside the artificial synaptic cleft is again akin to that produced at a disk nanoelectrode.[50] Hence:

$$\phi_{\text{cleft}} = 4D_{\text{cleft}} C_{\text{pore}}^{\text{exit}} R_{\text{pore}} \quad (A3)$$

The three fluxes in eqn (A1)–(A3) are equal under steady state conditions so that: $\phi_{\text{ves}} = \phi_{\text{pore}} = \phi_{\text{cleft}} = i/(2F)$ where i is the monitored amperometric current.[35] This allows solving the system in eqn (A1)–(A3), from which it follows that:

$$i_{\text{tube}} = 8FD_{\text{ves}} C_{\text{ves}}^{\text{rel}} R_{\text{pore}} / \left[1 + \chi_{\text{matrix}} \frac{D_{\text{ves}}}{D_{\text{cleft}}} \left(1 + \frac{4L}{\pi R_{\text{pore}}}\right)\right] \quad \text{(A4)}$$

For a toroid pore, eqn (A2) does not apply since $L \approx 0$ and $C_{\text{pore}}^{\text{enter}} = C_{\text{pore}}^{\text{exit}} = C_{\text{ves}}^{\text{exit}}/\chi_{\text{matrix}}$, so that one obtains:

$$i_{\text{tore}} = 8FD_{\text{ves}} C_{\text{ves}}^{\text{rel}} R_{\text{pore}} / \left(1 + \chi_{\text{matrix}} \frac{D_{\text{ves}}}{D_{\text{cleft}}}\right) \quad \text{(A5)}$$

Comparing eqn (A4) and A5 shows that the two fluxes are strictly proportional and differ only by a correcting factor, *viz.*:

$$i_{\text{tore}} = i_{\text{tube}}(1 + \Gamma_{\text{corr}}) \quad \text{(A6)}$$

where:

$$\Gamma_{\text{corr}} = \left(\frac{4L}{\pi R_{\text{pore}}}\right) \bigg/ \left(1 + \frac{D_{\text{cleft}}}{\chi_{\text{matrix}} D_{\text{ves}}}\right) \quad \text{(A7)}$$

so that upon using a tore-based analysis instead of a tube-based one, when this applies, the only bias incurred is that D_{ves} value would need to be corrected to $D_{\text{ves}}^{\text{tore}}(1 + \Gamma_{\text{corr}})$ where $D_{\text{ves}}^{\text{tore}}$ is that obtained through application of the tore-based model.

Though an upper boundary may be estimated for Γ_{corr}. Firstly, based on the results from Ewing *et al.*, which compared the mean neurotransmitter amounts released from fully swollen matrixes to those after normal exocytosis,[34] one may consider that a partially swollen matrix (*viz.*, still mostly constricted by its membrane) as considered here retains about 2/3 of its initial content (*i.e.*, leading to an amperometric current intensity lower than the baseline noise after 1/3 of its content has been released). This suggests that χ_{matrix} is at most a few units. Considering a toroid pore, our analysis showed that $D_{\text{cleft}}/D_{\text{ves}}$ was close to 100. Hence, the expression of Γ_{corr} simplifies to:

$$\Gamma_{\text{corr}} = \frac{4}{\pi} \chi_{\text{matrix}} \frac{D_{\text{ves}}}{D_{\text{cleft}}} \times \frac{L}{R_{\text{pore}}} \quad \text{(A8)}$$

On the other hand, TIRFM experiments[4] evidence that fusing bulk vesicles are fully illuminated. Since vesicles have a diameter (*ca.* 300 nm) comparable to the length of the evanescent wave field, this shows that L/R_{ves} is much smaller than unity. Hence, L/R_{pore} is presumably at most a few tens. Therefore, even if the fusion pore has a tubular shape rather than a toroid one, Γ_{corr} is at most commensurable to a few units.

This ensures that considering a toroid connection instead of a possible tubular one does not affect at all the principle of our analysis. Furthermore, since we resorted to an independent entry (*viz.*, from published patch-clamp data) for calibrating R_{pore}, any possible bias incurred if Γ_{corr} was not negligible would be shifted onto the κ value reported in the text, which would be underestimated with respect to reality, but whose effect would be null due to the compensating auto-coherence of our analysis. For this reason, in the absence of precise independent information on C_{ves}^{rel}, χ_{matrix}, D_{ves}/D_{cleft} and L/R_{pore}, we choose to rely on a tore-based model both for simplicity and to avoid the need to consider an additional unknown parameter (*viz.*, L/R_{ves} in our normalized framework).

II. Derivation of eqn(1) of the main text

The charge Q released by single vesicle and captured at the electrode surface is obtained upon integrating the amperometric spike current including the PSF. Hence, $Q = 4(2F)\pi R_{ves}^3 C_{ves}^{rel}/3$ (note that the electrooxidation of adrenaline is a two-electron process).[35] On the other hand, within the nanodisk analogy used here as a first analytical approximation, the current corresponding to the opened fusion pore is given by:[50] $i_{pore} = 4(2F)D_{ves}C_{ves}^{rel}R_{pore}$. A ratio of these two expressions readily affords:

$$Q/i_{pore} = \pi R_{ves}^3/(3 D_{ves} R_{pore}) = \pi/[3(D_{ves}/R_{ves}^2) \times (R_{pore}/R_{ves})] \quad (A9)$$

which provides eqn (1) upon noting that $\kappa = D_{ves}/R_{ves}^2$ and $\alpha_{pore}^{open} = (180/\pi)R_{pore}/R_{ves}$, where α_{pore}^{open} is in degrees, as everywhere in the text.

III. Variations of the fusion pore energy vs. its radius in a lipidic pore

For the sake of simplicity, in eqn (3) we explained the different contributions to the initial fusion pore energy by considering the case of a purely lipidic structure since, ultimately, this is the relevant situation based on the present experimental and theoretical data. Under such circumstances, noting that the main influence of $\Delta W_{edge}^{closure}$ is restricted to sub-nanometric radii, while for non-tense membranes $\Delta W_{tension}$ and ΔW_{edge}^{line} act mostly at larger distances; in first approximation, the potential energy well (eqn (3)) displays a minimum at $R_{pore}^{min} \approx (\gamma/2\pi\rho)^{1/2}$ with $\Delta W_{pore}^{min} \approx 2(2\pi\rho\gamma)^{1/2}$ (for $n = 1$), and a maximum at $R_{pore}^{max} \approx \rho/\sigma$ with $\Delta W_{pore}^{max} \approx \pi\rho^2/\sigma$. Depending on the relative magnitudes of ρ and σ, the pore energy may be rather shallow (Fig. 7-IA), relatively well defined (Fig. 7-IB) or may even disappear if $\sigma > (2\pi\rho^3)/\gamma)^{1/2}$, *i.e.*, when the surface tension is high enough (*i.e.*, whenever σ approaches and exceeds 1 mN m^{-1} for the ρ and γ values obtained in Fig. 5).

When the potential energy well displays a local minimum at R_{pore}^{min} and a local maximum at R_{pore}^{max}, the energy difference $\Delta W_{pore}^{\neq} = \Delta W_{pore}^{max} - \Delta W_{pore}^{min}$ represents the activation energy required to rupture the initial fusion pore. Therefore, whenever $\Delta W_{pore}^{\neq} \gg k_B T$, after dissipating its initial energy, a pore of initial radius smaller than R_{pore}^{max} should rest around its stable minimum without any need for biological constraints to enforce such stability. Conversely, a pore created with an initial radius larger than R_{pore}^{max} should proceed spontaneously to its exponential growth while dissipating its initial energy.

While the pore is opened, catecholamine cations which maintain the cohesion of the intravesicular polyelectrolyte peptidic matrix through charge compensation and several hydrogen-bonds[53] are replaced by small hydrated cations to preserve electroneutrality. This provokes a trend for the matrix to swell.[39] On the other hand, while the pore is nanometric, most of the matrix remains necessarily enclosed in its membrane and swelling is drastically refrained. This situation forces the vesicle to equilibrate its swelling pressure by an increase of the surface tension of its membrane; hence, the tension coefficient σ necessarily increases with time.[23,24] From eqn (3) it is understood that this provokes a progressive

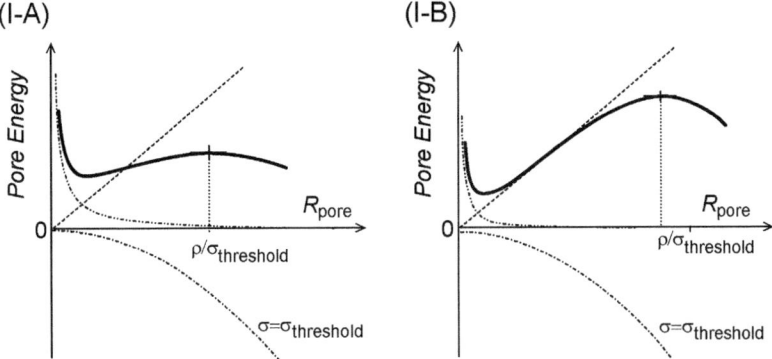

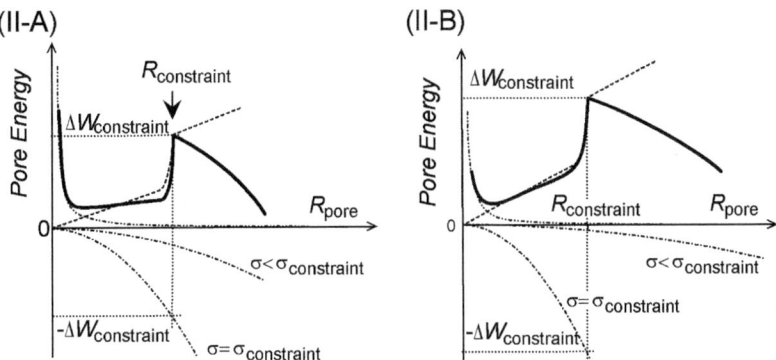

Fig. 7 Schematic potential energy of the initial fusion pore, for purely lipidic behavior (I), or when the pore is constrained by a biological structure or scaffolding (II); see text. In each case, situations featuring a low (A) or high (B) membrane tension relative to the pore edge one are represented. The overall energy is shown by the solid curve in each case together with its components: line edge energy (dashes), edge closure energy (dash–dot–dot), and surface tension energy (dash–dot); see text.

decrease of $\Delta W_{pore}^{max} \approx \pi \rho^2/\sigma$ and of $\Delta W_{pore}^{\neq} = \Delta W_{pore}^{max} - \Delta W_{pore}^{min}$. Therefore, provided it remains open for a sufficient time, any fusion pore will eventually reach a state in which $\Delta W_{pore}^{\neq} \lesssim k_B T$ or less so that it will irreversibly proceed to its exponentially expanding phase at a rate controlled by its energy dissipation[23,24,40,42] (see below, section V).

IV. Variations of the fusion pore energy vs. its radius in presence of biological constraints

Section III focused on a purely lipidic pore in which the line tension and surface tension energies scale as R_{pore} and R_{pore}^2, respectively.[23,24,40,42] However, in a biological membrane such as those considered here the situation may be far more complicated. Indeed, upon increasing its radius, the pore edge may meet some biological structure or scaffold which then should oppose its further expansion while the constraint holds. As a first approximation this can be represented as a sterical constraint applying for $R_{pore} \approx R_{constraint}$ and beyond.

The line tension energy may then be tentatively represented as being the sum of a linear contribution (viz., $\approx 2\pi\rho R_{pore}$) and a Heaviside function (viz., $\Delta W_{scaffold} \times \Theta[R - R_{pore}]$) so that $\Delta W_{edge}^{line} \approx 2\pi\rho R_{pore}$ when $R_{pore} < R_{constraint}$, while $\Delta W_{edge}^{line} \approx 2\pi\rho R_{pore} + \Delta W_{scaffold}$, when $R_{pore} \geq R_{constraint}$. Whenever $\Delta W_{constraint} = 2\pi\rho R_{constraint} + \Delta W_{scaffold}$ is not negligible vs. $k_B T$ this situation leads to a characteristic energy profile which drastically differs from the potential energy well shape predicted for a purely lipidic pore (compare Fig. 7-IIA and 7-IIB).

Such biological constraints may involve still-locked SNAREs, actin meshes of the cell membrane cytoskeleton, transmembrane proteins locally nailing together the two bilayers, etc. Assuming that for $R_{pore} < R_{constraint}$ the pore still obeys a purely lipidic energy profile, the activation barrier limiting the pore widening beyond $R_{constraint}$ is $\Delta W_{pore}^{\neq} \approx \Delta W_{constraint} - 2(2\pi\rho\gamma)^{1/2}$. Hence, the pore expansion is stalled around $R_{constraint}$ up to when either the constraint disappears (e.g., SNAREs unlocking, local reorganization of the membrane cytoskeleton,[11,12] etc.) or when the membrane surface tension energy increases sufficiently for the pore energy to overcome $\Delta W^{\neq}$ (see Fig. 7-II). Such a process may well repeat at a series of $R_{constraint}$ values when several constraints are met sequentially when the pore enlarges. This could explain the multi-step sequential opening of the initial fusion pore observed in this work (compare Fig. 4B,C).

V. Role of viscous dissipation on the kinetics of energy release

Initial fusion pores may be created only through overcoming the cell and vesicle membranes repulsion.[8,48] It is now widely accepted that the required energy is provided by the Ca^{2+}-mediated assembling of SNAREs complexes.[7] Seen in physicochemical terms, the situation is then akin to that of a chemical reaction which requires overcoming an activation barrier and then relaxation to a more stable state, viz., here, that featuring the opened pore. Consequently, when this transition state is overcome the system necessarily possesses an excess energy with respect to the final pore potential energy well.[54,55] In any classical physicochemical non-structured environment, radiative energy transfer into molecular rearrangements and vibration modes of numerous chemical bonds rapidly dissipate this excess energy within a few picoseconds. However, this cannot occur so fast in a highly structured and viscous system with so few potentially vibrating bonds as the one at hand. Here, dissipation may occur mostly through structural deformations involving sliding of membrane bilipidic layers with respect to each other and with respect to the structured electrolyte bathing it.[23,24,40,42]

Hence, the pore dynamics (viz., its ability to enlarge or close depending on its initial position vs. its local potential energy minimum and maximum, see below) is kinetically controlled by the rate at which its initial energy excess may be dissipated viscously. Denoting by η the two-dimensional viscosity of the system (note that the pore involves two curvature radii in orthogonal planes whatever its toroid or tubular shape), a purely lipidic pore thus experiences an algebraic growth of its radius given by:[23,24,40,42]

$$8\eta \frac{dR_{pore}}{dt} = -\frac{1}{2\pi} \frac{d\Delta W_{pore}}{dR_{pore}} = \sigma R_{pore} - \rho + \frac{\gamma}{2\pi R_{pore}^2} \quad (A10)$$

(using our above approximation, viz., $W_{edge}^{closure} = \gamma/R_{pore}$). Thus:

$$\frac{\mathrm{d}R_{\text{pore}}}{R_{\text{pore}}} = \frac{\sigma}{8\eta}\left\{1 - \frac{R_{\text{pore}}^{\max}}{R_{\text{pore}}}\left[1 - \left(\frac{R_{\text{pore}}^{\min}}{R_{\text{pore}}}\right)^2\right]\right\}\mathrm{d}t \qquad (A11)$$

where $R_{\text{pore}}^{\min} \approx (\gamma/2\pi\rho)^{1/2}$ and $R_{\text{pore}}^{\max} \approx \rho/\sigma$. Eqn (A11) shows that the pore should enlarge spontaneously towards $R_{\text{pore}}^{\min}$ when $R_{\text{pore}} < R_{\text{pore}}^{\min}$ or towards large values when $R_{\text{pore}} > R_{\text{pore}}^{\max}$, or decrease spontaneously towards $R_{\text{pore}}^{\min}$ when $R_{\text{pore}}^{\min} < R_{\text{pore}} < R_{\text{pore}}^{\max}$. However, these trends are regulated kinetically by the time constant $\tau = 8\eta/\sigma$, so they may be extremely slow in a non-highly-tense viscous membrane system as the one at hand ($\sigma \approx 0.25$ mN m^{-1}). For biological membranes $\eta \approx 10^{-6}$ N s m^{-1} appears a good average estimate for the two-dimensional viscosity,[42] so that τ is expected to fall in the range of several tens of milliseconds. Such a large time constant value with respect to the life-time of initial fusion pores is perfectly coherent with our observation that the initial fusion pores giving rise to PSF events did not relax during their life-time and remained distributed almost all over their potential energy well (compare Fig. 5A).

Acknowledgements

This work has been supported in part by the CNRS (UMR 8640), Ecole Normale Superieure (ENS, Paris), University Pierre and Marie Curie (UPMC), and the French Ministry of Research. The authors thank ANR (Chaire d'Excellence project "MicroNanoChem") and CNRS for financial support of this project in UMR 8640.

References and notes

1 R. D. Burgoyne and A. Morgan, *Physiol. Rev.*, 2003, **83**, 581–632.
2 A.-S. Cans, N. J. Wittenberg, R. Karlsson, L. Sombers, M. Karlsson, O. Orwar and A. G. Ewing, *Proc. Natl. Acad. Sci. U. S. A.*, 2003, **100**, 400–404.
3 See *e.g.*: (a) E. Neher, *Nature*, 1993, **363**, 497–498; (b) A. W. Henkel, H. Meiri, H. Horstmann, M. Lindau and W. Almers, *EMBO J.*, 2000, **19**, 84–93.
4 See *e.g.* among many TIRFM reports: (a) D. Zenisek, J. A. Steyer and W. Almers, *Nature*, 2000, **406**, 849–854; (b) J. G. Burchfield, J. A. Lopez, K. Mele, P. Vallotton and W. E. Hughes, *Traffic*, 2010, **11**, 429–439; (c) M. W. Allersma, L. Wang, D. Axelrod and R. W. Holz, *Mol. Biol. Cell*, 2004, **15**, 4658–4668.
5 L. J. Breckenridge and W. Almers, *Nature*, 1987, **328**, 814–817.
6 A. E. Spruce, L. J. Breckenridge, A. K. Lee and W. Almers, *Neuron*, 1990, **4**, 643–654.
7 R. Jahn and R. H. Scheller, *Nat. Rev. Mol. Cell Biol.*, 2006, **7**, 631–643.
8 D. Leckband and J. Israelachvili, *Q. Rev. Biophys.*, 2001, **34**, 105–267.
9 A. Gil, L. M. Gutierrez, C. Carrasco-Serrano, M. T. Alonso, S. Viniegra and M. Criado, *J. Biol. Chem.*, 2002, **277**, 9904–9910.
10 M. Criado, A. Gil, S. Viniegra and L. M. Gutierrez, *Proc. Natl. Acad. Sci. U. S. A.*, 1999, **96**, 7256–7261.
11 M. L. Gardel, J. H. Shin, F. C. MacKintosh, L. Mahadevan, P. Matsudaira and D. A. Weitz, *Science*, 2004, **304**, 1301–1305.
12 R. E. Guzman, P. Bolanos, A. Delgado, H. Rojas, R. DiPolo, C. Caputo and E. H. Jaffe, *Pfluegers Arch.*, 2006, **454**, 131–141.
13 J. Zimmerberg and K. Gawrisch, *Nat. Chem. Biol.*, 2006, **2**, 564–567.
14 J. B. Sorensen, *Annu. Rev. Cell Dev. Biol.*, 2009, **25**, 513–537.
15 E. Neher and B. Sakmann, *Nature*, 1976, **260**, 799–802.
16 O. P. Hamill, A. Marty, E. Neher, B. Sakmann and F. J. Sigworth, *Pfluegers Arch.*, 1981, **391**, 85–100.
17 E. Neher and A. Marty, *Proc. Natl. Acad. Sci. U. S. A.*, 1982, **79**, 6712–6716.
18 See *e.g.*: (a) E. Borroni, P. Ferretti, W. Fiedler and G. Q. Fox, *Cell Tissue Res.*, 1985, **241**, 367–372; (b) G. Q. Fox, *Cell Tissue Res.*, 1996, **284**, 303–316.
19 C. Amatore, S. Arbault, M. Guille and F. Lemaître, *Chem. Rev.*, 2008, **108**, 2585–2621.
20 M. Lindau and G. A. de Toledo, *Biochim. Biophys. Acta, Mol. Cell Res.*, 2003, **1641**, 167–173.
21 K. Lollike, N. Borregaard and M. Lindau, *J. Cell Biol.*, 1995, **129**, 99–104.

22 A. Albillos, G. Dernick, H. Horstmann, W. Almers, G. A. deToledo and M. Lindau, *Nature*, 1997, **389**, 509–512.
23 C. Amatore, Y. Bouret, E. R. Travis and R. M. Wightman, *Angew. Chem., Int. Ed.*, 2000, **39**, 1952–1955.
24 C. Amatore, Y. Bouret, E. R. Travis and R. M. Wightman, *Biochimie*, 2000, **82**, 481–496.
25 T. J. Schroeder, R. Borges, K. Pihel, C. Amatore and R. M. Wightman, *Biophys. J.*, 1996, **70**, 1061–1068.
26 F. Segura, M. A. Brioso, J. F. Gómez, J. D. Machado and R. Borges, *J. Neurosci. Methods*, 2000, **103**, 151–156.
27 (a) C. Amatore, A. I. Oleinick and I. Svir, *ChemPhysChem*, 2010, **11**, 149–158; (b) C. Amatore, A. I. Oleinick and I. Svir, *ChemPhysChem*, 2010, **11**, 159–174.
28 C. Amatore, S. Arbault, I. Bonifas, Y. Bouret, M. Erard, A. G. Ewing and L. A. Sombers, *Biophys. J.*, 2005, **88**, 4411–4420.
29 C. Amatore, S. Arbault, I. Bonifas, M. Guille, F. Lemaître and Y. Verchier, *Biophys. Chem.*, 2007, **129**, 181–189.
30 C. Amatore, S. Arbault, I. Bonifas and M. Guille, *Biophys. Chem.*, 2009, **143**, 124–131.
31 J. Crank, *The Mathematics of Diffusion*, 2nd edn, Oxford University Press, Oxford, 1975.
32 See e.g.: C. Amatore, Y. Bouret, E. Maisonhaute, J. I. Goldsmith and H. D. Abruña, *Chem.-Eur. J.*, 2001, **7**, 2206–2226.
33 B. Farell and S. J. Cox, *Bull. Math. Biol.*, 2002, **64**, 979–1010.
34 See e.g.: D. M. Omiatek, Y. Dong, M. L. Heien and A. G. Ewing, *ACS Chem. Neurosci.*, 2010, **1**, 234–245.
35 For the factor 2 (i.e., two-electron overall route during electrooxidation of adrenaline under amperometric conditions), see: E. L. Ciolkowski, K. M. Maness, P. S. Cahill, R. M. Wightman, D. H. Evans, B. Fosset and C. Amatore, *Anal. Chem.*, 1994, **66**, 3611–3617.
36 A. Meunier, O. Jouannot, R. Fulcrand, I. Fanget, M. Bretou, E. Karatekin, S. Arbault, M. Guille, F. Darchen, F. Lemaître and C. Amatore, *Angew. Chem., Int. Ed.*, 2011, **50**, 5081–5084.
37 R. E. Coupland, *Nature*, 1968, **217**, 384–388.
38 (a) D. T. Gillespie, *J. Phys. Chem.*, 1977, **81**, 2340–2361; (b) It may be surprising that this analysis evidenced that $\kappa = D_{ves}/R_{ves}^2$ was a constant while R_{ves} exhibits a large distribution[37] (hence also D_{ves}). However, a plausible rationale for this fact is that the larger the vesicle, the larger the volume allowed to the matrix partial swelling. Since the volume expansion dV of the matrix resulting from a variation dR of its radius (allowed by the pore opening and increase in the membrane tension) is $dV/dR = 4\pi R^2$, it seems reasonable to consider that the change in diffusion coefficient scales with R_{ves}^2. Indeed in the partially swollen matrix, neurotransmitters certainly mostly diffuse through the altered zones; see, e.g.: C. Amatore, R. S. Kelly, E. W. Kristensen, W. G. Kuhr and R. M. Wightman, *J. Electroanal. Chem.*, 1986, **213**, 31–42.
39 (a) C. Nanavati and J. M. Fernandez, *Science*, 1993, **259**, 963–965; (b) J. Zimmerberg, M. Curran, F. S. Cohen and M. Brodwick, *Proc. Natl. Acad. Sci. U. S. A.*, 1987, **84**, 1585–1589; (c) P. E. Marszalek, B. Farrell, P. Verdugo and J. M. Fernandez, *Biophys. J.*, 1997, **73**, 1160–1168; (d) P. E. Marszalek, B. Farrell, P. Verdugo and J. M. Fernandez, *Biophys. J.*, 1997, **73**, 1169–1183.
40 O. Sandre, L. Moreaux and F. Brochard-Wyart, *Proc. Natl. Acad. Sci. U. S. A.*, 1999, **96**, 10591–10596.
41 C. Taupin, M. Dvolaitzky and C. Sauteret, *Biochemistry*, 1975, **14**, 4771–4775.
42 Y. A. Chizmadzhev, P. I. Kuzmin, D. A. Kumenko, J. Zimmerberg and F. S. Cohen, *Biophys. J.*, 2000, **78**, 2241–2256.
43 T. J. Schroeder, J. A. Jankowski, K. T. Kawagoe, R. M. Wightman, C. Lefrou and C. Amatore, *Anal. Chem.*, 1992, **64**, 3077–3083.
44 The rise-times cannot be due to the diffusion between the initial fusion pore and the electrode surface, since in the "artificial synapse" configuration[16] the latter is placed within a few hundred nanometers of the cell membrane. Hence, diffusion of released neurotransmitters over the artificial synaptic cleft was faster than a few microseconds and could not lead to the observation of the recorded rising times. Similarly, the sampling time resolution was 25 µs so this could be an instrumental artifact.
45 However, it must be noted that very few papers report observation of initial fusion pores flickering in amperometric traces; see e.g.: (a) Z. Zhou, S. Misler and R. H. Chow, *Biophys. J.*, 1996, **70**, 1543; (b) R. G. W. Staal, E. V. Mosharov and D. Sulzer, *Nat. Neurosci.*, 2004, **7**(4), 341. Yet, it is not yet a clear case and, beyond the authors' claims, the possibility remains that such observations resulted from accidental spike superimposition (Zhou et al.) or could be a very specific feature pertaining to cell models different from chromaffin cells, producing sufficiently long-lasting initial fusion pores with long periods of opening and closing so as to be not filtered by diffusion (Staal et al.).

46 (a) A. Kusumi, Y. Umemura, N. Morone, T. Fujiwara, in *Anomalous Transport: Foundations and Applications*, ed. R. Klages, R. Radons and I.M. Sokolov, Wiley, New York, 2008; (b) A. Kusumi, C. Nakada, K. Ritchie, K. Murase, K. Suzuki, H. Murakoshi, R. S. Kasai, J. Kondo and T. Fujiwara, *Annu. Rev. Biophys. Biomol. Struct.*, 2005, **34**, 351–378; (c) A. Kusumi and Y. Sako, *Curr. Opin. Cell Biol.*, 1996, **8**, 566–574; (d) H. Murakoshi, R. Iino, T. Kobayashi, T. Fujiwara, C. Ohshima, A. Yoshimura and A. Kusumi, *Proc. Natl. Acad. Sci. U. S. A.*, 2004, **101**, 7317–7322.
47 L. D. Landau and E. M. Lifshitz. *Statistical Physics*, vol. 5, 3rd edn, Pergamon Press, Oxford, 1980.
48 J. Klein and E. Kumacheva, *J. Chem. Phys.*, 1998, **108**, 6996–7009.
49 C. Amatore, A. Oleinick and I. Svir, *ChemPhysChem*, 2009, **10**, 211–221.
50 C. Amatore, A. I. Oleinick and I. Svir, *J. Electroanal. Chem.*, 2006, **597**, 69–76.
51 C. Amatore, *Chem.-Eur. J.*, 2008, **14**, 5449–5464, and references therein.
52 C. Amatore, S. Arbault, Y. Bouret, M. Guille and F. Lemaître, *ChemPhysChem*, 2010, **11**, 2931–2941.
53 J. L. Barrat and J. F. Joanny, Theory of polyelectrolyte solutions in *Advances in Chemical Physics*, ed. I. Prigogine and S. Rice, Vol. XCIV, Wiley, New York, 1966, pp. 27–33.
54 R. Long, C.-Y. Hui, A. Jagota and M. Bykhovskaia, *J. R. Soc. Interface*, 2012, **9**, 1555–1567.
55 The Gaussian curvature (see: U. Seifert and R. Lipowsky, in Structure and Dynamics of Membranes, From Cells to Vesicles, ed. R. Lipowsky and E. Sackmann, Elsevier, Amsterdam, 1995, pp. 409–411), contributes also to the initial energy of the fusion pore. The modulus of Gaussian curvature energies is predicted to be about $300 k_B T$.

Faraday Discussions

RSC Publishing

PAPER

Potential-dependent single molecule blinking dynamics for flavin adenine dinucleotide covalently immobilized in zero-mode waveguide array of working electrodes

Jing Zhao,[a] Lawrence P. Zaino III[b] and Paul W. Bohn[*ab]

Received 12th February 2013, Accepted 6th March 2013
DOI: 10.1039/c3fd00013c

Single molecules exhibit a set of behaviors that are characteristic and distinct from larger ensembles. Blinking is one such behavior that involves episodic transitions between luminescent and dark states. In addition to the common blinking mechanisms, flavin adenine dinucleotide (FAD), a cofactor in many common redox enzymes, exhibits blinking by cycling between a highly fluorescent oxidized state and a dark reduced state. In contrast to its behavior in flavoenzymes, where the transitions are coupled to chemical redox events, here we study single FAD molecules that are chemically immobilized to the Au region of a zero-mode waveguide (ZMW) array through a pyrroloquinoline quinone (PQQ) linker. In this structure, the Au functions both to confine the optical field in the ZMW and as the working electrode in a potentiostatically controlled 3-electrode system, thus allowing potential-dependent blinking to be studied in single FAD molecules. The subset of ZMW nanopores housing a single molecule were identified statistically, and these were subjected to detailed study. Using equilibrium potential, E_{eq}, values determined from macroscopic planar Au electrodes, single molecule blinking behavior was characterized at potentials $E < E_{eq}$, $E \sim E_{eq}$, and $E > E_{eq}$. The probability of observing a reduced (oxidized) state is observed to increase (decrease) as the potential is scanned cathodic of E_{eq}. This is understood to reflect the potential-dependent probability of electron transfer for single FAD molecules. Furthermore, the observed transition rate reaches a maximum near E_{eq} and decreases to either anodic or cathodic values, as expected, since the rate is dependent on having significant probabilities for both redox states, a condition that is obtained only near E_{eq}.

Introduction

The effects of confinement on chemical reactivity are of intense interest, however these effects are difficult to study at the ensemble level where the

---

[a] Department of Chemical and Biomolecular Engineering, University of Notre Dame, Notre Dame, IN 46556, USA. E-mail: pbohn@nd.edu
[b] Department of Chemistry and Biochemistry, University of Notre Dame, Notre Dame, IN 46556, USA

magnitude—and even sign—of the effect may vary from system to system and may be confounded by extraneous physical and chemical characteristics of the immediate environment.[1–7] One problem in obtaining consistent results is developing a robust experimental protocol that allows the fundamental biophysical/chemical parameters to be isolated from confounding effects. In this regard, zero mode waveguides (ZMWs) offer some unique advantages. ZMWs were initially used to study single binding/catalysis events for systems with μM ligand concentrations.[8–11] Consisting of zeptoliter (1 zL = 10^{-21} L) volume cylindrical nanopores in an opaque film (typically Al[12,13]), ZMWs serve to confine the electromagnetic field inside the nanocylinder, thus enhancing the signal-to-background ratio in spectroscopic experiments. Owing to their small volume, excellent optical confinement, spatial localization, and signal enhancement, ZMWs have been used to study single molecule DNA sequencing,[14] protein–protein interactions,[15] and plasma membrane dynamics.[10,11,16] Furthermore, spectroscopy can be accomplished either by imaging a large number of parallel ZMWs simultaneously,[12,13,15] or by fluorescence correlation.[16,17]

Previously, we constructed Au-based ZMWs for studies of the enzyme monomeric sarcosine oxidase (MSOX),[18] a flavoenzyme in which the flavin adenine dinucleotide (FAD) cofactor exhibits blinking by cycling between a highly fluorescent oxidized state and a dark reduced state. In these experiments electron transfer was constrained to occur between solution-phase substrate and MSOX immobilized to the SiO_2 floor in the interior of the ZMW. The resulting observation of single molecule fluorescence signatures correlated to redox events raises the possibility of using the same strategy to observe direct heterogeneous electron transfer from metallic electrodes to single redox-active molecules. Thus, in contrast to its behavior in flavoenzymes, where the transitions are coupled to chemical redox events, here we study single FAD molecules that are chemically immobilized to a Au working electrode in a zero-mode waveguide (ZMW) array structure through a pyrroloquinoline quinone (PQQ) linker.[19]

Single molecule electrochemistry has been studied since the pioneering experiments by Bard and coworkers utilizing a recessed SECM tip to trap electroactive species, thus amplifying the electrochemical signal above the background by redox cycling and realizing the limit of electrochemical sample size.[20,21] Although these and subsequent experiments have observed the electrochemical behavior of single molecules, observing single electron transfer events is a much more challenging objective. To approach this goal, Lemay and coworkers used lithographically-prepared nanoelectrodes to observe single enzyme electrochemistry in structures producing fA-scale currents commensurate with the electron transfer activity of ~10 molecules.[22,23]

In trapped molecule experiments, assuming that the electrode kinetics are not rate-limiting, the electron transfer rate is given by an inverse transit time

$$\tau^{-1} = \frac{2D}{\langle x^2 \rangle} \quad (1)$$

where D is the diffusion coefficient and x is the separation between electrodes. In contrast, single molecule spectroelectrochemistry produces emitting states that, as long as they can be physically retained in the focal volume, turnover at a rate,

$$N = \frac{\sigma \phi_F P}{A h \nu} \quad (2)$$

where the turnover rate, N, is determined by the absorption cross-section, σ, and source flux, $P/Ah\nu$, as modified by the fluorescence quantum yield, ϕ_F. In trapped molecule electrochemistry with $D \sim 10^{-5}$ cm^2 s^{-1} and $d \sim 10$ nm, it is possible to achieve redox cycling rates of the order $\sim 10^7$ s^{-1}. In contrast, single molecule fluorescence, limited ultimately by photobleaching, can achieve turnover values as high as 10^4 s^{-1}. Furthermore, by careful manipulation of the optical environment, using total internal reflection excitation for example, it is possible to drastically reduce the background, so that single molecule fluorescence can be observed at S/N ratios as large as 100.[24,25] Thus, if redox events can be coupled to the probability that a chromophore emits, then excited state cycling can be used to replace redox cycling to identify and locate electron transfer events.

Recent studies have demonstrated amplification of electron transfer reactions by spectroscopic monitoring.[22,26–28] For example, the conversion between a fluorescent and non-fluorescent state can be monitored while modulating the potential of the cell. Barbara, Bard and coworkers showed that single molecule spectroelectrochemistry can be performed on polymer films coated on indium tin oxide with total internal reflectance fluorescence (TIRF) microscopy.[28] Spectroelectrochemical studies of single molecules have also been performed by single molecule spectroscopy under potential control, using confocal microscopy to monitor a focal volume adjacent to the working electrode surface.[27,29]

Flavoenzymes, containing a FAD cofactor based on the isoalloxazine chromophore, have been a particular target of single molecule fluorescence studies,[30,31] because the molecule is highly fluorescent in the oxidized state and dark in the reduced state.[32] Previous research from this laboratory used ZMWs to study the dynamics of single MSOX enzymes, which contain a FAD cofactor.[18] Thus, the combination of ZMWs (to isolate single FAD molecules and reduce the spectroscopic background) with a Au cladding (used as a working electrode to control the electrochemical potential) provides an ideal platform to investigate the relationship between single molecule fluorescence and single electron transfer events. In addition, because an array of ZMWs is imaged simultaneously, multiple single FAD molecules can be studied in parallel.

Experimental

ZMW fabrication

Zero-mode waveguides were fabricated by a procedure similar to that reported previously.[18] Briefly, a $\sim$170 μm thick fused silica coverslip was cleaned using freshly prepared piranha solution (3 : 1 H$_2$SO$_4$: H$_2$O$_2$ – *Caution: piranha is an extremely aggressive oxidant and should be used with great care*), rinsed with copious amounts of a series of solvents, including deionized (DI) water, acetone, and isopropanol, followed by a final DI water rinse, then dried using N$_2$. The substrate was then coated with a thin ($d \leq 100$ nm) layer of Au by thermal evaporation at $\sim$0.1 nm s^{-1}. The modified 1 cm $\times$ 1 cm area was then coated with 10 nm Cr to passivate the top surfaces against subsequent organomercaptan self-assembly. All evaporated film thicknesses were measured by a quartz crystal film-thickness monitor. A dual-source focused ion beam (FIB) instrument (Helios Nanolab 600, FEI Corp.) was used for milling and characterization. ZMWs were patterned in a 70 μm $\times$ 70 μm square array with a lattice constant of 2 μm by FIB ion milling. The diameters of the individual nanopores constituting the ZMWs were chosen to be between 60 and 150 nm. Finally, after FAD immobilization polydimethylsiloxane (PDMS) wells were placed on the coverslip to isolate individual arrays for convenient sample application.

FAD immobilization

The Au electrodes were cleaned by boiling in 2M KOH for 1 h followed by rinsing with DI water. They were stored in concentrated H_2SO_4 and then soaked for 10 min in concentrated HNO_3 and rinsed again with DI water. The clean Au electrodes were then soaked in a solution of 0.05 M cystamine in DI water for 1 h and then rinsed thoroughly with DI water after emersion. The cystamine-modified Au-electrodes were incubated in 3 mM PQQ in 0.1 M pH 7.2 HEPES-buffer in the presence of 5 mM 1-ethyl-3-[3-dimethylaminopropyl]carbodiimide hydrochloride (EDC) for 2 h. The PQQ-functionalized electrodes were subsequently reacted with 1 mM 3-aminophenylboronic acid in 0.1 M pH 7.2 HEPES buffer in the presence of 5 mM EDC for 2 h, followed by rinsing with DI water. The resulting electrode was treated with FAD, typically at a concentration of 10–30 nM in 0.1 M pH 7.0 phosphate buffer for 2 h at room temperature and then overnight at 4 °C followed by washing with DI water. The concentration was determined to yield an optimal distribution of ZMWs occupied with single FAD molecules.

Electrochemical measurements

Electrochemical measurements were performed using an electrochemical analyzer (Reference 600, Gamry Instruments). The measurements were carried out at an ambient temperature using FAD modified Au as a working electrode, an unmodified Au film as a counter electrode, and a saturated Ag/AgCl reference electrode. For CV measurements, FAD immobilization was accomplished as above, but using 1 mM FAD in 0.1 M pH 7.0 phosphate buffer to prepare macroscopic FAD-derivatized planar Au working electrodes. Cyclic voltammetry was performed in 0.1 M pH 7.2 phosphate buffer at 50 mV s^{-1}.

Fluorescence measurements

Fluorescence measurements were performed on an Olympus IX71 wide-field epi-illumination microscope. Laser radiation at 488 nm was reflected into the microscope and defocused to illuminate an area ~70 μm × 70 μm on the fused silica substrate side of the ZMW sample to directly excite single FAD fluorescence in individual ZMW nanopores. The resulting fluorescence was collected by the same oil immersion objective (100×, 1.4 NA) and projected onto a 512 × 512 pixel Andor charge-coupled device (CCD) camera (Andor Technology) with a 1.5× tube lens, achieving a net magnification of 150×. A dichroic mirror (Chroma Z488RDC) and an emission filter (Chroma HQ525/50m) were used to discriminate against scattered excitation radiation. Sequential image frames were acquired with 50 ms exposure for subsequent data analysis using Igor (Wavemetrics) software. Movies were processed by first identifying individual localized fluorescence regions of interest (ROIs), the intensity of which was obtained by integrating the signal over a 6 × 6 pixel area. Temporal fluorescence intensity trajectories were then calculated from each ROI for subsequent statistical analysis. Potential-dependent fluorescence trajectories were obtained by applying the stated potential *vs.*

Ag/AgCl to the entire ZMW array in 0.1 M pH 7.2 phosphate buffer. All solutions were sparged with N_2 prior to ZMW fluorescence experiments to remove O_2.

Results and discussion

Construction of electrochemical ZMW arrays

To obtain fluorescence readout of single molecule electron transfer events, it is necessary to isolate single electroactive molecules so that they may be addressed both electrochemically and optically. The strategy used here involves the fabrication of controlled-potential ZMW structures in which a subpopulation of ZMW nanopores contain a single molecule—the rest containing either zero or multiple molecules, *viz.* Fig. 1. The ZMWs containing single FAD molecules are identified by standard statistical tests applied to the fluorescence trajectories from all ZMWs.[33,34] The optical intensity through a cylindrical zero-mode waveguide can be expressed as

$$I(z) = \exp\left[-2z\sqrt{\frac{1}{\lambda_c^2} - \frac{1}{\lambda_m^2}}\right] \quad (3)$$

where $\lambda_c = 1.7d$ is the cut-off wavelength, d is the diameter for a cylindrical ZMW, and λ_m is the wavelength in the core medium, 367 nm in the present experiment. Thus, the nominal observation volume, *i.e.* the volume in which the radiation field is contained, for a 100 nm diameter ZMW is ~125 zL. Although the light-blocking efficiency of the Au is lower than the commonly used Al, Au was used to construct ZMWs in these experiments because Au enables a wide range of self-assembly surface chemistry which can be used to synthesize the immobilized

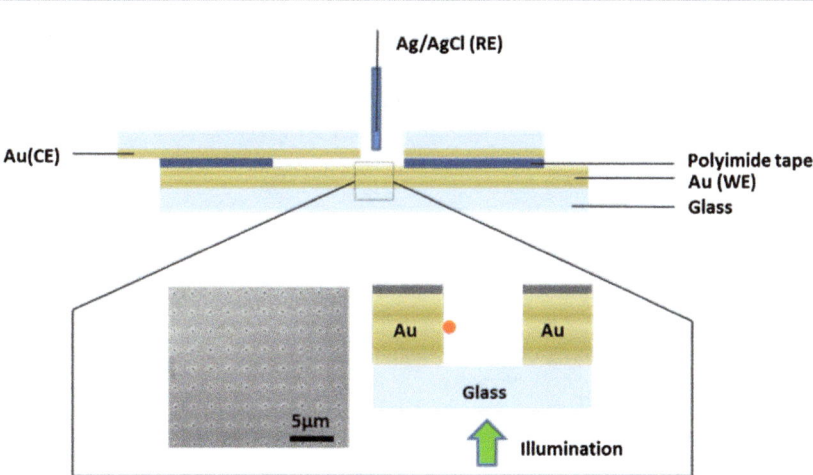

Fig. 1 Schematic (side view) diagram of the ZMW-electrochemical cell made between Au-coated fused silica containing ZMW working electrode (WE) arrays and a Au-coated coverslip counter electrode (CE). The structure uses a macroscopic Ag/AgCl reference electrode (RE). Each square ZMW array consists of ~1000 ZMWs with 2 μm period. The diameter of the ZMW nanopores is 100 nm. The laser illuminates the device from the fused silica side. (*Expanded view*) Left: SEM image showing a portion of the ZMW array. Right: Schematic diagram showing a typical location of an individual FAD/FADH$_2$ molecule (red) immobilized on the Au WE.

electroactive species. In this work, FAD molecules are immobilized on the Au sidewalls of the ZMW nanopores—rather than the fused silica bottom surfaces—specifically to access heterogeneous electron transfer processes. However, as indicated by eqn (3), there is a dispersion in the excitation intensity among single molecules in different ZMWs. Those near the bottom of the ZMWs are more strongly excited due to the intensity decay along the z-axis of the ZMWs.

In these experiments, uniform ZMW arrays containing ~10^3 ZMWs were prepared by fast FIB milling to allow parallel multiplexed observation of multiple single molecules at the same applied potential. Assuming a Poisson distribution governs molecular occupancy of the ZMWs, the concentration of FAD was adjusted in the range 10–30 nM (depending on the ZMW diameter) to optimize the yield of ZMWs with single FAD chromophores bound to the Au sidewalls within the observation volume. To ensure that electrochemically addressable immobilized fluorescent FAD were obtained, FAD molecules were isolated on the surface of the Au sidewalls using an organothiol-directed 3-step synthesis shown in Scheme 1. Following the molecular logic gate work of Pita and Katz,[19] the Au surface was prepared with a self-assembled monolayer presenting a terminal amine for derivatization with PQQ and subsequently with m-aminophenylboronic acid, both of which were carried out with EDC coupling chemistry.[35,36] PQQ has three possible sites for amide coupling, but these experiments were carried out without control over the three possible regioisomers. Finally, FAD was immobilized by coupling to the surface phenylboronic ester through the sugar moiety. The efficacy of this immobilization strategy was tested on macroscopic planar Au surfaces. The resulting structures demonstrate that the linker is competent to support electron transfer as supported by cyclic voltammetry and differential pulse voltammetry, the latter yielding a robust measure of surface coverage of 9.3×10^{-11} mol cm^{-2}, in good agreement with Pita and Katz.[19] Binding of FAD on large scale planar Au resulted in a peak ($E = -0.48$ V vs. Ag/AgCl) corresponding to the redox process for surface immobilized FAD. This peak position was used in subsequent ZMW electrochemistry experiments as an estimate of the position of the single molecule equilibrium potential, $E_{eq} = -0.48$ V vs. Ag/AgCl.

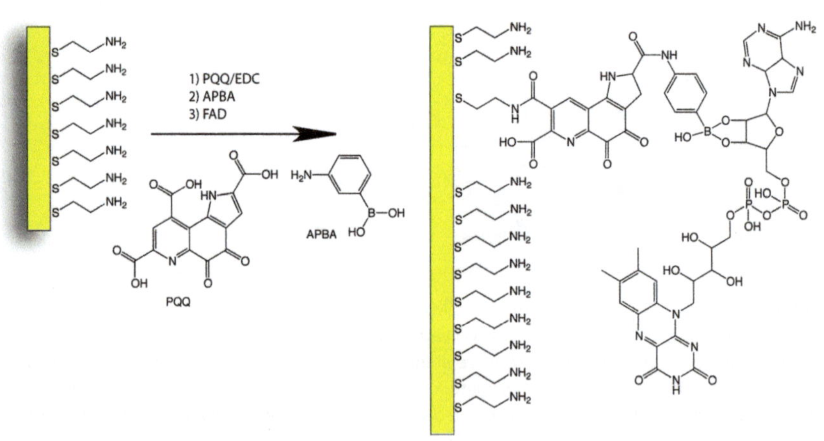

Scheme 1

In any surface immobilization scheme, non-specific adsorption is a potential problem. In addition, the existence of solution-phase FAD molecules diffusing in or out of the observation volume would also complicate recognition of the FAD–FADH$_2$ transitions occurring on Au sidewalls. To evaluate the influence of these potential sources of interference, the fluorescence from a ZMW array filled with free FAD solution was monitored in the absence of PQQ surface linker molecules using the same concentration as that used for immobilization, *i.e.* 30 nM. Under these conditions, the intensity of non-specifically adsorbed and freely diffusing FAD was found to be negligible in comparison to that of single molecules of covalently attached FAD within the same temporal observation window. Therefore, interferences from non-specific adsorption and desorption of FAD and from freely diffusing FAD were neglected.

Representative single molecule emission behavior

The wide-field imaging strategy implemented here allows multiplexed real-time observation of multiple ZMW nanopores, each containing a single FAD molecule. FAD derivatization solution concentrations in the range 10–30 nM were found to yield acceptable single molecule occupancies, although structures prepared from 10 nM FAD solutions were found to minimize contributions from multiply occupied ZMWs. To distinguish single FAD molecules from multiple molecules, fluorescence levels corresponding to all potential ROIs, *i.e.* ZMWs, were plotted as intensity histograms. The fluorescence levels of FAD molecules immobilized in the ZMW were determined by comparison with the fluorescence intensity of ZMWs containing zero molecules, *i.e.* ZMWs exhibiting no jumps in fluorescence level during the entire trajectory, which were regarded as defining the background. These ZMW ROIs produced intensity distributions with a single peak at the lowest count level for 50 ms acquisition, *i.e.* ∼5500 counts, allowing this level to be assigned to ZMWs containing zero molecules. ZMWs with episodic fluorescence intensities exceeding this background were assigned to occupancies $n \geq 1$. Analysis of the integrated fluorescence intensity histograms in these frames frequently yielded peaks at multiple positions, as shown in Fig. 2, but with variable intensities. The next two discrete histogram peaks can be assigned to $n = 1$ and 2, respectively. Variable intensity distributions from the same number of molecules in different ZMW nanopores is attributed to differential positioning of the fluorescent FAD molecules relative to the ZMW excitation field and to different orientations of the chromophores relative to the surface. Increased FAD loading was also observed with larger ZMW diameters.

In addition to the histograms, typical fluorescence intensity trajectories of single FAD molecules, obtained without potential control, exhibit blinking and eventual photobleaching. Under the same laser excitation and solution conditions used to acquire potential-dependent data, control experiments (without potential control) show that the photobleaching times of immobilized FAD molecules, $\tau_{\text{photobleach}} = 110 \pm 70$ s, are much longer than the average on-time duration (ranging from a few hundred milliseconds to a few seconds) in the ZMW fluorescence trajectories obtained under potential control. Also, open-circuit FAD measurements yield an on-time distribution that decays exponentially with time constant $\tau_{\text{blink}} = 9.7 \pm 0.6$ s, the timescale of which is compared below to the temporal characteristics of blinking under potential control. Given the much

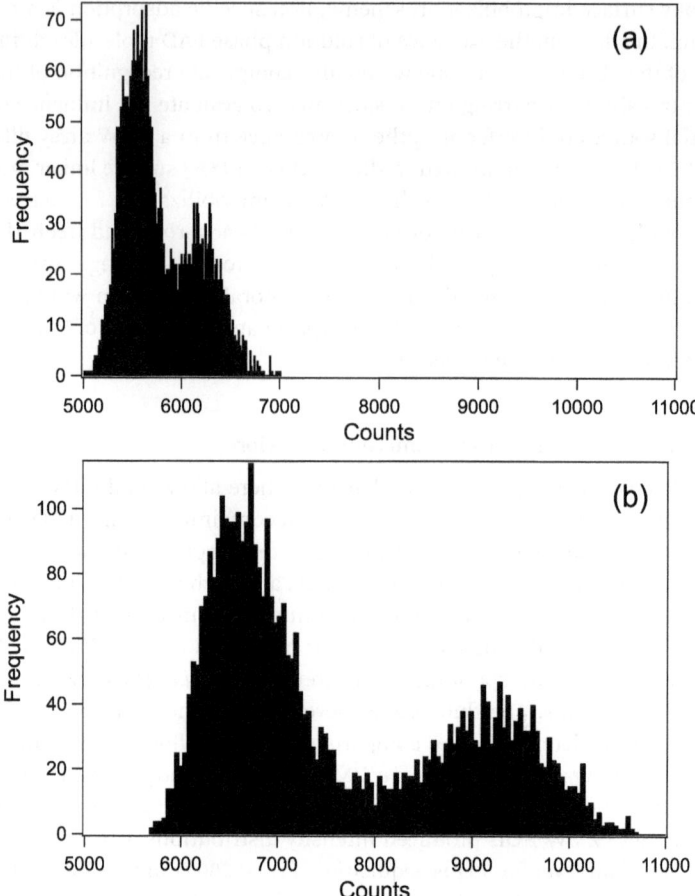

Fig. 2 Representative histograms of 6 min trajectories illustrating the relationship of the discrete fluorescence levels to molecular occupancy. (a) Single molecule occupancy exhibits an intensity peak near that of the background, ~5500 counts, and a single distinct peak at higher intensity. (b) Multiple molecule occupancy exhibits peaks at single molecule levels and higher.

longer time scale of photobleaching compared to potential-dependent blinking, this phenomenon can be safely neglected in analysis of potential-dependent single molecule fluorescence behavior. Furthermore, the characteristic blinking behavior under potential control can readily be distinguished from the open circuit behavior, so the intrinsic blinking does not confound interpretation of the potential-dependent single molecule behavior.

Potential dependent single molecule intensity distributions

When applying a potential to the ZMW working electrode array, movies of ZMW single molecule fluorescence show repetitive stochastic bursts at frequencies that are potential dependent. Typical data shown in Fig. 3 all exhibit stochastic two-state off–on signals, but the on- and off-time distributions vary dramatically as a function of potential. The digital, two-state nature of these fluorescence trajectories is consistent with the interpretation that each burst comes from a single

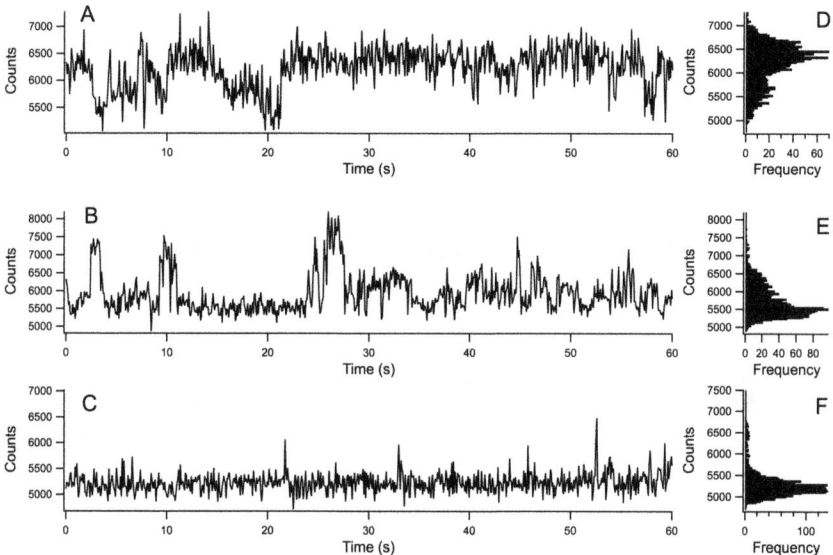

Fig. 3 *Left*: 60 s long fluorescence trajectories at applied potentials of (A) −0.2 V; (B) −0.48 V; (C) −0.8 V. Each point in the trajectory is a 50 ms integration acquired at 82 ms intervals. *Right*: Intensity histograms for the full 150 s acquisition at applied potentials of (D) −0.2 V; (E) −0.48 V; (F) −0.8 V.

FAD molecule. Were multiple molecules responsible, the trajectory would exhibit a variable number of intensity states, depending on the number of molecules, with measurable activity at the intermediate levels. Because the digital fluorescence trajectories observed in the absence of applied potential have completely different dynamics (*vide supra*), the single FAD fluorescence transitions are clearly related to electrode potential and, thus, to electron transfer events.

Fig. 4 shows the distribution of fluorescence intensities for three separate single FAD molecules at each of three potentials (−0.2 V, −0.48 V, −0.8 V vs. Ag/AgCl), chosen to bracket E_{eq}. At each potential, the molecules show behavior which is qualitatively similar, but the details of the distributions differ. In contrast, the behavior is distinctly different at the three different potentials, with the weight of the distributions shifting from mostly 'on' at −0.2 V to mostly 'off' at −0.8 V, with molecules displaying a distribution weighted between the two states near the macroscopic E_{eq} value. Differences in the intensity distributions are observed among individual single FAD molecules at the same potential, *i.e.* different single FAD molecules in different ZMW nanopores, which could arise from a number of possible sources. Most directly, E_{eq} is certainly expected to vary from molecule to molecule, reflecting variations in bonding, steric issues and the local nature of the electrode.[37]

Starting at the most anodic potential, −0.2 V vs. Ag/AgCl, the behavior for which is displayed in Fig. 4(A)–(C), molecules represented in Fig. 4(A) and (B) are clearly dominated by a preponderance of on-state behavior, as would be expected from the macroscopic electrochemistry of the FAD cofactor. In contrast, the molecule in Fig. 4(C) shows a larger fraction of off-state, suggesting that the local E_{eq} for this molecule is shifted positive relative to the macroscopic average value.

When the applied potential is near the macroscopic $E_{eq} = -0.48$ V, Fig. 4(D)–(F), the distributions are shifted to a larger fraction of off-times, as expected at this

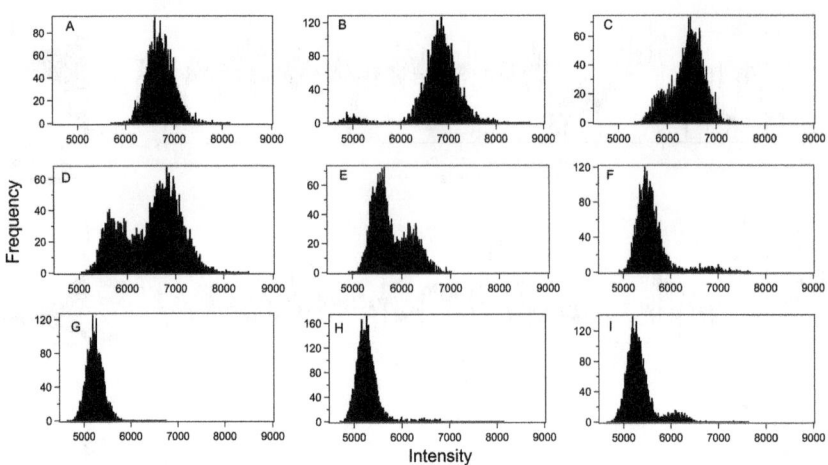

Fig. 4 Histograms of single molecule intensity distributions from 6 min trajectories for three representative molecules at each of three potentials. (A)–(C) −0.2 V; (D)–(F) −0.48 V; (G)–(I) −0.8 V. All potentials vs. Ag/AgCl reference.

more cathodic potential. Here, too, there are significant molecule-to-molecule differences consistent with local variations in E_{eq}, cf. Fig. 4(D) and 4(F). At −0.8 V, Fig. 4(G)–(I), the fraction of on-times in the distributions are negligible, except for the molecule in Fig. 4(I) where it is small but measurable. These observations, taken collectively, are consistent with the macroscopic behavior in which oxidized FAD (predominant at −0.2 V) is luminescent, while reduced $FADH_2$ (the dominant species at −0.8 V) is dark. To further illustrate potential control over electron transfer to single FAD molecules immobilized in ZMWs, histograms of fluorescence intensity over tens of ZMW spots were analyzed at multiple potentials. Just as the results for the three specific molecules in Fig. 3 indicate, the distribution of fluorescence intensity shifts from levels consistent with a single molecule in the on-state to an off-state level at more cathodic applied potentials.

In addition to the fluorescence intensity distributions, the rate of on → off and off → on transitions also changes with applied potential. At cathodic (anodic) potential, the intensity distributions indicate molecules in the off (on) state, and as can be seen in Fig. 3, transitions between levels are infrequent and the duration in the opposite state is short. However, as is evident in Fig. 3, the rate of transitions is larger near E_{eq}. Transition rates at specified potentials were obtained by fitting the histogram of on-time and off-time distributions. In one particularly long (10 min observation time) but otherwise representative fluorescence trajectory, obtained at E_{eq} = −0.48 V vs. Ag/AgCl, both on-time and off-time distributions exhibited single exponential decay with approximately the same decay constant, $k_{on} \approx k_{off} \approx 2.5$ s^{-1}, consistent with the reduction rate and the oxidation rate being comparable to each other near E_{eq}.

Potential dependence

We hypothesize that the probability of electron transfer is potential-dependent on a single molecule basis. The probability of observing a reduced state (p_{red}) should

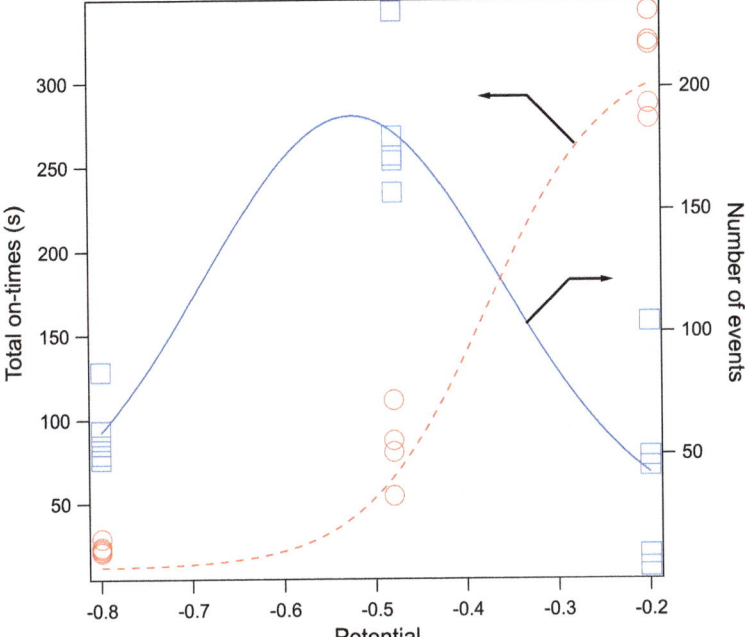

Fig. 5 Potential dependence of total on-time duration (left, red, circles) and the number of transition events (right, blue, squares) for 15 separate 6-min trajectories; 5 at each potential. The total on-time duration and number of transitions are fit to sigmoid (red, dashed) and Gaussian (blue, solid) functions, respectively.

approach 1 as the potential is scanned negative of E_{eq}, and similarly, the probability of observing an oxidized state (p_{ox}) should approach 1 as the potential is scanned positive of E_{eq}. Because the observed transition rate depends on having non-zero probabilities for both p_{red} and p_{ox}, the maximum should be reached near E_{eq}, falling off to either side, so the transition rate is expected to be a maximum when the potential is held near the ensemble-averaged value of E_{eq}. Comparison of the fluorescence trajectories of single FAD molecules at different potentials in Fig. 3 clearly indicates that the rate at which transitions between FAD and FADH$_2$ occur is potential dependent. Consistent with this hypothesis, Fig. 5 shows that the total on-time duration increases with increasingly anodic applied potential. In addition, Fig. 5 indicates that the rate at which transitions between luminescent (oxidized, FAD) and dark (reduced, FADH$_2$) states occur can be modulated with electrochemical potential, with the maximum rate occurring near E_{eq}. The curve is symmetric around E_{eq}, and consistent with expectations, the transition frequency is a maximum at E_{eq}.

Conclusions

We have explored the possibility of using electrochemically-active zero-mode waveguides to examine potential-dependent blinking behavior of single FAD/FADH$_2$ molecules covalently immobilized to the Au cladding layer of the ZMW. The optical fields are not as well-confined in Au-clad ZMWs as in those with

Al cladding, but the small difference is compensated by access to Au self-assembly chemistry, which is used here to fashion a molecular wire construct utilizing an intermediate PQQ linker. The efficacy of this linker strategy is well-demonstrated by the single molecule electrochemical results. Fluorescence temporal trajectories and intensity distributions clearly indicate molecule-to-molecule variations in behavior at the same nominal potential, likely illustrating variations in local E_{eq} that may reflect, among other things, variations in bonding, steric constraints and the local nature of the electrode. Despite these individual differences, characterization of several tens of molecules at each potential show definite potential-dependent behavior: (1) the weight of the fluorescence intensity distributions shifts from mostly 'on' at -0.2 V to mostly 'off' at -0.8 V, with molecules displaying a distribution weighted between the two states near the macroscopic E_{eq} value; and (2) the rate of on→off and off→on transitions also changes with applied potential, reaching a maximum near E_{eq}. These observations are consistent with behavior expected for single electron transfer events. Using the dark↔emissive transition as an indicator of the $FADH_2 \leftrightarrow FAD$ 2-electron redox process, strongly supports an interpretation of single redox events. The fluorescence intensity distributions scale with applied potential in exactly the same way as expected for the single molecule probabilities, p_{ox} and p_{red}. In addition, using a Fermi's Golden Rule formalism leads to the prediction that the rate of reduction reactions, for example, should scale with the probability that the molecule is initially oxidized, i.e. $p_{ox} = 1 - p_{red}$. This, in turn, predicts that the transition rate should be largest near E_{eq}, exactly as observed.

Given the success of these potential-controlled ZMW-immobilized single molecule spectroelectrochemistry experiments, we anticipate further exploiting this capacity to modulate the quantum efficiency of a redox-sensitive fluorophore by immobilizing single copies of FAD-containing redox-active flavoenzymes in order to explore individual electron transfer events at single enzyme molecules. In these single enzyme spectroelectrochemistry experiments, we expect to see even more dramatic molecule-to-molecule differences given the structure (surface orientation) dependent efficiency of electron transfer.

Acknowledgements

This work was supported by the National Science Foundation through the Center for Nanoscale Chemical-Electrical-Mechanical Manufacturing Systems (JZ) and grant 1111739 (LPZ). Fabrication of ZMW electrode arrays was partially supported by the Notre Dame Integrated Imaging Facility and the Notre Dame Nanofabrication Facility, and development of the ZMW structures was funded by the Department of Energy through DE FG02 ER0715851.

References

1 L. Homchaudhuri, N. Sarma and R. Swaminathan, *Biopolymers*, 2006, **83**, 477–486.
2 M. Jiang and Z. H. Guo, *J. Am. Chem. Soc.*, 2007, **129**, 730–731.
3 N. Kozer, Y. Y. Kuttner, G. Haran and G. Schreiber, *Biophys. J.*, 2007, **92**, 2139–2149.
4 N. Kozer and G. Schreiber, *J. Mol. Biol.*, 2004, **336**, 763–774.
5 Y. Y. Kuttner, N. Kozer, E. Segal, G. Schreiber and G. Haran, *J. Am. Chem. Soc.*, 2005, **127**, 15138–15144.
6 S. Zorrilla, M. A. Hink, A. J. W. G. Visser and M. P. Lillo, *Biophys. Chem.*, 2007, **125**, 298–305.

7 S. Zorrilla, G. Rivas, A. U. Acuna and M. P. Lillo, *Protein Sci.*, 2004, **13**, 2960–2969.
8 M. Levene, J. Korlach, S. Turner, M. Foquet, H. Craighead and W. W. Webb, *Science*, 2003, **299**, 682–686.
9 J. M. Moran-Mirabal and H. G. Craighead, *Methods*, 2008, **46**, 11–17.
10 J. M. Moran-Mirabal, A. J. Torres, K. Samiee, B. Baird and H. Craighead, *Nanotechnology*, 2007, **18**, 195101.
11 K. T. Samiee, J. M. Moran-Mirabal, Y. K. Cheung and H. Craighead, *Biophys. J.*, 2006, **90**, 3288–3299.
12 M. Foquet, K. Samiee, X. Kong, B. P. Chauduri, P. Lundquist, S. Turner, J. Freudenthal and D. B. Roitman, *J. Appl. Phys.*, 2008, **103**, 034301.
13 J. Korlach, P. J. Marks, R. L. Cicero, J. J. Gray, D. L. Murphy, D. B. Roitman, T. T. Pham, G. A. Otto, M. Foquet and S. Turner, *Proc. Natl. Acad. Sci. U. S. A.*, 2008, **105**, 1176–1181.
14 J. Eid, A. Fehr, J. Gray, K. Luong, J. Lyle, G. Otto, P. Peluso, D. Rank, P. Baybayan, B. Bettman, A. Bibillo, K. Bjornson, B. Chaudhuri, F. Christians, R. Cicero, S. Clark, R. Dalal, A. Dewinter, J. Dixon, M. Foquet, A. Gaertner, P. Hardenbol, C. Heiner, K. Hester, D. Holden, G. Kearns, X. X. Kong, R. Kuse, Y. Lacroix, S. Lin, P. Lundquist, C. C. Ma, P. Marks, M. Maxham, D. Murphy, I. Park, T. Pham, M. Phillips, J. Roy, R. Sebra, G. Shen, J. Sorenson, A. Tomaney, K. Travers, M. Trulson, J. Vieceli, J. Wegener, D. Wu, A. Yang, D. Zaccarin, P. Zhao, F. Zhong, J. Korlach and S. Turner, *Science*, 2009, **323**, 133–138.
15 T. Miyake, T. Tanii, H. Sonobe, R. Akahori, N. Shimamoto, T. Ueno, T. Funatsu and I. Ohdomari, *Anal. Chem.*, 2008, **80**, 6018–6022.
16 J. B. Edel, M. Wu, B. Baird and H. Craighead, *Biophys. J.*, 2005, **88**, L43–L45.
17 J. Wenger, D. Gerard, P.-F. Lenne, H. Rigneault, J. Dintinger, T. W. Ebbesen, A. Boned, F. Conchonaud and D. Marguet, *Opt. Express*, 2006, **14**, 12206–12216.
18 J. Zhao, S. P. Branagan and P. W. Bohn, *Appl. Spectrosc.*, 2012, **66**, 163–169.
19 M. Pita and E. Katz, *J. Am. Chem. Soc.*, 2008, **130**, 36–37.
20 F. R. F. Fan and A. J. Bard, *Science*, 1995, **267**, 871–874.
21 F. R. F. Fan, J. Kwak and A. J. Bard, *J. Am. Chem. Soc.*, 1996, **118**, 9669–9675.
22 A. J. Bard, *ACS Nano*, 2008, **2**, 2437–2440.
23 F. J. M. Hoeben, F. S. Meijer, C. Dekker, S. P. J. Albracht, H. A. Heering and S. G. Lemay, *ACS Nano*, 2008, **2**, 2497–2504.
24 W. E. Moerner and D. P. Fromm, *Rev. Sci. Instrum.*, 2003, **74**, 3597–3619.
25 H. Park, G. T. Hanson, S. R. Duff and P. R. Selvin, *J. Microsc.*, 2004, **216**, 199–205.
26 E. Cortes, P. G. Etchegoin, E. C. Le Ru, A. Fainstein, M. E. Vela and R. C. Salvarezza, *J. Am. Chem. Soc.*, 2010, **132**, 18034–18037.
27 C. H. Lei, D. H. Hu and E. J. Ackerman, *Chem. Commun.*, 2008, 5490–5492.
28 R. E. Palacios, F. R. F. Fan, A. J. Bard and P. F. Barbara, *J. Am. Chem. Soc.*, 2006, **128**, 9028–9029.
29 C. H. Lei, D. H. Hu and E. Ackerman, *Nano Lett.*, 2009, **9**, 655–658.
30 H. Lu, L. Xun and X. Xie, *Science*, 1998, **282**, 1877–1882.
31 W. Min, I. V. Gopich, B. P. English, S. C. Kou, X. S. Xie and A. Szabo, *J. Phys. Chem. B*, 2006, **110**, 20093–20097.
32 A. Tyagi and A. Penzkofer, *J. Photochem. Photobiol., A*, 2010, **215**, 108–117.
33 W. P. Ambrose, P. M. Goodwin, J. H. Jett, A. Van Orden, J. H. Werner and R. A. Keller, *Chem. Rev.*, 1999, **99**, 2929–2956.
34 J. C. Fister, S. C. Jacobson, L. M. Davis and J. M. Ramsey, *Anal. Chem.*, 1998, **70**, 431–437.
35 S. T. Plummer, P. W. Bohn, R. A. Stockton and M. A. Schwartz, *Langmuir*, 2003, **19**, 7528–7536.
36 Q. Wang and P. W. Bohn, *J. Phys. Chem. B*, 2003, **107**, 12578–12584.
37 G. K. Rowe, M. T. Carter, J. N. Richardson and R. W. Murray, *Langmuir*, 1995, **11**, 1797–1806.

Faraday Discussions

Networks of DNA-templated palladium nanowires: structural and electrical characterisation and their use as hydrogen gas sensors

Mariam N. Al-Hinai,[†ac] Reda Hassanien,[b] Nicholas G. Wright,[c] Alton B. Horsfall,[c] Andrew Houlton[a] and Benjamin R. Horrocks*[a]

Received 15th February 2013, Accepted 4th March 2013
DOI: 10.1039/c3fd00017f

Electroless templating on DNA is established as a means to prepare high aspect ratio nanowires *via* aqueous reactions at room temperature. In this report we show how Pd nanowires with extremely small grain sizes (<2 nm) can be prepared by reduction of $PdCl_4^{2-}$ in the presence of λ-DNA. In AFM images the wires are smooth and uniform in appearance, but the grain size estimated by the Scherrer treatment of line broadening in X-ray diffraction is less than the diameter of the wires from AFM (of order 10 nm). Electrical characterisation of single nanowires by conductive AFM shows ohmic behaviour, but with high contact resistances and a resistivity (~10^{-2} Ω cm) much higher than the bulk value for Pd metal (~10^{-5} Ω cm @20 °C). These observations can be accounted for by a model of the nanowire growth mechanism which naturally leads to the formation of a granular metal. Using a simple combing technique with control of the surface hydrophilicity, DNA-templated Pd nanowires have also been prepared as networks on an Si/SiO_2 substrate. These networks are highly convenient for the preparation of two-terminal electronic sensors for the detection of hydrogen gas. The response of these hydrogen sensors is presented and a model of the sensor response in terms of the diffusion of hydrogen into the nanowires is described. The granular structure of the nanowires makes them relatively poor conductors, but they retain a useful sensitivity to hydrogen gas.

1 Introduction

Controlled fabrication of materials with a specific shape or size remains an important topic in nanotechnology. Nanowires are of particular interest for the

---

[a]Chemical Nanoscience Laboratory, School of Chemistry, Newcastle University, Bedson Building, NE1 7RU, UK. E-mail: b.r.horrocks@ncl.ac.uk; Tel: +44(0)191 2225619
[b]Department of Science and Mathematics, Faculty of Education, Assiut University, New Valley Branch, El-Kharja 72511, Egypt
[c]School of Electrical and Electronic Engineering, Newcastle University, Merz Court, NE1 7RU, UK

† Current address: Department of Engineering, Sohar College of Applied Science, Oman

fabrication of functional electronic devices and duplex DNA has been shown to be highly effective for promoting the anisotropic growth of a variety of materials such as organic conducting polymers,[1-8] metals,[9-23] metal oxides,[24-26] semiconducting chalcogenides,[27-30] and superconducting alloys.[31] The self-assembly and molecular recognition abilities of DNA may also alleviate the problems of inter-element wiring and positioning at the nanometer scale.[32] The templating reaction proceeds by the formation of nuclei on the template and these nuclei grow and may eventually overlap.[33] The resulting structure is typically either in the form of a smooth nanowire or a series of beads on a string. DNA-templated nanowires have several advantages: ease of fabrication, deposition and controlled alignment on a substrate by molecular combing.[4,27] They are therefore of particular interest in sensing technology.

It is now widely recognised that hydrogen gas sensors are essential for safety reasons in any widespread use of hydrogen as a low-carbon energy source.[34,35] Hydrogen is highly flammable and becomes explosive when its concentration exceeds 4% in air.[36] The sensors are required to be rapid, sensitive, reliable, cheap, simple to operate and to have high selectivity. The required selectivity can be obtained by using the ability of palladium to sorb hydrogen gas.[37] Electronic sensors have advantages over optical and other devices with respect to the simplicity of the sensor structures and compatibility of the sensor fabrication process with miniaturization and the current IC process.[38] However, bulk Pd-based hydrogen sensors suffer from fundamental problems including long diffusion-limited response times and irreversible resistance changes resulting from the large internal stress caused by the expansion that occurs upon excess hydrogen absorption.[39] Sensors have therefore been designed using various Pd-based nanomaterials, including nanowires, as transducers.[40-54] Interest in nanowires for use in sensor technology has grown and therefore Pd/DNA nanowires may be attractive for hydrogen sensing. In particular, DNA-templated nanowires offer a simple and cost effective sensor fabrication process. In this report we describe the chemical and structural characterisation of Pd/DNA nanowires, their electrical behaviour and discuss the response of a Pd/DNA nanowire network as a hydrogen sensor.

2 Experimental

2.1 Preparation of Pd/DNA nanowires

Palladium nanowires (Pd/DNA) were prepared by templating the electroless reduction of Pd(II) on lambda DNA (λ-DNA) or calf thymus DNA (CT-DNA). λ-DNA was purchased from New England Biolabs (cat. no. N30011S Hitch, Herts, UK) at a concentration of 500 ng μL^{-1} in pH 8 buffer; it was used as supplied. CT-DNA was obtained from Sigma-Aldrich (highly polymerised, 6% sodium). All other reagents were obtained from Sigma-Aldrich and were of AnalaR grade or equivalent. Deionized water was obtained from a Barnstead nanopure™ purification train with nominal resistivity of 18.2 MΩ cm.

Our standard preparation of Pd/DNA nanowires:

10 μL of freshly-prepared K_2PdCl_4(aq) (3 mM) was added to 20 μL of λ-DNA (500 ng μL^{-1}) in the presence of 5 μL sodium citrate (aq) (0.5 mM). The Pd(II) ions were reduced to Pd(0) metal by addition of 5 μL aqueous dimethylaminoborane (DMAB, 10 mM) or $NaBH_4$ (10 mM). The solution was thoroughly mixed and allowed to react at room temperature for 2–24 h as indicated in the results section. 24 h reaction times were required to form smooth continuous nanowires. Use of

NaBH$_4$ or DMAB as reducing agent produced similar nanowires; some differences between the two are noted in the electrical characterisation below.

Slightly different preparation methods were used for UV-Vis spectroscopy and XRD analysis because these techniques require larger samples. CT-DNA was employed for reasons of cost.

For UV-Vis experiments:

2 mL aqueous CT-DNA (162.5 ng µL^{-1}) was mixed with 1 mL freshly-prepared K$_2$PdCl$_4$ (aq) (3 mM) in the presence of 0.5 mL sodium citrate (aq) (0.5 mM). 0.5 mL of aqueous dimethylaminoborane (DMAB, 10 mM) was added dropwise to the pale yellow solution.

For XRD experiments:

Half the quantities employed in the UV-Vis experiment were used. A black precipitate of Pd/DNA was formed and was collected by filtration and washed thoroughly with hot ethanol.

2.2 FTIR spectroscopy

FTIR spectra (in the range 800–1800 cm^{-1}) were recorded in transmission mode with a Bio-Rad Excalibur FTS-40 spectrometer (Varian Inc., Palo Alto, CA, USA) equipped with a liquid nitrogen cooled deuterated triglycine sulphate (DTGS) detector. 128 scans were co-added and averaged and the resolution was 4 cm^{-1}. Pd/DNA solution (8 µL) was deposited on a clean p-Si(100) chip (1 cm × 1 cm) and dried in air for 1 h prior to analysis. A clean p-Si(100) chip served as the background.

2.3 UV-Vis spectroscopy

Ultraviolet-visible (UV-Vis) absorption spectroscopy was employed to monitor the reaction progress. UV-Vis absorbance spectra of Pd nanowires before and after DMAB reduction were recorded *in situ* on a Thermo Spectronic GENESYS 6 spectrophotometer at room temperature (wavelength range: 250 to 600 nm).

2.4 X-ray photoelectron spectroscopy

An Axis-Ultra photoelectron spectrometer equipped with a monochromatic Al-Kα X-ray as the excitation source (1486.7 eV) with an operating power of 150 W (15 kV, 10 mA) was used to collect photoemission spectra of Pd–DNA samples. The photoelectrons were filtered by the hemispherical analyzer and recorded by multichannel detectors. The chamber pressure was 3.2×10^{-9} Torr. For the survey scan, the pass energy was 20 eV and the step size was 0.3 eV. The binding energies obtained in the XPS analysis were calibrated using C1s (284.6 eV) as a reference. Spectral peaks were fitted with mixed singlet functions or a Doniach–Sunjic doublet function after subtraction of a Shirley-type background using the WinSpec program developed by LISE laboratory, Belgium. The Pd/DNA samples were prepared by depositing 8 µL of solution on a clean Si(100) substrate and drying in air at room temperature in a laminar flow hood to minimize contamination (Model VLF 4B, Envair, Haslingden, Lancs, U.K.).

2.5 X-ray diffraction

The XRD analysis was recorded from a powder sample using X-ray diffractometer system (XPERT-PRO) with graphite monochromatized Cu Kβ radiation ($\lambda =$

0.15418 nm). The scanning rate of $0.033° \text{ s}^{-1}$ was applied to record the pattern in the 2θ range of 25° to 90°.

2.6 Atomic force microscopy, conductive AFM and scanning conductance microscopy

Tapping mode imaging of surface topography was performed in air on a Multimode Dimension Nanoscope V (Veeco Instruments Inc., Metrology Group, Santa Barbara, CA, USA) using TESP7 probes (n-doped Si cantilevers, Veeco Instruments Inc., Metrology Group), with a resonant frequency of 234–287 kHz and a spring constant of $20\text{–}80 \text{ N m}^{-1}$. Data acquisition was carried out using Nanoscope version 7.00b19 (Dimension Nanoscope V) software (Veeco Instruments Inc.). Vibrational noise was reduced with an isolation table/acoustic enclosure (Veeco Inc., Metrology Group). Si(100) chips (1 cm × 1 cm) cut from 100 mm diameter wafer (B-doped, p-type single-side polished; Compart Technology, Cambridge, UK) were used as substrates. The substrate was degreased by immersion in boiling organic solvents (acetone and 2-propanol) for 7 min, then rinsed with water and dried in a stream of nitrogen before silanization. Some chips were rendered hydrophobic by exposure to Me_3SiCl vapour for 5–7 min. This procedure reduces the number of nanowires that adhere to the substrate when a drop of Pd/DNA solution is placed on the chip. Low nanowire densities facilitate the study of individual nanowires and also the molecular combing process by which they are aligned as the drop of solution is dragged across the surface. Omission of the silanization treatment resulted in the formation of nanowire networks, which were also investigated.

Scanning conductance microscopy (SCM) measurements were carried out in air on the same Dimension Nanoscope V system using MESP probes (n-doped Si cantilevers, with a metallic Co/Cr coating, Veeco Instruments Inc., Metrology Group), with a resonant frequency of ca. 70 kHz, a quality factor of 200–260 and a spring constant of $1\text{–}5 \text{ N m}^{-1}$. In SCM experiments, a bias voltage is applied between the tip and substrate. No dc current flows because the substrate is coated with a dielectric, but an electrostatic field is produced between the tip and the substrate. The phase angle between the tip motion and the driving force is related to the force gradient at the tip and is sensitive to the conductance of samples resting on the dielectric as well as their polarizability. The reported SCM phase images show the phase of the tip oscillation at a set lift height above the substrate surface (10–100 nm typ.). The substrates comprised p++Si(100) chips with a thermally grown oxide layer (240 nm) as determined by a spectrometric thin film analyzer (Filmetrics F40).

For conductive AFM (c-AFM) measurements, a constant bias was also applied between the tip and the sample (the tip was grounded). We have previously shown that DNA-templated nanowires facilitate a simple means of making contacts to individual nanowires.[4] Briefly, a large amount of nanowires are deposited on the substrate by drop-casting; as the droplet dries, individual nanowires protrude from the main mass of nanowires and are aligned by the receding meniscus. Individual nanowires can be easily imaged by AFM at the edge of the dried mass. Electrical contact to the mass of nanowires was made by applying a drop of In/Ga eutectic to one corner of the chip and to the metallic chuck. c-AFM imaging was performed in contact mode, with an applied bias of 0.5 V. The imaged area was about 1 mm away from the In/Ga contact. The closed loop system of the Dimension V instrument makes it possible to reproducibly position the tip at a point of interest identified in the image of a single Pd/DNA nanowire and to record current–voltage (I–V) curves at that point. The conductance was estimated using the slope of the I–V curve at zero bias.

2.7 High-resolution transmission electron microscopy (HRTEM)

The samples for HRTEM and EDS (energy dispersive X-ray spectroscopy) were prepared by placing a drop of the Pd/DNA nanowires solution onto a carbon-coated Cu grid followed by slow evaporation of water under ambient conditions. The TEM images were acquired using a high resolution analytical field emission microscope (HR-FETEM, JEOL JEM-2100F) at Sultan Qaboos University (Oman) at an accelerating voltage of 200 kV and the EDS spectrum was recorded during the TEM experiment.

2.8 Electrical characterisation

Two-terminal conductivity measurements were performed using gold microelectrodes fabricated on clean oxidised silicon using a standard reverse photolithography process. The electrode pattern consists of 8 pairs of large gold pads with 2 µm-wide fingers. The gap between each pair of fingers ranged from none to several micrometers and the two pads in each pair are separated by $\simeq$ 80 µm. The two small gold fingers in each pair provide contacts to the DNA-templated Pd wires and the two large gold pads serve as electrical contacts for external measurement of the system. The Au microelectrodes were deposited in 100 nm-deep trenches produced by reactive ion etching of the SiO_2. The amount of deposited Au was controlled so that the surface of the Au microelectrodes is level with the surrounding oxide surface; this enables the alignment and AFM imaging of nanowires across the fingers. In order to study single nanowires, the oxide layer was treated with Me_3SiCl vapour to facilitate molecular combing. A 2 µL drop of an aqueous solution of Pd/DNA nanowires was placed on these electrodes and nanowires were aligned across the gap between the Au fingers by molecular combing. To obtain Pd/DNA nanowire networks, the Me_3SiCl treatment was omitted. Electrical measurements were made using a probing station (Cascade Microtech) and a B1500A semiconductor analyser (Agilent). All of the electrical measurements were carried out under dry nitrogen without light illumination. I–V curves at various temperatures were performed on the probe station using a thermal chuck system (Model ETC-200L, ESPEC, Japan).

2.9 Hydrogen sensing

To evaluate the Pd/DNA nanowire networks as hydrogen gas sensors, Pd/DNA nanowires were combed on clean Si/SiO_2 substrates on which microfabricated Au electrodes had previously been fabricated. The nanowires adhere strongly to the hydrophilic SiO_2 surface to form a network of interconnected nanowires. Such nanowire networks have the advantage of having multiple routes for current flow within the active sensor element, but, more importantly, between the nanowire network and the contact pads. This creates a significantly more robust and resilient device than a single nanowire sensor. The device was then connected to a sample holder by wire bonding and then placed in the chamber of the H_2 gas sensing system. The system was composed of a vacuum chamber with heatable sample holder, controlled by a Lakeshore 331 temperature controller, a Keithley

6487 picoammeter for *I*–*V* measurements and cylinders of pure N_2 and 2.5% H_2 in N_2 to modify the ambient. The chamber was initially evacuated to a base pressure of about 9×10^{-2} Torr and then N_2 was injected using a mass flow controller to create an inert ambient for control tests. Following this, the sensing gas (2.5% H_2 in N_2 mixture) was injected, such that the chamber pressure during the measurements was maintained at $\simeq 9$ Torr. Different H_2 in N_2 concentrations were achieved by varying the relative flow rates of the gases using the mass flow controllers. The hydrogen sensing performance of these network devices was evaluated by recording the electrical response upon exposure to cycles of N_2 flow and H_2 gas of different concentrations (600–2300 ppm) at 330–400 K.

2.10 Growth model

The growth of Pd on λ-DNA was modelled by a 2D lattice gas simulation. A square lattice (i, j) of 2000 × 200 lattice points was used to represent the Pd and the surrounding solution. Points at the boundary ($j = 0$) represented the template. Periodic boundary conditions were imposed in the direction parallel to the template. Lattice sites were either occupied by an atom of the solid, or by a dissolved metal ion or empty. At each time, a site was selected at random and the following rules applied based on the status of the site and a nearest neighbour, which was also chosen at random:

1. Solid sites are allowed a chance k_{diss} to dissolve into an empty neighbouring site to form a metal ion at that site.
2. Dissolved metal ions were allowed a chance k_{dep} to deposit and become a new solid site if a neighbouring site was solid.
3. Dissolved metal ions simply move to the neighbouring site if it is empty.

The values of k_{diss} and k_{dep} were calculated according to the number of solid neighbours of a given site (*n*). If a site has *w* solid neighbours, then dissolution was assumed to require the breaking of additional bonds and,

$$k_{diss} = k_{diss}^0 e^{-wn} \tag{1}$$

On the other hand, deposition was favoured by the additional bonds formed and k_{dep} is given by:

$$k_{dep} = k_{dep}^0 e^{wn} \tag{2}$$

k_{diss}^0, k_{dep}^0 and *w* are parameters input to the simulation. This feature of the model is important to allow for the possibility of a surface tension which is an important consideration for the growth process.[33] If $w = 0$, then the number of bonds between sites has no effect on the growth and surface tension effects are absent from the model. The initial state of the system was taken to be hemispherical nuclei of radius 20 sites on the *x*-axis, representing the DNA template. The simulations were run for 10^6–10^7 time steps.

3 Results and discussion

3.1 Chemical characterisation of Pd/DNA nanowires

Fourier transform infra-red (FTIR) spectroscopy was used to probe the interaction of Pd with DNA; spectra of the nanowires show shifts in the band positions and

Table 1 FTIR spectral assignments for λ-DNA and Pd/DNA nanowires in the 1000–1800 cm^{-1} 'fingerprint' region

Wavenumber (cm^{-1}) in λ-DNA	Wavenumber (cm^{-1}) in Pd/DNA	Assignment
1705	1662	guanine ring
1409	1395	N–H deformation
1060 (broad)	1074	P–O or C–O backbone stretch
	1112	PO$^-_2$ symmetric stretch
1235	1239	asymmetric PO$^-_2$ stretch

intensities in the fingerprint region that indicate intimate interactions between Pd species and DNA and that the prepared nanowires are not mere mixtures of Pd and DNA molecules (Table 1). Pd(II) is known to bind to the N7 atoms of the guanine and adenine bases[55] and also the N3 atoms of the thymine and cytosine bases.[14] However, less is known about the interaction of Pd(0) with DNA.

Fig. 1 shows the fingerprint region of the λ-DNA infrared spectrum. The most intense feature is a broad unresolved band due to PO$_2^-$ symmetric stretch modes and P–O or C–O stretches of the phosphate backbone and sugar centered near 1060 cm^{-1}. After Pd-templating, this feature is reduced in intensity, shifted to higher energy and two components are resolved; one at 1112 cm^{-1} which we assign to the symmetric phosphate mode and another at 1074 cm^{-1} due to the C–O stretching modes. The other significant changes are in the asymmetric PO$_2^-$ stretch near 1235 cm^{-1} (decreased intensity, shifted to 1239 cm^{-1}), the N–H deformation near 1400 cm^{-1} (shifted to 1395 cm^{-1}) and the modes associated with the nucleobases in the range 1500–1700 cm^{-1}. In particular, the guanine mode at 1705 cm^{-1} increases in intensity and shifts to 1662 cm^{-1} after metallisation. Taken together, these changes indicate that there is a strong interaction of the templated material with both the nucleobases and the phosphate backbone.

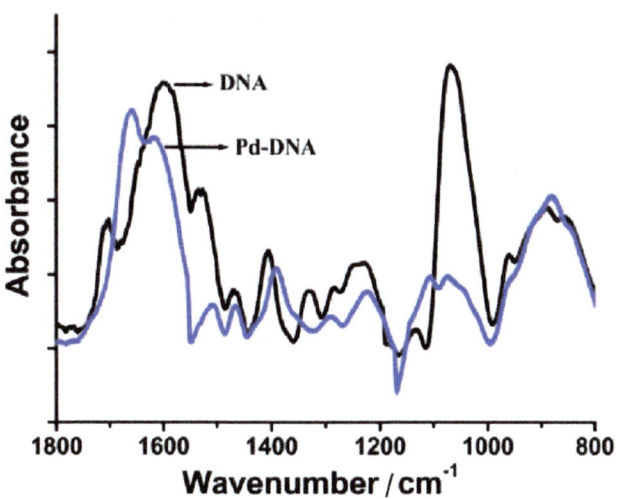

Fig. 1 FTIR spectrum of Pd/DNA nanowires (blue) and FTIR spectrum of bare λ-DNA (black). 128 scans co-added and averaged, 4 cm^{-1} resolution.

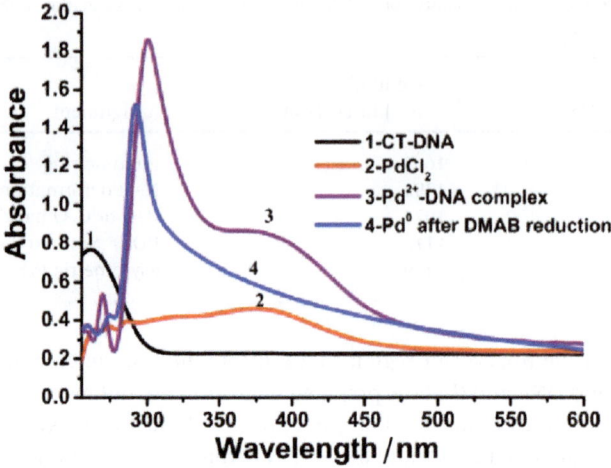

Fig. 2 UV-Vis absorption spectra at different stages of the synthesis process: absorption spectra of CT-DNA (curve 1); aqueous K_2PdCl_4 solution (curve 2); the Pd(II)-DNA complex before DMAB treatment (curve 3) and the Pd/DNA solution after DMAB reduction (curve 4) showing the loss of the Pd(II) band near 400 nm.

UV-Vis absorption spectra were employed to monitor the reduction of Pd(II) ions and the formation of metallic Pd(0) upon addition of the reducing agent. Electronic spectra of metal nanowires show absorption bands due to the excitation of the plasmon resonance. Noble metals that give a distinct plasmon peak, such as Au,[56] Ag[57] and Cu[58] are relatively easy to examine by optical spectroscopy. For Pd, the overlap of the DNA and bands related to the reducing agent and Pd(II) complicates the spectra. Fig. 2 shows the UV-Vis absorption spectra of the Pd/DNA solution at different stages of the template synthesis process. The aqueous calf-thymus (CT-DNA) solution exhibits the characteristic nucleic acid absorption band at 260 nm (curve 1); upon complexation with Pd(II) ions, the dominant absorption maximum shifted to about 300 nm (curve 3). The formation of a Pd(II)-DNA complex was supported by the red-shift of the Pd(II) band to 400 nm from the value observed in aqueous $PdCl_4^{2-}$ of 390 nm (curve 2). After DMAB reduction to produce Pd metal, the 300 nm peak for the Pd(II)-DNA complex was narrowed, blue-shifted to 290 nm and the 400 nm Pd(II) peak disappeared. A long absorption tail, decreasing with wavelength in the range 300–600 nm, remains (curve 4). The shape of this absorption is consistent with metallic Pd nanostructures.[59,60]

A high-resolution transmission electron microscopy (HRTEM) image and the EDS spectrum of a network of DNA/Pd nanowires is shown in Fig. 3. The EDS spectrum consisted mainly of Pd, C, Cu and P peaks. Cu peaks came from the carbon-coated Cu TEM grid, the Pd peak from nanowires, the C peak came from the TEM grid as well as DNA and the small P peak came from the DNA. The EDS analysis confirms directly the presence of Pd on the DNA template.

Fig. 4 shows the X-ray powder diffraction pattern for a sample of Pd/DNA nanowires. Peaks due to reflections from the (111), (200), (220) and (311) planes of metallic Pd at 2θ values of 39°, 46°, 68° and 82° were observed; these are consistent with a previous report.[59] The peaks are very broad and, upon fitting a pseudo-Voigt function to the most intense diffraction peak at $2\theta = 39°$, we

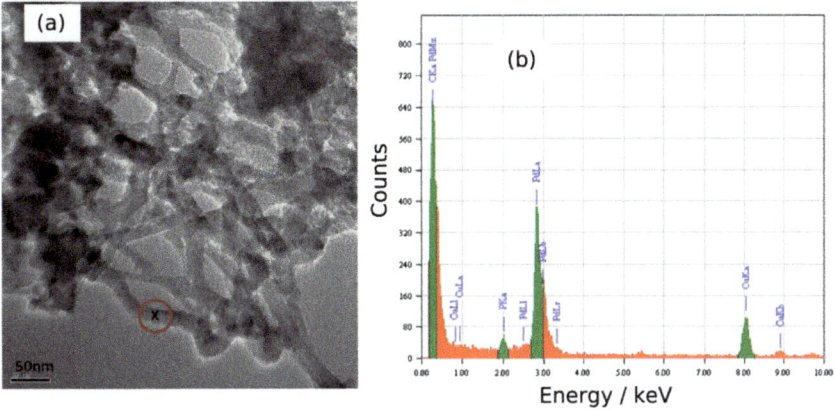

Fig. 3 (a) HRTEM image of Pd/DNA nanowire network on a carbon-coated Cu grid and (b) the EDS spectrum at the point "x" in the image corresponding to a nanowire.

estimate a crystallite diameter of 1.6 nm from the Scherrer equation. No peaks attributable to Pd oxides were observed, which rules out the presence of crystalline oxides.

XPS survey spectra indicated the presence of C, N, O, P (from the DNA template), Na (from the buffer or NaBH$_4$ reductant) and Pd and Cl (from PdCl$_4^{2-}$). Fig. 5 shows the Pd 3d spectrum. Peaks for both Pd(0) and Pd(II) oxidation states were observed. The doublet at 335.1 eV and 340.4 eV is assigned to Pd(0) and that at 336.4 eV and 341.7 eV to Pd(II) following previous reports.[59,61] The feature at 346.3 eV is most likely a plasmon loss band associated with the peak at 335.1 eV. The plasmon energy of 11.2 eV is larger than the 7 eV expected for a surface plasmon and smaller than the bulk plasmon energy of 14 eV,[62,63] but may be an effect of the small particle size. The Pd(II) state observed by XPS, but not XRD is interpreted as an amorphous oxyhydroxide coating the metal crystallite surfaces. We suggest this results from aquation of the PdCl$_4^{2-}$ followed by hydrolysis. PdCl$_4^{2-}$ is known to undergo exchange of Cl$^-$ ligands for H$_2$O in non-complexing acidic solutions, the pH of λ-DNA is buffered at about 8, but some aquation of

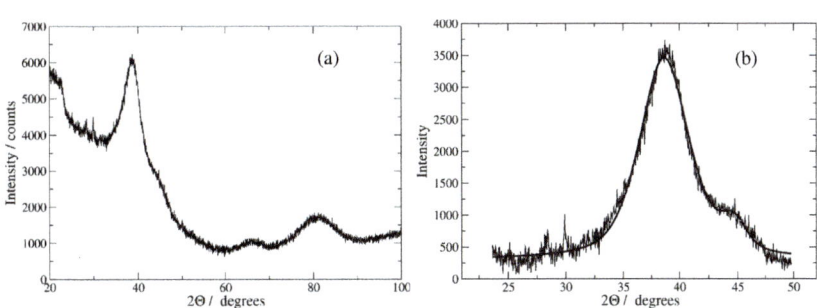

Fig. 4 (a) X-ray diffraction pattern of Pd–DNA powder; the diffraction peaks originate from the (111), (200), (220) and (311) planes of metallic Pd at 2θ values of 39°, 46°, 68° and 82°. (b) pseudo-Voigt functions fitted to the peaks at 39° and 46°. The crystallite size was estimated using the Scherrer equation as 1.6 nm.

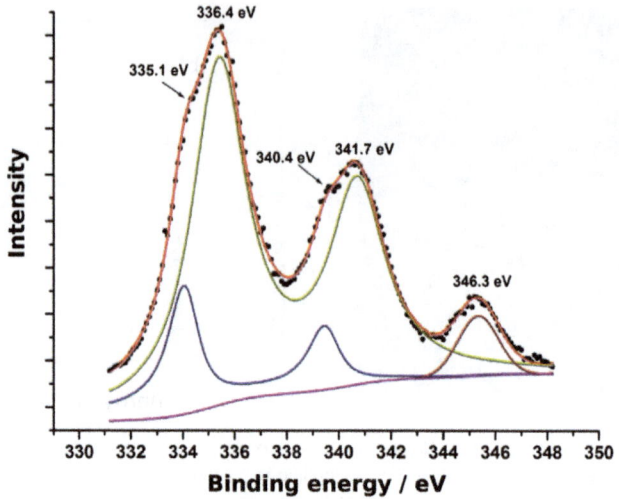

Fig. 5 The Pd 3d XPS spectrum of Pd/DNA nanowires. The experimental data are the black symbols and the components making up the theoretical fit are shown as solid lines.

$PdCl_4^{2-}$ is still possible.[64] $Pd(OH_2)_4^{2+}$ is unstable to hydrolysis at pH values above the $pK_a = 2.3$.[65]

3.2 Structural characterisation of Pd/DNA nanowires

Fig. 6 shows a selection of AFM images of the Pd/DNA nanowires aligned by molecular combing on Me$_3$SiCl-treated Si/SiO$_2$ substrates. Parallel Pd/DNA nanowires with diameters in the range 5–45 nm were observed. The height of bare λ-DNA, as measured by AFM, is well-known to be less than the crystallographic diameter of the double helix; values in the range 1–2 nm are typical. The nanowires appear smooth (Fig. 6a,b) and little Pd formed off the template is present. These AFM observations also revealed that the samples must be incubated at least 24 h before alignment in order to obtain smooth and continuous Pd nanowires over the whole length of λ-DNA. This was shown by imaging different samples which are aligned only a few hours after preparation and it was observed that in this case, the nanowires appeared with discontinuous parts along the DNA strands, as shown in Fig. 6d for the nanowires aligned after 2 h. The diameter distribution of the Pd/DNA nanowires was obtained by observing over 90 wires. Fig. 6c is a histogram of the average wire diameter 24 h after preparation. The average diameter of an individual wire was determined using section analysis along a 3 μm portion of the wire. The majority of the observed wires have diameters in the range 5–20 nm. It is clear that the Pd crystallite size determined by linewidth analysis of the XRD pattern is much less than the diameter of the nanowires, in contrast to the cases of Cu/DNA,[17] Cu$_2$O/DNA[24] and Fe$_3$O$_4$/DNA nanowires[26] where the AFM heights are comparable to the crystallite size from the Scherrer analysis. It is therefore also clear that the Pd metal does not form on the DNA template in the same way as previously studied systems, and a schematic illustration of the nanowire structure is presented in Fig. 7.

In order to rationalise the AFM and XRD results, we carried out lattice gas simulations of the growth process. The initial configuration was assumed to be a

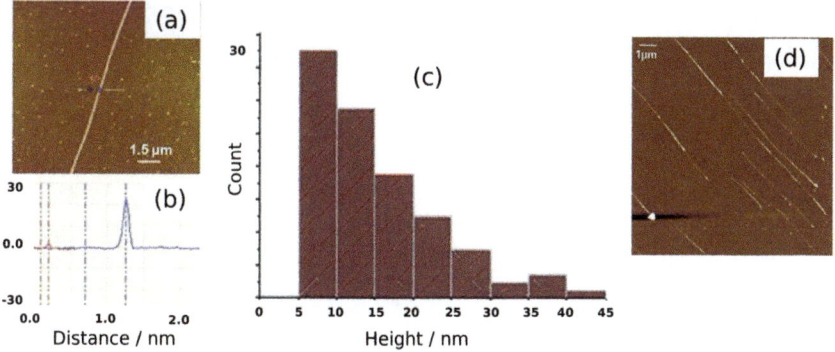

Fig. 6 AFM images of Pd/DNA nanowires. (a) AFM image of a single Pd/DNA nanowire (scale bar = 1.5 μm); (b) Section of the nanowire in (a) at the indicated points showing the height difference between the Pd/DNA nanowire (≃23 nm) and the bare DNA strand (2 nm). (c) Height distribution of 90 Pd/DNA nanowires, the heights were determined from AFM images. (d) Discontinuous nanowires aligned after a reaction time of 2 h. The scale bar is 1 μm and the height (colour) scale is 30 nm.

series of spherical nuclei on the DNA template (Fig. 8a). The solid sites are shown in yellow (colour scale values > 1) and the solution sites are shown as purple-black (colour scale values < 1). The solution starts supersaturated with respect to the solid, i.e., $\frac{c}{c_\infty} > 1$ where c is input to the simulation as a parameter, along with rate constants for dissolution and deposition of the solid, and c_∞ is the concentration of dissolved metal ions at equilibrium. The parameter w determines the additional barrier to dissolution because a site has bonds to its nearest neighbours via eqn (1) and for deposition via eqn (2). This leads to the solid having a bulk surface tension which acts to minimize the area of its surface. The supersaturation can be related to the rate constants, after taking appropriate macroscopic averages, by $\frac{\langle k_{dep} \rangle}{\langle k_{diss} \rangle} c$. We therefore ran simulations with different values of k_{diss}^0, as defined in eqn (1), in order to study the effect of the driving force on the templating reaction.

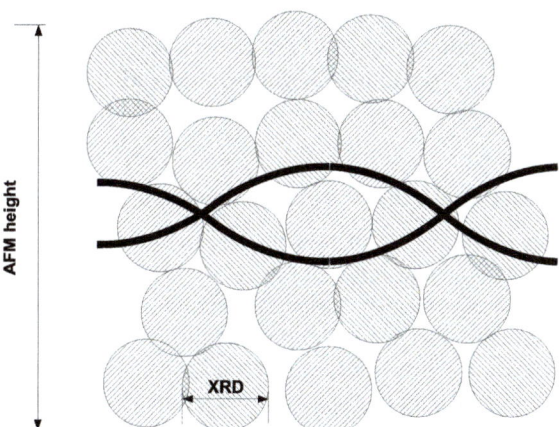

Fig. 7 Schematic representation of the structure of the prepared DNA/Pd nanowires. The Pd crystallite size (≃1.6 nm) is substantially smaller than the nanowire heights observed by AFM (5–45 nm).

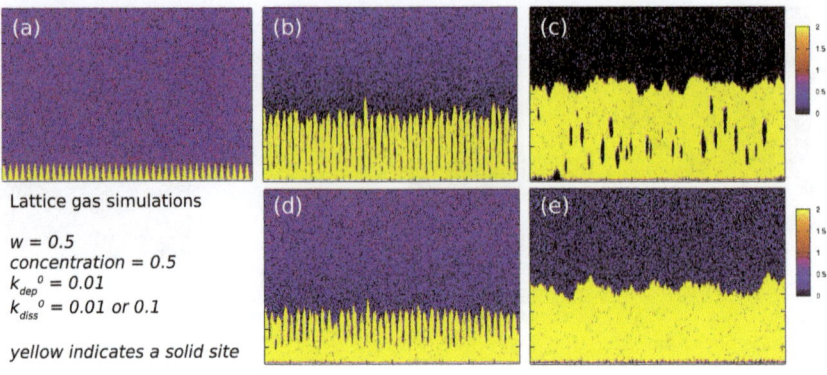

Fig. 8 Lattice gas simulations of the growth of Pd/DNA nanowires on a square lattice of 2000 × 200 sites. The template lies along the x-axis in each figure. The common parameters are $w = 0.5$, concentration of dissolved metal ions, $c = 0.5$ (per site) and $k^0_{dep} = 0.01$. (a) $t = 0$; (b) $k^0_{diss} = 0.01$ and $t = 10^4$ time-steps; (c) $k^0_{diss} = 0.01$ and $t = 10^7$ time-steps; (d) $k^0_{diss} = 0.1$ and $t = 10^4$ time-steps and (e) $k^0_{diss} = 0.1$ and $t = 10^7$ time-steps.

Unlike other materials we have studied, Pd deposition is expected to be much less reversible and therefore is modelled by choosing a larger driving force for deposition, i.e., a lower value for k^0_{diss}. Fig. 8b,d shows after 10^4 steps of the simulation the nuclei have grown; the growth is dominated by deposition of atoms at the step edges because of the large value of w chosen, and the crystallites are no longer spherical. When $k^0_{diss} = 0.01$ the crystallites are still separated from each other, but for $k^0_{diss} = 0.1$ there is already substantial overlap of neighbouring crystallites. As the growth continues, the overlap of neighbouring crystallites by dissolution and reprecipitation results in the appearance of a continuous nanowire (Fig. 8e) whose surface roughness has previously been modelled by a linear thermodynamic treatment.[33] However, when $k^0_{diss} = 0.01$, events leading to the coalescence of neighbouring crystallites are rare and, though the external surface of the nanowire may appear very similar, there remain internal voids even at long simulation times. Such a simulation cannot model all the details of the actual Pd deposition, but it does indicate that when the deposition process is irreversible, a complex internal structure is a natural consequence. In the actual chemical system, additional features related to the presence of grain boundaries, deposition of oxide and the possibility of further nucleation events will naturally produce a granular morphology as in Fig. 7.

3.3 Electrical characterisation of Pd/DNA nanowires

The conductivity of the Pd/DNA nanowires was assessed qualitatively by the non-contact scanned conductance microscopy (SCM) technique. The measurement is performed via lift mode, a two-pass imaging technique in which the tip oscillates normal to the surface and a dc bias is applied between the tip and substrate. SCM is able to detect conductive objects resting on a dielectric film because of the change in the energy stored in the tip-substrate capacitance after insertion of a nanowire between the tip and the substrate; this change in energy is equivalent to a change in the force constant of the cantilever and therefore of the phase of the tip motion.[66] Previous studies have revealed that conducting 1D structures such as single-wall carbon nanotubes have a negative phase shift with respect to background[67] and insulating nanostructures such as polymers or λ-DNA show a positive phase shift.[66,67] Fig. 9a–c shows a tapping mode AFM image and the corresponding scanned conductance images of Pd/DNA nanowires, when bias potentials of +6 V and −6 V were applied. The dark appearance of the nanowires in the phase images (Fig. 9b and c), which corresponds to a negative phase shift as the tip crosses the nanowire, and the quadratic variation of the phase shift with the bias voltage (Fig. 9d) are evidence that the nanowires are conductive and that the phase shift is not due to trapped charge effects. The thin, bright nanowires (positive phase shift) in the phase images are attributed to bare DNA strands because the height of these strands is in the range 1.5–2 nm. Moreover, the phase shift increases as the radius of the templated nanowires increases (Fig. 9e).

It should be noted that, although the majority of the tested nanowires are conductive, the measurements also revealed the existence of non-conductive nanowires in the sample. These findings are consistent with the results of the XPS experiments which indicate the presence of oxide in the sample.

Quantitative measurements of Pd/DNA nanowire conductance were obtained by depositing nanowires on chemically oxidised Si(100) substrates to measure I–V curves using c-AFM following a previously developed method.[4] The conductive AFM tip serves as one contact to a nanowire which protrudes from a dense network of nanowires to which an In/Ga eutectic is used to complete the circuit. Fig. 10 shows an SEM image of individual Pd/DNA nanowires stretched out from a

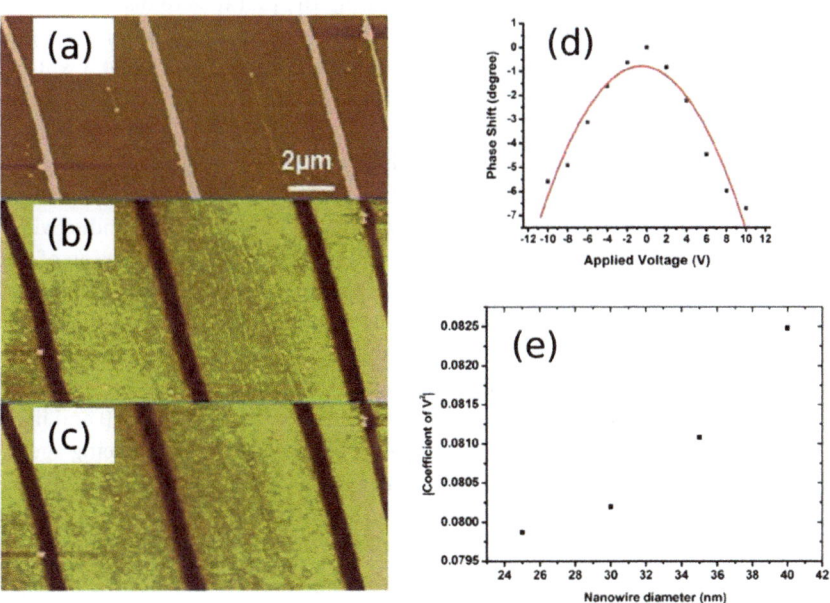

Fig. 9 Electrical characterisation of Pd/DNA nanowires by scanned conductance microscopy. (a) AFM height image of Pd/DNA nanowires (the data scale is 25 nm). (b) and (c) the corresponding SCM phase images of the same nanowires with applied bias voltages of +6 V and −6 V respectively (the data scale is 3°). (d) Phase shift *versus* applied voltage for a single Pd/DNA nanowire (20 nm diameter) at a lift height of 60 nm aligned on 240 nm thick SiO_2 on highly doped Si. (e) A quadratic was fitted to the plots of phase shift against bias voltage. The coefficients of V^2 are plotted against the nanowire diameter at a constant lift height of 60 nm.

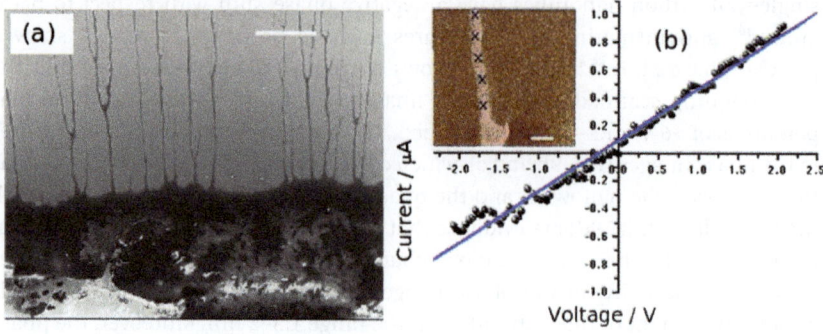

Fig. 10 c-AFM measurements of Pd/DNA nanowires. (a) Electron micrograph of the dense mass of Pd/DNA nanowires (lower third of image); single Pd/DNA nanowires are seen protruding from the mass (scale bar = 20 μm). (b) Current–voltage curve with the tip contacting a single nanowire (inset = cAFM current image, data scale = 500 nA, tip/sample bias = 5 V, scale bar = 1 μm).

dense mass of nanowires deposited from a dried droplet and the *I*–*V* curve of a single nanowire selected at the edge of such a dense mass. The dense mass of nanowires serves as the second contact to the nanowire. This technique allows the resistance of a single nanowire and the contacts to be separately determined; a series of *I*–*V* curves at different points on the nanowire are recorded and the resistance is then plotted as a function of distance along the nanowire. The conductivity is obtained from the slope of this plot and the measured height; the contact resistance is estimated by extrapolating the plot to zero distance.

Fig. 11 shows the plot of resistance against distance. The slope of the regression lines gives the nanowire resistance per unit length as ($5.1 \pm 0.2 \times 10^8$ Ω cm^{-1}). Using the diameter of the nanowire observed in the contact mode image ($\simeq$20 nm), the conductivity was determined as 1.6×10^4 S m^{-1}. This is the largest value of conductivity we observed for single Pd/DNA nanowires. This value is lower than both the bulk conductivity of Pd (9.5×10^6 S m^{-1})[68] and a previous report of a 50 nm diameter Pd/DNA nanowire of 2×10^6 S m^{-1}.[14] A conductivity of 3.33×10^5 S m^{-1} was also reported for a nanowire prepared by reduction of PdO (height 30 nm and length 450 nm[61]). The conductance of thin nanowires fabricated on DNA templates *via* Pd(II) reduction have also been reported; the average resistance of a single wire was 800 GΩ at 10 V.[68] Our c-AFM measurements on various nanowires revealed resistance values in the range of 0.4–0.8 GΩ for the DNA/Pd nanowires prepared by the reduction of Pd(II) ions with DMAB and in the range of 2–8 MΩ for the nanowires prepared using NaBH$_4$ as a reducing agent.

Fig. 11 also shows the effect of increasing the tip/nanowire contact force by increasing the set point voltage from 0 to 3 V. A clear decrease in the intercept on the resistance axis is observed at higher forces, which is expected if the tip/nanowire junction is the dominant contact resistance. This is expected based on considerations of the area and previous estimates of the resistance at the second contact of 50–200 Ω.[24] It is also clear that the contact resistance is very substantial and dominates the raw *I*–*V* data.

cAFM is inconvenient for temperature-dependent *I*–*V* measurement, therefore two-terminal Au microelectrode devices were fabricated in order to study this aspect. The DNA/Pd nanowires were transferred by the method of molecular combing onto clean Si/SiO$_2$ substrates in the gap between microfabricated Au

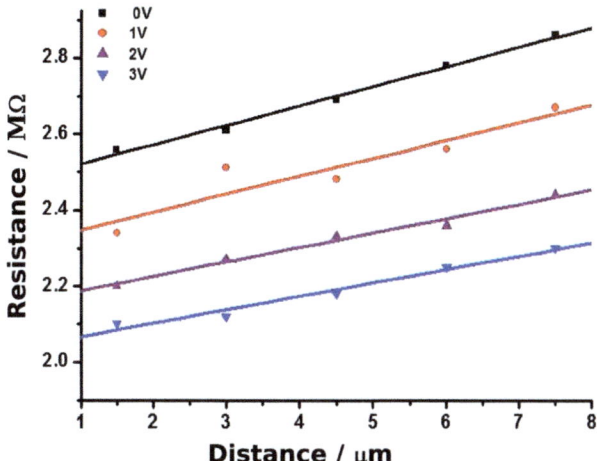

Fig. 11 cAFM measurements of resistance at different points along a single Pd/DNA nanowire. The distance is that between the point at which the tip contacts the nanowire and the dense mass shown in Fig. 10. The nanowire resistance per unit length is the slope of the plot and the intercept at zero distance is the contact resistance. Measurements obtained at different setpoints (tip/sample force) are shown in different colours.

electrodes. In contrast to the case of polymer nanowires,[4] the large contact resistance observed with Pd/DNA samples makes the I–V measurements of single Pd/DNA nanowires by two-terminal measurements difficult and non-linear characteristics were often observed (diode-like characteristics with a turn-on voltage of $\simeq 0.6$ V). Treatment of the single nanowires with reducing agents (gaseous hydrogen) did not reliably improve the behaviour. Instead we found that omitting the silanization of the SiO_2 resulted in networks of Pd/DNA nanowires with consistent linear I–V characteristics (Fig. 12). Such nanowire networks have the advantage of having multiple routes for current flow within the nanowires and between the nanowire network and the contact pads. The networks exhibited linear I–V curves with measured currents in the nA range compared to the leakage current of the electrodes alone ($\simeq 100$ fA). The conductance at zero bias increased with temperature over the range 233–373 K, which is opposite to the expected behaviour of a metal. However, similar observations[68] have been made and attributed to a conduction mechanism based on the concept of a granular metal. Such behaviour seems natural given the mechanism of growth discussed above and the small grain size observed.

The single Pd/DNA nanowires also showed increased currents at higher temperatures; the linear portion of the data above the turn-on voltage could be used to make a plot of $\ln G$ versus $1/T$ from which an activation energy of 0.35 eV was derived, but the interpretation of this value is not straightforward. The zero bias conductance of the network was simpler to analyse and the behaviour for a granular metal under conditions where the grains are separated by a tunneling barrier was obtained (Fig. 12). The regression line shows the fit of the Efros–Shklovskii law,[69,70] $\frac{G}{G_0} = \exp\left(-\sqrt{\frac{T_0}{T}}\right)$, to the data and $T_0 = 1850 \pm 120$ K or in terms of an energy, 0.16 ± 0.01 eV.

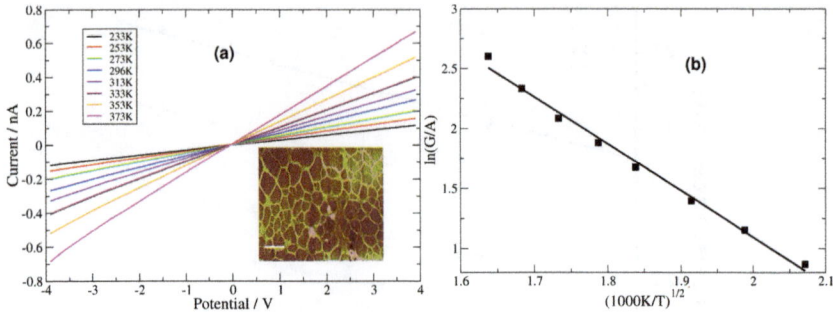

Fig. 12 (a) Two-terminal microelectrode current–voltage curves for a network of Pd/DNA nanowires at temperatures from 233–373 K. (Inset: An AFM height image of the network, the data scale is 20 nm, scale bar = 500 nm). (b) The zero bias conductances, G, from (a) were divided by mean cross-sectional area, A, and $\ln(G/A)$ was plotted against $T^{-1/2}$ to test the Efros–Shklovskii law.

3.4 Hydrogen sensing with Pd/DNA nanowire networks

The more reliable electrical behaviour of the nanowire networks makes for a significantly more robust and resilient sensor than our single Pd/DNA nanowires. We have previously described the sensor performance in more detail,[71] but give here a brief summary before discussing the modelling of the sensor response in terms of hydrogen diffusion in the nanowires. Fig. 13 shows the response of the resistance of the network to pulses of H_2/N_2 mixtures and pure N_2 gas. The increase in resistance is caused by adsorption of H_2 on the Pd surface dissociation to atomic hydrogen and in-diffusion to form the more resistive Pd hydride.[72]

The device signal, defined as $S = \dfrac{R_{H_2} - R_{N_2}}{R_{N_2}}$, is clearly reversible and the time for 90% of the signal to be achieved was about 85 s at 2300 ppm hydrogen. Recovery was slightly faster with a 90% recovery time of ≃60 s. The sensitivity is comparable to previous devices,[47,50] although the response times are rather slow by comparison with electrodeposited Pd nanowires[47] which have larger crystallite sizes. The maximum sensitivity of the sensor is increased from 0.09 to 0.26 with increasing temperature from 330 K to 400 K with good repeatability. Interestingly,

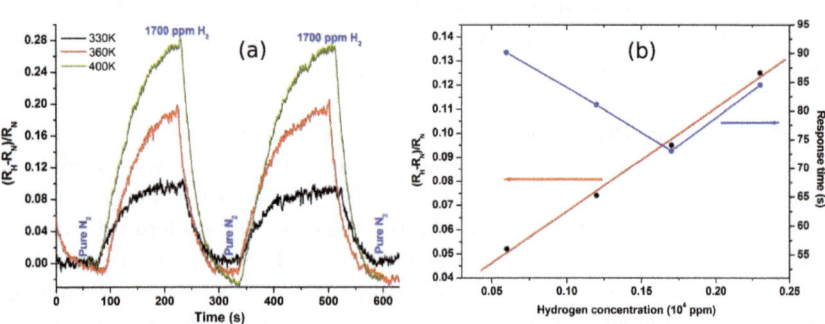

Fig. 13 (a) Fractional change in network resistance to pulses of 1700 ppm H_2 and pure N_2 at temperatures from 330 K to 400 K. (b) Fractional change in resistance and response time against hydrogen concentration at 330 K. Response time is defined as the time to achieve 90% of the steady-state resistance change.

this is in contrast to previous reports in which sensor response was observed to decrease with increasing temperature from 323 to 373 K,[46] although it should be noted that it has recently been shown that even the sign of the response can change with temperature and that there may be competing effects such as hydride formation and swelling of the material in a single device.[75]

The response of the network to hydrogen was modelled as that of an ensemble of identical nanowires. The sorption of hydrogen in the nanowires was treated as a standard diffusion problem with cylindrical symmetry:

$$\frac{\partial^2 C}{\partial R^2} + \frac{1}{R}\frac{\partial C}{\partial R} = \frac{\partial C}{\partial T} \qquad (3)$$

The boundary and initial conditions are $C(1,T) = 1$; $\frac{\partial C(0,T)}{\partial R} = 0$; and $C(R,0) = 0$ for $0 \leq R < 1$. The dimensionless groups are defined as $C = c(r)/c^*$; $R = r/a$ and $T = Dt/a^2$. D is the diffusion coefficient of hydrogen in the Pd/DNA nanowire, a is the radius of the nanowire and c^* is the concentration of hydrogen at the nanowire surface. r is the radial coordinate and t is the time. The term $\frac{\partial^2 C}{\partial Z^2}$ vanishes for a nanowire which is uniform along its length. We also assume that the process is limited by diffusion in the cylindrical nanowire and not by any barrier at the surface. The radial variation of the dimensionless concentration is given by

$$C(R,T) = 1 - 2\sum_{n=1}^{\infty} \frac{J_0(\lambda_n R)}{\lambda_n J_1(\lambda_n)} e^{-\lambda_n^2 T} \qquad (4)$$

where J_0 and J_1 are Bessel functions of the first kind of order zero and one and the λ_n are the zeros of J_0 on the positive real axis. In order to model the change in conductance, we assume that the conductivity of the Pd/DNA nanowire is determined by the local hydrogen concentration, the fractional change is small and the relationship is linear:[76]

$$\sigma(r) = \Delta\sigma_H C(R) + \sigma_0 \qquad (5)$$

The fractional change in conductance (G) of the nanowire can then be expressed in terms of the integral of $2\pi r\sigma$ in eqn (5) from $R = 0$ to $R = 1$:

$$\frac{\Delta G}{G_0} = 2\frac{\Delta\sigma_H}{\sigma_0}\int_0^1 C(R)R dR \qquad (6)$$

Substituting eqn (4) into eqn (6) gives the final result which can be used to model the data for conductance or resistance changes upon exposure to hydrogen:

$$\frac{\Delta G}{G_0} = -\frac{\Delta R}{R_0} = \frac{\Delta\sigma_H}{\sigma_0}\left[1 - 4\sum_{n=1}^{\infty}\frac{e^{-\lambda_n^2 T}}{\lambda_n^2}\right] \qquad (7)$$

The system is linear, so the corresponding equation for the diffusion of hydrogen out of the nanowires upon exposure to nitrogen is simply:

$$\frac{\Delta G}{G_0} = 4\frac{\Delta\sigma_H}{\sigma_0}\sum_{n=1}^{\infty}\frac{e^{-\lambda_n^2 T}}{\lambda_n^2} \qquad (8)$$

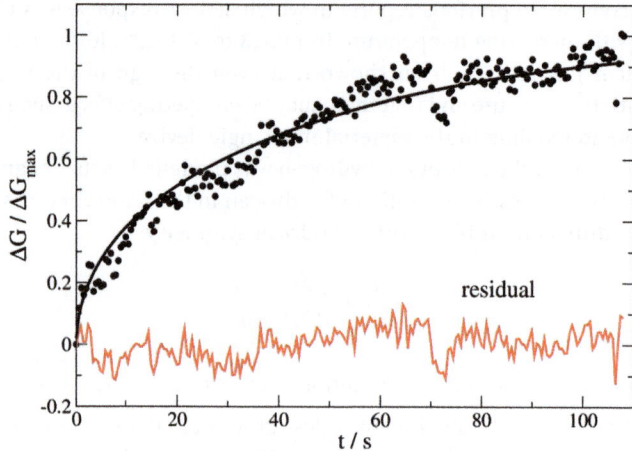

Fig. 14 Change in nanowire conductance as a fraction of the steady-state value (ΔG_{max}). The temperature was 330 K, the mean nanowire radius was 20 nm and the hydrogen diffusion coefficient determined using eqn (7) as a regression model was $1.4 \pm 0.1 \times 10^{-14}$ cm^2 s^{-1}.

Eqn (7) and (8) were used as the regression models for the experimental data at each measured temperature (Fig. 14). The sums were truncated after λ_{20} which was found to give sufficient precision. We estimated the hydrogen atom diffusion coefficient in Pd/DNA at 330 K as $1.4 \pm 0.1 \times 10^{-14}$ cm^2 s^{-1}. This is much smaller than the reported diffusion coefficient in bulk Pd of 3.8×10^{-7} cm^2 s^{-1} at room temperature[37] and that obtained in a recent electrochemical study of 3.2×10^{-7} cm^2 s^{-1}.[77] Previous reports also revealed that the hydrogen diffusion coefficient values in Pd thin films (50 nm to 1.34 μm thick) are 2–3 orders of magnitude smaller than of the bulk Pd at 298 K[73] and in Pd thin films with thickness of 22.5 nm at 330 K which were reported as 6.9×10^{-11} cm^2 s^{-1}.[74] Other workers have

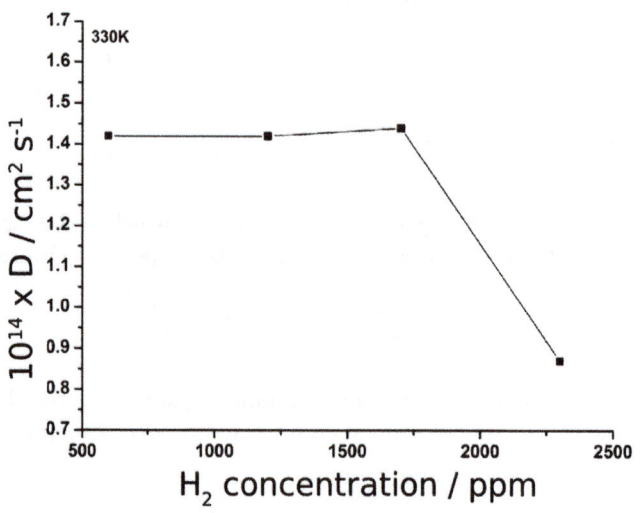

Fig. 15 The change in diffusion coefficients obtained from the sensor response with increasing [H$_2$] at 330 K.

studied H/D exchange in Pd using molecular beams; the exchange takes place on a timescale of order 10 s for 7 nm diameter Pd nanoparticles at 280 K.[78] Our results are therefore not unprecedented for Pd nanostructures. The Pd/DNA nanowires here consist of small crystallites and therefore the low diffusion coefficient may be caused by grain boundary traps.[74] We did observe a small difference between response and recovery, but with the recovery process slightly faster (Fig. 15). This suggests that diffusion is not the sole process limiting the response rate.

The diffusion coefficient is plotted as a function of hydrogen concentration at 330 K (Fig. 15). The results show that D_H is roughly constant ($\sim 1.4 \times 10^{-14}$ cm^2 s^{-1}) over the range 600–1800 ppm but decreases to 0.85×10^{-14} cm^2 s^{-1} when [H$_2$] increased to 2300 ppm. This may indicate the onset of the α to β phase transition in Pd which has been reported to occur at concentrations of order 10^4 ppm.[50]

4 Conclusions

Pd/DNA nanowires formed by aqueous DNA-templating reactions have a substantially different structure to many other DNA-templated materials. A comparison of X-ray diffraction and AFM measurements shows that the apparently smooth wires observed in AFM comprise very small grains (1.6 nm) unlike Cu/DNA,[17] Cu$_2$O/DNA[24] and Fe$_3$O$_4$/DNA nanowires,[26] which have been shown to have a beads-on-a-string morphology in which the diameter of the wire matches the crystallite size estimated by diffraction. The nanowires are substantially less conductive than bulk Pd, but when prepared in the form of networks on Si/SiO$_2$ substrates by molecular combing, they show stable I–V characteristics and their resistance increases in a reversible manner upon exposure to hydrogen gas. These networks exhibit response and recovery times of the order of 1 min and the fractional change in resistance was 26% at 400 K and 2300 ppm hydrogen. The hydrogen diffusivity estimated by modelling the sensor response was substantially smaller than the previous reported values for bulk Pd and thin films, which is likely due to grain boundary effects as a result of their granular nature. Nevertheless, these nanowire networks show comparable sensitivity to single Pd nanowire sensors and are very simple to fabricate.

Acknowledgements

Dr Issa Al Amri from Sultan Qaboos University, Oman is thanked for help with the electron microscopy. ONE is thanked for funding research into nanowires at Newcastle.

References

1 Y. Ma, J. Zhang, G. Zhang and H. He, *J. Am. Chem. Soc.*, 2004, **126**, 7097–7101.
2 L. Dong, T. Hollis, S. Fishwick, B. A. Connolly, N. G. Wright, B. R. Horrocks and A. Houlton, *Chem.–Eur. J.*, 2007, **13**, 822.
3 J. Hannant, J. H. Hedley, J. Pate, A. Walli, S. A. Farha Al-Said, M. A. Galindo, B. A. Connolly, B. R. Horrocks, A. Houlton and A. R. Pike, *Chem. Commun.*, 2010, **46**, 5870.
4 R. Hassanien, M. Al-Hinai, S. A. Farha Al-Said, R. Little, L. Šiller, N. G. Wright, A. Houlton and B. R. Horrocks, *ACS Nano*, 2010, **4**, 2149.
5 S. M. D. Watson, J. H. Hedley, M. A. Galindo, S. A. Farha Al-Said, N. G. Wright, B. A. Connolly, B. R. Horrocks and A. Houlton, *Chem.–Eur. J.*, 2012, **18**, 12008.

6 P. Nickels, W. U. Dittmer, S. Beyer, J. P. Kotthaus and F. C. Simmel, *Nanotechnology*, 2004, **15**, 1524.
7 S. A. F. Al-Said, R. Hassanien, J. Hannant, M. A. Galindo, S. Pruneanu, A. R. Pike, A. Houlton and B. R. Horrocks, *Electrochem. Commun.*, 2009, **11**, 550.
8 S. Pruneanu, S. A. F. Al-Said, L. Q. Dong, T. Hollis, M. A. Galindo, N. G. Wright, A. Houlton and B. R. Horrocks, *Adv. Funct. Mater.*, 2008, **18**, 2444.
9 E. Braun, Y. Eichen, U. Sivan and G. Ben-Yoseph, *Nature*, 1998, **391**, 775.
10 K. Keren, M. Krueger, R. Gilad, G. Ben-Yoseph, U. Sivan and E. Braun, *Science*, 2002, **297**, 72.
11 K. Keren, R. S. Berman and E. Braun, *Nano Lett.*, 2004, **4**, 323.
12 W. E. Ford, O. Harnack, A. Yasuda and J. M. Wessels, *Adv. Mater.*, 2001, **13**, 1793.
13 J. Richter, R. Seidel, R. Kirsch, M. Mertig, W. Pompe, J. Plaschke and H. K. Schackert, *Adv. Mater.*, 2000, **12**, 507.
14 J. Richter, M. Mertig, W. Pompe, I. Mönch and H. K. Schackert, *Appl. Phys. Lett.*, 2001, **78**, 536.
15 Z. X. Deng and C. D. Mao, *Nano Lett.*, 2003, **3**, 1545.
16 C. F. Monson and A. T. Woolley, *Nano Lett.*, 2003, **3**, 359.
17 S. M. D. Watson, N. G. Wright, B. R. Horrocks and A. Houlton, *Langmuir*, 2010, **26**, 2068.
18 E. Braun and K. Keren, *Adv. Phys.*, 2004, **53**, 441.
19 H. A. Becerril, R. M. Stolenberg, D. R. Wheeler, R. C. Davis, J. N. Harb and A. T. Woolley, *J. Am. Chem. Soc.*, 2005, **127**, 2828.
20 F. Patolsky, W. Weizmann, O. Liobasheveski and I. Willner, *Angew. Chem., Int. Ed.*, 2002, **41**, 2323.
21 T. Nishinaka, A. Takano, Y. Doi, M. Hashimoto, A. Nakamura, Y. Matsushita, J. Kumaki and E. Yashima, *J. Am. Chem. Soc.*, 2005, **127**, 8120.
22 H. Yan, S. H. Park, G. Finkelstein, J. H. Reif and T. H. LaBean, *Science*, 2003, **301**, 1882.
23 H. Kudo and M. Fujihara, *IEEE Trans. Nanotechnol.*, 2006, **5**, 90.
24 R. Hassanien, S. A. Farha Al-Said, L. Šiller, R. Little, N. G. Wright, A. Houlton and B. R. Horrocks, *Nanotechnology*, 2012, **23**, 075601.
25 D. Sarkar and M. Mandal, *J. Phys. Chem. C*, 2012, **116**, 3227.
26 H. D. A. Mohamed, S. M. D. Watson, B. R. Horrocks and A. Houlton, *Nanoscale*, 2012, **4**, 5936.
27 L. Dong, T. Hollis, B. A. Connolly, N. G. Wright, B. R. Horrocks and A. Houlton, *Adv. Mater.*, 2007, **19**, 1748.
28 W. U. Dittmer and F. C. Simmel, *Appl. Phys. Lett.*, 2004, **85**, 633.
29 L. Levina, V. Sukhovatkin, S. Musikhin, S. Cauchi, R. Nisman, D. P. Bazett-Jones and E. H. Sargent, *Adv. Mater.*, 2005, **17**, 1854.
30 J. L. Coffer, S. R. Bigham, R. F. Pinizzotto and H. Yang, *Nanotechnology*, 1992, **3**, 69.
31 D. S. Hopkins, D. Pekker, P. M. Goldbart and A. Bezryadin, *Science*, 2005, **308**, 1762.
32 A. Houlton, A. R. Pike, M. A. Galindo and B. R. Horrocks, *Chem. Commun.*, 2009, 1797.
33 S. M. D. Watson, A. Houlton and B. R. Horrocks, *Nanotechnology*, 2012, **23**, 505603.
34 D. W. Keith and A. E. Farrell, *Science*, 2003, **301**, 315.
35 R. Moy, *Science*, 2003, **301**, 47.
36 M. Khanuja, S. Kala, B. R. Mehta and F. E. Kruis, *Nanotechnology*, 2009, **20**, 015502.
37 T. B. Flanagan and W. A. Oates, *Annu. Rev. Mater. Sci.*, 1991, **21**, 269.
38 J. S. Noh, J. M. Lee and W. Lee, *Sensors*, 2011, **11**, 825.
39 E. Lee, J. M. Lee, J. H. Koo, W. Lee and T. Lee, *Int. J. Hydrogen Energy*, 2010, **35**, 6984.
40 Y. Hu, D. Perello, U. Mushtaq and M. Yun, *IEEE Trans. Nanotechnol.*, 2008, **7**, 693.
41 Y. Im, C. Lee, R. P. Vasquez, M. A. Banger, N. V. Myung, E. R. Menke, R. M. Penner and M. H. Yun, *Small*, 2006, **2**, 356.
42 R. K. Joshi, S. Krishnan, M. Yoshimura and A. Kumar, *Nanoscale Res. Lett.*, 2009, **4**, 1191.
43 V. La Ferrara, B. Alfano, E. Massera and G. Di Francia, *IEEE Trans. Nanotechnol.*, 2008, **7**, 776.
44 E. C. Walter, F. Favier and R. M. Penner, *Anal. Chem.*, 2002, **74**, 1546.
45 F. Yang, D. K. Taggart and R. M. Penner, *Nano Lett.*, 2009, **9**, 2177.
46 N. Tasaltin, S. Ozturk, N. Kilinç and Z. Ziya Ozturk, *Appl. Phys. A: Mater. Sci. Process.*, 2009, **97**, 745.
47 F. Yang, S. C. Kung, M. Cheng, J. C. Hemminger and R. M. Penner, *ACS Nano*, 2010, **4**, 5233.
48 F. Favier, E. C. Walter, M. P. Zach, T. Benter and R. M. Penner, *Science*, 2001, **293**, 2227.
49 S. Mubeen, T. Zhang, B. Yoo, M. A. Deshusses and N. V. Myung, *J. Phys. Chem. C*, 2007, **111**, 6321.
50 X. Q. Zeng, M. L. Latimer, Z. L. Xiao, S. Panuganti, U. Welp, W. K. Kwok and T. Xu, *Nano Lett.*, 2011, **11**, 262.

51 S. Yu, U. Welp, L. Z. Hua, A. Rydh, W. K. Kwok and H. H. Wang, *Chem. Mater.*, 2005, **17**, 3445.
52 K. J. Jeon, M. Jeun, E. Lee, J. M. Lee, K.-I. Lee, P. von Allmen and W. Lee, *Nanotechnology*, 2008, **19**, 495501.
53 D. Ding, Z. Chen and C. Lu, *Sens. Actuators, B*, 2006, **120**, 182.
54 Y. Sun and H. H. Wang, *Adv. Mater.*, 2007, **19**, 2818.
55 G. B. Onoa, G. Cervantes, V. Moreno and M. J. Prieto, *Nucleic Acids Res.*, 1998, **26**, 1473.
56 H. Nakao, H. Shiigi, Y. Yamamot, S. Tokonami, T. Nagaoka, S. Sugiyama and T. Ohtani, *Nano Lett.*, 2003, **3**, 1391.
57 L. Berti, A. Alessandrini and P. Facci, *J. Am. Chem. Soc.*, 2005, **127**, 11216.
58 I. Lisiecki and M. P. Pileni, *J. Phys. Chem.*, 1995, **99**, 5077.
59 S. Kundu, K. Wang, D. Huitink and H. Liang, *Langmuir*, 2009, **25**, 10146.
60 J. A. Creighton and D. G. Eadon, *J. Chem. Soc. Faraday Trans.*, 1991, **104**, 6767.
61 K. Nguyen, M. Monteverde, A. Filoramo, L. Goux-Capes, S. Lyonnais, P. Jegou, P. Viel, M. Goffman and J.-P. Bourgoin, *Adv. Mater.*, 2008, **20**, 1099.
62 V. D. Vankar and R. W. Vook, *Surf. Sci.*, 1983, **131**, 463.
63 P. Legare, Y. Holl and G. Maire, *Solid State Commun.*, 1979, **31**, 307.
64 L. I. Elding, *Inorg. Chim. Acta*, 1972, **6**, 683.
65 B. I. Nabivanets and L. V. Kalabina, *Russ. J. Inorg. Chem.*, 1970, **15**, 818.
66 C. Staii, A. T. Johnson and N. J. Pinto, *Nano Lett.*, 2004, **4**, 859.
67 M. Bockrath, N. Markovic, A. Shepard, M. Tinkham, L. Gurevich, L. P. Kouwenhoven, M. W. Wu and L. L. Sohn, *Nano Lett.*, 2002, **2**, 187.
68 J. Lund, J. Dong, Z. Deng, C. Mao and B. A. Parviz, *Nanotechnology*, 2006, **17**, 2752.
69 A. L. Efros and B. I. Shklovskii, *J. Phys. C: Solid State Phys.*, 1975, **8**, L49.
70 B. Abeles, P. Sheng, M. D. Coutts and Y. Arie, *Adv. Phys.*, 1975, **24**, 407.
71 M. N. Al-Hinai, N. G. Wright, A. B. Horsfall, R. Hassanien, B. R. Horrocks and A. Houlton, *IEEE Sensors*, 2011, 1.
72 B. D. Kay, C. H. F. Peden and D. W. Goodman, *Phys. Rev. B*, 1986, **34**, 817.
73 H. Hagi, *Mater. Trans., JIM*, 1990, **31**, 954.
74 Y. Li and Y. T. Cheng, *Int. J. Hydrogen Energy*, 1996, **21**, 281.
75 D. Yang, L. Valentín, J. Carpena, W. Otaño, O. Resto and L. F. Fonseca, *Small*, 2013, **9**, 188.
76 R. Smith and D. Otterson, *J. Phys. Chem. Solids*, 1970, **31**, 187.
77 B. C. M. Martindale, D. Menshykau, S. Ernsta and R. G. Compton, *Phys. Chem. Chem. Phys.*, 2013, **15**, 1188.
78 W. Ludwig, A. Savara, R. J. Madix, S. Schauermann and H.-J. Freund, *J. Phys. Chem. C*, 2012, **116**, 3539.

Faraday Discussions

DISCUSSIONS

General discussion

DOI: 10.1039/c3fd90028b

Professor Kornyshev opened the discussion of the paper by Professor Wang: Have you been working on destructive micropropeller rockets, not only those that are physically drilling, but may be caring 'chemical weapons' such as chemotherapy drugs?

Professor Wang responded: Yes, drug-loaded microrockets are the ultimate goal (*à la* the movie *Fantastic Voyage*); yet, we still face grand challenges for *in vivo* applications, and these will be better realized using externally-powered (magnetically or ultrasound-based) micromachines.

Professor Kornyshev continued: Concerning the motion of micro-propelled rockets: what is the achievable precision of their trajectories?

Professor Wang responded: Reproducible propulsion behavior is achieved under reproducible rocket fabrication and operational conditions (*e.g.*, fuel level).

Dr Kataky asked: Nanomachines for *in vivo* applications are exciting with tremendous possibilities. My concern is that these nanomachines must be hydrophobic; as such they will be retained by cell walls and tissues as foreign bodies. This may lead to other health issues. Can you comment on this?

Professor Wang replied: There are indeed many challenges for *in vivo* application of nanomachines, including powering and navigating them in biological fluids, along with biocompatibility and performing complex tasks at predetermined body locations.

Dr Tschulik commented: In your presentation you have demonstrated impressively a variety of possible applications of self propagated micrometer-scaled polymer-based rockets. One of the applications you suggested is using these micro rockets to transport specific target molecules or objects to desired places within lab-on-a-chip devices or other micro-channel based analytical systems. For these applications rockets of small dimensions seem to be desirable. What is the minimum size at which you can produce functioning micro- or even nano rockets by your production method?

Professor Wang answered: Using the template approach, the smallest microtube rockets that we have made are 4 micrometers long and 0.2 micrometers in outer diameter. These fit nicely within common microchannels of 3–5 micrometers width.

Professor Pumera added a comment: The smallest electrochemically deposited microjet engines have a diameter of ~200 nm.

Professor Wang noted: I agree, for template-prepared rockets. I recall a smaller one prepared by a rolled-up fabrication, described in *Chem. Rec.*

Professor Amatore said: Thank you very much for your stimulating presentation. In order to 'pick up' cells you use magnetic field to drive your nanomachines. This is elegant, but requires human- or computer-recognition and decision, and in particular that one has access to microscopic views of the cell arrangement to drive the nanomachine. How will you do this in a real biomedical application?

Professor Wang responded: A good and important point. We envision that future cell-capture biomedical applications will involve the parallel 'swimming' of multiple magnetically aligned multiple receptor-functionalized motors within the small (microliter) sample reservoir of a microchip system, and will not require manual intervention. For a recent review see: J. Wang, *Lab. Chip.*, 2012, **12**, 1944.

Professor Amatore commented: You envisioned essentially 'externally-imposed directionality' to your micro-machines, but could it be feasible to think about using chemotaxis analogs, *i.e.*, relying on the direction of specific chemical gradients, for example, as do many microbes or bacteria?

Professor Wang answered: Yes, bio-inspired following of concentration gradients (*e.g.*, pH gradient around cancer tumors) is the ultimate – yet highly challenging – goal of biomedical (*in vivo*) applications of synthetic micomotors. More practical, at this early stage, for such *in vivo* applications are externally powered (magnetic or ultrasound) micromotors.

Professor Kornyshev asked: Could the propelled nano-rockets be used for cholesterol removal in the disease angina?

Professor Wang responded: Yes, such rockets (and other magnetically-propelled microswimmers) have already been used as nano-drills and may eventually be used to remove cholesterol-based plaques from clogged blood vessels.

Mr Abbas asked: What are the design limitation/constrains of the nano-rocket; does the ratio between the height and the diameter of the nano-rocket affect its speed and mobility?

Professor Wang responded: Yes, the propulsion is dependent upon various geometrical factors, as described by Mei's team in the theoretical model paper and its predictions: 'Dynamics of catalytic tubular microjet engines: Dependence on geometry and chemical environment' (J. Li *et al.*, *Nanoscale*, 2011, **3**, 5083).

Dr Batchelor-McAuley opened the discussion of the paper by Professor Amatore: The vesicles released by the chromaffin cells used within your study are relatively large as compared to other biological systems of interest, such as neurons. Given this size difference, how generally applicable are your results?

Professor Amatore replied: You are perfectly right in terms of the vesicle sizes (*ca.* 150 nm radius for chromaffin cells *vs. ca.* 20–30 nm for neurons); but the internal structures which hold the neurotransmitter cations within the vesicles differ. Chromaffin cells have a dense core matrix (*viz.* they contain a densely packed anionic polyelectrolyte, chromogranin A) while electron microscopy absorbance shows that the vesicles of neurons contain either a fluid or a swollen polyelectrolyte. Hence, rates of diffusion are expected to depend on these factors and are certainly not to be transposed.

However, chromaffin cells are considered as very good "work-horses" by neurobiologists and we chose them for this reason. In this respect, since we do not make any assumption (except for the initial spherical shape of the vesicles), our strategy and model should be adaptable to any type of cell/vesicle for which precise patch-clamp data are available. For example, in collaboration with Professor Andrew Ewing (University of Göteborg, Sweden) we tested PC12 cells and the model worked perfectly.

Now, concerning the initial fusion pore, it is agreed among neurobiologists that this system should not drastically differ among different cells, though the vesicle size and content may differ greatly. This is not unrealistic since initial fusion pore radii (*ca.* 1 nm) are extremely small *vs.* those of any vesicle. Under such conditions our results should be conserved except maybe for the value of the surface tension energy parameter σ determined here. Indeed, if σ is mostly influenced by the cell membrane properties it should remain within the range found here for most cells. Conversely, if this is mostly influenced by the vesicle membrane tension, one may expect large variations between cells due to the vesicle radius and content, since both factors necessarily influence the Laplace tension.

In this respect, since the σ value we determined here is as expected for a poorly tense cell membrane (while a much smaller vesicle is expected to have a much larger surface tension) we are inclined to favor that this parameter is mostly regulated by the cell membrane tension. Hence, our results should be valid for any cell obeying the same machineries.

Dr Walcarius opened the discussion of the paper by Professor Schuhmann: The size of the UME tip you used is in the micrometric range (10 μm), *i.e.*, the same range as the studied cells (around 50 μm). Do you feel it could be useful to use nanoelectrodes to get some higher resolution mapping of the observed events? Does it make sense from the biological point of view (heterogeneous oxygen consumption at the cell surface)?

Professor Schuhmann responded: It is important to establish a tiny gap between the microelectrode and the cell which is depleted in oxygen due to the metabolic action of the cell. Thus, one needs an object with similar dimensions as the cell. However, it would be advantageous to have a smaller active electrode surface embedded in a large glass insulating sheath (*e.g.* an RG value of 100; a

1-μm Pt disk surrounded by 100 μm of glass). However, the time of planar diffusion would then be even shorter and it would be even more difficult to detect the oxygen concentration in the gap. Assuming optimal, high-sensitivity amplifiers a positioning to a gap of 100 nm with the use of a 100 nm electrode would be optimal, since the hemispherical diffusion zone would not overlap substantially with the cell. However, the extremely low currents make the experiment very difficult.

Professor Amatore asked: Thank you very much for providing such an elegant solution to this problem which has remained unresolved in the literature up to now. However, I wish to comment on a point of semantics. As you know, SECM originates from the work that Wightman and myself did for probing the concentration near an active object (tissue, a cell, another electrode, *etc.*), and was then implemented for monitoring surface activity by scanning, before finally receiving its final form as SECM in Bard's hands. I am not recalling this to legitimatise any pretence of paternity, but just to stress that in our work we intended that the collector would not interfere with the generator, *i.e.* no feedback would occur, to measure the real generator-emitted flux. Conversely, in the hands of Bard and of the SECM-school, feedback became important to characterize the electroactivity of a surface. In my view this is the real contribution of Bard in SECM and the reason why I do not call my work in the area SECM. Yet, now you are using SECM and avoiding feedback to measure real gradients without interference by the SECM tip... So, could you tell us if you feel the topic of your talk belongs to SECM or not?

Professor Schuhmann responded: Historically I agree that the birth of the name "SECM" belongs to Alan Bard, based on previous work by Engstrom. In my opinion, Bard recognized in the work from Engstrom, and based on his own experiments with electrochemical STM, that the change in a feedback current can be the source of a new scanning probe microscopy. Since your work with Wightman using a generator/collector configuration was before 1989, SECM was initially very much focused on feedback effects, however, it was later extended to all types of modes including the generator/collector mode, local impedance modes, AC-SECM, redox competition, local material deposition and etching, *etc.*, especially through the series of publications by Bard in *Anal. Chem.* Thus, in my opinion SECM is used presently for methods employing Faradaic currents at microelectrodes for generating information about the local site underneath the tip. In this context I want to refer to our paper M. Etienne *et al.*, *Anal. Chem.*, 2006, **78**, 7317–7324 "Feedback-independent Pt nanoelectrodes for shearforce-based constant-distance mode scanning electrochemical microscopy" in which we deliberately prepared nanoelectrodes which could not approach the surface due to feedback interaction.

Essentially, I would like to extend the discussion by the following thought: A lot of work was dedicated to determine the local kinetic constant at a given surface site using z-approach curves without scanning the tip in the x,y-direction. Is this SECM? Is the word "microscopy" used accurately in this context. For example, the first paper with the words "SECM" and "nanoelectrodes" in the title showed only z-approach curves and no scanning at all (I could add many more examples).

Now, close to 25 years after the invention of the name "SECM", including previously known findings, I would suggest that we agree that the elucidation of phenomena using positioned micro- and nanoelectrodes and Faradaic or capacitive currents measured at or invoked by the microelectrode should be summarized under the heading "SECM". Although this is does not reflect the historical truth, I would put a generator/collector experiment into this group of experiments.

Professor Mount commented: Do the products of oxygen reduction (*e.g.* peroxide) affect the cell lifetime? Does the time-gating you use have the advantage of reducing or removing this effect?

Professor Schuhmann replied: In our experiments we reduce oxygen within the small gap between the tip and the cell for a short time, after which the tip-potential is switched back to a no-effect potential. The tip is then retracted, repositioned with respect to x,y and approached again. Using this method, the species produced are liberated into the bulk and the overall interaction time as well as the amount of the produced reactive oxygen species stay as small as possible. Thus, we are convinced that the proposed method is the best that can be achieved addressing the question raised.

Professor Barbero commented: The data you presented clearly shows that the SECM electrode could reduce drastically the oxygen content of the whole cell. While this a problem for the analytical measurement of changes in oxygen concentration the cell, it could be an excellent way to control rapidly the oxygen content in individual cells. The subject is very important to assess the mechanism that cells use to survive anoxia and/or to trigger collective mechanisms in immobilized tissue.

Professor Schuhmann responded: I have not yet considered the possibility of using a SECM tip to decrease locally the oxygen concentration in the environment and also inside a single cell. For this, studies have to be performed looking at the survival of the cell during a prolonged decrease of the internal oxygen concentration. I agree that this would be an interesting further step in research based on the proposed findings.

Professor Amatore noted: It may look surprising that so much oxygen is present in a cell for its release to overcompensate the negative feedback due to the cell bulk. However this is fully compatible with some former experiments from my laboratory (see, *e.g.*, S. Arbault *et al.*, *Anal. Chem.*, 1995, **67**, 3382–3390) in which we showed that when most cells (*i.e.*, except for neurons) are placed suddenly under anaerobic conditions they continue to be able to produce oxidative stress, *viz.* ROS and RNS whose synthesis by the cell requires oxygen, for *ca.* 10 min. This is also compatible with the fact that the body can tolerate *ca.* 30 min anaerobic conditions, for example after a heart attack. For neurons the oxygen storage appears much more limited and that is the reason why brain areas where a stroke occurs are significantly damaged after a few minutes.

Professor Kanoufi asked: The principle of the detection of O_2 consumption has been made in the redox competition mode, as illustrated in Fig. 6 of your paper. There, a first electrochemical pulse is generated to inject O_2 from water splitting oxidation. During this pre-electrochemical step a large amount of H^+ could be formed. Have you evaluated the amount of O_2 or H^+ generated during this step? And is such a pre-electrochemical treatment invasive for the cell?

Professor Schuhmann responded: In this specific case, unlike the previously published redox-competition mode (see: K. Eckhard, *et al.*, Redox-competition mode of scanning electrochemical microscopy (SECM) for visualisation of local catalytic activity, *Phys. Chem. Chem. Phys.*, 2006, **8**, 5359–5365), no oxygen-injection pulse is performed and hence the cell will not feel a substantial pH value change.

Professor Kanoufi continued: Your Fig. 7 shows how you could image the spatio-temporal evolution of O_2 in the vicinity of the cell. This imaging strategy is very promising and demonstrates the potential of the technique. It confirms that at least two processes are operating: O_2 consumption and O_2 permeation from the cell. The results are presented for two times, $t = 20$ ms and $t = 490$ ms; did you record the current evolution during the whole pulse ($t = 0 \rightarrow 500$ ms)? Could you characterize the dynamics of both phenomena (consumption and permeation)?

Professor Schuhmann answered: During the duration of the oxygen detection pulse (500 ms) 100 current-values were recorded. Thus, we have a time resolution of 5 ms and we could evaluate the oxygen decay and then the oxygen increase curve in the gap using the recorded data. This should be complemented by experiments at different distances (*e.g.* 100 nm, 200 nm, 300 nm, ...) and with different-sized electrodes to establish a model for the overall process.

Dr Kataky enquired: Can the SECM tip monitor lateral diffusion of oxygen from the cells?

Professor Schuhmann replied: The modulation of the O_2 concentration by the cell metabolism is very slow and easily compensated by the fast diffusion of O_2 in the bulk. In addition, one has to keep in mind all aspects of convolution of topography and electrochemical activity in SECM experiments. Taking this into account, I feel that the only feasible way is to position the SECM tip in a known and short distance from the cell, from above.

Following on from the discussion on the variation of the pH value within the gap between the positioned microelectrode and the cell as well as the counter ion movement to keep electric charge compensation, I would like to point out the approximately 4 times faster diffusion of protons due to the so-called super-exchange mechanism (Grothuss mechanism) in which the proton does not have to diffuse, but only hydrogen bonds are capped over. By this one may anticipate that the pH decrease in the gap between tip and cell is smaller than initially expected.

Professor Barbero continued: I agree that the effectiveness of electrochemical actuators involving pH changes must take into account the larger diffusion coefficient of protons ($ca.$ 10^{-4} cm^2 s^{-1}) and hydroxyls ($ca.$ 5×10^{-5} cm^2 s^{-1}) with respect to other ions (smaller than 1×10^{-5} cm^2 s^{-1}) which can reduce the pH-effect inside the gap or pore. However, in small, long pores the faster diffusion of proton and hydroxyl makes water dissociation a faster mechanism for charge compensation than ion movement. Therefore, at short times it is possible to generate local pH changes (or increase a pH change) which could influence the course of the reaction.

Dr Tschulik addressed Professor Schuhmann: In order to detect the oxygen consumption of the living cells you investigate in your paper, you measure the time-response of the oxygen reduction reaction using an SECM setup. Since during these measurements your working electrode is very close (about 100 nm) to the cell surface any motion of the oxygen consuming cell, especially in the perpendicular-to-the-electrode direction, could strongly affect the measured time response. Does such cell-motion occur within the time scale of your measurement?

Professor Schuhmann responded: This is a very good question. Typically, after shear-force-based positioning of the tip in close proximity to the cell, we switch off the electrode vibration and we immediately invoke the potential pulse profile with a typical duration for the oxygen-reduction pulse of 500 ms. Since the correct determination of the oxygen consumption of the cell is obtained at times after the decline of the charging current (*e.g.* > 20 ms) we do not feel that cell swelling or movement plays a role.

Professor Amatore added: This is not really a question but a comment. Based on the constancy of your data with the Fe^{2+} probe I am fully convinced that the cell topography is followed perfectly by the SECM probe. Hence the deviations observed for the O$_2$ reduction current above the cell cannot be an artifact related to any change in collection efficiency due to a change in the electrode–cell distance. Hence, they are clearly indicative of the cellular release of O$_2$ under the electrochemical constraints imposed by the electrode.

Miss Hirani communicated: Given that at longer pulse lengths the tip is inducing transmembrane diffusion of oxygen, how does this impact on the current measured at the next x,y-grid point?

Professor Schuhmann communicated in reply: The time between two x,y-grid points is chosen to be long enough to establish equilibrium before the next measurement is performed. For this, the potential at the tip is switched to a no-effect potential, the electrode is raised up, moved to the new x,y-location and approached using a shear-force positioning mode. Once the new measuring point is reached the new pulse profile is invoked. Usually the duration at a potential low enough for oxygen reduction at the tip is 500 ms, which is not sufficient to completely empty the cell.

Professor Mount communicated: Fig. 4 appears to show an asymmetric PC12 cell topography. Could this be a potential effect of shear force distance control, scan direction and/or cell membrane flexibility? Evidence against an effect of scan direction would be that the observed topography were independent of scan direction. Is this the case?

Professor Schuhmann communicated in reply: I think one has to assume the function of the detection of topography by means of shear force interactions. We have to assume a disk electrode, in this case a single carbon fibre of about 7 μm diameter insulated with a polymer layer of about 2–3 μm thickness. Shear force interaction will occur at the point of closest approach between this tip and the surface. I would interpret the shown topography as caused by a change in the predominant shear force interaction points during scanning. Between an x-position of 10 to 40 interaction occurs with the Petri dish. Then, when approaching the cell sideways most probably a new interaction point becomes determining at the right side of the tip. Once the electrode has travelled across the surface, an interaction point at the left side of the tip will feel the cell. Since these interaction points are changing and taking into account the diameter of the tip, the linescan shows a convolution of the size of the cell plus that of the tip.

Professor Kanoufi returned to the discussion of the paper by Professor Amatore: The information that could be obtained from those electrochemical measurements is really amazing. It is particularly interesting to observe the sequential opening of a pore, and to characterize its growth within few nanometres (<2 nm in Fig. 4). This Figure is also very interesting for its kinetic information. It actually shows the characteristic time rise for the pore opening, $\alpha_{pore}(t)$, until it reaches a steady aperture, α_{pore}^{open}. I imagine the next step in the investigation is to depict statistically the kinetics of $\alpha_{pore}(t)$. However, it is interesting to note that for the three kind of pores presented in Fig. 4, all the openings detected proceed within a similar timescale on the order of 5–10 ms. Does this help to identify the kinetically limiting process beyond the opening phenomenon?

Professor Amatore answered: Thank you, dear Frédéric, you understood perfectly! In fact we have already obtained data showing that the initial fusion pore does not expand completely (*i.e.*, 'full' fusion) as believed previously by most neurobiologists and ourselves. Conversely, the pore expands so that it represents an angular opening of *ca.* 10 degrees and then its expansion is abruptly stopped. We believe that the pore expansion 'bumps' into some biological structure. This may be the cell surface cytoskeleton or a protein of the dynasore family wrapped around the neck of the vesicle/cell membranes connection. At this stage we do not know, but plan to perform experiments (*e.g.* dual fluorescent probe marking observed by TIRFM) to identify this biological structure in the future.

Concerning the part of your question related to the rates of expansion, one should note that for such systems the energy released during the expansion must be dissipated. Basically there are two modes and both are controlled by the viscous friction of each leaflet of the bilipidic membranes *vs.* each other (this may be made more difficult by the presence of transmembrane proteins, cholesterol, *etc.*) or *vs.* the solutions on each side and possibly the cytoskeletal mesh. The shapes of the opening functions during the stalled full fusion stage agree qualitatively with such a view as we established previously, though our previous analysis assumed improperly (but in agreement with the current views at the time) that 'full' fusion was complete (*i.e.* led to a final angular opening of 180 degrees). Our present results, indicating a maximum angular opening of *ca.* 10 degrees, force us to treat anew the $\alpha_{pore}(t)$ functions measured within this new framework.

Finally concerning your comment about the time-scales in Fig. 4 (right column) I believe that one should be careful about their significance. First the time-scale is not as homogeneous as you understood, but ranged from a few milliseconds to a few tens of a millisecond. In this respect, again viscous dissipation is crucial and may even be hampered by still-locked or partially locked SNARE complexes. The role of viscosity is particularly evident through the pore energy curve shown in Fig. 5A. Indeed, should viscous dissipation be fast, all pore radii would fall around the energy well minimum (with a statistical distribution related to the fluctuation of thermal quanta due to the fact that these are single events). Yet we observe a near-even distribution of pore sizes corresponding to a range of energies exceeding *ca.* 5 thermal quanta. I take this as showing that the pore energy dissipation is so slow, due to high viscosity, that the pores remain almost poised at their initial opening.

Professor Bond opened the discussion of the paper by Dr Bohn: What is the significance and meaning of the term 'reversible' or 'equilibrium potential' when applied to a single molecule attached to an electrode surface as described in your paper? Given the wide range of possible conformations in the oxidised and reduced forms you mention, the different locations of electroactive centres relative to the electrode surface and the need for charge neutralisation *via* ingress or egress of electrolyte ions, could values for each molecule be very different? Could differences be in the range of say 50 to 100 mV or even larger?

Dr Bohn replied: Indeed the term 'equilibrium potential' refers to a population in normal parlance, while single-molecule experiments necessarily probe individual members of the ensemble from which the macroscopic equilibrium potential is measured. Thus, the interpretation of terms like 'equilibrium potential' in the context of a single molecule must be based on probabilities. As an ansatz, we assign an equilibrium potential for a single molecule to be the applied potential at which the probabilities for oxidation and reduction are equal. As Professor Bond notes, there are a number of local factors applying both to the molecule as well as to the electrode and its surface site distribution which can be expected to influence the single-molecule equilibrium potential. In our experience, the variations among estimated values for single molecules which are all part of the same zero mode waveguide array can exceed 100 mV.

Professor Schuhmann asked: A quite complicated binding chemistry was chosen, using PQQ as linker. PQQ has a quinone-type redox conversion at substantially higher formal potential and three carboxylic acid residues which cannot be addressed selectively. Assuming that the PQQ does not take part in the electron-transfer process, why you did not choose a dicarboxylic acid for binding of the boronic acid derivative which later binds the ribose of FAD?

Dr Bohn answered: Initially, we were concerned that the single FAD molecules would not be able to undergo efficient electron transfer across the cystamine monolayer, and the work by Willner shows that PQQ bound to FAD in the manner described constitutes a competent redox mediator, outside of the enzyme. We could certainly have used a dicarboxylic acid to link the FAD molecules to the Au ZMW surface, and in fact are going to do these experiments. Our expectation is that we will see electron transfer just as Willner *et al.* did when they bound FAD to the electrode surface through the phenylboronic ester.

Professor Amatore continued: I wish to come back to your discussion with Alan Bond. Your 'single events' data are necessarily statistical, and hence chaotic within the probabilities imposed by the driving force. So, have you tried to perform statistical treatments of the data obtained at each potential? I am interested in the sense that, for example (as you well know), Marcus theory involves statistical fluctuations of the inner and outer spheres to allow the starting and final energy wells to intersect where electron transfer occurs. This must apply to each of your single events, though in a chaotic mode. Could then your data provide quantitative views on the statistical distribution of Marcus reorganization energies?

Dr Bohn replied: This is a very interesting observation. The challenges to applying our single molecule spectroelectrochemistry approach to the measurement of statistical distributions of reorganization energies would be to identify a system where a single electron transfer pathway is dominant, and then to acquire enough single molecule trajectories to make a meaningful statistical comparison. We are working to improve our apparatus, so that we can acquire parallel trajectories more efficiently, which will address the second issue. There are two subtle problems here – first, the molecule must avoid photobleaching for long enough to generate a trajectory that can be analyzed with good confidence, and then one must collect enough of these long trajectories. We are also looking into different linker chemistries which can direct electron transfer to occur *via* a single dominant pathway. Thus, there are no fundamental hurdles to using single molecule spectroelectrochemistry to map out distributions of reorganization energies – only experimental challenges.

Professor Mount said: Can you see enough molecules to measure an ensemble of lifetimes in the pores? If so, is the mean and distribution the same as/similar to the lifetime observed for the molecule in solution?

Dr Bohn answered: I assume by 'lifetime' you mean fluorescence lifetime. If so, the answer is that we are not currently able to measure single molecule lifetimes with our apparatus. Nevertheless, we would not expect the lifetime to be the same as free FAD in solution. If nothing else, the proximity to the Au electrode would be expected to give rise to some excitation transfer to the plasmons of the metal, thus affecting the observed lifetime. It would, however, be very interesting to see if the ZMW-bound single FADs exhibit a single (as in the protein) or double (as in free solution) exponential decay.

Professor Schuhmann commented: In the scheme shown from earlier work of Willner *et al.* using the PQQ–boronic acid–FAD binding chemistry the FAD was integrated within glucose oxidase and hence could not access the electrode surface. In the presented configuration, the FAD can bend down towards the electrode and electron transfer may occur from the FAD to the electrode directly. This is in addition much more likely because at the applied electrode potentials the PQQ was kept in its reduced state and hence an electron-transfer cascade from FADH2 to PQQ and finally to the electrode cannot occur. Thus, as I already mentioned in my question above, the reasoning for using PQQ in the electrode architecture remains unclear.

Dr Bohn responded: First, we note that our data do not address the question of electron transfer mechanism for the bound FAD. Having said this, we agree that the direct (through the cystamine monolayer) electron transfer is viable, but we do not believe it possible to completely discount a PQQ mediated process based on the available evidence. First, as is consistent with the lifetime data of Zhong *et al.* there is a hinge motion capable of bringing the isoalloxazine ring in close proximity with the PQQ ring. In addition, it is at least worth considering whether the potential of the PQQ/PQQH$_2$ couple is quite different from its bulk value, at least in some cases. Such individual differences could render the mediated electron transfer active. Again we emphasize that our data are silent about these questions.

Professor Kornyshev opened the discussion of the paper by Dr Horrocks: Why do metal nanowires not fuse? van der Waals forces will tend to cause fusion, unless you charge them, as in the case of functionalized nanoparticles. So, how do you protect nanowires against fusion?

Dr Horrocks replied: We do not attempt to protect the Pd nanowires described in the paper against fusion. This is not necessary for the measurements described because they are carried out after the wires are deposited on a substrate and dried; at this point they are immobile. However, it is certainly correct that the nanowires will aggregate in solution given enough time, but the process is slow because of the relatively low concentrations involved. In previous work (with conductive polymers and metal oxides) we have observed this phenomenon (R. Hassanien *et al.*, *Nanotechnology*, 2012, **23**, 075601 and S. Pruneanu *et al.*, *Adv. Funct. Mater.*, 2008, **18**, 2444), which is interesting in itself because the nanowires form rope-like structures and the twisting of the nanowires about each other may be right or left handed.

Professor Barbero commented: The grainy nature of the Pd nanowires should introduce an additional effect on the hydrogen sensing related to changes in the grain-to-grain contact due to Pd grain expansion. Do you find evidence of that effect?

Dr Horrocks replied: Previous workers have shown that there are at least two possible origins for the response of Pd sensors to hydrogen partial pressure. The first type of sensor are those in which the sorption of hydrogen increases the resistivity of the material because PdHx has a higher resistivity than Pd. Ref. 72 and 76 in the paper are examples. The second type of sensor is that which shows a decrease in resistivity because the swelling of the crystallites that occurs when Pd sorbs hydrogen and results in the closing of some gaps in the structure. A recent report has in fact shown that an individual sensor can show both increases and decreases of resistivity as the temperature changes, ref. 75 in our paper. Our devices have only exhibited the first type of response, *i.e.*, an increase in resistivity and therefore we have only modelled such a mechanism in the paper. We don't see any evidence that changes in grain-to-grain contact are ever the dominant factor in our devices, though of course we cannot rule it out completely. Other possibilities would be, *e.g.*, reduction of PdO by hydrogen, but such an effect would probably not be as reversible as we observe. As far as we can tell, the main effect of the grain structure on the devices is to produce a resistance that decreases with increasing temperature as shown in Fig. 12 in our paper, and perhaps to account for the low values of diffusion coefficient that we estimate.

Professor Pumera said: What is the device-to-device reproducibility of the analytical signal?

Dr Horrocks replied: The device-to-device reproducibility is about 20% in relative terms for the signal. However, we have not yet prepared enough devices to have good enough statistics for a more precise estimate.

Dr Batchelor-McAuley returned to the discussion of the paper by Dr Bohn: Earlier the question of "what the formal potential of a single molecule actually means" was raised. I would like to follow on from the comment by Professor Schuhmann, who said that he believed that an electron-hopping mechanism for the electron transfer is not likely to be operative due to the differences in the formal potential of the linker quinone and the flavin. In light of the earlier comments on the formal potentials of single molecules, I would like to ask Professor Schuhmann if such large ensemble thermodynamic arguments for electron hopping are still fully valid when discussing such systems? Is it more kinetically favorable for the electron to tunnel across the linker chain, the molecule to bend towards the electrode surface to allow electron transfer, or the linker quinone to provide a lower energy pathway for electron hopping (*cf.* the STM work on long range electron transfer by Jens Ulstrup, *Proc. Natl. Acad. Sci. U. S. A.*, 2005, **102**(45), 16203–16208).

Dr Bohn replied: Ulstrup *et al.* posit nuclear configurational fluctuations as the source of local fluctuations in the energy levels of the bridge (azurin in their case) that bring the oxidized bridge orbital below the Fermi level of the tip, where it can facilitate resonant tunneling. The same mechanism could certainly operate in our PQQ-FAD system, although there is significantly greater structural flexibility than in the corresponding cofactor–apoenzyme reconstruction or in the azuring system studied by Ulstrup *et al.* This, in turn, would argue for greater configurational effects and would leave open the possibility of tunneling through this level. Experiments to compare the two possible pathways (through-bond *vs.* through space) would need to rigorously define the distance dependences in order to permit a realistic comparison.

Professor Amatore addressed Dr Batchelor-McAuley: The rate of electron transfer is certainly influenced by fluctuations in bending/extension of the chain linking the redox center to the surface since this controls the intensity of the electronic coupling, so you should not consider that all redox centers are poised at a given distance from the surface. Furthermore the PQQ moiety, even if not expected to be electroactive in your potential range, may contribute to favoring the electronic coupling depending on its orientation *vs.* the redox center and the surface. Thus, though your data are certainly exciting they cannot provide any firm conclusion. Could you think of using rigid linkers as do, for example, photochemists and (bio)photophysicists when investigating single events?

Professor Schuhmann added: Dr Batchelor-McAuley suggested that the formal potential of individual molecules may differ substantially from the thermodynamic equilibrium and hence it may be likely that some PQQ molecules may be able to still transfer electrons from $FADH_2$. On the one hand, I feel that the use of the expression "formal potential" for an individual molecule is not appropriate since the Nernst equation is not defined with a single molecule. In addition, even if one assumed a substantial variation of the potential for the redox conversion of individual $PQQ/PQQH_2$ couples a potential difference of 400 mV would decrease this probability by 7 to 8 orders of magnitude (59 mV per decade).

Dr Batchelor-McAuley answered both Professor Amatore and Professor Schuhman: I fully agree with these comments. The challenge is the possible need to reconcile these views with some significant experimental work on charge transfer through DNA. A good example of this work comes from Barton *et al.* (*Nat. Chem.*, **3**, 228–233 (2011)) where – assuming the validity of the results (P. Salvatore *et al.*, *J. Am. Chem. Soc.*, 2012, **134**, 19092–19098) – electron transfer is found to occur efficiently across 100-mer dsDNA (34 nm). Importantly, the redox group being used (Nile Blue) does not exhibit a potential close to that of the DNA bases and hence on the basis of the above argument one would assume that this long-range charge transfer would not be feasible.

Dr Bohn replied: We agree that the use of a 'formal potential' is not the best way to describe single molecule behavior. A better approach would be to describe the potential at which the probabilities for reduction and oxidation are equal. We know from the spread in our transition rate and total on-time data (which increase at more anodic potentials), that these can vary dramatically from molecule-to-molecule. However, for a given molecule we do not know what this is until we measure the distributions.

Professor Wang addressed Dr. Horrocks: What are the advantages of the DNA-based Pd hydrogen gas sensor over carbon nanotubes decorated with Pd particles?

Dr Horrocks responded: I'm not sure that CNTs decorated with Pd particles would produce devices in which the current flows through the Pd. When we compare our Pd/DNA sensors with examples from the literature we find that their sensitivity is comparable to the best devices reported and I believe this is because the current path is through the Pd. The aspect of their performance which most needs improvement is the response time – a few other sensors have been reported which are notably faster, *e.g.*, ref. 47 in our paper.

Professor Chen asked: You have fabricated and evaluated the Pd/DNA nanowire networks as a hydrogen gas sensor. Is your sensor reusable?

Dr Horrocks answered: Yes, the sensor can be cycled through many exposures of hydrogen/nitrogen without obvious degradation in response.

Professor Schuhmann addressed Dr. Bohn: The redox conversion of PQQ could be visualized by means of cyclic voltammetry to substantially more anodic potentials (*e.g.* + 300 mV *vs.* Ag/AgCl 3 M Cl$^-$). The question is, if one could see an enhanced electron-transfer *via* the PQQ if the authors had applied an electrode potentials at which PQQ was in its oxidized form and by this could facilitate the transfer of electrons from FAD? In cyclic voltammetry of multiple combined electrodes the PQQ/PQQH$_2$ redox peak should be visible. Was this observed? And could the authors detect an enhanced electron transfer at higher electrode potentials?

Dr Bohn answered: This is an interesting experiment, in principle. However, we cannot measure the electron transfer events by the current generated. First, the pores in the ZMW arrays are not individually addressable, and even if they were we would need to measure single molecule currents from 30–50 single molecules in a given experiment (the other 950 pores containing either 0 or multiple molecules). Nevertheless, one would expect the electron transfer efficiency for FADH$_2$ oxidation to increase exactly as the question posits. The only way we have at present to approach the experiment is to measure the off-time distributions (which reflect the lifetimes of the FADH$_2$ species) as a function of applied potential as we scan (very slowly) to more positive potentials. We are in fact working on this experiment now.

Professor Williams communicated: Your method shows that the $E(1/2)$ values of different molecules may be different, and indeed could extract the current–potential curve for an individual molecule from the potential dependence of the transition probabilities. It would be interesting to know just how different individual molecules might be. Derived from this idea comes the question of how the ensemble current–potential curve, as determined on a macroscopic electrode or on the full array, depends on the distribution of $E(1/2)$.

Professor Bohn communicated in reply: Indeed, the $E_{1/2}$ values are likely quite different. Our data do not address this directly, although the spread in individual molecule on- and off-state distributions, *e.g.* near the macroscopic equilibrium potential of 0.48 V, shows that the there is likely to be significant spread in the individual molecule values. However, the work of Bard and Barbara (R. E. Palacios *et al.*, *J. Am. Chem. Soc.*, 2006, **128**, 9028–9029) using single molecule emission to elucidate the behavior of single redox centers in poly(9,9dioctylfluorene-co-benzothiadiazole) shows variations in $E_{1/2}$ values for different redox centers covering 100s of mV. Although the comparison is not strict, since the PQQ-FAD construct is an immobilized single molecule as opposed to single redox center in an extended material, we believe the underlying behaviors are likely to be similar.

Faraday Discussions

PAPER

Anodic TiO$_2$ nanotubes: double walled vs. single walled

Ning Liu,[a] Hamed Mirabolghasemi,[a] Kiyoung Lee,[a] Sergiu P. Albu,[a] Alexei Tighineanu,[a] Marco Altomare[ab] and Patrik Schmuki*[a]

Received 20th February 2013, Accepted 18th March 2013
DOI: 10.1039/c3fd00020f

Electrochemical formation of self-organized TiO$_2$ nanotube layers has been a highly active research field for more than 10 years. In the present manuscript we investigate the formation of two distinctly different anodic TiO$_2$ nanotube morphologies, 'single walled' and 'double walled' tubes, which are formed mainly depending on the nature of the anodization electrolyte. While the widest used electrolytes are ethylene glycol (EG) based, forming double walled structures, tubes formed in dimethyl sulfoxide (DMSO) based electrolytes show a single tube walled morphology. Here we provide reasons for the formation of double walled tubes, characterize tubes for their composition, structure and certain properties, and give measures to suppress or minimize double wall formation. Except for the fact that in DMSO single walled tubes are formed, we also show that they grow sufficiently slowly to allow partial crystallization of the tubes during growth – this, in turn drastically influences their electronic properties. Finally we discuss the effects and potential consequences of double or single wall growth for TiO$_2$ nanotube applications.

1 Introduction

Over the past decade the growth of one-dimensional self-organized anodic oxide structures has attracted great scientific interest. A particular focus has been the formation of self-aligned TiO$_2$ nanotube arrays. This is to a large extent due to the high expectations in view of applications, such as in dye-sensitized solar cells, biomedical or photocatalytic coatings (for an overview, see e.g. ref. 1–3).

Experimental procedures to grow ordered TiO$_2$ tubular structures by anodization are based on dilute fluoride containing electrolytes. In early experiments water based electrolytes[4-6] were used which then became replaced by glycerol or ethylene glycol based electrolytes, as the latter allow for a much better definition of the tubes (smooth surface, high degree of organization and considerably higher aspect ratios).[1] Over the years other organic solvents or fluoride-free electrolytes have also been explored – but usually such fluoride free electrolytes lead to less organized

---

[a]Department of Material Science WW-4, LKO, University of Erlangen-Nuremberg, Martensstrasse 7, 91058 Erlangen, Germany. E-mail: schmuki@ww.uni-erlangen.de; Fax: +49-9131-85 275 82; Tel: +49-9131-85 275 75
[b]On leave from Department of Chemistry, University of Milan, Via Golgi 19, I-20133 Milano, Italy

structures.[7–9] In the last five years, most work in the field has focused on correlations between a characteristic tube feature, such as tube length and diameter, or tube crystal structure to a specific functional property. In the great majority these works consider the tubes to be uniform over their wall thickness and length, and disregard the potential effect of electrolyte or detailed formation conditions to induce morphological or composition gradients along or across tube sections.

In the present work we focus on a feature that hardly any attention is paid to in tube formation literature but that is inherent to any anodic oxidation process. Namely that due to the coupling of cation and anion migration during oxide growth, a double layer oxide is formed. For tube formation the double layer finally turns into an outer- and inner-shell in the morphology. These layers can affect critically not only the geometry but also the properties of TiO_2 nanotube layers.[10–12]

Fig. 1a illustrates the formation of a double layer structure in anodic oxide formation. The anodic potential causes oxidation of metal, in our case metallic Ti to Ti^{4+}. Under the applied field the Ti^{4+} ions migrate across the oxide outward to the oxide/electrolyte interface. Concurrent O^{2-} ions from the electrolyte (water) are inserted in the oxide lattice and migrate towards the oxide/metal interface. In principle, new oxide can thus be formed at either interface, depending on the cation to anion transfer numbers t_+/t_- in the oxide. For TiO_2, oxide growth occurs at both interfaces (as typically the transfer number (t_+) for Ti^{4+} is 0.4).[13] In other words, Ti^{4+} ions are comparably mobile and when arriving at the outer interface they can react with H_2O or O^{2-} to form oxide (or oxyhydroxide). This outer layer typically shows pick-up of electrolyte anions or other electrolyte species as molecules adsorbed at the oxide/electrolyte interface can simply be overgrown by ejected Ti^{4+} species reacting with H_2O. Moreover, ions from the solution may be

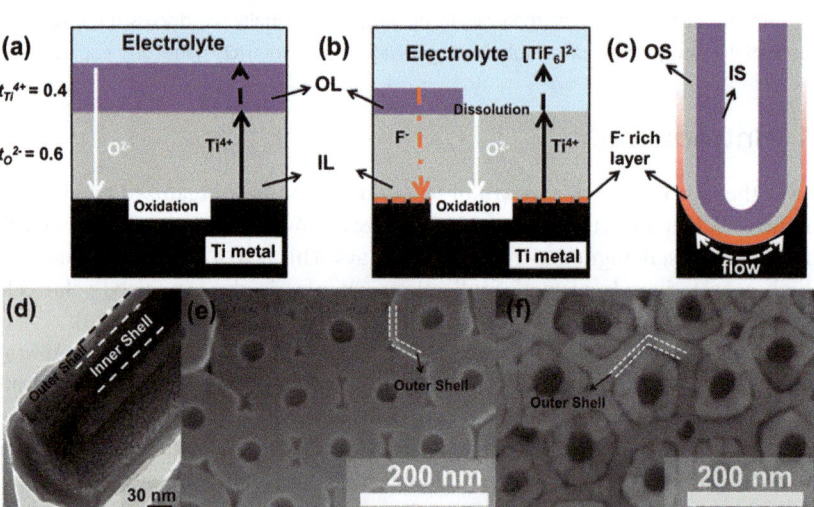

Fig. 1 (a, b) Schematic representation of formation of a double layer structure in anodic oxide in the absence and presence of fluoride ions; (c) diagram of the flow mechanism pushing oxide and fluoride layer up the cell walls by viscous flow; (d) TEM image of double wall morphology for TiO_2 nanotubes of 15 μm thickness formed in EG with addition of 0.1 M NH_4F and 1 M H_2O at room temperature; (e, f) SEM images of double walled TiO_2 nanotubes of 15 μm thickness formed in EG with addition of 0.1 M NH_4F and 1 M H_2O at room temperature before (e) and after annealing (f) at 450 °C.

incorporated by competing inward migration. In typical anodization cases the inner oxide layer (at the metal side) is of a higher quality (less electrolyte contaminated) than the outer layer, as plain overgrowth of adsorbed species can only take place by oxide growth at the outer interface. The driving force for ion migration (with a flux of J) is the effective field (F) across the oxide layer thickness (t), $J = F/t$. Under potentiostatic conditions and for an insoluble oxide, J is permanently decreased by oxide thickening (*i.e.* field lowering), and thus reaches a final very low value (practically, oxide formation is stopped after some time).

The key to nanotube formation is to maintain a steady-state situation of oxide formation and dissolution[14] which establishes a steady reduced thickness of the oxide and thus allows for a permanent formation of new high field oxide (Fig. 1b). In case of anodic TiO_2, dissolution can be achieved by: (i) F^- ions in the electrolyte that capture arriving F^- at the oxide/electrolyte interface and solvatize the Ti ions as stable TiF_6^{2-}; and (ii) by plain chemical dissolution of the oxide or oxyhydroxide as TiF_6^{2-}. Both dissolution effects result in a lower layer thickness which allows steady high field conditions and thus continuous oxide production to be maintained. The oxide is then (due to local instabilities and stress effects[14]) cracked open, pushed side ways, and finally upwards to a 3D morphology. Under a set of conditions self-organization is stabilized and highly organized 3D structures (tubes) form where the oxide is permanently pushed up the wall (as illustrated in Fig. 1c).[14-16] Therefore in the case of nanotube growth the steady state oxide thickness of the tube bottom determines the strength of the ion driving field ($F = \Delta U_{app}/t_{ox}$) and thus the overall growth rate. Also it can be perceived that for conditions that minimize the formation of the outer contaminated layer (which becomes the tubes' inner shell [IS]), higher quality oxide tubes (maximized outer shell [OS]) can be formed. In this context most noteworthy is that for tubes grown in fluoride containing ethylene glycol (EG) electrolytes, which is by far the most used electrolyte, a very significant formation of an inner layer and strong uptake of carbon from the electrolyte is observed.[10] The as-formed TiO_2 nanotubes are typically amorphous. In this state typically only TEM can reveal clearly the double wall nature of such tubes [Fig. 1d]. However, to make tubes useful for many applications, they need to be annealed to a crystalline material (*e.g.* anatase, which is reported to be the most efficient TiO_2 polymorph for electron transport based applications such as solar cells[17]). In the annealing process the double nature of the tube wall can become very clearly evident, as shown by the comparison of a plain view of TiO_2 nanotubes formed in EG before and after annealing in Fig. 1e and f.

In the present paper, we discuss key factors that affect the formation of these inner contamination shells and how to minimize inner layer formation (leading to 'single walled' tubes), and finally we discuss some potential impact on applications. We start out with the findings of a number of preliminary experiments and some prior work[10,12] showing that typically the most used tube formation electrolyte, based on EG, has a tendency to form multiwalled tubes, while particularly dimethyl sulfoxide (DMSO) containing electrolytes aid in the formation of single walled tubes.

2 Experimental

To grow TiO_2 nanotubes, titanium foil (99.6% purity, 0.1 mm, Goodfellow, England) was used. Prior to anodization the Ti foils were cleaned by sonication in acetone and ethanol followed by rinsing in deionized water (DI) and drying under a nitrogen stream. Anodization was carried out in a two electrode configuration with platinum counter electrode using a high-voltage potentiostat (Jaissle IMP 88 PC). For tube growth we used the following electrolyte compositions at different temperature (room temperature [around 20 °C], and 60 °C): (i) ethylene glycol (EG, Sigma–Aldrich, containing less than 0.2 wt% H_2O) with addition of 0.1 M NH_4F (Sigma–Aldrich, 98%) and 1 M DI water; (ii) dimethyl sulfoxide (DMSO, Sigma–Aldrich, containing less than 0.2 wt% H_2O) with 2 wt% HF (Sigma–Aldrich, 40%).

After the anodization, the samples were immersed in ethanol for 1 h and then left to dry in an N_2 stream. Some samples were annealed at 450 °C in air for 1 h using a Rapid Thermal Annealer (Jipelec JetFirst 100). For morphological characterization field-emission scanning electron microscopy (Hitachi FE-SEM S4800) and conventional TEM (Philips CM30 TEM/STEM were used. XRD patterns were collected using an X'pert Philips MPD diffractometer with a Panalytical X'celerator detector, using graphite-monochromatized Cu-Kα radiation ($\lambda = 1.54056$ Å). For the chemical analysis, an EDX (EDAX Genesis, fitted to SEM chamber) was used.

Solid-state conductivity measurements were carried out by Ag epoxy applied through a 2.1 mm wide opening in a mask onto the nanotube surface. After preheating the samples for water evaporation, resistivity values were then obtained from I–V curves taken at room temperature in an N_2 glove box.

The photocurrent measurements were performed in 0.1 M Na_2SO_4 at +500 mV vs. Ag/AgCl in a three-electrode electrochemical cell equipped with a quartz glass window. Photocurrent spectra were recorded in the range of 300–600 nm using an Oriel 6356 Xe lamp and Oriel Cornerstone 7400 1/8m monochromator.

3 Results and discussion

First we would like to point out how to judge properly the the presence or absence of an inner (IS) layer after tube formation. Fig. 2a shows tubes formed in an EG electrolyte at room temperature. Shown are the cross section of an approx. 20 μm thick tube layer and SEM plane view images taken at different tube heights. These images show that towards the tube bottom, the thickness of the tube inner shell strongly increases, i.e. the profile indicates that at the tube top the inner layer has been more strongly dissolved under the permanent mild etching of the anodizing solution ($TiO_2 \rightarrow TiF_6^{2-}$) than at the bottom. This results in an approximate V-shaped inner tube profile (in Fig. 2b) and implies that the inner tube shell is significantly more prone to chemical etching compared with the outer shell. This in turn can be ascribed to the higher degree of impurity incorporation in the inner shell (Fig. 2c). After annealing, the top part of such a tube still shows a single wall morphology while clearly at the bottom a double layer structure is apparent (in Fig. 2d). This illustrates that proper judgment of single or double wall formation can only be done close to the bottom of the tubes – most reliably by TEM or by SEM after annealing the tube layer (shown in Fig. 2d). It may be noteworthy that the inner layer shows, under most annealing conditions, a distinct granular morphology – this can be ascribed to the extremely high carbon content in these tubes (≈ 20 at% from EDX in Fig. 2c). When these carbon (hydrocarbon) compounds are evaporated as CO_2 during annealing this leads to the voids in the structure.

The fact that the solubility of the inner layer in a given electrolyte, to a large extent, determines the thickness of this layer is evident from tubes grown at

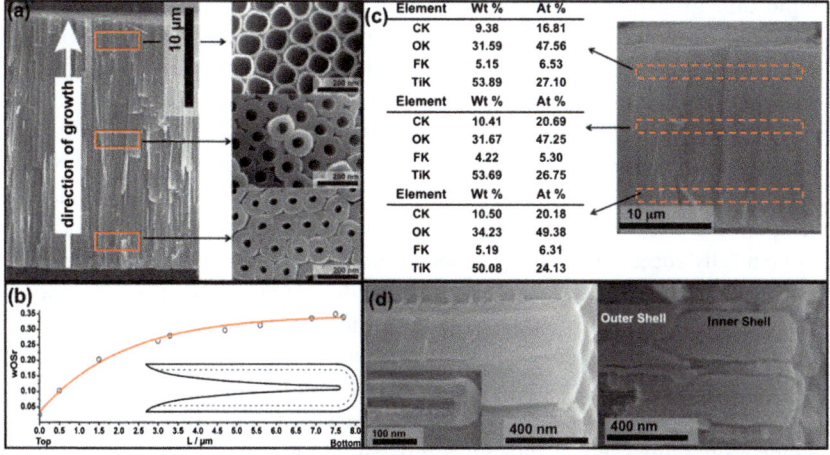

Fig. 2 (a) SEM images and (b) illustration of the gradient in the tube-wall thickness (wOSr = wall to outer shell ratio) of TiO$_2$ nanotubes formed in EG with addition of 0.1 M NH$_4$F and 1 M H$_2$O at room temperature taken at the top, from the fractures in the middle, and close to the bottom of a tube layer; (c) SEM image with EDX analysis for the double walled tubes of 15 μm thickness at 3 different locations (inserted boxes); (d) SEM images of cross-sections of double walled tubes before (left) and after (right) annealing at 450 °C.

different temperatures. Fig. 3 shows SEM views and I–t curves for tubes grown in an EG electrolyte at different temperatures. From the SEM images it is clear that with an elevated temperature the inner layer is dissolved more rapidly and thus thinned down. The finding of thinner layers at elevated temperatures is consistent with a strongly increased current flow at higher temperatures and a more

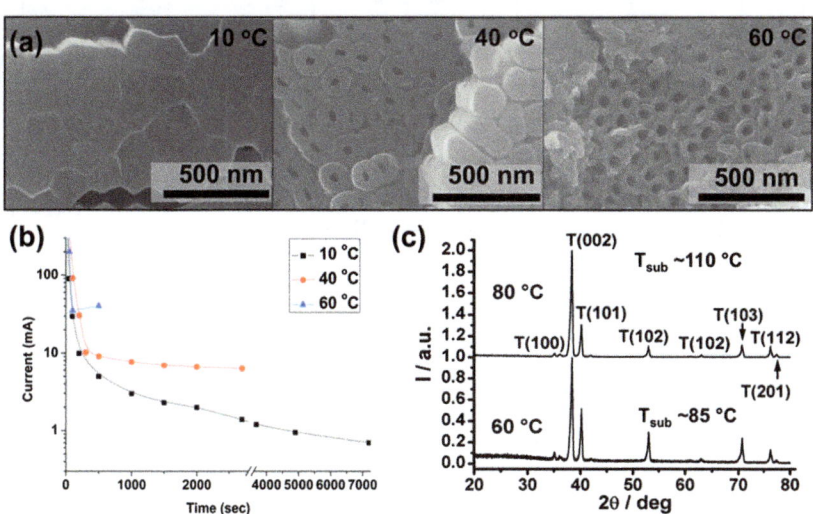

Fig. 3 (a) SEM top-view images of TiO$_2$ nanotubes formed in EG with addition of 0.1 M NH$_4$F and 1 M H$_2$O at different temperatures; (b) typical current-time (I–t) curves for anodization at different temperatures; (c) XRD spectra for as-formed TiO$_2$ nanotubes of 15 μm thickness anodized at different temperatures.

rapid tube growth. It is remarkable that at sufficiently low temperatures essentially the entire inner core of the tubes can be filled with the inner tube shell (Fig. 1). In turn, one could assume that in view of producing single walled (outer shell only) tubes, a sufficiently high electrolyte temperature could be a successful strategy. However, it turns out that in all our experiments, simply using elevated temperatures did not lead to a complete removal of the inner layer – at the bottom an inner layer was always clearly present. Also, extended waiting at elevated temperature after anodization (*i.e.* under open-circuit conditions) was not found to be a fully successful strategy. Tube layers that are exposed for a while under OCP conditions to the electrolyte tend to delaminate from the substrate by preferential etching at the tube oxide/metal interface. Also it is noteworthy that for EG tubes, even after anodization at comparably high electrolyte temperatures, no trace of crystallization of the oxide could be found (Fig. 4e).

Most remarkable, however, is that a change in the electrolyte can result in single walled tubes. Fig. 4 shows a comparison of EG tubes with tubes produced in a DMSO-based fluoride electrolyte. From the *I–t* curves (in Fig. 4a and b) it is clear that the EG electrolyte at room temperature delivers much higher current densities than DMSO; as a result TiO_2 nanotube growth is extremely slow in DMSO. However, if the temperature is elevated, the growth rate of the nanotubes can be accelerated strongly in both electrolytes. For DMSO at 60 °C the rate still does not reach EG at room temperature. However, DMSO tubes show a distinct

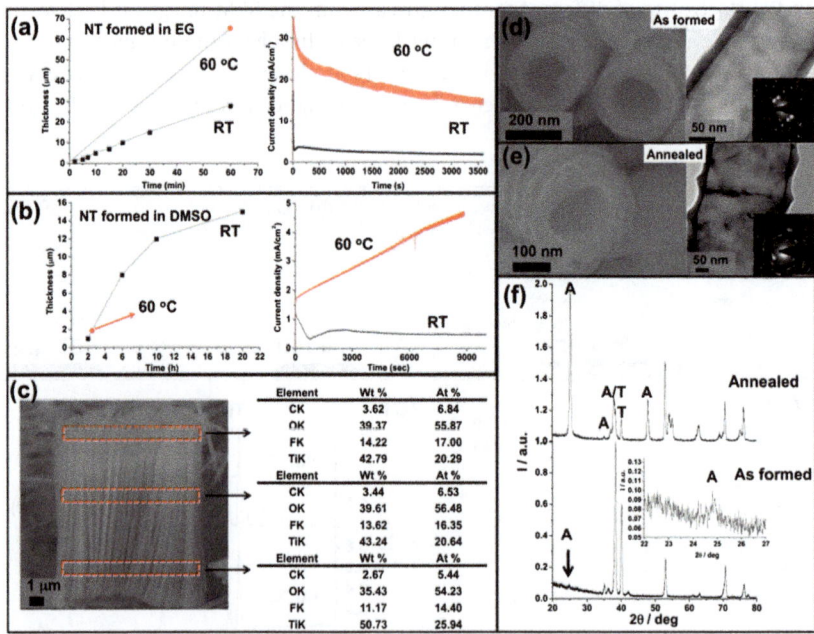

Fig. 4 (a, b) TiO_2 nanotube-layer thickness with anodization time and *I–t* curves for different electrolytes; (c) SEM image with EDX analysis for single walled tubes of 15 μm thickness formed in DMSO with addition of 2 wt% HF at room temperature at 3 different locations (inserted boxes); (d, e) SEM and TEM images of single walled tubes of 15 μm thickness formed in DMSO with addition of 2 wt% HF at room temperature before (d) and after (e) annealing at 450 °C; (f) XRD spectra for TiO_2 nanotubes of 15 μm thickness before and after annealing at 450 °C.

morphological feature that, in contrast to the EG electrolyte, in DMSO under all anodization conditions a single walled structure was obtained. Fig. 4c and d show typical SEM and TEM images for tubes formed in DMSO that demonstrate the absence of an inner shell. EDX measurements taken at different tube heights (Fig. 4c) confirm a uniform tube composition over the entire length. The detected carbon level is at the usual natural contamination level of nanostructured TiO_2.

Basically, from a comparison of the I–t curves one may assume that the slow ion flux in DMSO at room temperature is the cause of the single wall nature (assuming that the chemical dissolution process of the inner tube wall is rate determining and, due to a lower ion flux through the oxide, a sufficient solubility of the inner layer prevents precipitation). Nevertheless, when comparing results in pure DMSO, where at elevated temperature similar current densities as in EG at room temperature can be reached, it becomes clear that also in DMSO under comparable ion flux conditions no inner layer is formed.

An explanation for the different behavior in EG and DMSO may be given in terms of voltage induced radical and hole formation and capture as illustrated in Fig. 5. If an n-type semiconductor oxide, such as TiO_2, is polarized increasingly in the anodic direction, the Fermi level is lowered (towards higher redox-potentials) until it reaches the valence band edge (this when $E_{appl} \approx U_{fb} + E_g$, where E_{appl} is the effective applied voltage, U_{fb} is the flatband potential (for $TiO_2 \approx -0.4$ V vs. Ag/AgCl at neutral pH, and E_g is the bandgap of TiO_2 of ≈ 3 eV [i.e. at around 3–4 V vs. Ag/AgCl]); this leads to valence band ionization.[18] If the semiconductor is as highly doped (defective) as the anodic TiO_2 film, tunneling breakdown of the Schottky junction becomes possible.[19] The result is surface generated valence band holes (as in a photocatalytic case) which can react with the environment either directly with any species in the electrolyte, or e.g. with H_2O to form radical species ($H_2O \rightarrow OH$). For EG based electrolytes the decomposition products of radical or hole capture are a range of EG oxidation products and oligomerized EG.[20] These, in form of anions, will strongly adsorb on anodic TiO_2 and thus be incorporated by overgrowth in the OS. DMSO, on the other hand, is reported to capture radicals at least a factor of 10 faster than EG. Most importantly, after radical capture the reaction products differ significantly from the EG case[21–24] – please note that no sulfur can be detected in the EDX of Fig. 4c.

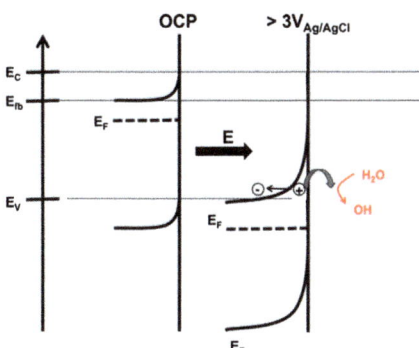

Fig. 5 Schematic illustration of the energetic situation at the semiconductor (TiO_2)/electrolyte interface. Sufficient voltage-induced band-bending creates radical and hole formation and capture.

Another interesting aspect is that in DMSO the as-formed tubes, anodized at room temperature, show some crystallization to anatase – this is evident from the SAD patterns shown in Fig. 4d, inset, and from the XRD in Fig. 4f. In XRD there is clearly a peak at ≈26°, this anatase peak is much weaker than for a sample fully annealed at 450 °C in air that shows complete crystallization, nevertheless it is clearly detectable. The origin of partial crystallinity clearly must be ascribed to the extremely slow growth of the DMSO tubes (3 days to grow a 20 μm thick layer), as tubes grown at 60 °C in DMSO to a comparable length do not show any trace of crystallinity.

The partial crystallinity of the slowly growing DMSO tubes has a strong effect on the electrical properties. Fig. 6 shows a comparison of conductivity measurements carried out with EG formed tubes and DMSO formed tubes before and after annealing. Clearly a difference in the conductivity of the as-formed tubes of ≈2 decades can be measured. After annealing, a less pronounced difference is measured but the single walled tubes obtained in DMSO still show a higher conductivity – this reflects a 'better' quality oxide obtained in DMSO. Supporting evidence for the better quality oxide can be obtained, if the photoelectrochemical response of the different tubes is tested. Meanwhile, for both annealed tube types the typical band-gap of anatase (≈3.1 eV) is obtained, and a clear difference in the photocurrent transients can be seen.

Fig. 7 shows a plot for the determination of an indirect band-gap for tubes formed in EG and DMSO, and single photocurrent transients. While the steady state magnitude of the photocurrent for both tubes is similar and the shape of the transients in both cases is characteristic of a trap-filling mechanism as, *e.g.*, described in ref. 25, a faster photocurrent increase/decrease kinetics of the DMSO-formed tubes is evident. The faster saturation and decay upon switching light on and off is indicative of a lower density of trap-states for the DMSO-formed

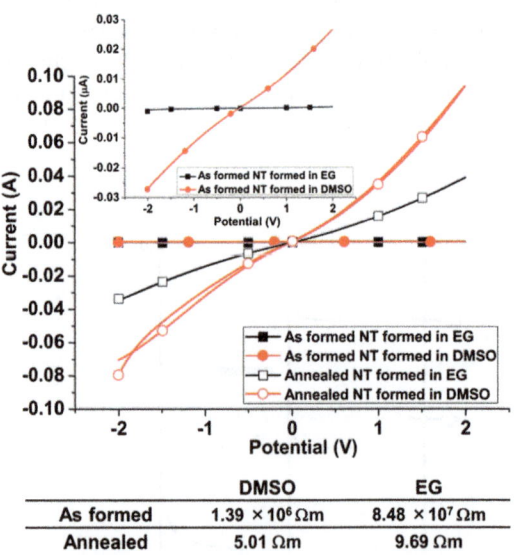

	DMSO	EG
As formed	$1.39 \times 10^6 \, \Omega m$	$8.48 \times 10^7 \, \Omega m$
Annealed	$5.01 \, \Omega m$	$9.69 \, \Omega m$

Fig. 6 Comparison of the electrical resistance of the single wall nanotubes of 15 μm thickness formed in DMSO with addition of 2 wt% HF, and double wall nanotube layers formed from EG with addition of 0.1 M NH$_4$F and 1 M H$_2$O at room temperature, before and after annealing at 450 °C.

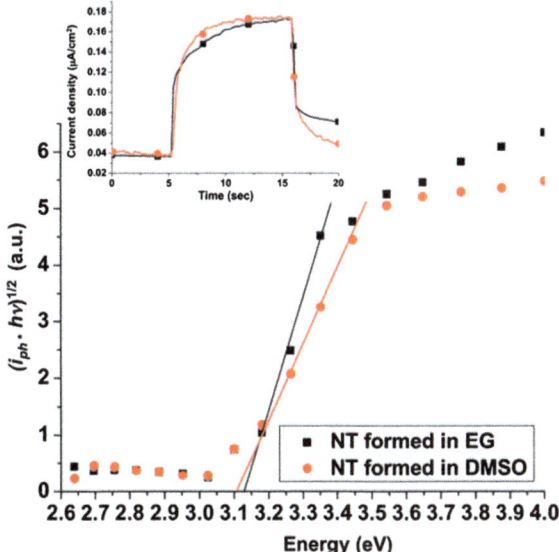

Fig. 7 Band-gap evaluation from photocurrent spectra for the single wall nanotubes of 15 μm thickness formed in DMSO with addition of 2 wt% HF and double wall nanotubes formed in EG with addition of 0.1 M NH_4F and 1 M H_2O at room temperature after annealing at 450 °C. Inset: photocurrent transients.

tubes. Both findings, electrical conductivity and phototransients, indicate that removal of the inner contamination layer leads to beneficial electrical properties in the tubes. This bears potential for adjusting the electronic properties for any application of TiO_2 nanotubes, where charge transport within the tubes is crucial, such as solar cells, photocatalysis, and sensing. Moreover, the fact that single walled tubes represent geometrically a better-defined tube has an impact on applications such as drug delivery systems or size exclusion membranes.

4 Conclusions

The present work investigates the finding that anodic TiO_2 nanotubes can consist of an inner and an outer shell of an oxide layer ('double walled'), or can consist of the outer shell only ('single walled'). We also show how to influence the extent of double layer formation by solvent, temperature and additives, and describe approaches to reach a single wall tube morphology. The origin of the double walled tubes is conceived to be inherent to the TiO_2 anodization process – *i.e.* the formation of a 'pure' oxide layer by O^{2-} inward migration and a layer grown by Ti^{4+} outward migration. The latter layer, in EG based electrolytes, is highly loaded with carbon species from electrolyte decomposition. In DMSO electrolytes, the formation of this contaminated layer can be fully suppressed. This is ascribed to the much lower tendency of DMSO to form deposition products that can be incorporated in the anodic oxide and the complete removal of the outward migration layer. In DMSO, tube growth at room temperature is very slow – *i.e.* it is sufficiently slow that partial crystallization of the tubes can occur. This has a drastic effect on the electrical conductivity of the tubes. In fact, one may speculate to what extent the conductivity of the tubes in general may be adjusted by adjusting the growth rate. Usually TiO_2 nanotubes layers are reported to be amorphous, however the present findings are in line with early work on TiO_2 anodization reporting the feasibility to induce crystallinity by slow growth and that fluorides can influence significantly the crystallization behavior of TiO_2 layers.[26] Overall, the ability to form more defined TiO_2 nanotubes (contamination layer free) clearly is a step towards improved properties and functionality of self-organized nanotube layers – not only in view of electron-transport dominated applications but also simply because a much more defined inner tube diameter becomes accessible.

Acknowledgements

The authors would like to greatly acknowledge DFG and the DFG Cluster of Excellence (EAM), for financial support.

References

1 P. Roy, S. Berger and P. Schmuki, *Angew. Chem., Int. Ed.*, 2011, **50**, 2904.
2 I. Paramasivam, H. Jha, N. Liu and P. Schmuki, *Small*, 2012, **8**, 3073.
3 S. Bauer, P. Schmuki, K. Mark and J. Park, *Prog. Mater. Sci.*, 2013, **58**, 261.
4 V. Zwilling, M. Aucouturier and E. Darque-Ceretti, *Electrochim. Acta*, 1999, **45**, 921.
5 G. K. Mor, O. K. Varghese, M. Paulose and C. A. Grimes, *Adv. Funct. Mater.*, 2005, **15**, 1291.
6 R. Beranek, H. Hildebrand and P. Schmuki, *Electrochem. Solid-State Lett.*, 2003, **6**, B12.
7 K. Nakayama, *et al.* at the 208th ECS Meeting, California, Abstract 819 and 843, 2005.
8 R. Hahn, J. M. Macak and P. Schmuki, *Electrochem. Commun.*, 2007, **9**, 947.
9 R. Hahn, T. Stergiopoulus, J. M. Macak, D. Tsoukleris, A. G. Kontos, S. P. Albu, D. Kim, A. Ghicov, J. Kunze, P. Falaras and P. Schmuki, *Phys. Status Solidi RRL*, 2007, **1**, 135.
10 S. P. Albu, A. Ghicov, S. Aldabergenova, P. Drechsel, D. LeClere, G. E. Thompson, J. M. Macak and P. Schmuki, *Adv. Mater.*, 2008, **20**, 4135.
11 S. E. John, S. K. Mohapatra and M. Misra, *Langmuir*, 2009, **25**, 8240.
12 H. Mirabolghasemi, N. Liu, K. Lee and P. Schmuki, *Chem. Commun.*, 2013, **49**, 2067.
13 J. W. Schultze and M. M. Lohrengel, *Electrochim. Acta*, 2000, **45**, 2499.
14 K. R. Hebert, S. P. Albu, I. Paramasivam and P. Schmuki, *Nat. Mater.*, 2011, **11**, 162.
15 P. Skeldon, G. E. Thompson, S. J. Garcia-Vergara, L. Iglesias-Rubianes and C. E. Blanco-Pinzon, *Electrochem. Solid-State Lett.*, 2006, **9**, B47.
16 K. G. Singh, A. A. Golovin and I. S. Aranson, *Phys. Rev. B: Condens. Matter Mater. Phys.*, 2006, **73**, 205422.
17 K. Zhu, N. R. Neale, A. Miedaner and A. J. Frank, *Nano Lett.*, 2007, **7**, 69.
18 Y. Y. Song, P. Roy, I. Paramasivam and P. Schmuki, *Angew. Chem., Int. Ed.*, 2010, **49**, 351.
19 S. M. Sze, *Physics of Semiconductor Devices*, John Wiley & Sons, 1981.
20 H. Asoh and S. Ono, *Mater. Sci. Forum*, 2003, **419–422**, 957.
21 L. M. Dorfman and G. E. Adams, *Reactivity of the Hydroxyl Radical in Aqueous Solutions*, U.S. Dept. Commerce, Natl. Bureau Standards #46, NSRDS-Washington, D.C.,1973.
22 W. Rosenblum, *Ann. N. Y. Acad. Sci.*, 1983, **411**, 110–119.
23 B. Wieland, J. P. Lancaster, C. S. Hoaglund, P. Holota and W. J. Tornquist, *Langmuir*, 1996, **12**, 2594.
24 K. S. Raja, T. Gandhi and M. Misra, *Electrochem. Commun.*, 2007, **9**, 1069.
25 R. P. Lynch, A. Ghicov and P. Schmuki, *J. Electrochem. Soc.*, 2010, **157**, G76.
26 J. Kunze, A. Seyeux and P. Schmuki, *Electrochem. Solid-State Lett.*, 2008, **11**, K11.

Faraday Discussions

PAPER

The simplest model of charge storage in single file metallic nanopores

Alexei A. Kornyshev*

Received 4th March 2013, Accepted 22nd March 2013
DOI: 10.1039/c3fd00026e

The problem of voltage controlled accumulation of ions in a narrow nanopore, which can accommodate just one row of ions of an ionic liquid and is filled with ions when the electrode is unpolarised, is mapped on an exactly solvable one-dimensional two state Ising model. Analytical solution of this, presumably simplest, statistical mechanical model reveals the dependence of the electrical capacitance on voltage, pore radius, and temperature. The voltage dependence of capacitance has the character of a smeared resonance, whose position and height is affected by a tiny change of the pore radius. Consequently, even the slightest dispersion of pore radii in the whole electrode, unavoidable in any real system, softens the voltage dependence.

Introduction

Electrochemistry has matured as a physical discipline from the studies of electrical double layer and electrode kinetics at well defined electrodes. In the first half of the 20th century these were liquid metals, such as Hg, with a pure, smooth, self healing atomically flat surface.[1] Such metals were not of great interest for practical applications, and the next big step forward, in the 1970s, was moving to another class of well characterised electrodes – single crystal electrodes, predominantly of noble metals, such as Pt, Ag, and Au.[2] This is how 'electrochemical surface science' was born[3,4] and a number of elementary processes at the electrode–electrolyte interface were understood.

On the other hand, because all the events in any industrially important electrochemical processes normally take place at the electrode–electrolyte interface, one generally tends to maximise the interfacial area. In electrochemical engineering, since the middle of the previous century, the use of 'volume filling' surfaces, achievable for highly porous electrodes, was mainstream for batteries, fuel cells, or supercapacitors.[5–8] Later, the discovery of various forms of carbon has opened new horizons for engineering nanoporous electrodes of well controlled volume-filling structures, with the surface area many orders of magnitude larger than the projection area.[9,10] For instance, using such electrodes

Department of Chemistry, Faculty of Physical Sciences, Imperial College London, London SW7 2AZ, United Kingdom. E-mail: A.Kornyshev@imperial.ac.uk

for supercapacitors, where the electrical energy is stored in a form of a charge accumulated in electrical double layers, allows to scale up energy storage proportionally to the hugely enhanced interfacial area.[11,12]

A standard way to increase the electrode–electrolyte interface is to decrease the characteristic size of the pores, still accessible to the electrolyte, simultaneously decreasing the thickness of pore walls – altogether increasing the porosity of the electrodes. It was recently found, moreover, that the capacitance *per true unit surface area* increases with pore size, down to the size below which ions can no longer penetrate pores. The effect has been observed both for solvated ions in organic solvents[13] and solvent-free ionic liquids.[14] Although the existence of this effect has been recently questioned[15] on the grounds of possible inaccuracies of the determination of the true internal surface area,[16] the common opinion is that, although there may be some variance in the strength of this effect, qualitatively it will remain valid, *i.e.* the experiments of ref. 13 and 14 demonstrate a real phenomenon, but not an artifact [*cf.* also ref. 17 and 18].

Explanation of this effect was suggested in ref. 19 and 20 on the basis of the following idea. The Coulomb interaction inside a nanoscale pore in an electronically conducting material is exponentially screened, with the decay length proportional to the pore width or radius. The exact laws are different for different pore geometries, *e.g.* cylindrical or slit-like, but the qualitative consequences of such enhanced screening are the same. Pair interaction energies for large ionic liquid ions (or solvated inorganic ions) in a pore of a diameter that is only slightly larger than the diameter of an ion (plus its solvation shell, if applicable) will be just about a couple of k_BT for the nearest neighbors. Interaction of the next nearest neighbors will be less important, and negligible for narrow nanopores. Such suppression of Coulomb interactions allows unbinding of cation–anion pairs packing inside the pore of counterions of predominantly one sign, when polarising the electrode. The findings of ref. 19 based on a mean-field theory and obtained for slit pores have been verified by computer simulations[20] performed in the same geometry of the pores. In the simulations, some features of the capacitance–voltage dependence appeared to be smoothened, but otherwise these simulations approved the qualitative consequences of the enhanced screening in metallic nano-confinement. Some of those findings (effect of pore radius) have been recently confirmed by more involved atomistic molecular dynamic simulations.[21]

The state of ions in the pore in which their Coulomb interactions are replaced by Yukawa-like interactions[22,23] with the decay range proportional to but a few times smaller than the pore width was called a '*superionic state*'.[19,20] The term depicted the ability of ions of the same sign to stay close to each other due to the interaction with polarisable metallic pore walls. Consequently, such a state allows packing into a pore more ions of the same sign for a given voltage, enhancing thereby the capacitance of the pore per unit surface area. With the increase of the pore width, the Coulomb interactions become less screened and the capacitance goes down, as in the experiments of ref. 14.

If one further increases the pore size, a second layer of counterions will be able to sneak into the pore, and this will increase the capacitance. However, further increasing the width of a slit or radius of a cylindrical pore before letting a third layer/line to build up inside the pore will tend to decrease the capacitance. This is because the screening of electrostatic interactions will, as explained above, get weaker and the capacitance will start to go down again. And off these alternations go for a while, leading to decaying 'oscillations' of capacitance as a function of the pore size, as was first found in ref. 24–26.

The voltage dependence of the specific capacitance of nanoporous electrodes is a separate important issue that was first analyzed theoretically in ref. 19, 20 and 27. The results revealed a dramatic dependence of the capacitance–voltage curves on the nanopore width. This effect was later shown to have important consequences for the choice of the pore size for maximising energy storage.[28]

On the other hand, the power of the capacitor is determined by how fast we can charge/discharge the pore. Energy capacity and the speed of charging are generally competing with each other.[29] Energy storage scales up with the surface area. The surface area increases with the length of the pores and with the decrease of their radius, as one may then have more pores per projection surface area. But the longer the pores, the more time it will take to transport the ions to fully charge them. Energy storage devices are characterised by Ragone plots: the deliverable power *vs.* the storable energy, and relative to batteries, modern supercapacitors lie on this diagram towards higher power, but lower energy. In batteries, the rate-limiting step is, in most cases, the elementary act of electrochemical reaction. Because of that, they charge and discharge much slower than supercapacitors, in which the reaction stage is absent and where the end task is gathering the counterions at the corresponding electrode; in the case of a porous electrode, inside that electrode. With that picture in mind, the baseline for the development of supercapacitors is the increase of the energy storage per unit volume/weight of the device at, however, minimal losses in the power delivery.

Leaving the kinetics of nanoporous electrodes for a separate study[30,31] let us concentrate on the electrical capacitance of porous electrodes, important for the energy capacity.[29,28] Earlier reported models and simulations[19–21,24–28] were instrumental for understanding the capacitance behavior of nanoporous electrodes, but the studies based on them are numerical in implementation. Can one work out some simple model that could lead to a *transparent, analytical, tutorial textbook-like formula*, which would illuminate the main features of the capacitance dependence on voltage and the pore-size dependence? Such a model is possible for a particular case which has its own, although limited, relevance. Namely, for the capacitance of ultra narrow quasi single-file pores, in which the ions may be assumed to settle at or close to the main axis of a pore that can accommodate just one row of ions, whose arrangement in the pore can be considered as one dimensional, as sketched in Fig. 1. We present this model for the case of a pure, solvent-free ionic liquid, which will be restricted even further, subject to the assumptions formulated below.

The model

Basic assumptions

1. There is a propensity for cations and anions to occupy the pore in the electrode, even when the latter is not polarised. When it becomes polarised, co-ions will tend to be replaced by counterions.[32]

2. This propensity for ions to fill the pore is so strong that a non-polarised pore is densely packed with ions, with no empty voids between them. This simplification may not generally be valid; the more general case is mathematically more sophisticated and is to be studied.

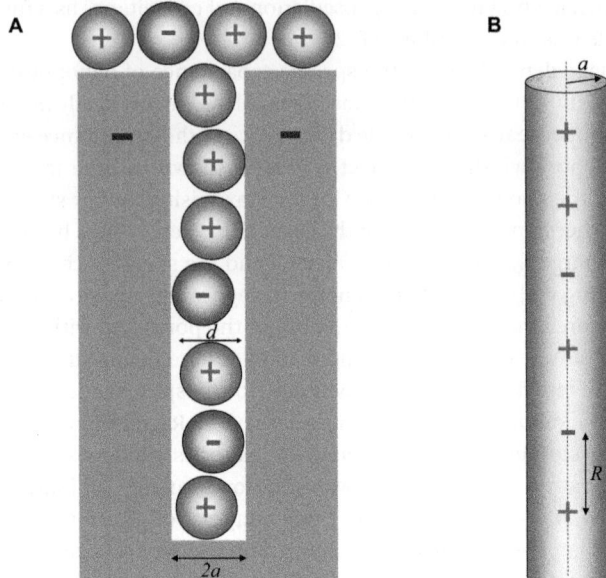

Fig. 1 A 'single-file' pore in a (negatively) polarised electrode: (A) a schematic cross-section, (B) a one dimensional lattice representation of distribution of ions. The pore is assumed to accommodate, roughly, one line of ions, ($d < 2a < 2d$).

3. Electrostatic interactions of ions in the pore are screened by the electronic polarisation of pore walls, and only the interactions of the nearest neighbors must be taken into account.

The study of this simple case results in a compact analytical formula for capacitance, which has a number of lessening consequences. Before formulating the model mathematically, we need to substantiate and discuss the last assumption.

Interionic interaction potential

Electrostatic interaction between charges in a cylindrical pore inside a metal is a classical problem of electrostatics, solved long ago (see *e.g.* a textbook[33]). For two unit point charges $U(R)$ sitting on an axis of a pore of radius a at a distance R from each other, the interaction energy scaled to the thermal energy, $k_B T$, is given by

$$\frac{U(R)}{k_B T} = \pm \frac{L_B}{R} \left\{ 1 - \frac{2}{\pi} \int_0^\infty \frac{dx}{I_0(x)^2} \frac{\sin[(R/a)x]}{x} \right\}. \quad (1)$$

Here $\pm$ corresponds to the interaction of ions of the same or opposite sign, respectively. L_B is the Bjerrum length; in Gaussian units used throughout the paper, $L_B = e^2/\varepsilon k_B T$, where e is the elementary charge and ε the effective dielectric constant inside the pore due to electronic polarisability of ions (if we assume the latter to be ≈ 2, at room temperature $L_B \approx 28$ nm). $I_0(x)$ is the modified cylindrical Bessel function of zero order. Using residue theory, the integral in the r.h.s. was calculated exactly to give the result in the form of a series,

$$\frac{U(R)}{k_B T} = \pm 2 \frac{L_B}{a} \sum_{m=1}^{\infty} \frac{e^{-\frac{k_m R}{a}}}{k_m [J_1(k_m)]^2}. \quad (2)$$

Here $J_1(k_m)$ is the first order cylindrical Bessel function and k_m are zeros of Bessel function of zero order, $J_0(k_m) = 0$; hence $k_1 = 2.4; J_1(y_1) = 0.52; k_2 = 5.52, J_1(y_2) = -0.34, \ldots$[34] Notably, for $R > a$ it is sufficient to keep only the first term of the series,

$$\frac{U(R)}{k_B T} \underset{R>a}{\approx} \pm 3.08 \frac{L_B}{a} \exp\left\{-\frac{R}{(a/2.4)}\right\}. \quad (3)$$

Fig. 2 shows that this approximation works amazingly well. Thus, interaction in a cylindrical pore with metallic walls decays exponentially with the decay length $a/2.4$.

It is also interesting to compare the interaction potentials in a cylindrical pore and the one in a flat gap (employed in the theory of charge accumulation in slit pores[19,20]). Fig. 3 shows this comparison using the asymptotic laws, perfectly sufficient for this study.

In both cases the screening is exponential, but in a cylindrical pore it is stronger than in a slit pore, because in a cylindrical pore the interacting charges are surrounded by metal 'from all sides'.[35] In a cylindrical pore, the decay range is $1/4.8$ of the pore diameter, whereas in a slit pore it is $1/3.14$ of the pore width.

Exponential screening in a nanogap or nanopore is a result of the infinite set of image charges that counterbalance the charges inside the nanopore. The importance of this effect in the theory of adsorption at a flat electrode has been studied decades ago (for review see ref. 1); their role in the double layer theory at

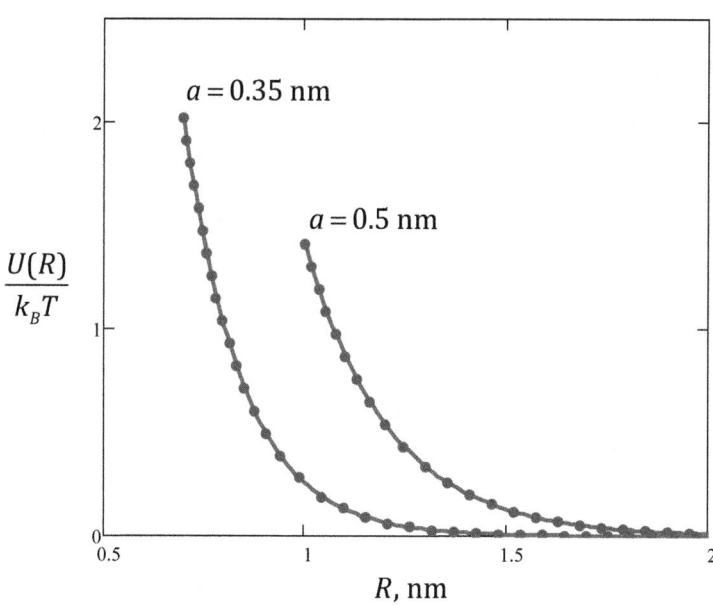

Fig. 2 Interaction potential scaled to thermal energy: comparison between the exact eqn (1) (lines), and its asymptotic form (dots), eqn (3). The graphs were plotted for two indicated pore radii, a, with the absolute values determined by $L_B \approx 28$ nm. For each of the two cases the results are shown down to distances of the pore diameter, $R = 2a$.

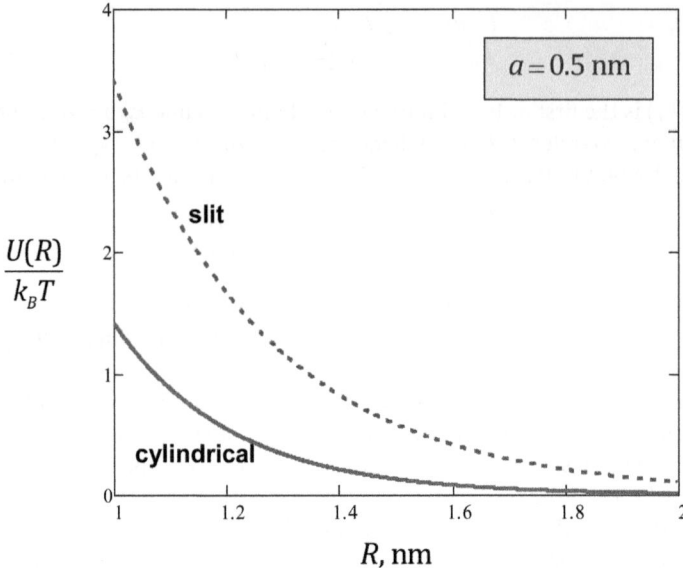

Fig. 3 Comparison of the interaction potentials in a cylindrical pore of radius a and a slit pore of half-width a. The plots calculated using eqn (3) for cylindrical and $\frac{U(R)}{k_BT} \approx 2\frac{L_B}{\sqrt{aR}} \exp\left\{-\frac{R}{2a/\pi}\right\}$ for slit pore[19] geometry.

flat interfaces has recently been emphasised and explored in ref. 36 and 37. Note that the effect of image charges in gap geometry is much stronger than at a flat interface. Indeed, a charge located at distance a near a flat interface, creates one mirror image of opposite sign in the metal at point $-a$; the charge and its image create together a potential which at long distances $R > a$ takes the form of an effective dipole-type potential $e\frac{2a^2}{\varepsilon R^3}$ instead of the Coulomb potential $e\frac{1}{\varepsilon R}$. It is the multiple set of image charges that emerge in the metal in a pore geometry that provide a much stronger, exponential screening.

Effects of the field penetration into the massive metal plates, which we did not touch above, has been intensively studied in the past both at the interface and in the flat gap geometry (see e.g. ref. 1 and 38–40). Unless $\varepsilon \gg 1$, all these effects effectively extend distance a, but otherwise they will lead to very nontrivial consequences.[38,41,42] In our problem, $\varepsilon \approx 2$ and such consequences will be of no or minor importance.

Mapping the problem on a 1-dimensional Ising model

Considering the dense packing of cations and anions of the same size, we may assume the 1d lattice constant to be equal to their diameter. Had they been of different size, the lattice constant would be voltage dependent. For instance, in a typical case of ionic liquids with cations larger than anions, when the pore is occupied predominantly by cations, the lattice constant will be larger than when the pore is packed with anions. We will not consider this complication in this study.

The idea of the model is sketched in Fig. 4. It uses the lexicon of spin models, standard in statistical mechanics.[43] When the site i of the 1d lattice is occupied by a cation it is assigned the value of 'spin' $S_i = 1$, when it is an anion, $S_i = -1$ (as mentioned in the introduction of the model, the sites are assumed to be always occupied by either cations or anions).

The Hamiltonian of the system (in the units of $k_B T$) can be written in terms of a one-dimensional Ising model with nearest neighbour interaction between 'spins' in the presence of an external 'field'[41]

$$\frac{H}{k_B T} = \sum_i \left\{ \frac{\alpha}{2}(S_i S_{i+1} + S_i S_{i-1}) + u S_i \right\}, \quad (4)$$

where the coupling constant characterising interactions of the neighbouring 'spins'

$$\alpha = |U(d)|/k_B T \quad (5)$$

with d, the ion diameter, taken the lattice constant.

Subject to eqn (3) $U(2d) \approx U(d)\exp\{-2.4\ d/a\}$. Since in the case we are considering in our model, $d < 2a < 2d$, i.e. the pore can accommodate ions but only one row of ions, the assumption of the nearest neighbour interactions is well justified.

The first item under the sum in eqn (4) favors cations and anions to neighbour each other. The last term, u, is the electrostatic potential drop between the electrode and the bulk of electrolyte, again taken in the units of $k_B T/e$ (at room temperature $k_B T/e \approx 25.6$ mV). The value of u is constant across and along the pore. This term favours occupation of the sites in the pore by counterions and expulsion of co-ions from the pore.

Recall that in the theory of magnetism, α (often denoted there by letter J) is the so called spin-coupling parameter, which characterises interactions of real magnetic spins of atoms; $\alpha > 0$ favours antiferromagnetic order, whereas $\alpha < 0$ corresponds to ferromagnetic order; u characterises the interaction of spins with the external magnetic field.[43] In our problem $\alpha > 0$, because the anion–cation

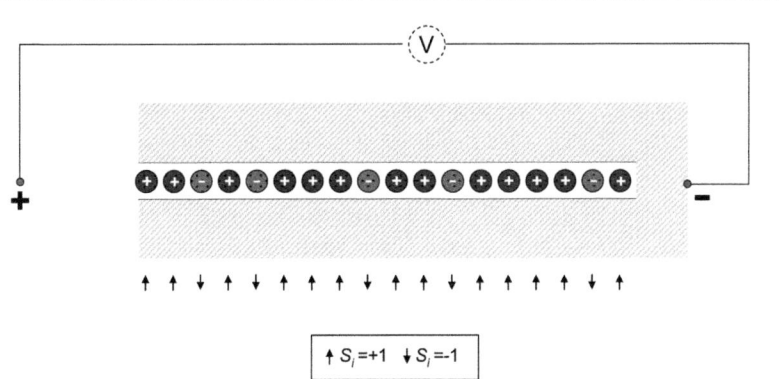

Fig. 4 1d spin description of the occupation of a cylindrical pore. Each site of a 1d lattice is occupied either by a cation (the state with spin +1) or anion (the state with spin −1). Voltage is controlled relative to the exterior of the pore. Nonzero polarisation of the electrode favors one of the spin-states over the other for each site. In this sketch the electrode is negatively polarised and spin +1 states are more favourable. The temperature tends to smear up this trend favoring equal distribution of spin +1 and spin −1 states.

pairs of neighbours are favoured (analogy with anti-ferromagnetism); a nonzero voltage u competes with that trend, at very large voltage the pore will contain only ions of one sign – the counterions. Note that in our problem u is the voltage, not electric field.

If the ions have different propensity to settle in an unpolarised pore due to specific interaction with the walls depending on the sort of ions, u must be replaced by

$$u \Rightarrow \tilde{u} = u - w \tag{6}$$

where $-w$ is half of the difference between the free energy of transfer of a cation and of an anion from the bulk into the interior of the pore, scaled to $k_B T$. $w > 0$ would mean preferential adsorption of cations.

The value of w will be affected by the difference in the size of cations and anions, but it may be nonzero even if they are of the same size, but of different chemical nature. Consideration of ions of different size will require a more sophisticated model, because the size of the ions enters the model only through the voltage-independent lattice constant. The simplest extension of the model will be making the lattice constant voltage-dependent, as the distance of closest approach between cations, between anions, and between cations and anions may be different. The corresponding extension of the model will be considered elsewhere. Here, we will still adopt correction (6) reminding ourselves that the value of w may be, generally, nonzero, even for the simplest case considered below.

Exact solution

General equations of the model. The expression for differential capacitance

The 1d Ising model with nearest neighbour interactions in an external field has a simple exact solution described in textbooks. Namely, the statistically averaged value of 'spin' is given by,[44]

$$\langle S_i \rangle = -\frac{\sinh(u-w)}{\sqrt{[\sinh(u-w)]^2 + e^{4\alpha}}} \tag{7}$$

The derivative

$$\chi(u) = \frac{d\langle S_i \rangle}{du} = -\frac{e^{-2\alpha}\cosh(u-w)}{\left\{1 + e^{-4\alpha}[\sinh(u-w)]^2\right\}^{3/2}} \tag{8}$$

characterises the response function. Whereas in magnetism χ is proportional to magnetic susceptibility, in our system this derivative characterises the response to the electrode polarisation. This response function determines the voltage-dependent differential capacitance. Given per its unit surface area it is called *specific* differential capacitance, and it is related to $\chi(u)$ as[45]

$$C(u) = -\frac{\varepsilon L_B}{2\pi ad}\chi(u). \tag{9}$$

Following eqn (3) and (5), the coupling constant can be approximated by

$$\alpha \simeq 3.08 \frac{L_B}{a} e^{-2.4d/a}. \tag{10}$$

Linear/low voltage response

As a particular case of the above expression for capacitance, one can obtain its value close to the voltage when the row of ions in the pore is electroneutral as a whole:

$$C(u)|_{u=w} \equiv \overline{C} = \frac{\varepsilon L_B}{2\pi a d} e^{-2\alpha} \simeq \frac{\varepsilon L_B}{2\pi a d} \exp\left\{-6.16 \frac{L_B}{a} e^{-2.4d/a}\right\}. \tag{11}$$

When $w \neq 0$, but the electrode is not polarised

$$C(u)|_{u=0} = \overline{C} \frac{\cosh(w)}{\left\{1 + e^{-4\alpha}[\sinh(w)]^2\right\}^{3/2}}. \tag{12}$$

The two expressions coincide when there is no preferential adsorption of ions into the pore ($w = 0$).

Understanding absolute values (scaling factor estimates)

If all the lengths in the dimensional pre-factor $\frac{\varepsilon L_B}{2\pi a d}$ in eqn (9) or (11) are taken in nm, in order to get the capacitance in µF/cm² one must multiply the result by 11.11.

To get a feeling about an order of magnitude of this prefactor, let us take $\varepsilon = 2$, $L_B = 28$ nm, $a = 0.35$ nm, and $d = 0.7$ nm. We then obtain 36.4 nm^{-1} or 404.37 µF/cm². This example corresponds to $\alpha = 2.04$, and the exponential term $e^{-2\alpha}$ will contribute a factor 0.0169 reducing the capacitance to 6.82 µF/cm², but a slight increase of α will make this value much smaller.

It is interesting to compare $C_0^{\text{(cylindrical)}} = C(u)|_{u=w=0} = (\varepsilon L_B/2\pi a d)e^{-2\alpha}$ with the linear response specific compact layer capacitance of a flat electrode, in Helmholtz approximation,[1] $C_0^{\text{(flat)}} = \varepsilon/(2\pi d)$. Their ratio, $C_0^{\text{(cylindrical)}}/C_0^{\text{(flat)}} = [L_B/a]e^{-2\alpha}$, for the above set of parameters comprises 1.4. For larger pore radii, α will increase, and the ratio will rapidly become much smaller than 1.

Thus the nanopore capacitance is extremely sensitive to the interionic interaction parameter, affected by the radius of the pore and screening properties of the pore walls if the latter are not ideally metallic.[1] On the contrary, the limit of the classical Helmholtz model is valid for the case of a highly concentrated electrolyte, which corresponds to 'zero' Gouy length;[1] there, everything is screened (electric field is zero beyond the compact layer), and ion–ion interactions do not enter the expression for the capacitance.

Results and discussion

Pore size effect and the voltage dependence

The shape of the function $\chi(u)$ is determined by the coupling parameter α, the whole function being symmetric about the point $u = w$. In the illustration below we show the case of $w = 0$, but if different, the zero point on the abscissa would correspond to $u = w$.

Following eqn (3) and (5), for $d = 0.7$ nm, and $a = 0.35$ nm, $\alpha = 2.04$; with the pore radius increasing just by 0.2 nm the coupling constant increases more than twice, to $\alpha = 4.622$. The larger the value of α, the higher the voltages required to charge the pore. The graphs in Fig. 2 illustrate this. We see that when the pore diameter is just a little larger than the diameter of an ion, the capacitance becomes almost negligible for small voltages. The reason for this is clear: it requires some voltage to start accepting the charge into the pore.

Non-monotonic capacitance–voltage dependence lies in two competing trends dominating in different voltage domains. With initial increase of the voltage, one unbinds cation–anion pairs, progressively facilitating charge accumulation, and the capacitance increases. However, at further voltage increase the majority of co-ions will get expelled outside of the pore and be replaced by counterions. Indeed, in eqn (7), at $u \to \pm\infty$, $\langle S_i \rangle \to \mp 1$, and, following eqn (8) and (9), the response function $\chi(u)$ and capacitance $C(u)$ vanish $\propto \exp(-2|u|)$, i.e. approaching saturation, further charging becomes increasingly difficult. The two regimes are separated by the maxima of capacitance, which lie close to $u_{\max} \approx w \pm (2\alpha + \frac{1}{2}\ln 2)$; at this voltage $\langle S_i \rangle \approx \mp 1/\sqrt{2}$, i.e. $\approx 50\%$ of the maximal charge is already accumulated in the pore.

Recall again that in Fig. 2 we have shown the case $w = 0$. Had it been different, the curves would have been shifted by the value of w (i.e. to the right when $w > 0$, or to the left when $w < 0$).

Temperature dependence

Temperature appears in the capacitance expression (eqn (8) and (9)) in several places: the values of voltage, u is scaled to the, so called, 'thermal voltage' $k_B T/e$ (25.6 mV at room temperature), whereas the energy of interionic interaction, α, and the free energy of preferential adsorption, w, are both scaled to thermal energy $k_B T$. To make things even more complicated, the temperature dependence is also inherent in the free energy of preferential adsorption through the entropic term. If, however, we neglect the latter (which is admissible for the case of strong specific adsorption of both cations and anions, the keystone of the two state model), the resulting trends appear to be simple (cf. Fig. 6). With the temperature increase the capacitance peaks become lower and broaden; the capacitance at low polarisation increases, because the structures that the voltage needs to break down in order to charge the pore get softer. The overall effect is noticeable but is not and cannot be large, because in the nanopore the pertinent 'energies' are larger than $k_B T$. The saturation regimes (the wings) are unaffected.

The effect of pore radii distribution

The sharp maxima in Fig. 5 and 6 refer to a single pore capacitance. But even the slightest dispersion of pore radii has a dramatic effect on the capacitance plots, due to the resonance character of a single pore capacitance–voltage dependence. Fig. 7 and 8 show that for any physically meaningful dispersion of pore radii, the values of the capacitance in the maxima reduce from anomalously large to more 'reasonable' values, i.e. closer to the observable ones. The second message is that the wings decrease much less steeply, and the overall capacitance–voltage dependence becomes less 'dramatic'.

Pore size and temperature effects on the linear response capacitance

It is remarkable that in a single file pore, the capacitance close to $u = w$ (or $u = 0$ when $w = 0$) appears to be very small, unless the interionic interactions are very strongly screened. The effect of the pore size that eqn (11) predicts (Fig. 9) is qualitatively similar to that experimentally observed,[14] but at pore radius only slightly larger than the ion radius, the single file capacitance practically vanishes. To the best of our knowledge, such small values of specific capacitance have never been reported. This is different in the case of slit pores,[19,20] where non-polarised pores delivered substantial capacitance. Thus this can mean that in

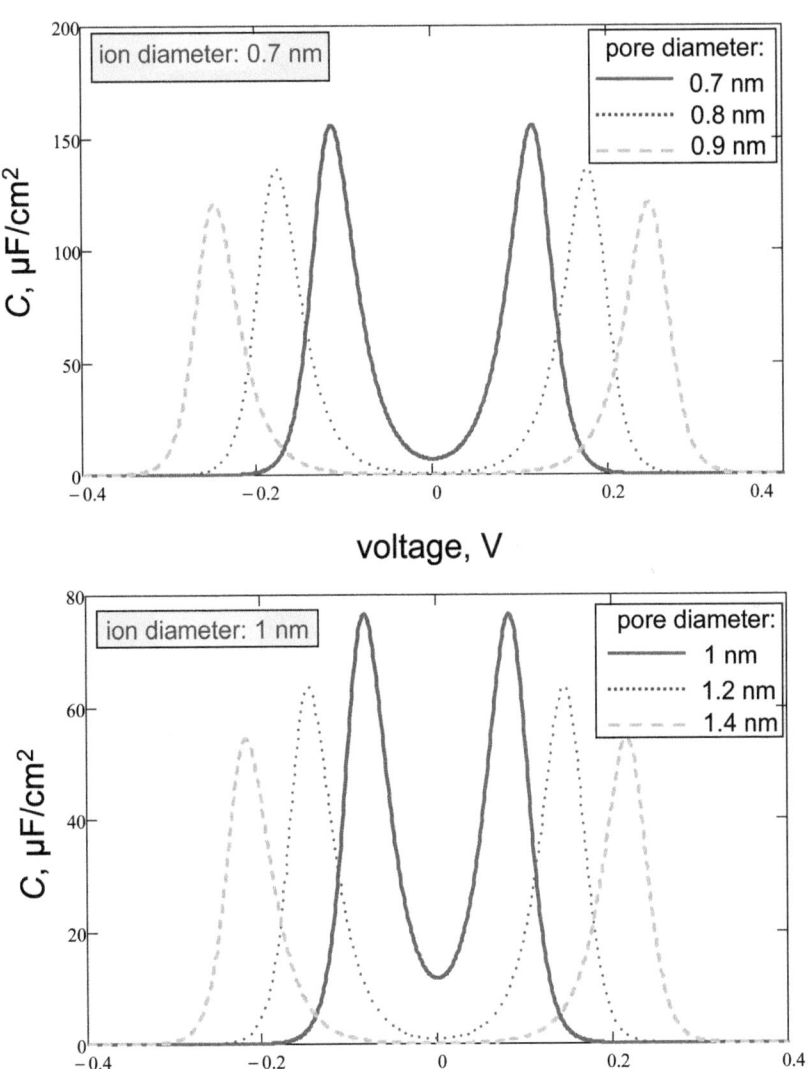

Fig. 5 Ising model results for the specific voltage-dependent differential capacitance of a single pore per its unit surface area, calculated via eqn (8)–(10). The effect of the diameter of the pore, shown for ion diameter 0.7 nm (upper panel) and 1 nm (lower panel); the effective dielectric constant of the pore interior, $\varepsilon = 2$. Remarkably, the capacitance vanishes very quickly with voltage because of the saturation of charge density in a single file geometry.

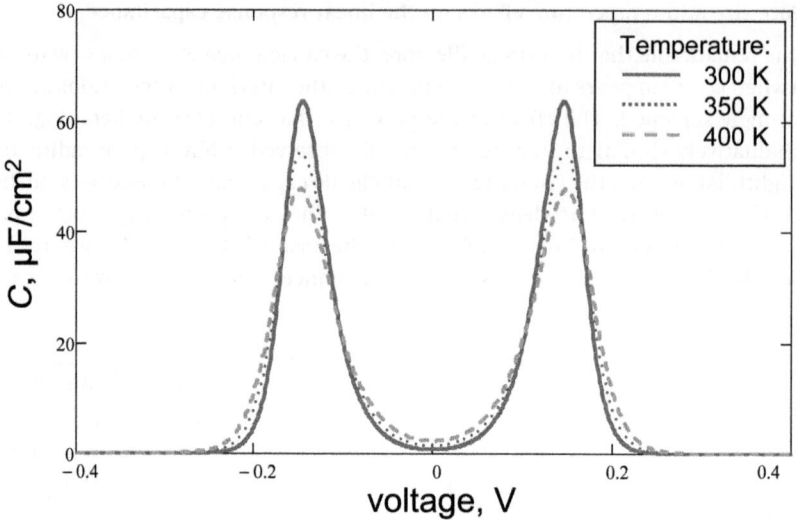

Fig. 6 Temperature effect on the voltage dependence of the specific differential capacitance of a single pore: calculated via eqn (8)–(10), shown for ion diameter 1 nm and pore diameter 1.2 nm. As in Fig. 5, $\varepsilon = 2$.

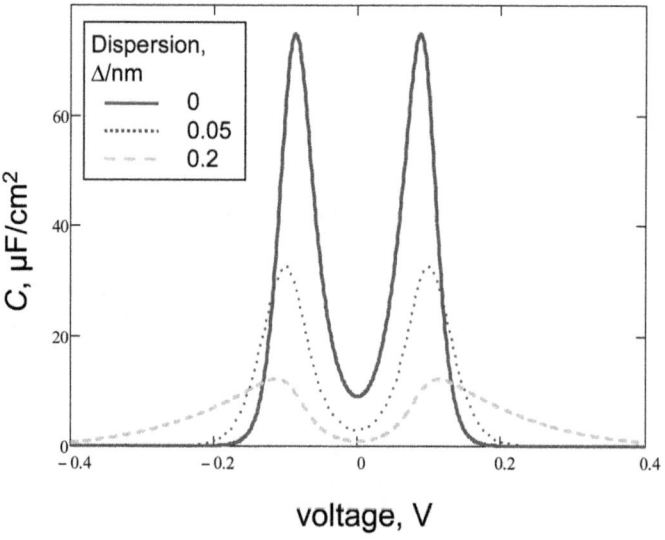

Fig. 7 Specific differential capacitance of a narrow cylindrical pore at dense packing of ions, just about to accommodate one row of ions on its axis, weighted over narrow Gaussian pore size distributions. Ion diameter $d = 0.7$ nm, pore diameter $2a = 0.72$ nm; curves correspond to the indicated pore-radii dispersion; $\varepsilon = 2$. Already for minor pore-radii dispersion, the saturation regions move to higher voltages, and the capacitance curve displays moderate values.

experiments reported so far, we may not deal with single-file pores. An alternative explanation may be that there is a preferential adsorption of one sort of ions at a nonpolarised electrode, and the values of $u = 0$ do not actually correspond to the dip between the peaks. Last, but not least, if the pore contains not only ions, but

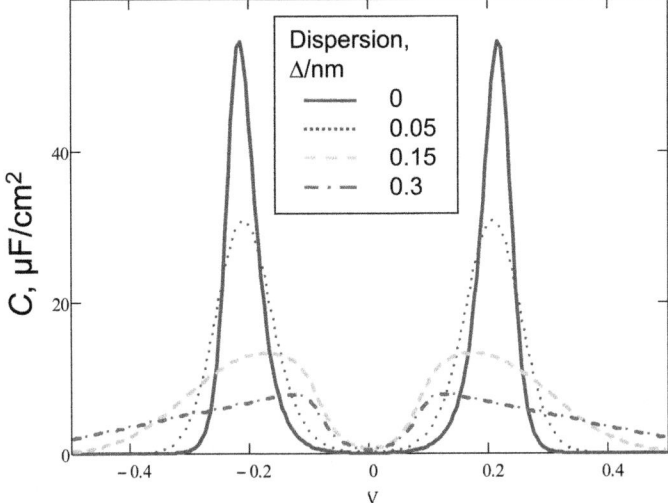

Fig. 8 The same as in Fig. 7 but shown for a larger average diameter of pores, $2a = 1.4$ nm (for the same size of ions, $d = 0.7$ nm), a larger spread of indicated pore-radii dispersion, and slightly larger voltage range. Broadening of pore-radii distribution does not make the linear response capacitance smaller, unlike the examples shown in Fig. 7. There, the range of smaller pores simply cannot accommodate ions, as taken into account in the calculation, and that decreased the capacitance. The role of the latter effect is minor in the curves of this figure.

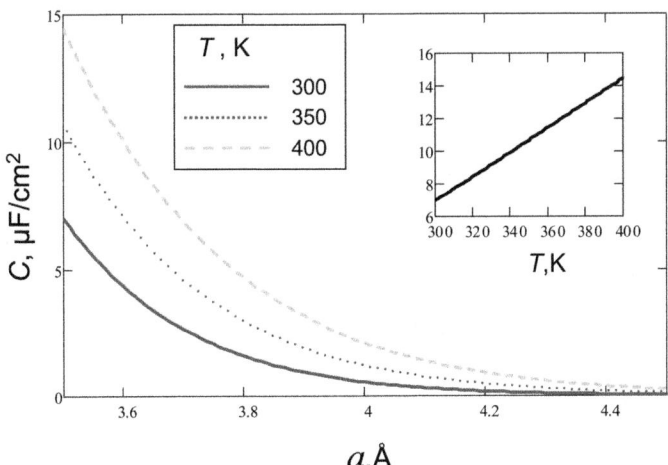

Fig. 9 Differential 'linear response' capacitance of a nanopore as a function of pore radius at different temperatures. Curves are calculated for the packing of ions of 0.7 nm diameter. The inset is the capacitance temperature dependence shown for the pore diameter equal to ion diameter. Other parameters are as in Fig. 2.

empty voids, the two state Ising model will not be applicable, whereas its extension to the three state model may give a different result. All these options remain to be studied.

The temperature dependence of the linear response capacitance is also worth noting. Outlining explicitly the temperature dependence in eqn (10), we get

$$C(u)|_{u=w} \simeq \frac{\varepsilon L_B^{(r)}}{2\pi a d} \frac{T_r}{T} \exp\left\{-6.2\frac{L_B^{(r)}}{a} e^{-2.4d/a} \frac{T_r}{T}\right\} \quad (13)$$

where T and T_r (=300) are, respectively, running and room temperatures, $L_B^{(r)}$ is the Bjerrum length at $T = T_r$; for $\varepsilon = 2$, $L_B^{(r)} \simeq 28$ nm. The inset in Fig. 9 shows such dependence. Due to competing trends in the pre-exponential and exponential factors, it *linearly* grows with temperature in the relevant interval of temperatures.

Concluding remarks

Ultraporous electrodes with single-file pores may themselves not be practical for supercapacitors because of large ohmic power losses accompanying the charging–discharging process, as swapping co-ions with counterions in a single file may have substantial kinetic hindrance. For the same reason, accurate measurements of the equilibrium differential capacitance in such ultra narrow pores are difficult, and the author is not aware of any data on the voltage dependence reported so far. Still, theoretical results may be useful for understanding what, in principle, ultra nanoporous electrodes could deliver for energy storage.

Formulation of the model presented above was inspired by ref. 47. The latter suggested mapping of the electrical double layer problem in ionic liquids at a semi-infinite flat electrode onto a 1d-three state spin model in an external field with long range Coulomb interactions, for which the authors have obtained a sophisticated exact solution. The model of the present paper and its solution is much simpler, because in the metallic pore electrostatic interactions are exponentially screened, *i.e.* are short range, but most importantly the 1d spin-analogy is literal here.

The resulting 'toy' model of single-file charge storage allows the rationalisation of several interesting features. It shows the interplay between the voltage, temperature and the electrostatic interactions between electrolyte ions inside the nanopore, screened by free electrons of the electrode. The existence of an exact analytical solution for this model reported in this paper, describes all these dependences in the most simple terms. It may trigger new experimental electroanalytical studies of nanotemplated electrodes with ionic liquids, in order to verify the predicted trends. In designing such experiments one should keep in mind the two main limitations of the model:

(i) The description of the electrode as an 'ideal metal' may be fairly good for many metallic electrodes, but it can be questioned for carbon materials. Recent investigation[46] has shown that the reduction of screening will be noticeable with the corresponding effect on α, and subsequently on the capacitance. A similar effect is produced by a slight increase of the radius of the pore, except the latter also increases the denominator in eqn (8), thus affecting not only $\chi(u)$ through the coupling constant α, but also the pre-factor. Hence the main effect of such refinement will be the shift of the maxima to higher voltages. Accounting for a modified screening pattern can be easily implemented through correcting the input value of α.

(ii) We assumed that pores of a nonpolarised electrode are stuffed with ions (the stuffing is neutral, if there is no preferential adsorption of ions of one sign). We assumed furthermore that both sorts of ions are strongly adsorbed in a pore of a nonpolarised electrode. For the latter to be true there must be a strong energetic drive to compensate for the losses of entropy incurred when squeezing the ion into a narrow nanopore. Thus no empty space in the pore was allowed. This made it possible to reduce the description to a *two state model*: each site of the lattice can be in 'spin-up' (occupied by cation) or 'spin-down' (occupied by anion) states. To take into account a balance between cations, anions, and empty voids, one must develop a theory beyond the two state model, namely a three state model in which the 'spin' can also acquire the value 0 to account for the possibility of empty sites in the lattice, call them voids.[47] Such theory is more complicated and leads to a more cumbersome set of equations, providing more sophisticated capacitance curves, which will be reported elsewhere. The above considered 'tutorial' case is, however, not only valuable because it is simpler; it may as well correspond to physical reality, in the limit of strong adsorption of ions.

With all these reservations spelled out, it still made sense to present this model, as simple as it is, as a starting point for future theoretical and experimental developments.

Acknowledgements

I am thankful to my collaborators and co-workers – Ralph Colby, Maxim Fedorov, Yuri Gogotsi, Slavko Kondrat, Alpha Lee, Carlos Perez, Gunnar Pruessner, and Chris Rochester – for many useful discussions, and Annina Sartor for critical reading of the manuscript. Special thanks are due to enlightening conversations with the authors of ref. 47, Ron Horgan and David Dean. The work is a part of the project on ionic liquids in nano-confinement supported by the Engineering and Physical Sciences Research Council, Grant EP/H004319/1, awarded within the framework of the EPSRC/NSF – UK/USA international cooperation program.

References

1 A. A. Kornyshev, E. Spohr and M. A. Vorotyntsev, Electrochemical interfaces: at the border line, in *Encyclopedia of Electrochemistry*, ed. A. Bard *et al.*, Wiley-VCH, New York, 2002, vol. 1, pp. 33–132.
2 A. Hamelin, Double-layer properties at sp and sd metal single-crystal electrodes, in *Modern Aspects of Electrochemistry*, ed. B. E. Conway, R. E. White and J. O'M. Bockris, Plenum Press, New York, 1985, vol. 16, pp. 1–102.
3 M. J. Weaver and X. Gao, *In situ* Electrochemical Surface Science, *Annu. Rev. Phys. Chem.*, 1993, **44**, 459–494.
4 D. M. Kolb, Electrochemical Surface Science, *Angew. Chem., Int. Ed.*, 2001, **40**, 1162–1181.
5 Yu. A. Chizmadjev, V. S. Markin, M. P. Tarasevich and Yu. G. Chirkov, *Macro-kinetics of processes in porous media*, Nauka, Moscow, 1971.
6 D. Pletcher, N. A. Hampson and A. J. S. McNeil, Electrochemistry of porous electrodes, *Specialist Periodical Reports: Electrochemistry*, Royal Society of Chemistry, London, 1984, vol. 9, pp. 1–65.
7 J. S. Newman, *Electrochemical Systems*, II Edition, Prentice Hall, Englewood, NJ, 1991.
8 A. D. Carbó, *Electrochemistry of Porous Materials*, CRC Press –Taylor and Francis, 2007.
9 Y. Gogotsi (ed.) *Carbon Nanomaterials*, CRC Press – Taylor & Francis, 2006.
10 F. Beguin and E. Frackowiak (ed.), *Carbon Materials for Electrochemical Energy Storage Systems*, CRC Press, 2009.
11 Yu. M. Volfkovich and T. M. Serdyuk, Electrochemical Capacitors (review), *Russ. J. Electrochem.*, 2002, **38**, 1043–1068.

12 P. Simon and Yu. Gogotsi, Materials for electrochemical capacitors, *Nat. Mater.*, 2008, **7**, 845–854.
13 J. Chmiola, G. Yushin, Y. Gogotsi, C. Portet, P. Simon and P. L. Taberna, Anomalous increase in carbon capacitance at pore sizes less than 1 nanometer, *Science*, 2006, **313**, 1760–1763.
14 C. Largeot, C. Portet, J. Chmiola, P. L. Taberna, Y. Gogotsi and P. Simon, Relation between the ion size and pore size for an electric double-layer capacitor, *J. Am. Chem. Soc.*, 2008, **130**, 2730–2735.
15 F. Stoeckli and T. A. Centeno, Pore size distribution and capacitance in microporous carbons, *Phys. Chem. Chem. Phys.*, 2012, **14**, 11589–91.
16 F. Stoeckli and T. A. Centeno, On the determination of surface areas in activated carbons, *Carbon*, 2005, **43**, 1184–1190.
17 E. Raymundo-Pinero, K. Kierzek, J. Machnikowski and F. Beguin, *Carbon*, 2006, **44**, 2498–2507.
18 C. O. Ania, J. Pernak, F. Stefaniak, E. Raymundo-Pinero and F. Beguin, *Carbon*, 2006, **44**, 3126–3130.
19 S. Kondart and A. A. Kornyshev, Superionic state in double layer capacitors with nanoporous electrodes, *J. Phys.: Condens. Matter*, 2011, **23**, 022201, 1–5; corrigendum: 2013, **25**, 119501.
20 S. Kondart, N'Georgi, M. V. Fedorov and A. A. Kornyshev, Superionic state in nano-porous double layer capacitors: insights from Monte Carlo simulations, *Phys. Chem. Chem. Phys.*, 2011, **13**, 11359–11366.
21 C. Merlet, B. Rotenberg, P. A. Madden, P.-L. Taberna, P. Simon, Y. Gogotsi and M. Salanne, On the molecular origin of supercapacitance in nanoporous carbon electrodes, *Nat. Mater.*, 2012, **11**, 306.
22 H. Yukawa, On the interaction of elementary particles, *Proc. Phys. Math. Soc. Japan*, 1935, **17**, 48.
23 G. E. Brown and A. D. Jackson, *The Nucleon-Nucleon Interaction*, North-Holland Publishing, Amsterdam, 1976.
24 P. Wu, J. Huang, V. Meunier, B. G. Stumper and R. Qiao, Complex capacitance scaling in ionic-liquids-filled nanopores, *ACS Nano*, 2011, **5**, 9044–9051.
25 D. Jiang, Z. Jin and J. Wu, Oscillation of capacitance inside nanopores, *Nano Lett.*, 2011, **11**, 5373–5377.
26 G. Feng and P. T. Cummings, Supercapacitor capacitance exhibits oscillatory behaviour as a function of nanopore size, *J. Phys. Chem. Lett.*, 2011, **2**, 2859–2864.
27 K. Kiyohara, T. Sugino and K. Asaka, Phase transitions in porous electrodes, *J. Chem. Phys.*, 2011, **134**, 154710.
28 S. Kondrat, C. R. Pérez, V. Presser, Y. Gogotsi and A. A. Kornyshev, Effect of pore size and its dispersity on the energy storage in nanoporous supercapacitors, *Energy Environ. Sci.*, 2012, **5**, 6474–6479.
29 B. E. Conway, Optimisation of energy density and power density, in *Electrochemical Supercapacitors*, Kluwer, New York, 1999.
30 S. Kondrat and A. Kornyshev, Charging dynamics and optimisation of nanoporous supercapacitors, *J. Phys. Chem. C*, 2013, **117**, 12399–12406.
31 A. A. Lee, S. Kondrat, G. Oshanin, and A. A. Kornyshev, Charging a quasi-one-dimensional metallic nanopore, in preparation.
32 Dealing in this study exclusively with equilibria, we do not discuss the kinetics of the charging process, which may be slow in such narrow pores, but still warranted by some degrees of freedom that we do not consider explicitly (such as phonons in/undulations of the walls, fluctuations of orientations of intrinsically anisotropic ions, internal conformations of ions, *etc*).
33 W. Panovsky and M. Phillips, *Classical electricity and magnetism*, Addison-Wesley, Reading, Massachusetts, II edition, vol. 1, 1969.
34 M. Abramovich and I. Stegun, *Handbook of Mathematical Functions*, National Bureau of Standards – U.S. Government Printing Office, Washington, DC, 10th Edition, 1974.
35 Note that these screened potentials are not exactly Yukawa-like. Indeed, the Yukawa potential in additional to the screening exponential function contains R in the denominator. The general forms of interaction potentials both for cylindrical and slit geometries also recover $1/R$ law, but only at $R \ll a/2.4$, $a/1.57$, as they, of course, should, because at such short distances the interaction must approach the bare Coulomb law. However, this is not present in the asymptotic expressions, valid in the opposite limit, which is only interesting for this study.
36 M. S. Loth, B. Skinner and B. I. Shklovskii, Non-mean-field theory of anomalously large double layer capacitance, *Phys. Rev. E: Stat., Nonlinear, Soft Matter Phys.*, 2010, **82**, 016107.

37 M. S. Loth, B. Skinner and B. I. Shklovskii, Anomalously large capacitance of an ionic liquid described by the restricted primitive model, *Phys. Rev. B: Condens. Matter Mater. Phys.*, 2010, **82**, 056102.
38 M. A. Vorotyntsev and A. A. Kornyshev, Electrostatic Interaction at the metal/dielectric interface, *Zh. Eksp. Teor. Fiz.*, 1980, **78**, 1008–1019; *Sov. Phys. JETP*, 1980, **51**, 509–514.
39 A. M. Gabovich, I. C. Il'chenko, E. A. Pashitskii and Yu. A. Romanov, Electrostatic energy and screened charge interaction near the surface of metals with different Fermi surface shape, *Surf. Sci.*, 1980, **94**, 179–203.
40 A. M. Gabovich, V. M. Rosenbaum and A. I. Voitenko, Dynamical image forces in three-layer systems and field emission, *Surf. Sci.*, 1987, **186**, 523–549.
41 M. A. Vorotyntsev and S. N. Ivanov, Statistical-mechanics of an ion ensemble adsorbed on a metal-dielectric interface, *Zh. Eksp. Teor. Fiz.*, 1985, **88**, 1729–1737; *Sov. Phys. JETP*, 1985, **61**, 1028–1032.
42 A. A. Kornyshev and W. Schmickler, On the coverage dependence of the partial charge transfer coefficient, *J. Electroanal. Chem.*, 1986, **202**, 1–21.
43 K. Huang, *Statistical Mechanics*, Wiley & Sons, New York, 1987.
44 R. J. Baxter, *Exactly solved models in statistical mechanics*, Academic Press, London, 1982.
45 It is straightforward to show that the so-called volumetric capacitance, *i.e.* the capacitance per total volume of the electrode including its pore volume, reads: $C_V = -(\varepsilon L_B P/\pi a^2 d)\chi(u)$, where P is the porosity of the electrode. Presenting this expression for an interested reader, we will not exploit it in this paper.
46 C. Rochester, A. A. Lee, G. Pruessner, and A. A. Kornyshev, Interionic interactions in electronically conducting nanoconfinement, *ChemPhysChem*, in press.
47 V. Demery, D. S. Dean, T. C. Hammant, R. R. Horgan and R. Podgornik, Overscreening in a 1D lattice Coulomb gas model of ionic liquids, *Europhys. Lett.*, 2012, **97**, 28004.

Faraday Discussions

RSC Publishing

PAPER

Carbon nanotube based electrochemical sensor for the sensitive detection of valacyclovir

Badal Shah, Todd Lafleur and Aicheng Chen*

Received 22nd February 2013, Accepted 19th April 2013
DOI: 10.1039/c3fd00023k

An electrochemical sensor for the sensitive detection of valacyclovir has been developed, which is based on single-walled carbon nanotube (SWCNT)-modified glassy carbon electrodes. The electrochemical oxidation of valacyclovir at the SWCNT-modified glassy carbon electrodes has been investigated using cyclic voltammetry and differential pulse voltammetry. Our experimental results show that the SWCNT-modified glassy carbon electrode possesses high activity toward the electrochemical oxidation of valacyclovir. In a 0.1 M phosphate buffer (pH = 7.4), valacyclovir exhibited an irreversible oxidation peak at ~0.91 V. The effects of pH of and the amount of SWCNT deposited on the glassy carbon electrode on the activity of the sensor have also been studied. Under optimized conditions, the sensor demonstrates a linear response range from 5×10^{-9} to 5.5×10^{-8} M valacyclovir. The detection and quantification limits were found to be 1.80×10^{-9} M and 6.02×10^{-9} M, respectively. The selectivity, stability and reproducibility of the proposed sensor were examined as well. To validate its real world application, the electrochemical sensor has been successfully utilized in the detection of valacyclovir in human blood plasma and pharmaceutical samples. Thus, the electrochemical sensor developed in this study has strong potential to be employed in the quality control testing of pharmaceutical products and also for therapeutic drug monitoring in hospitals.

1. Introduction

Substituted purines represent an important category of compounds that are actively used as therapeutic agents against viral infection. Valacyclovir is the drug of choice for the treatment of herpes zoster and cold sores. It is also effective for the treatment or suppression of genital herpes in immunocompetent individuals and for the suppression of recurrent genital herpes in HIV infected individuals. The chemical structure of valacyclovir is shown in Scheme 1. Valacyclovir is the L-valyl ester and prodrug of the antiviral drug acyclovir. Subsequent to absorption, valacyclovir is rapidly and almost completely hydrolyzed to acyclovir and L-valine, an essential amino acid, *via* first-pass metabolism. This hydrolysis is mediated

---

Department of Chemistry, Lakehead University, 955 Oliver Road, Thunder Bay, Ontario P7B 5E1, Canada. E-mail: aicheng.chen@lakeheadu.ca; Fax: +1 807 346 7775; Tel: +1 807 343-8318

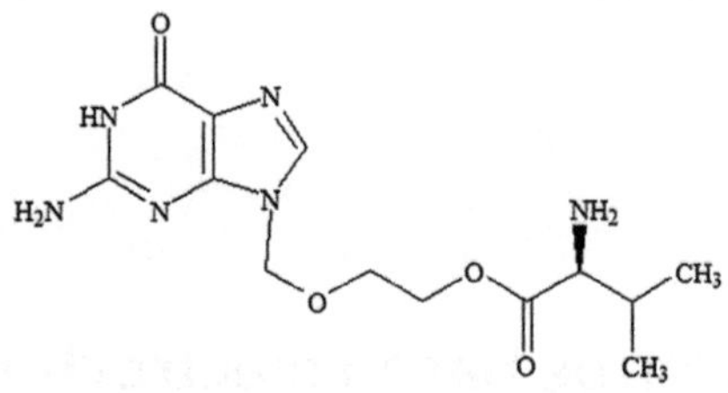

Scheme 1 Chemical structure of valacyclovir.

primarily by the enzyme valacyclovir hydrolase and occurs predominantly in the liver. Because of the low bioavailability of acyclovir, its prodrug (valacyclovir) is preferred for therapeutic treatment, through which it plays an important role in the therapy of viral diseases. This drug provides significant therapeutic benefits in the treatment of viral diseases, and no serious side effects associated with its use have been reported in its monograph. Because of the wide use of valacyclovir in the treatment of different viral diseases, its quantitative analysis has become very important and is under extensive study.

Various analytical methods such as HPLC,[1,2] RP-HPLC,[3,4] spectrophotometric methods[5,6] and LC-MSMS[7] have been proposed for the selective detection of valacyclovir in human serum plasma and pharmaceutical formulations. The simultaneous determination of valacyclovir and acyclovir was also reported using HPLC,[8] liquid chromatography/positive-ion electrospray ionization mass spectrometry (LC–ESI-MS/MS),[9] LC-MS-MS[10] and micellar electrokinetic chromatography methods.[11] Even though these methods are sensitive and selective, they are tedious and time consuming due to the requisite pretreatment of samples and optimization of chromatogram conditions. In contrast to these methods, electrochemical techniques coupled with carbon nanotube (CNT) modified electrodes are facile and cost-effective. CNTs are molecular-scale tubes of graphitic carbon, which are reported to possess exceptional electrocatalytic properties.[12] Single-walled carbon nanotubes (SWCNTs) consist of single rolled-up sheets of graphene with either open or closed ends, depending on the fabrication methodology, which have outstanding strength and a very large surface area per unit mass.[13]

Because valacyclovir is comprised of biologically important compounds such as polynucleic acids, whose electrochemical and enzymatic oxidation follow similar mechanistic pathways, knowledge of the voltammetric behavior of valacyclovir is of biological significance.[14–16] Carbon nanotube modified electrodes represent a class of nanomaterials that have been extensively used in the fabrication of sensors and biosensors due to their enhanced catalytic activity.[17–21]

The objective of this work is to establish a convenient, cost-effective and sensitive method for the determination of valacyclovir in pharmaceutical formulations and human bodily fluids based on the electrocatalytic activity of CNTs. In the present study, the electrochemical oxidation of valacyclovir on SWCNT-modified glassy carbon electrodes (GCEs) has been investigated for the first time. The effective detection of valacyclovir in human blood plasma and pharmaceutical samples shows that the electrochemical sensor proposed in this study might be adopted in clinical and quality control laboratories as well as for therapeutic drug monitoring in hospitals.

2. Materials and methods

2.1. Reagents and materials

Valacyclovir hydrochloride hydrate, SWCNTs and human serum (from human male AB Plasma) were purchased from Sigma-Aldrich. Generic valacyclovir 500 mg tablets were obtained from the Thunder Bay Regional Health Sciences Center pharmacy. All other reagents were of analytical grade and were used as supplied. All experiments were performed in a 0.1 M phosphate buffer solution (PBS) with different pH values that were maintained with K_2HPO_4 and KH_2PO_4. The electrolyte was subjected to continuous stirring during the experiments to maintain a homogeneous analyte concentration. Double distilled water, purified with a Nanopure® water system was used in the preparation of all solutions. A previous publication reported that valacyclovir was partially hydrolyzed to acyclovir in an aqueous solution following storage at +4 °C for several weeks.[3] For this reason, all valacyclovir solutions were freshly prepared and used within several hours to avoid hydrolysis.

2.2. Instrumentation

All electrochemical measurements were performed using a CHI 660 electrochemical workstation (CH Instrument Inc., USA). Electrochemical experiments were carried out utilizing a conventional three-electrode system including a GCE, having a diameter of 3 mm (modified and unmodified) as the working electrode, a Pt coil electrode as the counter electrode and a 3 M KCl saturated Ag/AgCl electrode as the reference electrode. The pH of the solution was measured with an Oakton® pH6 Acorn Series pH meter. A field-emission scanning electron microscopy (FE-SEM) (Hitachi SU-70) equipped with energy dispersive X-ray spectroscopy (EDS) was employed for the characterization of the SWCNT-modified GCE surface. All experiments were performed at room temperature and the potentials quoted are *versus* an Ag/AgCl electrode.

2.3. Fabrication of electrode

Prior to modification, the surface of the GCE was carefully polished with alumina powder using micro cloth pads and then rinsed thoroughly with double distilled water until a mirror-like surface was obtained. It was then cleaned *via* a brief ultrasonic exposure in order to remove any adsorbed substances on the surface and dried at room temperature. A 0.5 mg ml^{-1} suspension of SWCNTs in *N,N*-dimethylformamide was prepared under ultrasonic agitation. The GCE surface was subsequently coated with a known (10–70 μl) volume of this suspension and allowed to dry at room temperature.

2.4. Optimization of proposed sensor

The effects of the amount of carbon nanotubes coated on the GCE and the pH of the electrolytes on the performance of the sensor were studied. The SWCNTs-modified GCEs were tested under various pH conditions ranging from pH 5 to 9. Differential pulse voltammograms were recorded for each pH condition at a valacyclovir concentration of 15 nM. In order to assess the influence on sensor performance of the amount of SWCNT that was coated on the GCE, various volumes of a SWCNT suspension (0.5 mg mL^{-1}), ranging from 10 μL to 70 μL, were coated on the GCE and permitted to dry at room temperature. The prepared electrodes were tested by recording a differential pulse voltammogram of 20 nM valacyclovir in 0.1 M phosphate buffer (pH 7.4).

2.5. Concentration study

The SWCNT-modified GCE electrodes were employed as a working electrode in a three-electrode assembly. A calculated quantity of valacyclovir stock solution was added to the cell containing pH 7.4, 0.1M phosphate buffer to arrive at a valacyclovir concentration of 5 nM. After an accumulation time of 180 s, a differential pulse voltammogram was recorded at between −0.1 V and 1.1 V, with a pulse width of 0.2 s, pulse period of 0.5 s and potential increment of 4 mV. The identical procedure was followed to record a differential pulse voltammogram of various concentrations of valacyclovir, ranging from 5 to 55 nM. For comparison, differential pulse voltammograms were recorded at a GCE and the SWCNT-modified GCE for a valacyclovir concentration of 30 nM in a 0.1 M phosphate buffer. Cyclic voltammograms at a scan rate of 50 mV s^{-1} were also recorded for the same concentration.

2.6. Stability and reproducibility study

To test the stability of the sensor, ten differential pulse voltammograms were recorded using the same electrode at an identical valacyclovir concentration for several hours. In order to verify reproducibility, five different electrodes were prepared and differential pulse voltammograms were recorded for each.

2.7. Analysis of pharmaceutical samples in human serum

Ten valacyclovir tablets were crushed into a powder using a mortar. Human serum samples, purchased from Sigma were stored frozen until assayed. Human serum plasma was diluted with a phosphate buffer to remove any matrix interference. An aliquot volume of sample was fortified with the valacyclovir tablet powder to achieve a final concentration of 1.0×10^{-6} M, and treated with 0.4 mL acetonitrile. A volume of 5 mL was extracted from the same serum sample and then sonicated for two minutes. The mixture was then centrifuged for ten minutes at 4000 rpm to purge any protein residue. Appropriate volumes of the clean supernatant solution were analyzed in a voltammetric cell containing a pH 7.4, 0.1M phosphate buffer. A calibration curve method was used for quantification.

3. Results and discussion

3.1. Characterization of the fabricated electrode and electrochemical oxidation of valacyclovir

SEM was utilized to study the surface characteristics and morphology of the fabricated SWCNT-modified GCE. Fig. 1A shows that the GCE substrate was uniformly covered with the carbon nanotubes, which linked together to form a nanoporous network structure. Energy dispersive X-ray spectroscopy was employed to investigate the purity of the SWCNTs. Aside from the primary carbon peak, several small peaks originated from the impurities, for instance Fe, Al, Pt, S, Cl and Cr, were observed in Fig. 1B. Our EDS analysis also showed that the

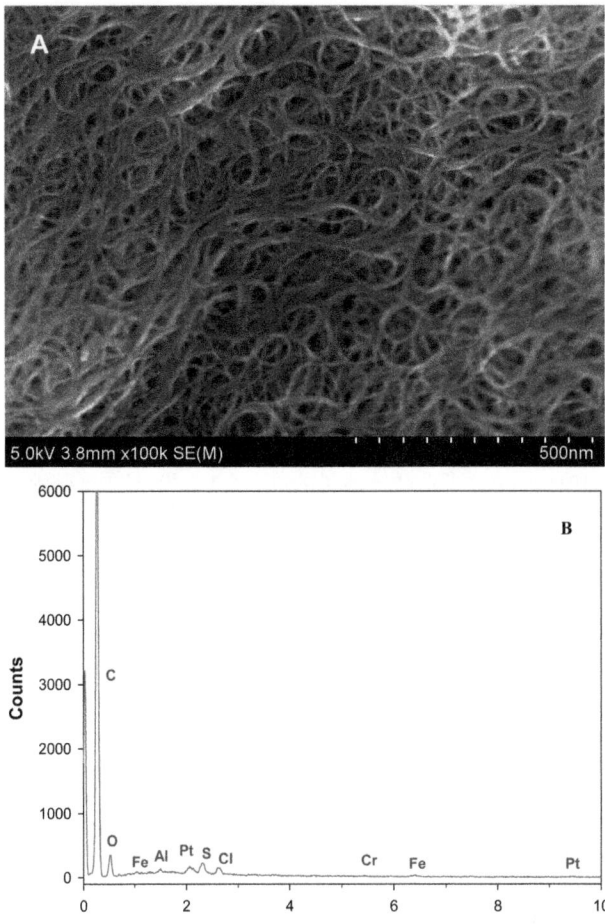

Fig. 1 (A) SEM image and (B) EDX spectrum of the SWCNT-modified glassy carbon electrode.

distribution of the impurities was not uniform. Some other impurities such as Mo and Zn were detected in some areas.

In order to explore the properties and potential of the SWCNT-modified GCE towards the detection of valacyclovir, the electrochemical behaviors of 30.0 nM valacyclovir at a bare GCE and the SWCNT-modified GCE were characterized using cyclic voltammetry. As shown in Fig. 2, during the anodic scan from −0.1 to 1.1 V, the electrochemical oxidation of valacyclovir on the SWCNT-modified GCE occurred when the electrode potentials were higher than 0.8 V. However, no reduction peak was observed during the cathodic sweep from 1.1 to −0.1 V, revealing that the electrochemical oxidation of valacyclovir is irreversible. Based on a previous study[16] and the electrochemical behaviour of guanine at a carbon-based electrode,[22–25] one can predict that the occurrence of the oxidation process is at a guanine moiety within the molecule. The electrochemical oxidation of valacyclovir involves a two-electron and two-proton transfer process to form the intermediate 8-oxoguanine, as illustrated in Scheme 2.

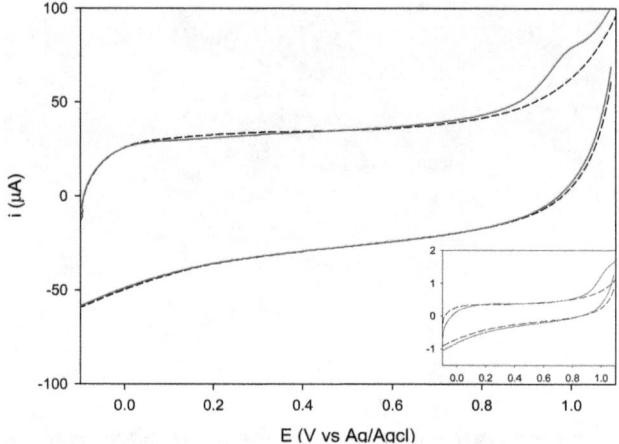

Fig. 2 Cyclic voltammogram (CV) of blank (dashed line) and 30 nM valacyclovir in pH 7.4, 0.1M phosphate buffer (solid line) and SWCNT-modified glassy carbon electrode and glassy carbon electrode. (Insert: CV at a bare glassy carbon electrode).

For comparison, the CV curves at the bare GCE are displayed as an insert of Fig. 2. Under identical experimental conditions, the peak current of the electrochemical oxidation of valacyclovir at the SWCNT-modified GCE was over 30 times higher than that at the bare GCE, whereas the peak potential was lower. Those results indicate that the modification of the GCE with SWCNTs significantly enhances the electrochemical oxidation of valacyclovir. In addition to their large surface area, some seminal work from the Compton group has deciphered the origins of the noteworthy enhancement in the electrocatalytic activity of CNTs.[26–29] It has been recognized that in most cases, the electrochemical activity of carbon nanotubes is due to the edge plane-like site defects, which may occur at the nanotube ends as well as along the tube axis. In addition, as described in a recent critical review conducted by Pumera,[31] there are several other possible causes that underlie the improvement of the catalytic activity. For instance, metallic impurities within CNTs strongly influence the redox properties of solutes, which may be responsible for a large portion of the observed "electrocatalytic" behavior of CNTs. Nanographite impurities within CNT samples could have a potent influence on the electrochemistry of many solutes. Oxygen-containing groups on CNTs and the structures of CNTs might have profound effects on their electrochemical properties as well. Pumera *et al.* have further demonstrated that even 100 ppm (0.01%wt.) of iron based impurities in carbon nanotubes is still active and that the

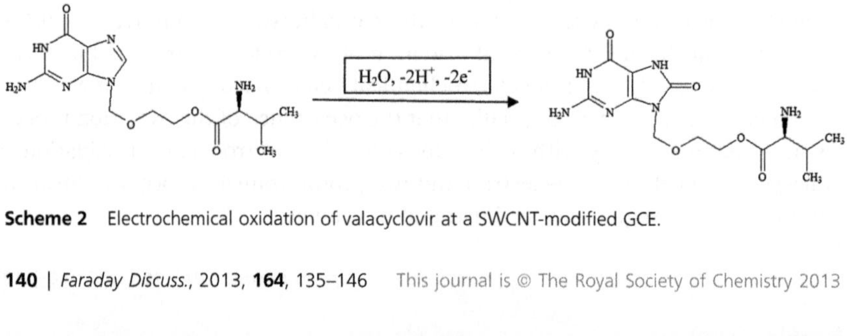

Scheme 2 Electrochemical oxidation of valacyclovir at a SWCNT-modified GCE.

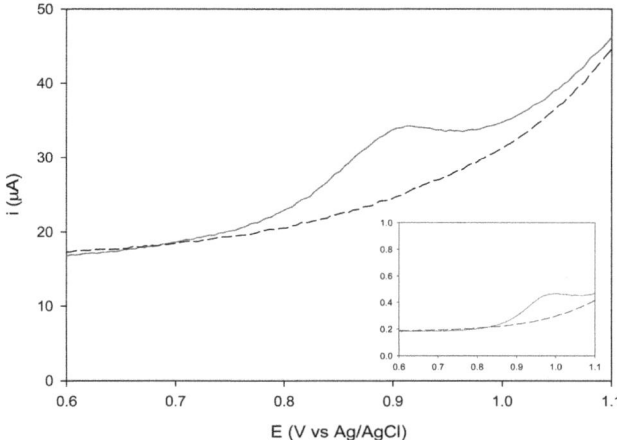

Fig. 3 Differential pulse voltammogram of blank (dashed line) and 30 nM valacyclovir in pH 7.4, 0.1M phosphate buffer (solid line) at (A) a glassy carbon electrode and (B) a SWCNT-modified glassy carbon electrode.

metallic impurities are difficult to remove, even when washed in HNO_3.[32,33] Thus, we can attribute the high catalytic activity of the SWCNT-modified GCE in the electrochemical oxidation of valacyclovir to its large surface area and impurities as well as the oxygen-containing groups that exist in the SWCNTs as shown in Fig. 1B.

The electrochemical behaviors of 30.0 nM valacyclovir at a bare GCE and a SWCNT- modified GCE were also compared by differential pulse voltammetry, with the results shown in Fig. 3. When a bare GCE was used as the working electrode, a small peak due to valacyclovir oxidation appeared at ~0.98 V as seen in the insert of Fig. 3. After being modified with the SWCNTs, the peak potential shifted to ~0.91 V and the peak current was increased by over 50 times, further confirming the significant enhancement in electrocatalytic activity. A comparison of Fig. 3 and Fig. 2 shows that, aside from the peak being sharper and better defined at the same concentration of valacyclovir than those obtained by cyclic voltammetry, it also has a lower background current, resulting in improved resolution. For these reasons differential pulse voltammetry rather than cyclic voltammetry was selected as an analytical method for the quantitative analysis of valacyclovir in the present study.

3.2. Optimization of proposed sensor

The effect of pH on the performance of the sensor was tested in the pH range from 5 to 9. As depicted in Fig. 4, the peak current of valacyclovir oxidation increased in alignment with the elevation of pH from 5.0 to 7.4, but then decreased with a further increase of the pH to 9.0, revealing that pH 7.4 is optimal. The influence of the volume of the SWCNT coating on the peak current was also investigated. For this purpose various volumes of a SWCNT suspension (0.5 mg mL^{-1}), ranging from 10 to 70 μL, were coated onto the GCE. It was observed that the increase in SWCNT deposition (from 10 to 50 μL) led to an augmentation of the peak current response of 30 nM of valacyclovir in a pH 7.4 0.1M phosphate buffer. At a SWCNT

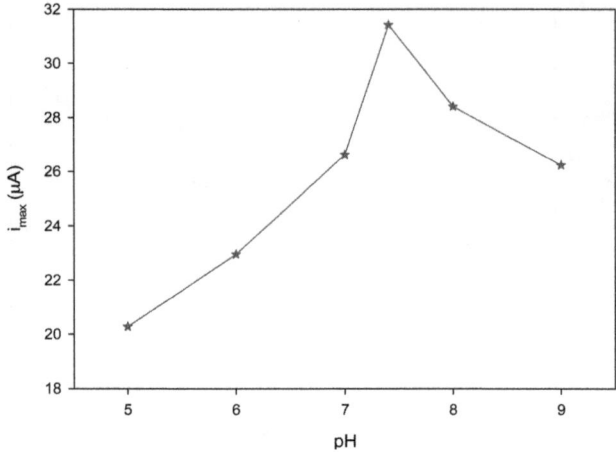

Fig. 4 Influence of pH on the peak current of 15 nM valacyclovir in 0.1 M phosphate buffer pH 7.4.

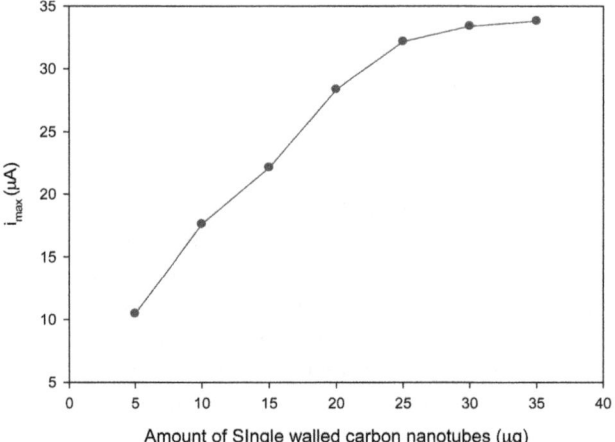

Fig. 5 Influence of volume of SWCNT-coating on the glassy carbon electrode on peak current of 20 nM valacyclovir in 0.1M phosphate buffer pH 7.4.

volume of more than 50 μL, the peak currents of alternating current voltammetry became almost constant, as shown in Fig. 5. Thus it was concluded that the optimum amount of SWCNT required to catalyze the oxidation of valacyclovir was 50 μL. Based on this finding, a 0.1M phosphate buffer (pH 7.4) and coating of a 50 μl SWCNT suspension was used in subsequent experiments.

3.3. Concentration study

To establish a methodology for the determination of valacyclovir, differential pulse voltammograms were recorded using increasing amounts of the compound. A pH 7.4 0.1 M phosphate buffer was selected as the supporting electrolyte and the peak at 0.91 V was considered for the analysis. Fig. 6A shows that the peak current increased linearly with an increase in the concentration of valacyclovir. Utilizing the optimized conditions described above, a calibration curve was

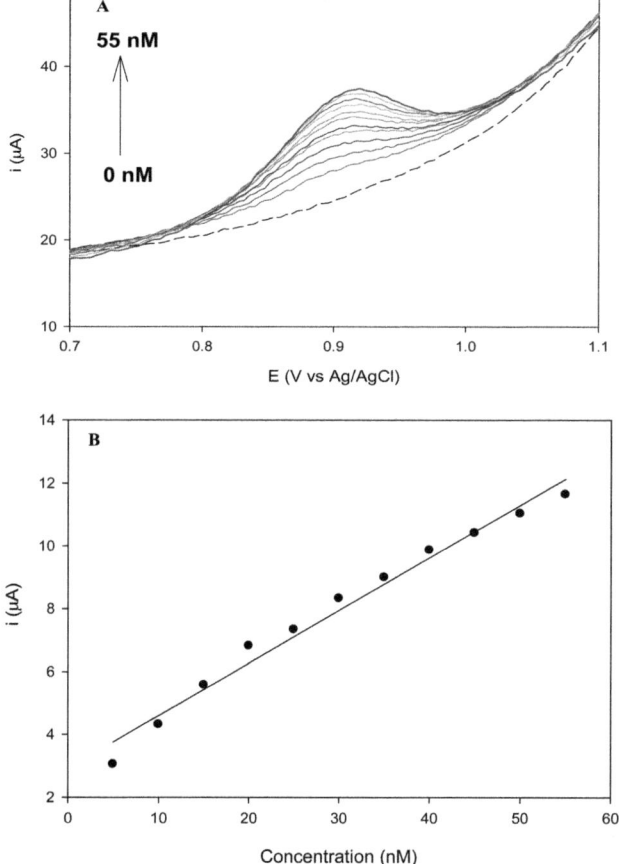

Fig. 6 (A) Differential pulse voltammograms of valacyclovir recorded in a 0.1 M pH 7.4 phosphate buffer solution at the SWCNT-coated glassy carbon electrode; and (B) the determined calibration curve of valacyclovir in the range 5 to 55 nM.

obtained in the linearity range of 5×10^{-9} to 5.5×10^{-8} M, as shown in Fig. 6B and 6C. The limit of detection (LOD) and limit of quantification were calculated using the following formula:

$$\text{LOD} = 3 \text{ S m}^{-1}$$

$$\text{LOQ} = 10 \text{ S m}^{-1}$$

where S is the standard deviation of the peak current of blank ($n = 5$) and m is the slope of the calibration curve. LOD and LOQ were found to be 1.81×10^{-9} M and 6.02×10^{-9} M, respectively. The LOD and LOQ calculated for the present sensor are much lower than those in other reported work,[1,4,11,16] which are summarized in Table 1.

To investigate the stability of the fabricated electrochemical sensor, five differential pulse voltammograms were recorded at an interval of several hours at a 30 nM valacyclovir concentration. The percentage of relative standard deviation

Table 1 Comparison of the performance of various instrumental methods for valacyclovir determination

Method	Least measurable quantity/M	Limit of detection/M	Ref.
HPLC	2.5×10^{-6}	8.33×10^{-7}	1
Micellar electrokinetic chromatography	2.7×10^{-7}	8.8×10^{-8}	11
RP-HPLC	1.4×10^{-8}	1.05×10^{-9}	4
Electrochemical DPV	5.0×10^{-9}	1.80×10^{-9}	This work

Table 2 Recovery tests of valacyclovir 500 mg generic tablet in human serum plasma

Concentration spiked/M	Concentration detected/M	% Recovery
0.0	0.0	—
1.5×10^{-8}	1.4385×10^{-8}	95.90
2.0×10^{-8}	1.9937×10^{-8}	99.69
2.5×10^{-8}	2.5030×10^{-8}	100.12
3.0×10^{-8}	3.0177×10^{-8}	100.59

(RSD%) was found to be 1.61%, revealing that the fabricated electrodes exhibit very stable performance. To investigate the reproducibility of the electrode preparation process, five different electrodes were prepared and differential pulse voltammograms were recorded using same concentration under the optimized conditions. The RSD% was found to be 1.18%.

3.4. Analysis of pharmaceutical formulation in human serum plasma

To validate the application of the proposed sensor in a real world application, the analysis of a pharmaceutical formulation in human serum plasma was carried out. Generally used excipients in tablet formulations include microcrystalline cellulose, magnesium stearate, colloidal silicon dioxide, titanium dioxide, macrogol, polysorbate 80, carnauba wax *etc.*, which may cause interference during analysis. In human serum, proteins constitute the major interfering substances. Treatment with acetonitrile was used to precipitate proteins, which were then removed *via* centrifugation. As shown in Table 2, the recovery tests of valacyclovir ranging from 1.5×10^{-8} to 3.0×10^{-8} M were carried out using differential pulse voltammetry. Recovery studies were carried out after the addition of a known volume of the drug (valacyclovir 500 mg tablet) to human serum plasma. The recoveries in different samples were found to reside in the range of from 95.90% to 100.59%. No additional peak was found due to excipients in the potential range of from −0.1 V to 1.1 V, which substantiate the selectivity of this method for valacyclovir.

4. Conclusions

In this work, SWCNT-modified GCEs were fabricated and studied as a new electrochemical sensor for the detection of valacyclovir. The influence of the pH and the volume of SWCNTs deposited on GCE were studied to optimize the performance of the proposed sensor. SWCNT-modified electrodes exhibited much higher electrocatalytic activity, when compared with unmodified GCEs, which were characterized by the significantly higher peak current and lower peak potential of valacyclovir oxidation. The high catalytic activity of the SWCNT-modified GCE may be attributed to its large surface area as well as inherent impurities and the oxygen-containing groups that resided in the SWCNTs. Our testing results reveal that the proposed sensor has a lower detection limit in contrast to the reported instrumental analysis. The selectivity, stability and reproducibility of the proposed sensor have also been investigated with satisfactory results. The developed electrochemical sensor was further employed to determine valacyclovir concentrations in a pharmaceutical formulation in human serum plasma to demonstrate its capacity for a real world application. The proposed sensor is sensitive, selective and economical in comparison to other instrumental methods, which makes it a more suitable candidate for utilization in the quality control testing of pharmaceutical products, as well as for therapeutic drug monitoring in hospitals and clinical research organizations.

Acknowledgements

This work was supported by a Discovery Grant from the Natural Sciences and Engineering Research Council of Canada (NSERC). We thank Dr Prashnat Jani and Mr Jeff Chen (RPh) for their help in acquiring valacyclovir tablets. B. S. acknowledges the Ontario Trillium Scholarship. A. C. acknowledges NSERC and the Canada Foundation for Innovation (CFI) for the Canada Research Chair Award in Materials & Environmental Chemistry.

References

1 A. S. Jadhav, D. B. Pathare and M. S. Shingare, *J. Pharm. Biomed. Anal.*, 2007, **43**, 1568.
2 M. L. Palacios, G. Demasi, M. T. Pizzorno and A. I. Segall, *J. Liq. Chromatogr. Relat. Technol.*, 2005, **28**(5), 751.
3 P. N. Rao, K. R. Rajeswari and V. J. Rao, *Asian J. Chem.*, 2006, **18**(4), 2552.
4 A. Savasera, C. Ozkana, Y. Ozkana, B. Uslu and S. Ozkan, *J. Liq. Chromatogr. Relat. Technol.*, 2003, **26**(11), 1755.
5 D. V. P. Reddy, B. M. Gurupadayya, Y. N. Manohara and V. V. Bhaskar, *Asian J. Chem*, 2007, **19**(4), 2797.
6 M. Ganesh, C. V. Narasimharao, A. Saravarakumar, K. Kamalakannan, M. Vinoba, H. S. Mahajan and T. Sivakumar, *E-J. Chem.*, 2009, **6**(3), 814.
7 J. J. Sasanyaa, A. M. M. Abd-Allab, A. G. Parkerb and A. Cannavan, *J. Chromatogr., B: Anal. Technol. Biomed. Life Sci.*, 2010, **878**, 2384.
8 C. Pham-Huy, F. Stathoulopoulou, P. Sandouk, J. M. Scherrmann, S. Palombo and C. Girre, *J. Chromatogr., Biomed. Appl.*, 1999, **732**(1), 47.
9 M. Yadav, V. Upadhyay, P. Singhal, S. Goswamia and P. S. Shrivastav, *J. Chromatogr., B: Anal. Technol. Biomed. Life Sci.*, 2009, **877**, 680.
10 R. Kanneti, R. Rajesh, J. Raj and P. Bhatt, *Chromatographia*, 2009, **70**, 407.
11 K. M. Al Azzama, B. Saada, A. Makahleaha, H. Y. Aboul-Eneinb and A. A. Elbashir, *Biomed. Chromatogr.*, 2010, **24**, 535.
12 J. P. Metters and C. E. Banks, *Vacuum*, 2012, **86**, 507.
13 S. Gullapalli and M. S. Wong, *Chem. Eng. Prog.*, 2011, **107**, 28.
14 G. C. Visor, S. E. Jackson, R. A. Kenley and G. C. Lee, *J. Pharm. Sci.*, 1985, **74**, 1078.
15 A. M. Oliveira-Brett and F. M. Matysik, *J. Electroanal. Chem.*, 1997, **429**, 95.
16 B. Uslu, S. A. Ozkan and Z. Senturk, *Anal. Chim. Acta*, 2006, **555**, 341.
17 L. Agui, P. Yanez-Sedeno and J. M. Pingarron, *Anal. Chim. Acta*, 2008, **622**, 11.
18 A. Chen and S. Chatterjee, *Chem. Soc. Rev.*, 2013, DOI: 10.1039/c3cs35518g.
19 A. Chen and B. Shah, *Anal. Methods*, 2013, **5**, 2158, DOI: 10.1039/c3ay40155c.

20 L. Siegert, D. K. Kampouris, J. Kruusma, V. Sammelselg and C. E. Banks, *Electroanalysis*, 2009, **21**, 48.
21 J. Kruusma, V. Sammelselg and C. E. Banks, *Electrochem. Commun.*, 2008, **10**, 1872.
22 R. N. Goyal and A. Dhawan, *J. Electroanal. Chem.*, 2006, **591**, 159.
23 A. M. Oliveira-Brett, V. Diculescn and J. A. P. Piedade, *Bioelectrochemistry*, 2002, **55**, 61.
24 H. S. Wang, H. X. Ju and H. Y. Chen, *Anal. Chim. Acta*, 2002, **461**, 243.
25 S. Chatterjee and A. Chen, *Electrochem. Commun.*, 2012, **20**, 29.
26 R. R. Moore, C. E. Banks and R. G. Compton, *Anal. Chem.*, 2004, **76**, 2677.
27 C. E. Banks, T. J. Davies, G. G. Wildgoose and R. G. Compton, *Chem. Commun.*, 2005, 829.
28 C. E. Banks, R. R. Moore, T. J. Davies and R. G. Compton, *Chem. Commun.*, 2004, 1804.
29 C. E. Banks, A. Crossley, C. Salter, S. J. Wilkins and R. G. Compton, *Angew. Chem., Int. Ed.*, 2006, **45**, 2533.
30 B. Sljukic, C. E. Banks and R. G. Compton, *Nano Lett.*, 2006, **6**, 1556.
31 M. Pumera, *Chem. Rec.*, 2012, **12**, 201.
32 M. Pumera, *Chem.–Eur. J.*, 2009, **15**, 4970.
33 M. Pumera and Y. Miyahara, *Nanoscale*, 2009, **1**, 260.

Faraday Discussions

RSCPublishing

PAPER

Electroanalysis using modified hierarchical nanoporous carbon materials†

Rusbel Coneo Rodriguez, Angelica Baena Moncada, Diego F. Acevedo, Gabriel A. Planes, Maria C. Miras and Cesar A. Barbero*

Received 15th February 2013, Accepted 6th March 2013
DOI: 10.1039/c3fd00018d

The role of the electrode nanoporosity in electroanalytical processes is discussed and specific phenomena (slow double layer charging, local pH effects) which can be present in porous electrode are described. Hierarchical porous carbon (HPC) materials are synthesized using a hard template method. The three dimensional carbon porosity is examined using scanning electron microscopy on flat surfaces cut using a focused ion beam (FIB-SEM). The electrochemical properties of the HPC are measured using cyclic voltammetry, AC impedance, chronoamperometry and Probe Beam Deflection (PBD) techniques. Chronoamperometry measurements of HPC seems to fit a transmission line model. PBD data show evidence of local pH changes inside the pores, during double layer charging. The HPC are modified by *in situ* (chemical or electrochemical) formation of metal (Pt/Ru) or metal oxide (CoOx, Fe_3O_4) nanoparticles. Additionally, HPC loaded with Pt decorated magnetite (Fe_3O_4) nanoparticles is produced by galvanic displacement. The modified HPC materials are used for the electroanalysis of different substances (CO, O_2, AsO_3^{-3}). The role of the nanoporous carbon substrate in the electroanalytical data is evaluated.

Introduction

Carbon materials are widely used in electroanalytical chemistry.[1] Among them, glassy carbon (GC) is used as electrode substrate in electroanalytical research, by depositing sensitive materials onto flat GC surfaces.[2,3] Graphite and related materials are the most commonly used component of carbon paste electrodes.[4] Novel forms of carbon, such as carbon nanotubes,[5] and graphene,[6] have been shown to be useful in electrochemical sensing. Besides the low cost and extensive availability of carbon materials, their novel forms allow nanostructuring of the

---

Chemistry Department, National University of Rio Cuarto (UNRC), 5800-Rio Cuarto, ARGENTINA. E-mail: cbarbero@exa.unrc.edu.ar; Fax: +543584676233; Tel: +543584676233

† Electronic supplementary information (ESI) available: SEM of SiO_2 opal; HRTEM of PrRuNP-sPC; CV of methanol on PtRuNP-sPC; CV and DEMS of adsorbed CO on Pt nanoparticles deposited inside non hierarchical porous carbon; CV of PtMagNP-dPC to detect proton reduction. See DOI: 10.1039/c3fd00018d

surface, producing materials with novel properties. The main advantage is the creation of a large active surface area in small devices.

On the other hand, there are different methods to produce nanoporous GC. Since the chemical and electrochemical properties of GC are well known, it is possible to produce plain GC or modified electrodes with large surface areas. The synthetic methods allow tuning of the porosity, creating not only micropores ($d < 2$ nm) but also mesopores ($2 < d < 50$ nm) and macropores ($d > 50$ nm). In this way, not only the surface area but the mass transport of analytes can be controlled.[7]

Using nanoporous microparticles, large surface area electrodes could be built or small microelectrodes with reasonable areas can also be fabricated. A porous carbon ($A_{sp} = 500$ m^2 g^{-1}) microsphere ($r = 10$ µm) adsorbed on a 20 µm diameter ultramicroelectrode will have a surface area of ca. 0.02 cm^2. Therefore, for a reversible one electron transfer of a redox couple (Co = 1 µM), a faradaic current of ca. 0.36 mA (at 100 µs) will be observed. On the other hand, a compact carbon sphere will have a surface area of 1.25×10^{-5} cm^2 (at 100 µs). Therefore, the faradaic current will be ca. 0.2 µA. In this case, the faradaic current increases 1800 times by using the porous material, while maintaining the same geometrical area of the electrode.

We have extensively studied the synthesis of porous carbons by pyrolysis of porous polymeric (resorcinol/formaldehyde) resins (RF).[8,9] The resin gels maintain their nanoporosity during conventional drying trough stabilization of resin nanoparticles by cationic supramolecular species. The carbon source (precursor) is subjected to a heat treatment (pyrolysis) at high temperature in the absence of oxygen to produce the carbon.

Porous carbon can be obtained using templates of nanometric size.[10,11] Since precursor materials are organic polymers, containing hydrogen and oxygen besides carbon, the pyrolysis involve mass loss and volume contraction. Soft templates like polymers and surfactants are spontaneously eliminated during burning of the carbon material, therefore post treatments are not necessary. On the other hand, hard templates like silica and metal oxides survive to high temperatures, and must be removed after pyrolysis by chemical etching. In this process, the space initially occupied by the templates is transformed into the pores in the resulting carbon materials. The whole process results in a reverse copy of the template.[12]

We have shown that the micrometric fibers also stabilize the nanopores during drying.[13] In that way, a spontaneous hierarchical carbon material (bearing macropores and mesopores) could be produced. Since the main research goal was to produce electrode materials for supercapacitors,[14] both the capacitance and response rate should be maximized.

While capacitance depends mainly on the surface area, the response rate is related to the mass transfer and/or potential effects inside the pores, which depends on the length and tortuosity of the pores. One way to overcome such problems involves the fabrication of hierarchical porous carbons, having both macropores and meso/micropores. The presence of macropores makes the length of the meso/micropores small (<10 µm) and could make the mass transfer fast enough.[15]

Hierarchical porous carbon,[16-18] can be obtained using two templates in a somewhat complex processes. On the other hand, we have recently shown that the volume contraction of the carbon around a remaining hard template during pyrolysis induces the formation of additional mesoporosity, which besides the pores defined by the hard template, is able to create a hierarchical carbon material.[19] In that way, hierarchical porous carbon can be obtained in a single pyrolysis step.

Porous materials (*e.g.* carbon) are usually characterized by measuring the adsorption isotherm of inert gases (*e.g.* N_2). By modelling the adsorption data, the surface area, pore volume and pore size distribution can be evaluated. However, in electrochemical applications only the surface area accessible to the electrolyte is important. It is well known that small micropores cannot be filled with electrolyte and do not contribute to the electrochemical active area but are measured by the gas adsorption isotherms.[20] Therefore, *in situ* measurements of ion adsorption phenomena such as differential capacitance,[21] or Probe Beam Deflection (PBD),[22] render more useful data of the carbon texture. PBD has shown to be very useful to study porous materials,[23,24] including carbon aerogels.[25] The application of the technique allows one to distinguish between systems with fast mass transfer (where all the charging/discharging phenomena occur in a negligible time, called a discontinuous process) and slow mass transfer (where the charging/discharging phenomena occur during the whole measurement). Such a distinction is quite important for the electroanalytical application of the materials.

In the present work, we used the technique to study the changes on ion concentration occurring inside the pores when the state of charge of the carbon materials is altered. An important point here is that, unlike planar electrodes, porous electrodes could develop local ion concentration gradients inside the pores (*e.g.* H^+) which are different from bulk values. Those local phenomena could heavily affect the electrocatalytic activity.

On the other hand, a direct way to measure textural properties of disordered porous carbon is still missing. We have used a nanotomography method based on the cutting of thin slices of the porous material using a focused ion beam (FIB) of Ga ions. Then, microscopic images of the exposed surfaces are taken with a field emission scanning electronic microscope (FE-SEM). The images are reconstructed onto a 3D image of the solid.[26] The slices (*z* axis) can be as thin as 10 nm and the resolution of the FE-SEM is *ca.* 1 nm. We have used the technique to study porous carbon materials.[27] In the present work, we used it to study the macroporosity of hierarchical porous carbon.

Metal nanoparticles have been extensively used as electrocatalysts in fuel cells research, and have potential use in electroanalysis.[28] PtRu catalysts are the most active for methanol oxidation,[29] likely due to the formation of Ru oxides in the surface which are able to oxidize adsorbed poisons like CO. Such behaviour would allow the oxidization of other organic species of analytical interest (*e.g.* glucose[30]) which do not present reversible redox couples but have oxygenated groups or C–H bonds which can be irreversibly oxidized. Different electrode systems, such as electrochemically deposited Ru on Pt, bulk PtRu alloys and electrodeposited PtRu alloys, have been studied.[31-33] We have investigated the electrooxidation of methanol at mesoporous (MP) Pt and Pt–Ru electrodes.[34] Pure metal mesoporous catalysts (mesoporous Pt modified by adsorbed Ru) show improved surface activities, with values as high as 0.82 A cm^{-2}. However, differential electrochemical mass spectrometry (DEMS) measurements with pure metal catalysts reveal that only a half of the current obtained during methanol electrooxidation produces CO_2 while in commercial catalysts (20% Pt/Ru 1 : 1 on carbon, E-TEK®) all the current produces CO_2 (conversion efficiency ca. 100%). It is likely that such differences in catalytic activity towards methanol oxidation are due to diffusion effects taking place in the catalytic film. In the present work we show the successful deposition of PtRu inside hierarchical porous carbon (dPC), by a chemical method and the application of the electrode (PtRu-dPC) to detect carbon monoxide and methanol.

Redox metal oxides are widely used in electrocatalysis.[35,36] Therefore, they have also been used in electroanalysis.[37] Cobalt oxide nanoparticles (CoOxNP) could be easily deposited electrochemically onto flat GC electrodes.[38] They have been used for determination of hydrogen peroxide,[39] and arsenic ions.[40] In the present work we show the successful deposition inside hierarchical porous carbon (CoOxNP-dPC) and its use to detect arsenic ions and oxygen.

Additionally, we shown that it is possible to deposit magnetite (Fe_3O_4) nanoparticles (MagNP) inside hierarchical porous carbon, by a well-known chemical method,[41] to produce an electrocatalytic electrode material (MagNP-dPC). This material is used to detect arsenite ions using differential pulse voltammetry. By galvanic displacement,[42] the supported magnetite nanoparticles are decorated with Pt (PtMagNP-dPC). The electrode is able to reduce oxygen in the presence of a large concentration of methanol.

In summary, we use different examples of the current work in our group to discuss some points which we feel are important in the use of nanoporous electrode materials in electroanalysis:

— that large electrochemically active areas could improve the signal/noise ratio of electroanalytical measurement, but the length of the pores should be kept at a minimum to be able to discern the faradaic current from the slow double layer charging.

— that hierarchical carbon systems are interesting materials for electroanalytical applications since they possess both large surface areas and good accessibility of solution analytes to the electrode surface.

— that direct characterization techniques of the carbon porous structure are necessary to understand the role of the electrode material in the electroanalytical responses.

— that local pH gradients could easily develop inside the pores, in neutral solution, due simply to double layer charging coupled with slow mass transfer. Such pH changes could affect the electrocatalysis and/or chemistry coupled to the electrochemical reaction.

— that the modification of porous carbon with metal or metal oxide nanoparticles, a method extensively explored in fuel cell research, could be used with profit in electroanalysis.

Experimental

All reagents are of analytical grade and Millipore® quality water is used.

Materials

Synthesis of the hierarchical porous carbon. Hierarchical porous carbon is synthesized by a hard template approach, using SiO_2 nanospheres as moulding agent (Scheme 1). The SiO_2 nanoparticles (SiO_2-NP) of different diameters (typical 400 nm) were synthesized by the Stober method, that is TEOS hydrolysis in basic media.[43] The diameter of SiO_2-NP were determined by dynamic light scattering (DLS, Malvern 4700 with goniometer and 7132 correlator) with an argon-ion laser operating at 488 nm. All measurements were made at the scattering angle of 90°. The nanoparticle dispersity (measured by dynamic light scattering) is quite low. The SiO_2-NP were then deposited by gravitational sedimentation in the bottom of a vial.[17]

After solvent evaporation, the resulting opal was taken out of the vial and then treated at 1000 °C for 4 h to form a connected matrix of SiO_2 nanospheres (ESI Fig. S1†). While there is no long range order in the opal, it is clear that could be used as sacrificial template for a different solid which could be synthesized inside the interstitial space between spheres. In our case, the opal was impregnated with a carbon precursor, a mixture of resorcinol (1 g), formaldehyde (1.6 ml), and sodium carbonate (catalyst, 0.4 ml of 0.1 M solution). The impregnated SiO_2 opal was dried in an oven and heated at (105 °C) in air for 12 h to polymerize the resorcinol/formaldehyde mixture and obtain an opal impregnated resin (SiO_2-RF). To obtain the final carbon the RF resin should be carbonized by heating in inert atmosphere.[44] Based on previous data,[19] two methods of post-processing could be used: (i) the SiO_2 is removed before pyrolysis by treatment with hydrofluoric acid solution (Single Pore Carbon, sPC). (ii) The composite (SiO_2-RF) is pyrolyzed and the SiO_2 removed afterwards by treatment with hydrofluoric acid solution (Double Pore Carbon, dPC). The pyrolysis of both samples was carried out at 850 °C for a 24 h. The corresponding morphological characterizations were performed by means of SEM, HRTEM and FIB-SEM.

Deposition of PtRu nanoparticles inside porous carbon (PtRuNP-sPC). The carbon (sPC) was loaded (20% metal) with Pt/Ru nanoparticles using a variation of the formic acid method.[45] Due to the nature of the material to be impregnated, a porous solid with particle size around 25 μm, the order in which reactants are added to the carbon suspension was modified. In the conventional method, the

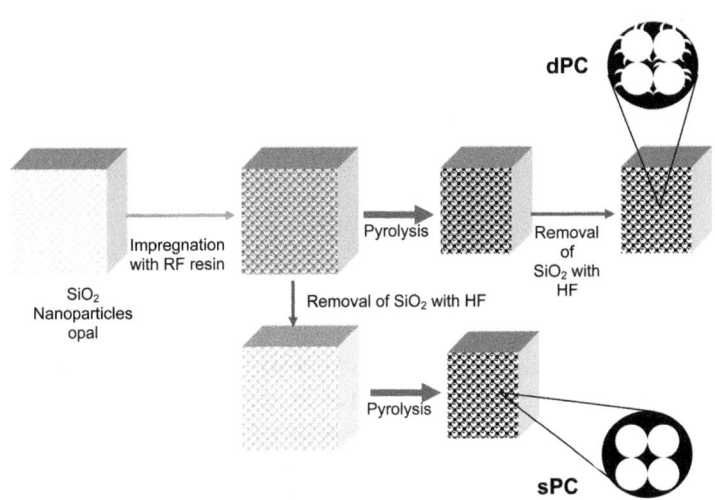

Scheme 1 Synthetic method to produce double pore carbon (dPC) or single pore carbon (sPC) using SiO_2 nanoparticles as hard template of resorcinol/formaldehyde resin.

last step involves the addition of the metallic precursors to be reduced. However for the porous carbon microparticles, the metallic precursors must reach the nanoparticle cores before the reduction starts. For this reason, the first step here is the addition, under vigorous stirring, of metal precursors to the HPC. This mixture is kept under stirring for 12 h, then the process is completed by the addition of an excess of formate solution (preceded by adjusting the pH to 10). Finally, the PtRu-sPC is washed with copious quantities of water, filtered and dried for its storage.

Deposition of cobalt oxide nanoparticles inside porous carbon (CoOxNP-dPC). Cobalt oxide nanoparticles were deposited electrochemically as described before.[37] A GC electrode modified by a porous carbon layer was cycled (50 mV s^{-1}, between -1.1 V and 1.25 V$_{Ag/AgCl}$) in a solution containing 0.01 M CoCl$_2$ and 0.1 M CH$_3$COONa. In the buffered solution, the Co^{II} ions are soluble but the Co^{III} species form an insoluble deposit. The amount of material was controlled by the number of deposition cycles (15 to 30 cycles).

HPC carbon loaded with magnetite (Fe$_3$O$_4$) nanoparticles (MagNP-dPC). To synthesize *in situ* magnetite (Fe$_3$O$_4$) nanoparticles inside the porous carbon, we used a procedure described before for the synthesis of free nanoparticles.[46] Powdered porous carbon (100 mg) is mixed with 25 ml of a solution consisting of 0.2 M ferric chloride (FeCl$_3 \cdot$6H$_2$O) and 0.1 M ferrous chloride (FeCl$_2 \cdot$6H$_2$O). The Fe/C ratio used is *ca.* 20% w/w. The mixture is subjected to magnetic stirring for 20 h, to incorporate the solution inside the carbon. After that, 25 ml of NaOH solution (0.1 M) are added dropwise. A black precipitate is formed instantaneously. The excess reactants are washed out with distilled water while the magnetite containing material is held at the bottom of the tube with a magnet.

Synthesis of Pt/magnetite nanoparticles on HPC (PtMagNP-dPC). The Pt/magnetite nanoparticles were synthesized by the galvanic displacement method,[47] using magnetite nanoparticles as the sacrificial material. Carbon loaded with magnetite nanoparticles, synthesized as described above, was mixed with hexachloroplatinic acid solution (0.01 M). The presence of Pt was detected by the presence of a hydrogen reduction current (in acid media) below 0.1 V$_{RHE}$.

Deposition of porous carbon particles on GC. A carbon ink is prepared by dispersing the powdered material in a solution Nafion®/water/ethanol. The ink was prepared by dispersing 4 mg of carbon in 1 ml of distilled water with 15 µl of 10% solution Nafion® in ethanol. The mixture was dispersed using ultrasound. To prepare the deposit, 10 ml of the ink were placed onto a flat GC surface and the solvent was evaporated in still air.

Methods

Scanning electron microscopies. Carbon morphology was observed using a field emission scanning electron microscope. A dual beam workstation (FEI Helios Nanolab 600) equipped with a FIB column employing a gallium liquid metal ion source, combined with a field emission gun scanning electron microscope (SEM) was used. The SEM column was used for general imaging and the FIB column was used for cross-section preparation.[48]

Electrochemical characterization of the HPC support. The electrochemical performance of the carbon electrodes was analyzed with a three electrode configuration. N$_2$ was bubbled through the solution to avoid dissolved oxygen.

The working electrodes are glassy carbon rods (3 mm diameter) inserted in Teflon, where carbon ink was deposited (see below). A carbon aerogel (Maketech, geometrical area 5 times of the working electrode) and a silver/silver chloride electrode (saturated KCl) were used as counter and reference electrode. Cyclic voltammetry (CV) and the AC impedance measurements were performed using a PC4 Potentiostat-Galvanostat-ZRA (Gamry Instruments, Inc).

Cyclic voltammetry. Cyclic voltammetry was carried out to test the electrochemical active area of both samples. An electrochemical evaluation of the carbon area has been preferred here due our interest in using these materials as supporting materials for fuel cell catalysts,[49] but can also be useful in other electrochemical devices for energy production and storage,[17] like electrochemical double layer capacitors,[50,51] and carbon/graphite composed materials for Li ion batteries.[52,53]

This measurement gives an accurate idea of the accessible surface for any electrochemical process in aqueous media. The differential capacitance was calculated as:

$$Cd = \frac{dQ}{dE} = \frac{dQ}{dt}\frac{dt}{dE} = i\,v \qquad (1)$$

where Q is the charge, E the potential, t the time, i the current and v the scan rate.

Using the current in amperes and the scan rate in volts per second, the differential capacitance is obtained in farads. Assuming that the capacitance surface density is constant, the capacitance gives a comparative idea of the extension of the electroactive area. However, it is known that surface redox groups (like quinone) could contribute to the capacitance value. Since these groups are active in acid media, the capacitance is measured in neutral media. The current in the resulting cyclic voltammograms was divided by the scan rate and the electrode mass to obtain specific capacitance *vs.* voltage profiles.

AC impedance. The AC impedance measurements were performed using a computerized potentiostat (GAMRY PC4) and CM 300 impedance software. The cell configuration has three electrodes with a counter electrode of carbon aerogel (Maketech, geometrical area 5 times of the working electrode) situated parallel to the working electrode and a reference electrode of SCE. The carbon aerogel was used as counter electrode to assure that its electrochemical active area was bigger than that of the working electrode. The measurements were made with a sinusoidal voltage perturbation of 1 mV and a resting time at each potential of 30 min. The circuit simulation and fitting were made using the analysis software in Excel (Microsoft) provided with the equipment.

Probe beam deflection. Probe Beam Deflection is a technique that measures the concentration gradient in front of the electrode by monitoring the refractive index gradient with a light beam.[22] The electrochemical charging of the double layer could be accompanied by a ion fluxes due to diffusion and migration. In a binary electrolyte both modes of mass transfer are necessarily coupled and a single binary diffusion coefficient describes the flux. The ion concentration in the solution changes, creating a gradient of refractive index normal to the electrode surface. A beam traveling parallel to the surface suffers a deviation proportional to the concentration gradient, therefore proportional to the extent and direction of ion flux. A positive beam deflection (away from the electrode) corresponds to incorporation of ions into the double layer while negative deflection (towards the electrode) implies release of ions to the solution. The Probe Beam Deflection arrangement was similar to the one described before.[54] The basic components of the PBD system are a 5 mW He-Ne laser (Melles Griot, 05 LHP11) and a bicell position sensitive detector (UDT PIN SPOT/2D). The laser beam is focused by a 50 mm lens to a diameter of roughly 60 µm in front of the planar electrode. The electrochemical cell, an optical glass cuvette with $2 \times 2 \times 4$ cm dimensions (1 cm of path length), is mounted on a 3 axis tilt table (Newport). A micrometric translation stage allows for controlled positioning of the sample with respect to the laser beam in 10 µm steps. The position sensitive detector is placed 25 cm behind the electrochemical cell and has a sensitivity of 3 mV µm^{-1}, which resulted in a deflection sensitivity of 1 mrad V^{-1}. All parts of the system are fixed on an optical rail and the whole set-up is mounted on an optic table (Melles Griot). The deflection signal was processed using a position monitor (UDT 201 DIV). Due to the fact that the PBD signal has to be monitored for long times (>50 s), the whole system was warmed up for 24 h before each measurement to eliminate thermal fluctuations. The signal of the two photo-diodes making the bicell detector were subtracted and normalized to the overall signal to eliminate laser intensity fluctuations. The chronodeflectometric pulses were fitted using the nonlinear fitting routine of Origin 7.0 (MicroCal). The glass cell contains a counter electrode of aerogel (with geometrical area 5 times the working electrode) and a Ag/AgCl (3 M NaCl) miniature reference electrode (BAS) separated from the solution with a Vycor diaphragm. The working electrodes were carbon composites plates (width 2 mm) attached onto Teflon plates with sides and back sealed and the active side unpolished. The electrochemical control of Probe Beam Deflection experiments was performed using a potentiostat (AMEL 2049). The set-up was controlled by a PC through a LabPC AD/DA card running in homemade software performed with LabView 5.1 (National Instruments). The deflection and the electrochemical signals were saved jointly in the PC.

Electronalytical measurements

Carbon monoxide oxidation. Electrochemical measurements were performed in 1 M CH_3OH–1 M H_2SO_4 at 60 °C, with a PC controlled Autolab PGSTAT30 potentiostat–galvanostat and a thermostatized three-electrodes electrochemical cell. We use a hydrogen reference electrode (RHE) in the electrolyte solution as the reference.

Oxygen reduction. Oxygen reduction measurements were performed in an O_2 saturated electrolyte produced by gently bubbling ultrapure (99.99) O_2 gas trough the solution by 15 min.

Arsenic ion oxidation. A mother solution of arsenic ions is prepared by dissolving weighted amounts of arsenic oxide (Aldrich, 99.9%) in 0.1 M NaOH solution. An aliquot of the mother solution is then mixed with phosphate buffer solution (pH = 7).

Results and discussion

Theory and numerical calculations

High surface area materials should be useful in electroanalysis because the faradaic signal scales with the active area. However, since the background signal in electrochemical measurement is due to double layer capacitance of the base electrode, its magnitude also scales with the active area. Moreover, small pores (such as micropores $d < 2$ nm) will contribute to double layer capacitance but are less accessible to mass transfer of analytes from solution. A usual method to eliminate capacitive background signal is to use pulses as the perturbation signal. For a small planar electrode, the capacitive charge (related to the background signal) decays exponentially with time while the faradaic charge (related to the analyte signal) decays with the square root of time. Therefore, after a perturbation pulse, the background signal will decay faster than the analytical signal. In that way, measuring at longer times increases the signal/background ratio. Such an approach is the basis of pulse voltammetries (normal and differential) and square wave voltammetry.[55] In differential pulse voltammetry (DPV), a small amplitude (10 to 100 mV) potential pulse is superimposed on a slowly changing base potential. The current is measured at two points for each pulse, the first point just before the application of the pulse and the second at the end of the pulse. These sampling points are selected to allow the decay of the nonfaradaic (charging) current. The difference between current measurements at these points for each pulse is determined and plotted against the base potential.[56]

However, the larger electrode area will increase not only the faradaic current but also the charging current, which is related the double layer capacitance. The charging current $i_{cap}(t)$ follows the expression

$$i_{cap}(t) = \frac{\Delta E}{Rs} e^{\left(-\frac{t}{Rs\ Cd_{sp}\ A}\right)} \qquad (2)$$

where ΔE is the potential pulse width, Rs is the cell resistance, A is the electrode area and Cd_{sp} is the aerial capacitance density.

On the other hand, for a simple reversible redox, at a potential where all reactant is converted at the electrode surface, the current will follow Cottrell's equation:

$$i_{far}(t) = nFACo\sqrt{\frac{Do}{\pi t}} \qquad (3)$$

where n is the number of electron exchanged, F is Faraday constant, A is the area and Do is the diffusion coefficient of the redox species.

Using eqn (2) and 3, it is possible to calculate the current dependence on time (Fig. 1).

$\Delta E = 0.1$ V; $Cd_{sp} = 20$ μFcm^{-2}, $Do = 1\ 10^{-5}\ cm^2 s^{-1}$, $Co = 1\ 10^{-6}$ M, $n = 1$. (a) $A = 0.1\ cm^2$, (b) $A = 10\ cm^2$.

As it can be seen, for small electrodes ($A = 0.01\ cm^2$, Fig. 1a) the capacitive current is larger than the faradaic current, at small times. At longer times, the faradaic current is larger than the capacitive current. Therefore, the approach of DPV is justified. The question is about large electrodes. For large electrodes ($A = 10\ cm^2$, Fig. 1b) the faradaic current is always larger than the capacitive current. It should be borne in mind that it is possible to fabricate an electrode with large active area and small geometrical area (<1 mm^2) depositing a porous carbon layer on a small flat GC electrode. Using only 1 mg porous carbon (specific area 500 m^{2}g^{-1}), it is possible to produce an electrode with ca. 500 cm^2 of active area.

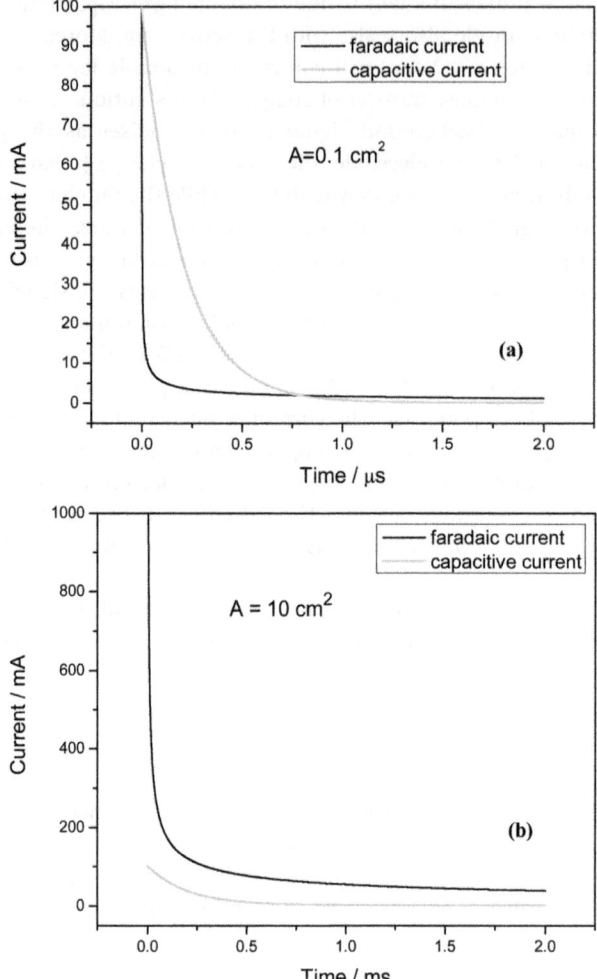

Fig. 1 Calculated capacitive and faradaic current measured after a potential pulse is applied to the electrode.

Such an electrode will show higher faradaic than capacitive currents at all times. Additionally, there is no need to wait until the double layer charging has decreased to a low value (as it is the case with small electrodes), only the width of the pulse should be kept constant.

The discussion above assumes that a porous electrode is only a planar electrode of extended area. However, it was suggested, based on PBD data, that ion transport in monolithic porous carbon occurs in a continuous process where the response scales with $\sqrt{t}$.[23] It can be thought that ions have to move inside the pores to compensate the charge inside the porous matrix. If the pores are tortuous and/or of small section, a small diffusion ($D < 10^{-7}$ $cm^2 s^{-1}$) through a relatively thick layer (>200 μm) will mean that 2000 s are necessary to charge the electrode; pure electrochemical measurements also predict a dependence of the fluxes with $\sqrt{t}$. To maintain electroneutrality, charge transport is linked to ion transport. It seems paradoxical for double layer charging to be controlled by mass transport, but that is the case in porous materials. Therefore, the current is controlled by the slow ion transport.

An alternative mechanism obeys a transmission line model (TLM) to account for electric potential profiles inside the pore.[57] In this case, the time profile of the charging current is given by eqn (4).[58]

$$i(t) = \Delta E \sqrt{\frac{Cd_{sp} A}{R_{pore} \pi t}} \left[1 + 2 \sum_{n=1}^{\infty} (-1)^{n+1} e^{(-n^2 \tau / t)} \right] \quad (4)$$

where ΔE is the potential step, R_{pore} is the ionic resistance, Cd_{sp} is the specific double layer capacitance and

$$\tau = R_{pore} C_{sp} A = \rho \left(\frac{L}{A S_{pore}} \right) Cd_{sp} A \quad (5)$$

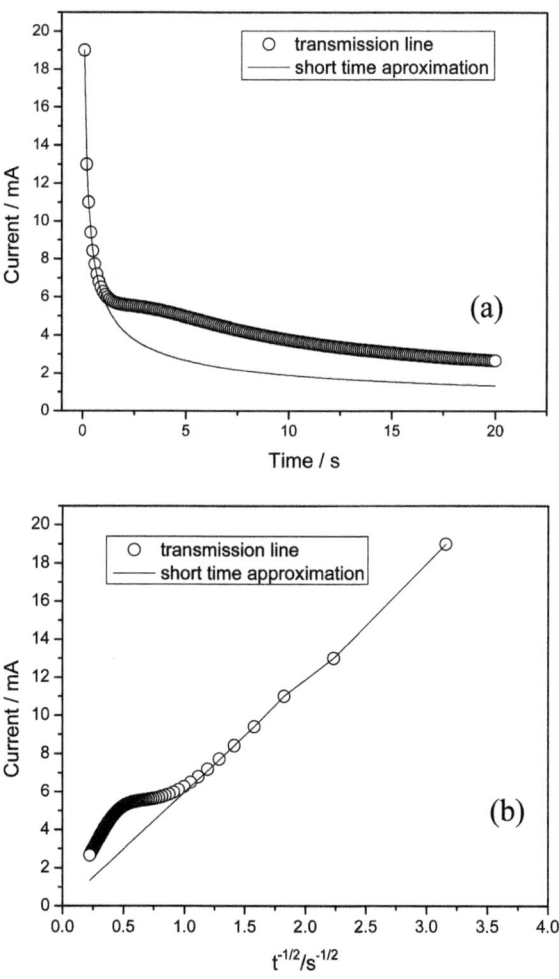

Fig. 2 Simulated time profile of capacitive current for a porous electrode following the full transmission line expression (eqn (3), open circles) or the short time approximation (eqn (4), full line). (A) current-time profile. (B) Cottrell plot. $R_{pore} = 18$ ohms. $Cd_{sp} = 20$ μF cm^{-2}, $A = 100$ cm^2.

where ρ is the resistivity, L is the electrode thickness and S_{pore} is the section of the pore.

If $L \to \infty \Rightarrow \tau \gg t$, therefore the term between brackets tends to 1, and the current obeys eqn (6):

$$i(t) = \Delta E \sqrt{\frac{Cd_{sp} A}{R_{pore} \pi t}} \qquad (6)$$

which defines a semi-infinite transmission line.

As it can be seen in Fig. 3, for small concentrations the faradaic current will be smaller than the capacitive current.

The slow charging behaviour precludes, or at least complicates the interpretation, of pulse potential data (including DPV), at least for small concentrations of redox species where the faradaic current is small. Therefore, experimental conditions should be set to assure that capacitive charging has subsided before the current measurement is taken. This is possible using short mesopores, such as those present in hierarchical carbons where the macropores limit the length of the solid mesopores.

On the other hand, in amperometric sensors usually large time pulses (>10 s) are used to allow "stationary" mass transport of the analytes and effective mass transfer of analytes at low concentrations. In such experimental conditions, the decay of the capacitive current is probably complete and only the faradaic current is measured. Obviously, if the mass transfer inside the pores allows the most of the pore area to be electrochemically active, the large electrode area will render a large current signal and improved sensitivity to a low concentration of analytes. However, it should be borne in mind that a commonly used technique to test the electrode activity, cyclic voltammetry, at usual scan rates (>10 mV s^{-1}) will be affected by the charging current and show a completely different picture of the system under study.

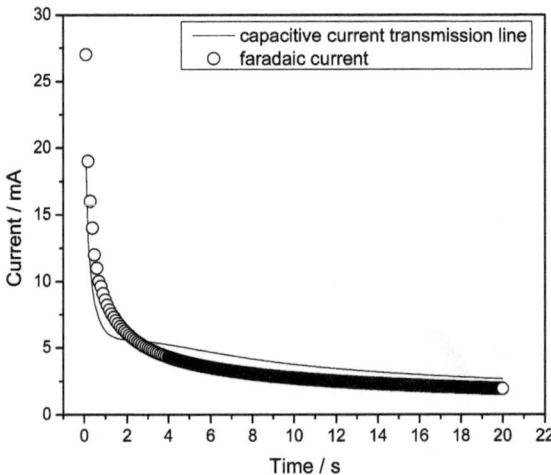

Fig. 3 Simulated responses of the capacitive current (eqn (3), full line) and faradaic current (eqn (2), open circles) of a porous electrode. The capacitive current follows the full transmission line expression with R_{pore} = 18 ohms, Cd_{sp} = 20 μF cm^{-2}, A = 100 cm^2. The faradaic current follows Cottrell equation with Do = 1 10^{-5} cm^2 s^{-1}, Co = 5 10^{-9} M, n = 1, A = 100 cm^2.

It should be mentioned that another technique useful to study porous electrodes is Probe Beam Deflection.[22] In this technique two extreme conditions of the double layer charging can be easily detected experimentally:

— *Discontinuous process*: where all the charge (and ions) are stored/released in a time span much smaller than the measurement. This condition is related to an RC circuit.

— *Continuous process*: where charge (and ions) are stored/released along the whole measurement. This condition is related to a semi-infinite transmission line.

The condition of a finite transmission line is more complex and has not been yet theoretically analyzed. Each kind of process shows a distinctive chronodeflectometric profile, making PBD a useful technique for porous electrode investigation.

The discussion suggests that hierarchical porous carbon should be a material of interest for electroanalytical applications possessing large surface areas and relatively fast double layer charging. Since carbon is a poor electrocatalyst for the electrochemical reaction of different substances, the carbon surface should be modified by incorporation of electrocatalysts. Porous carbon provides a large surface to adsorb electrocatalytic particles and also could protect the material from erosion by hiding the nanoparticles inside the pores. However, the particles must be of nanometric size to be able to decorate the inner pore surface and not be deposited only in the outer surface.

Synthesis of the hierarchical porous carbon

To produce carbon with and open structure, we used silica nanoparticles as hard template. The volume contraction during pyrolysis of RF resin is easily observable by comparison between the piece size before and after thermal decomposition of the precursor. However, when the carbon sample contains a rigid template, the observed contraction in the macroscale is very low due to the presence of a non compressible skeleton. In our case, this rigid support is composed of an array of nanometric SiO_2 nanoparticles with close contact between them. The SEM image of these SiO_2 nanoparticles (ESI, Fig. S1†) reveals the existence of almost monodisperse spheres with diameter of *ca.* 390 nm. The observed size agrees with the results of the DLS experiments which report an SiO_2 NP diameter around 400 nm, with a polydispersity index of ~3%. Such low dispersion favours the packing of the nanoparticles into a close packed structure. As is shown in Scheme 1, the hard template can be eliminated before or after pyrolysis. We have shown,[19] that elimination before pyrolysis produces a carbon with one pore size (sPC) which is directly determined by the template nanoparticle size but takes into account the 20–30% contraction of the material upon pyrolysis. On the other hand, if the hard template is left during pyrolysis, the contraction of the solid material has to take place in the interstitial spaces between particles. In this way, the solid becomes microfractured and the carbon has two pore sizes (dPC), one directly related to the hard template size (*e.g.* 400 nm) and another, much smaller, on the order of mesopores (<50 nm). Additionally, the material presents sintered beads of carbon linked into a matrix through their necks, which has been proposed as the building block of the RF resin. Those beads, after precipitation, form the condensed monoliths.[59,60] Such a structure is clearly observed in the fracture (Fig. 4a) cut of a dPC piece. On the other hand, in the flat FIB cut it is possible to observe the open three dimensional nature of the carbon (Fig. 4b).

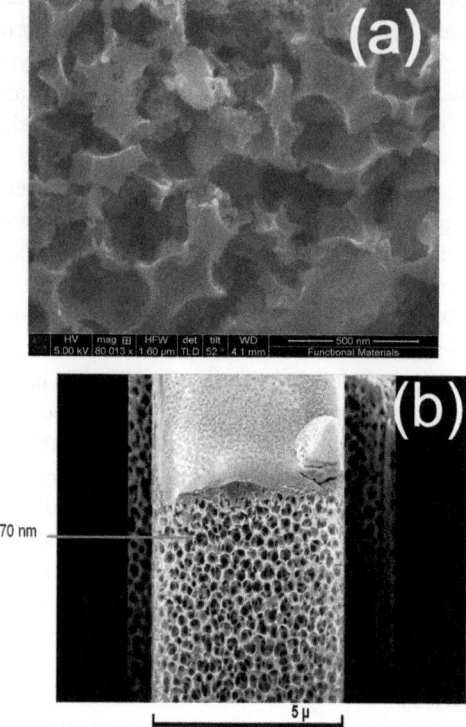

Fig. 4 FE-SEM micrographies of dPC synthesized using an opal of SiO$_2$ nanoparticles (400 nm diameter) as hard template. (a) Fracture surface. (b) Flat cut made using a FIB (Focused Ion Beam) of Ga.

The macropore structure seems quite open, allowing the analyte to diffuse easily to the electrode surface.

Electrochemical characterization of the HPC support

Measurements of double layer capacitances give a semiquantitative value of the extension of the electroactive area and are quite important to assess the relative contribution of the faradaic and capacitive signal. The electrodes made with dPC, where the silica template was removed after pyrolysis, show a CV curve with a large differential capacitance (130 F g^{-1} of carbon), associated with a large surface area (Fig. 5).

This capacitance value is quite similar to that reported before for the other porous carbon.[61] From these values, a specific surface electroactive area of ca. 650 m^2 g^{-1} can be roughly estimated (based on a specific capacitance of 20 μF cm^{-2}).[62] The value is larger than that expected for the exposed area of the holes left by 400 nm template nanoparticles. Therefore, the contraction of the carbon should have produced nanopores (meso and micropores) in the macropore inner surfaces.

In the case of a carbon material with only one pore size (sPC), where the template of SiO$_2$ nanoparticles template is removed before pyrolysis (sPC), the cyclic voltammogram (not shown) shows specific capacitance values reduced to ca. 30 Fg^{-1}. Such values correspond to specific surface areas pn the order of 150 m^{2}g^{-1}.

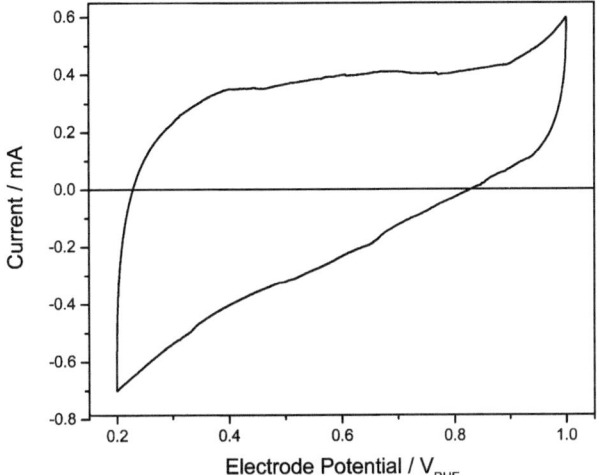

Fig. 5 Cyclic voltammetry of porous carbon (400 nm diameter) in H_2SO_4 1 M. Scan rate = 5 mV s^{-1}. Specific capacitance = 132 Fg^{-1}.

Chronoamperometry

As was discussed before, the chronoamperometric signal of a porous carbon will follow a transmission line model, which tends to $1/\sqrt{t}$ at short time span (semi-infinite condition). A plot of current as a function of $1/\sqrt{t}$ should be linear (Cottrell plot).

This is the case with thick (250 μm) monolithic porous carbon (Fig. 6, full squares), indicating that in a long time span (1000 s) the current is controlled by ion transport. On the other hand, for HPC (Fig. 6, open circles) the linearity only occurs up to 400 s. After that, the full transmission line regime is operative (as shown in Fig. 2).

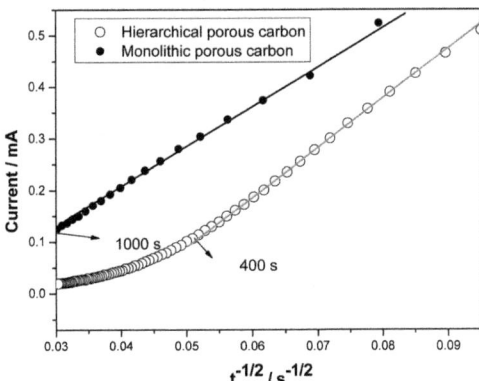

Fig. 6 Cottrell plots obtained by applying a potential pulse (between 0.0 and 0.35 V$_{SCE}$) to a monolithic porous carbon electrode (full circles) and a hierarchical porous carbon electrode (open circles). Both chronoamperometric data were obtained in 1 M H_2SO_4. The full lines are LSQ linear fits of the linear region of each plot.

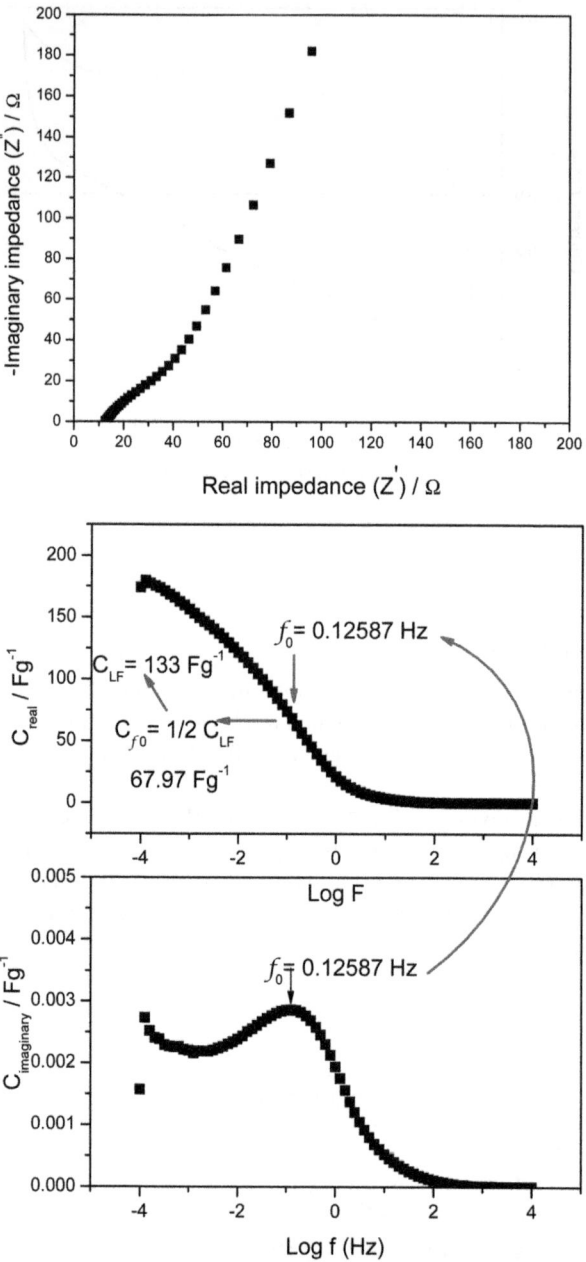

Fig. 7 Impedance dependence with frequency. (a) Nyquist plot (of a hierarchical porous carbon (400 nm) between 0.1 mHz and 10 kHz at 0.65 V_{RHE} in 1 M H_2SO_4. (b) Evolution of the real part of the capacitance vs. the frequency (above) and evolution of the imaginary part of the capacitance vs. the frequency (below).

AC impedance

While chronoamperometry gives a fairly reasonable account of the HPC electrode, it involves a large potential excursion. On the other hand, AC impedance uses a

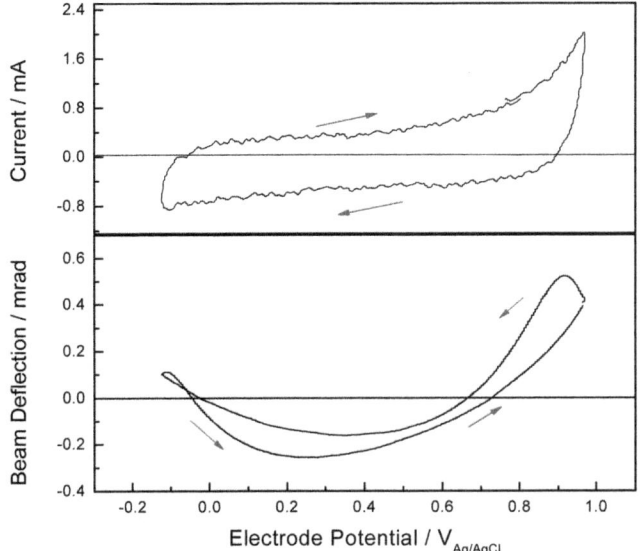

Fig. 8 Cyclic voltammetry (upper box) and cyclic deflectometry (lower box) of a HPC (400 nm) electrode in a 0.01 M KCl solution (scan rate 20 mV s^{-1}).

small potential amplitude assuring that the measurement is made on a linear system.[63] A typical Nyquist plot of an HPC is shown in Fig. 7.a

To determine the real capacitance, the real and imaginary capacitance are plotted as a function of frequency (Fig. 7.b), where:

$$C = C_{real} - j\, C_{imaginary}$$

$$C_{real}(\omega) = \frac{-Z_{imaginary}(\omega)}{\omega |Z(\omega)|^2}$$

$$C_{imaginary}(\omega) = \frac{-Z_{real}(\omega)}{\omega |Z(\omega)|^2}$$

In the plot, f_0 corresponds to $C_{real} = C_{LF}/2$; f_0 separates the capacitive behavior ($C_{real} > C_{LF}/2$) and the resistive behavior ($C_{real} < C_{LF}/2$) of the capacitor. f_0 corresponds to the maximum energy dissipation and $\tau_0 = 1/f_0$. Therefore Cd = 133 F g^{-1} which is in agreement with the capacitance measured by cyclic voltammetry.

The response at high frequencies show a deformed semicircle while the response at low frequencies does not show a line parallel to the y axis, suggesting that the system does not behave as a simple double layer capacitance. The deviation at low frequency usually shows the influence of mass transport, related to the microporous structure of the interface between electrode and electrolyte.[64]

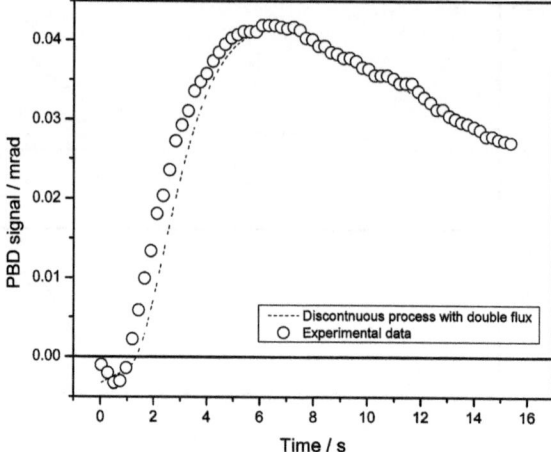

Fig. 9 Chronodeflectometry of a dPC in 0.1 KCl. The potential is pulsed between 0 and 0.9 $V_{Ag/AgCl}$. The dashed line present the simulation using two discontinuous process of opposite sense.

Probe beam deflection

To evaluate the ion exchange coupled with the double layer charging, we use Probe Beam Deflection (PBD). The cyclic voltammetry (Fig. 8. upper box) show a nearly constant current, typical of double layer charging. The PBD signal decreases at lower potential ($E < 0.3 \text{ V}_{Ag/AgCl}$) and increases at higher potential. This is coherent with a double layer with a potential of minimum charge‡ of ca. 0.25–0.35 $V_{Ag/AgCl}$).

To evaluate the dynamic behavior, the PBD signal is monitored during time after the potential is switched between 0.0 and 0.9 $V_{Ag/AgCl}$ (Fig. 9).

Distance beam-electrode = 100 μm. $D_1 = 2.58 \times 10^{-6}$ cm^{2}s^{-1}, $D_2 = 3 \times 10^{-5}$ cm^{2}s^{-1}. $C_1 = 1 \times 10^{-3}$ M. $C_2 = 1 \times 10^{-4}$ M.

The double layer process should show only one ion exchanged (e.g. Cl$^-$):

$$C(-) + Cl^- \rightarrow C(+)Cl^- + n\ e^- \quad (I)$$

where C(−) represents the carbon at potentials more negative than the potential of minimum charge (pmc) and C(+) represent the carbon at potentials more positive than the pmc. In that case, only one peak will be present.

It is noteworthy that the chronodeflectometry profile is fitted with a discontinuous process since the pore length in the hierarchical porous carbon is short enough to allow double layer charging in a negligible time compared with the span of the measurement (>16 s).

However, a prepeak is observed in the chronodeflectometric signal. A possible explanation is the fact that, inside the pore is faster to dissociate water and adsorb OH$^-$ instead of bringing Cl$^-$ from the solution, because protons are faster ($D = 8.1\ 10^{-5}$ cm^{2}s^{-1}) than Cl$^-$ ($D = 1.33\ 10^{-5}$ cm^{2}s^{-1}).[65] In that way, at short times,

---

‡ The potential of minimum charge (pmc) is the potential where a balance exists between negative and positive charges. It is the polycrystalline equivalent of the potential of zero charge (pzc) which is valid for single crystals.

water is dissociated and OH⁻ is adsorbed, then H⁺ are expelled creating a counterflux to the anion intake:

short time

$$C(-) + H_2O \rightarrow C(+)OH^- + H^+ + n\,e^- \quad \text{(II)}$$

long time

$$C(-) + Cl^- \rightarrow C(+)Cl^- + n\,e^- \quad \text{(III)}$$

$$C(+)OH^- + H^+ + Cl^- \rightarrow C(+)Cl^- + H_2O + n\,e^- \quad \text{(IV)}$$

This is observed as a prepeak in the chronodeflectometry signal. Inside the pore, the local pH is higher than the bulk pH. At longer times, anions are inserted, OH⁻ ions are desorbed and protons are inserted making the internal pH equal to the external pH. Such local pH could affect the solution chemistry and/or electrochemistry of the analyte and should be taken into account when using porous electrodes.

The diffusion coefficients used for the fitting (Fig. 9) correspond to the binary diffusion coefficient of KCl (main peak) and HCl (prepeak) since in a binary electrolyte like KCl, the electroneutrality maintenance links the mass transfer of anions and cations.[66]

sPC impregnation with PtRu

Previously, we have shown that catalytic metal nanoparticles (*e.g.* Pt) could be *in situ* synthesized inside porous carbon by the formic acid method.[67]

With aim of testing the utility of HPC as a support for the catalyst, a sample of dPC was loaded with 20% (w/w) of Pt/Ru (1 : 1) as described in the Experimental section. A transmission electron microscopy image of the PtRuNP-sPC (ESI, Fig. S2†) shows good nanoparticle dispersion, with a nanoparticle size around 2 nm.

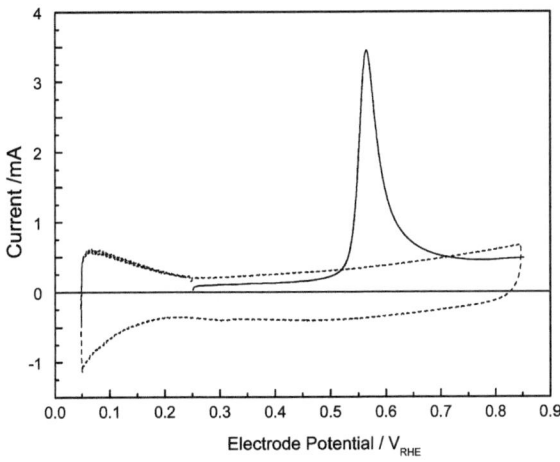

Fig. 10 CO stripping at 25 °C (full line) for PtRuNP-sPC. CV in clean electrolyte (dotted line). $E_{ad.} = 0.25V$, $t_{ad} = 10$ min. $v = 20$ mV s⁻¹.

CO and methanol electrooxidation on PtRuNP-sPC

The cyclic voltammogram in 1 M H_2SO_4 electrolyte (black dotted line in Fig. 10) shows the characteristic features of carbon supported PtRu materials.[68]

CO adsorption and stripping experiments were performed at 25 °C. The adsorption was performed by exposition of the electrode to a CO saturated solution (0.92 mM)[69] during 10 min, keeping the electrode potential fixed at 0.25 V. After that, clean electrolyte (CO free) was flushed for an additional 20 min in the presence of bubbling N_2, to prevent the presence of dissolved CO during the stripping. The CV shows a clear peak due to CO oxidation to CO_2.

$$CO + H_2O \rightarrow CO_2 + 2 H^+ + 2e^- \quad\quad (V)$$

It is noteworthy that oxidation of CO adsorbed on Pt nanoparticles loaded inside a monolithic porous carbon does not show the current due to CO oxidation.[67] However, simultaneous differential electrochemical mass (DEMS) spectroscopy measurements show that CO_2 was being produced in the same potential range, indicating that CO was being oxidized (ESI, Fig. S4†). It seems that slow double layer charging, present in non hierarchical porous carbon, does not allow the distinction between charging and faradaic current during cyclic voltammetry experiments. Since DEMS only detect volatile species, is insensitive to charging current effects and clearly show that CO electrochemical oxidation was occurring. In HPC electrodes, the open structure of the material diminishes the effect of slow charging and allows the CO oxidation current to be easily detected. From the CO adsorption and stripping experiments, it is possible to measure an important parameter: the electrochemically active area of the metal nanoparticles. The CO oxidation charge in Fig. 10 is 9.15 mC. Therefore, assuming 420 μC cm^{-2} for CO stripping, the electroactive surface area of the PtRuNO-sPC is 21.8 cm^2. Since the film contains 40 µg of nanoparticles, the calculation gives an area of 54.5 m^2g^{-1} for the surface area of the PtRu nanoparticles. The value is reasonable for small and highly distributed PtRu nanoparticles. The data show that PtRuNP-sPC is active to oxidize adsorbed redox species and cyclic voltammetry is able to distinguish between double layer charging and adsorbed species. The electrode is also able to oxidize methanol, (ESI, Fig. S3†) showing promise for electroanalysis of organic molecules which do not have reversible redox groups.

Oxygen reduction on CoOxNP-dPC

A possible way to electrocatalyze oxygen reduction is to use redox catalysts, such as redox metal oxides.[70] Cobalt oxide nanoparticles were easily deposited inside dPC by cyclic voltammetry (as described in the Experimental section). The electrode was tested for oxygen reduction (Fig. 11).

The voltammogram of the modified electrode in oxygen saturated (0.24 mM[71]) basic (0.1 M KOH) solution (Fig. 11) clearly shows two peaks due to the successive 2 electron transfer reactions occurring during oxygen reduction. By comparison, the electrode shows featureless changing current behaviours in deareated (bubbling N_2 during 15 min) solution.

The process at ca. -0.4 $V_{Ag/AgCl}$ involves the 1-electron reduction of oxygen which is mediated by quinone groups (Q) present at the carbon surface.[72]

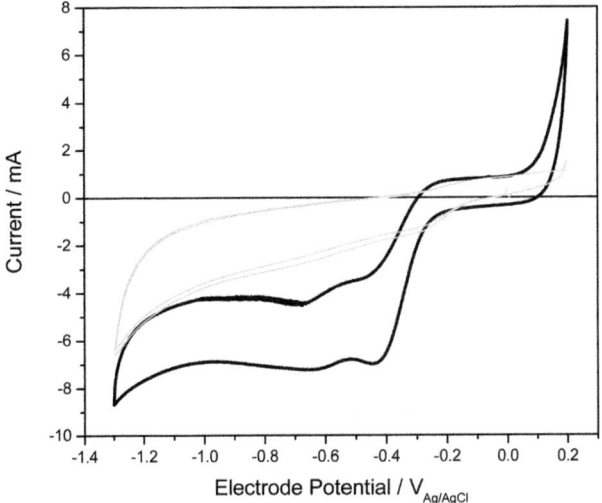

Fig. 11 Cyclic voltammetry of a porous carbon modified with CoOx nanoparticles electrode in saturated O_2 solution (black line) and saturated with N_2 (grey line). Electrolyte = 0.1 M KOH, scan rate = 10 mV s^{-1}).

$$Q + e^- \rightarrow Q^- \qquad (VI)$$

$$Q^- + O_2 \rightarrow Q + O_2^- \qquad (VII)$$

The O_2^- ion disproportionate into O_2 and HO_2^- (VIII) or is reduced to HO_2^- (IX)

$$2O_2^- + H_2O \rightarrow HO_2^- + O_2 + OH^- \qquad (VIII)$$

$$O_2^- + H_2O + e^- \rightarrow HO_2^- + OH^- \qquad (IX)$$

The peroxide radical anion could also be produced by direct reduction at the carbon surface, but at a slower rate.

The peroxide anion decomposes into oxygen and OH^-

$$2HO_2^- \rightarrow O_2 + 2OH^- \qquad (X)$$

Such reaction is catalyzed by redox hydroxides (like $Co(OH)_2$) which can be reversibly converted into more oxidized species:

$$Co(OH)_2 + HO_2^- \rightarrow CoOOH + OH^- + H_2O \qquad (XI)$$

to release O_2 by a coupled chemical reaction:

$$4CoOOH + 2H_2O \rightarrow 4Co(OH)_2 + O_2 \qquad (XII)$$

or to be electrochemically reduced back to the hydroxide

$$CoOOH + e^- + H_2O \rightarrow Co(OH)_2 + OH^- \qquad (XIII)$$

The metal oxide nanoparticles have a large surface area where those reactions can occur. Its reduced size makes the electron transport inside the particle, towards the base carbon electrode, quite fast. Therefore, it is likely that reaction (XIII) dominates over (XII).

The peroxide anion is electrochemically reduced to hydroxide at ca. -0.62 $V_{Ag/AgCl}$.

$$H_2O_2^- + e^- \rightarrow 2OH^- \qquad (XIV)$$

Such reaction could also be catalyzed by cobalt oxide species.

The results show that dPC modified with cobalt oxide nanoparticles are able to detect oxygen.

Arsenite oxidation on MeOx NP/dPC

Another relevant analyte is arsenic, a common contaminant of surface waters,[73,74] present as different ionic species. Metal oxides adsorb easily arsenic anions,[75] whose oxidation/reduction can be catalyzed by redox oxides. Hierarchical porous carbon, modified with *in situ* produced cobalt oxide nanoparticles (CoO$_x$NP-dPC) was used to detect arsenic ions in neutral solution (Fig. 12). The cyclic voltammogram measured in the presence of As ions (full line) shows a clear oxidation peak which is not present in the CV measured in the absence of As ions (dashed line). The peak is likely due to the oxidation of As^{III} to As^{V} species, catalyzed by the cobalt oxide:

$$2Co(OH)_2 \rightarrow 2CoOOH + 2e^- \qquad (XV)$$

$$2CoOOH + AsO_3^{-3} + 2H_2O \rightarrow 2Co(OH)_2 + AsO_4^{-3} + 2OH^- \qquad (XVI)$$

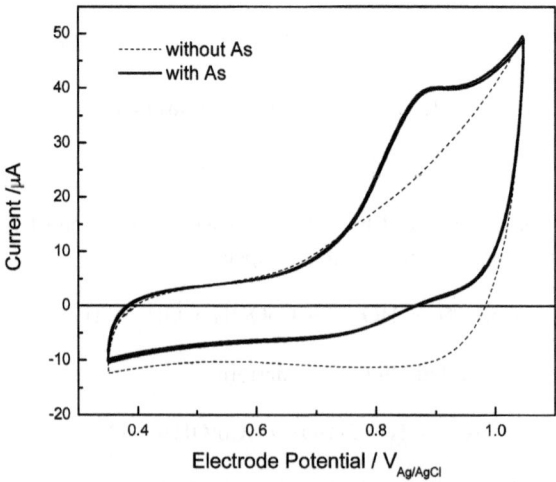

Fig. 12 Cyclic voltametry of a porous carbon electrode modified with CoO$_x$ nanoparticles in the absence (dashed line) and presence (full line) of AsO$_3^{-3}$ (187.5 ppm) ions. Electrolyte = 0.1 M phosphate buffer (pH = 7). Scan rate = 5 mV s^{-1}.

It is likely that reaction (XVI) occurs with the anion adsorbed in the nanoparticle surface. The metal oxide nanoparticles have a large surface area where such adsorption can occur.

As it can be seen, low levels (<200 ppm) of arsenic can be easily detected using cyclic voltammetry on modified HPC.

Iron oxides can also be used to detect arsenic. In Fig. 13 are shown the differential pulse voltammograms of a dPC electrode modified with chemically produced magnetite (Fe_3O_4) nanoparticles (MagNP-dPC).

The presence of the arsenite ions clearly changes the DPV profile from a single peak (due to Fe^{II}/Fe^{III} redox couple) to two peaks, likely due to different iron oxide species with surface-bonded arsenite. The mechanism is likely to be similar to the one observed with cobalt hydroxide. The peaks are superimposed on a broad current background.

The data show that DPV could be used with HPC electrodes but the low definition of the peaks (against background current) suggest that charging effects are not completely eliminated by the pulse technique. This is likely to be related to slow double layer charging which makes that the charging current still significant at the end of the potential pulse.

Oxygen reduction on PtMag-dPC

Oxygen is an interesting analyte in complex environmental samples.[76] In those applications, high sensitivity to the analyte should be combined with low sensitivity to the matrix. Both platinum,[77] and FePt,[78] are effective electrocatalysts of oxygen reduction. We synthesize Pt decorated magnetite nanoparticles inside dPC by means of galvanic displacement of Fe_3O_4 nanoparticles with H_2Cl_6Pt. The presence of Pt on the particles was confirmed by the presence of proton reduction currents at potentials more cathodic than 0.0 V_{RHE} (ESI, Fig. 5†).

Then, we tested the reduction of oxygen on the material. As it can be seen (Fig. 14) the peak due to immobilized redox oxide (dashed line in Fig. 14) shifted

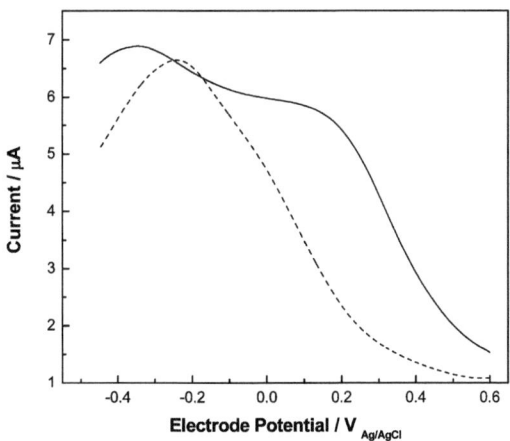

Fig. 13 Differential pulse voltammetry (DPV) of a dPC modified with Fe_3O_4 in the presence (full line) of $NaHAsO_3$ (300 ppm). The dashed line show the DPV of the electrode in absence of As. Scan rate = 20 mV s^{-1}. ΔE_{pulse} = 10 mV. t_{pulse} = 10 ms.

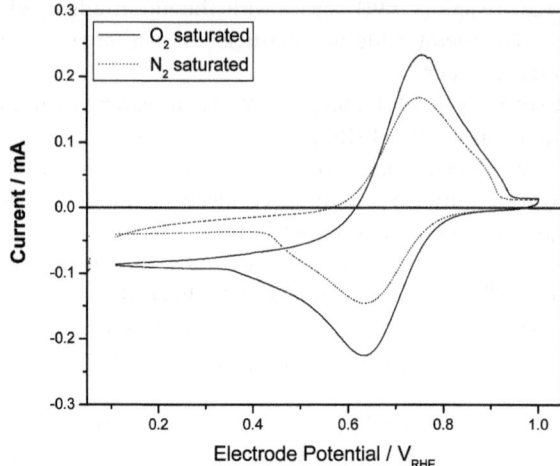

Fig. 14 Oxygen (saturated solution) reduction on PtMagNP-dPC electrodes in the presence of 0.5 M methanol. Scan rate = 5 mV s^{-1}. Solution = 0.1 M H$_2$SO$_4$ solution saturated with O$_2$ (15 min bubbling gas).

below the zero current axis due to the presence of a catalytic current for oxygen reduction. A likely catalytic mechanism involves the oxidation of Fe^{II} species by oxygen:

$$3Fe(OH)_2 + 1/2 O_2 = Fe_3O_4 + 3H_2O \qquad (XVII)$$

This reaction is catalyzed by the presence of Pt which allows breaking of the O$_2$ molecule.

The reduced species is regenerated by electrochemical reduction of the magnetite nanoparticles:

$$Fe_3O_4 + 2H^+ + 2e^- + 2H_2O = 3Fe(OH)_2 \qquad (XVIII)$$

The signal due to oxygen remains active in the presence of 0.5 M methanol (Fig. 14), showing a good capacity of rejecting the interference. Rotating disk voltammetry measurements show a current which increases with rotation rate, confirming the presence of mass transport controlled faradaic reactions.

Conclusions

Numerical calculations show that ideal large area electrodes could be used in electroanalysis disregarding the effect of charging currents. On the other hand, real porous electrodes could present slow double layer charging, compatible with a transmission line model (TLM), which could interfere with the faradaic measurements.

One way to produce porous electrodes with fast charging involves synthesizing hierarchical porous carbon (HPC) electrodes, where nanopores (micro and mesopores in IUPAC terminology) have short lengths due to the presence of macroporosity.

The synthesis of resorcinol–formaldehyde resins in the interstitial space of an opal made of silica nanoparticles allows the fabrication of HPC. By removing the hard template before or after pyrolysis, it is possible to produce a double pore (sPC) or single pore carbon (dPC), respectively. SEM characterization of FIB prepared samples shows an open three-dimensional structure. The specific double layer capacitance (measured by cyclic voltammetry and AC impedance) is in the order of 130 F g^{-1}, suggesting that the material has a large (ca. 650 m^2 g^{-1}) specific surface. Chronoamperometric measurements of monolithic porous carbon (without macropores) shows slow double layer charging compatible with a semi-infinite TLM. On the other hand, both dPC and sPC show a finite TLM, making them less affected by charging current effects.

Chronodeflectometric measurements of dPC in neutral media show a defined pre-peak which seems related to an increase of local pH (inside the pore) due to restricted diffusion of ions.

Both sPC and dPC were modified by *in situ* (chemical or electrochemical) synthesis of metal or metal oxide nanoparticles. sPC modified with PtRu nanoparticles is able to oxidize efficiently CO and methanol. CO monolayer oxidation current was easily detected, unlike previous results with modified non-hierarchical porous carbon.[67]

dPC electrodes modified with electrochemically grown cobalt oxide nanoparticles seem able to detect molecular oxygen or arsenic ions.

dPC modified with magnetite nanoparticles, chemically grown inside the carbon, could be used to detect arsenite by differential pulse voltammetry. dPC modified with magnetite nanoparticles, decorated with Pt by a galvanic displacement reaction, show a clear signal for oxygen reduction even in the presence of 0.5 M methanol.

Hierarchical porous carbon, modified with electrocatalytic nanoparticles, seems to be a promising material for electroanalysis.

Acknowledgements

R. C. R and A. B. M. thank CONICET and FONCYT (respectively) for graduate fellowships. D. F. A., G. A. P. and C. A. B. are permanent research fellows of CONICET. Funding by FONCYT, CONICET, MinCyT (Cordoba) and SECYT-UNRC is gratefully acknowledged. F. Soldera and F. Mucklich are thanked for the FIB-SEM measurements. R. C. R. thanks NanoCom Project (IRSES-UE) for funding his stay at U. Saarlandes.

References

1 R. L. McCreery, Carbon Electrodes: Structural Effects on Electron Transfer Kinetics, in *Electroanalytical Chemistry*, ed. A. J. Bard, Dekker, NY, 1991, Vol. 17, pp. 221–374.
2 M. Marti Villalba and J. Davis, *J. Solid State Electrochem.*, 2008, **12**, 1245.
3 E. Majid, S. Hrapovic, Y. Liu, K. B. Male and J. H. T. Luong, *Anal. Chem.*, 2006, **78**, 762.
4 I. S. Vancara, K. Vytras, K. Kalcher, A. Walcarius and J. Wang, *Electroanalysis*, 2009, **21**, 7.
5 G. A. Rivas, M. D. Rubianes, M. C. Rodrıguez, N. F. Ferreyra, G. L. Luque, M. L. Pedano, S. A. Miscoria and C. Parra, *Talanta*, 2007, **74**, 291.
6 Y. Shao, J. Wang, H. Wu, J. Liu, I. A. Aksay and Y. Lin, *Electroanalysis*, 2010, **22**, 1027.
7 A. Walcarius, *TrAC, Trends Anal. Chem.*, 2012, **38**, 79.
8 M. M. Bruno, N. G. Cotella, M. C. Miras, T. Koch, S. Seidler and C. Barbero, *Colloids Surf., A*, 2010, **358**, 13.
9 M. M. Bruno, N. G. Cotella, M. C. Miras and C. A. Barbero, *Colloids Surf., A*, 2010, **362**, 28.

10 Y. Deng, C. Liu, T. Yu, F. Liu, F. Zhang, Y. Wan, L. Zhang, C. Wang, B. Tu, P. A. Webley, H. Wang and D. Zhao, *Chem. Mater.*, 2007, **19**, 3271.
11 Z. Wang, F. Li, N. S. Ergang and A. Stein, *Chem. Mater.*, 2006, **18**, 5543.
12 B. Sakintuna and Y. Yürüm, *Ind. Eng. Chem. Res.*, 2005, **44**, 2893.
13 M. M. Bruno, N. G. Cotella, M. C. Miras and C. A. Barbero, *Chem. Commun.*, 2005, 5896.
14 G. Wang, L. Zhang and J. Zhang, *Chem. Soc. Rev.*, 2012, **41**, 797.
15 J. Balach, M. M. Bruno, N. G. Cotella, D. F. Acevedo and C. A. Barbero, *J. Power Sources*, 2012, **199**, 386.
16 A. F. Gross and A. P. Nowak, *Langmuir*, 2010, **26**, 11378.
17 O. D. Vele and A. M. Lenhoff, *Curr. Opin. Colloid Interface Sci.*, 2000, **5**, 56.
18 F. Ruo-wen, L. Zheng-hui, L. Ye-ru, L. Feng, X. Fei and W. Ding-cai, *New Carbon Mater.*, 2011, **26**, 171.
19 A. M. Baena-Moncada, G. A. Planes, M. S. Moreno and C. A. Barbero, *J. Power Sources*, 2013, **221**, 42.
20 E. Frackowiak and F. Beguin, *Carbon*, 2001, **39**, 937.
21 B. E. Conway. *Electrochemical Supercapacitors: Scientific Fundamentals and Technological Applications.* Springer, Berlin, 1999.
22 G. Lang, C. A. Barbero, *Laser Techniques for the Study of Electrode Processes*, Springer, Berlin, 2012.
23 C. A. Barbero, *Phys. Chem. Chem. Phys.*, 2005, **7**, 1885.
24 G. García, M. M. Bruno, G. A. Planes, J. L. Rodriguez, C. Barbero and E. Pastor, *Phys. Chem. Chem. Phys.*, 2008, **10**, 6677.
25 G. A. Planes, M. C. Miras and C. A. Barbero, *Chem. Commun.*, 2005, 2146.
26 E. L. Principe, High-density FIB-SEM 3D nanotomography: with applications of real-time imaging during FIB milling in *Focused Ion Beam Systems. Basics and Applications*, ed. N. Yao, Cambridge University Press, Cambridge, 2007.
27 J. Balach, F. Miguel, F. Soldera, D. F. Acevedo, F. Mücklich and C. A. Barbero, *J. Microsc.*, 2012, **246**, 274.
28 C. M. Welch and R. G. Compton, *Anal. Bioanal. Chem.*, 2006, **384**, 601.
29 A. S. Aricò, S. Srinivasan and V. Antonucci, *Fuel Cells*, 2001, **1**, 133.
30 A. Habrioux, E. Sibert, K. Servat, W. Vogel, K. B. Kokoh and N. Alonso-Vante, *J. Phys. Chem. B*, 2007, **111**, 10329.
31 G. García, J. A. Silva-Chong, O. Guillén-Villafuerte, J. L. Rodríguez, E. R. González and E. Pastor, *Catal. Today*, 2006, **116**, 415.
32 C. E. Lee and S. H. Bergens, *J. Phys. Chem. B*, 1998, **102**, 193.
33 H. A. Gasteiger, N. Markovic, P. N. Ross Jr. and E. Cairns, *J. Phys. Chem.*, 1994, **98**, 617.
34 G. A. Planes, G. García and E. Pastor, *Electrochem. Commun.*, 2007, **9**, 839.
35 Z. Chen, D. Higgins, A. Yu, L. Zhang and J. Zhang, *Energy Environ. Sci.*, 2011, **4**, 3167.
36 L. Trotochaud, J. K. Ranney, K. N. Williams and S. W. Boettcher, *J. Am. Chem. Soc.*, 2012, **134**, 17253.
37 U. Yogeswaran, S.-M. Chen and S.-H. Li, *Electroanalysis*, 2008, **20**, 2324.
38 C. Barbero, G. A. Planes and M. C. Miras, *Electrochem. Commun.*, 2001, **3**, 113.
39 A. Salimi, R. Hallaj, S. Soltanian and H. Mamkhezri, *Anal. Chim. Acta*, 2007, **594**, 24.
40 A. Salimi, H. Mamkhezria, R. Hallaj and S. Soltanian, *Sens. Actuators, B*, 2008, **129**, 246.
41 L. H. Reddy, J. L. Arias, J. Nicolas and P. Couvreur, *Chem. Rev.*, 2012, **112**, 5818.
42 C.-L. Lee and C.-M. Tseng, *J. Phys. Chem. C*, 2008, **112**, 13342.
43 W. Wang, B. Gu, L. Liang and W. Hamilton, *J. Phys. Chem. B*, 2003, **107**, 3400.
44 G. M. Jenkins, K. Kawamura, *Polymeric Carbons: Carbon Fibre, Glass and Char*, Cambridge University Press, Cambridge, 2011.
45 W. H. Lizcano-Valbuena, V. A. Paganin and E. R. González, *Electrochim. Acta*, 2002, **47**, 3715.
46 M. Mikhaylova, D. K. Kim, N. Bobrysheva, M. Osmolowsky, V. Semenov, T. Tsakalakos and M. Muhammed, *Langmuir*, 2004, **20**, 2472.
47 J. Zhang, Y. Mo, M. B. Vukmirovic, R. Klie, K. Sasaki and R. R. Adzic, *J. Phys. Chem. B*, 2004, **108**, 10955.
48 A. Velichko, C. Holzapfel and F. Mücklich, *Adv. Eng. Mater.*, 2007, **9**, 39–45.
49 B. Fang, J. H. Kim, M. Kim and J.-S. Yu, *Chem. Mater.*, 2009, **21**, 789.
50 E. Frackowiak and F. Béguin, *Carbon*, 2001, **39**, 937.
51 W. Xing, S. Z. Qiao, R. G. Ding, F. Li, G. Q. Lu, Z. F. Yan and H. M. Cheng, *Carbon*, 2006, **44**, 216.
52 L. Shen, C. Yuan, H. Luo, X. Zhang, K. Xu and F. Zhang, *J. Mater. Chem.*, 2011, **21**, 761.
53 Y. Wu, Z. Wen and J. Li, *Adv. Mater.*, 2011, **23**, 1126.
54 F. Garay and C. Barbero, *Anal. Chem.*, 2006, **78**, 6740–6746.
55 A. J. Bard, L. R. Faulkner, *Electrochemical Methods: Fundamentals and Applications*, Wiley, New York, 2000.

56 Samuel P. Kounaves, Voltammetric Techniques in Handbook of Instrumental Techniques for Analytical Chemistry, ed. Frank A. Settle, Prentice Hall PTR, New York, 1997.
57 R. de Levie, *Electrochim. Acta*, 1964, **9**, 1231.
58 X. Jin, L. Zhuang and J. Lu, *J. Electroanal. Chem.*, 2001, **519**, 137.
59 R. W. Pekala, *J. Mater. Sci.*, 1989, **24**, 3221.
60 R. W. Pekala, C. T. Alviso and J. D. LeMay, *J. Non-Cryst. Solids*, 1990, **125**, 67.
61 F. Lufrano and P. Staiti, *Int. J. Electrochem. Sci.*, 2010, **5**, 903.
62 X. Zhao, H. Tian, M. Zhu, K. Tian, J. J. Wang, F. Kang and R. A. Outlaw, *J. Power Sources*, 2009, **194**, 1208.
63 J. R. Macdonald, *Ann. Biomed. Eng.*, 1992, **20**, 289.
64 M. G. Sullivan, B. Schnyder, M. Bärtsch, D. Alliata, C. Barbero, R. Imhof and R. Kötz, *J. Electrochem. Soc.*, 2000, **147**, 2636.
65 C. A. Wraight, *Biochim. Biophys. Acta, Bioenerg.*, 2006, **1757**, 886.
66 F. Garay and C. A. Barbero, *Anal. Chem.*, 2006, **78**, 6733.
67 M. M. Bruno, G. A. Planes, M. C. Miras, C. A. Barbero, E. Pastor Tejera and J. L. Rodriguez, *Mol. Cryst. Liq. Cryst.*, 2010, **521**, 229.
68 J. R. C. Salgado, F. Alcaide, G. Alvarez, L. Calvillo, M. J. Lázaro and E. Pastor, *J. Power Sources*, 2010, **195**, 4022.
69 D. A. Wiesenburg and N. L. Guinasso, *J. Chem. Eng. Data*, 1979, **24**, 356.
70 Y. Wang, D. Zhang and H. Liu, *J. Power Sources*, 2010, **195**, 3135.
71 J. Emsley, *Oxygen in Nature's Building Blocks: An A–Z Guide to the Elements*. Oxford University Press. Oxford, 2001.
72 J. Xu, W. H. Huang and R. L. McGreery, *J. Electroanal. Chem.*, 1996, **410**, 235; T. Ohsaka, L. Q. Mao, K. Arihara and T. Sotomura, *Electrochem. Commun.*, 2004, **6**, 273; K. Vaik, D. J. Schiffrin and K. Tammeveski, *Electrochem. Commun.*, 2004, **6**, 1; K. Tammeveski, K. Kontturi, R. J. Nichols, R. J. Potter and D. J. Schiffrin, *J. Electroanal. Chem.*, 2001, **515**, 101.
73 K. Mandal and K. T. Suzuki, *Talanta*, 2002, **58**, 201.
74 A. Davis, D. Sherwin, R. Ditmars and K. A. Hoenke, *Environ. Sci. Technol.*, 2001, **35**, 2401.
75 S. R. Chowdhury and E. K. Yanful, *Water Environ. J.*, 2011, **25**, 429.
76 F. X. Simon, Y. Penru, A. R. Guastalli, J. Llorens and S. Baig, *Talanta*, 2011, **85**, 527.
77 I. E. L. Stephens, A. S. Bondarenko, U. Grønbjerg, J. Rossmeis and I. Chorkendorff, *Energy Environ. Sci.*, 2012, **5**, 6744.
78 C. Song, J. Zhang in PEM Fuel Cell Electrocatalyst and Catalyst Layer. Fundamentals and Applications, ed. J. Zhang, Springer, Berlin, 2008.

Faraday Discussions

RSC Publishing

PAPER

Pd@Au core–shell nanocrystals with concave cubic shapes: kinetically controlled synthesis and electrocatalytic properties

Ling Zhang,[ab] Wenxin Niu,[ab] Jianming Zhao,[ab] Shuyun Zhu,[ab] Yali Yuan,[ab] Tao Yuan,[ab] Lianzhe Hu[ab] and Guobao Xu*[a]

Received 15th February 2013, Accepted 22nd March 2013
DOI: 10.1039/c3fd00016h

A new type of concave cubic Pd@Au core–shell nanocrystals is synthesized through a kinetically controlled growth process. Pd nanocubes of 56 nm are used as the inner core, and CTAC and Br⁻ are used as the capping agent and selective adsorbent, respectively. A suitable ratio of $HAuCl_4$ and cubic Pd seeds and the presence of Br⁻ anions are critical to the growth of the concave cubic Pd@Au core–shell nanocrystals. The fast deposition rate on the corners of the cubic Pd seeds promotes the overgrowth of the Au outer shell along the <111> direction, leading to the formation of concave cubic nanostructures. The reduction process is monitored by the surface plasmon resonance spectra of the nanocrystals, and the extinction band became broader and red shifted as the nanocrystals became larger. The electrocatalytic properties of the concave cubic Pd@Au core–shell nanocrystals were investigated with the cathodic electrochemiluminescence reaction of luminol and H_2O_2. A possible electrocatalytic mechanism was proposed and analyzed.

1 Introduction

Noble metal nanocrystals (NCs) have been an important topic of research in recent years, for their wide applications in catalysis, sensing and imaging, therapy, and drug delivery.[1–10] The morphology of noble metal NCs significantly affects their properties by changing the binding energy and electromagnetic fields.[11–17] Noble metal NCs with concave surfaces exhibit interesting properties in catalysis, surface-enhanced spectroscopy, and optics, owing to their high-energy surface, concavities, and sharp corners/edges.[18–20] Site-specific etching and galvanic replacement are common methods to obtain NCs with concave surfaces or hollow interiors; however, sacrificial templates are essential.[20–23] Recently, noble metal NCs with concave surfaces have been synthesized through an

---

[a]*State Key Laboratory of Electroanalytical Chemistry, Chinese Academy of Sciences, Changchun 130022, China. E-mail: guobaoxu@ciac.jl.cn; Fax: +86 431 85262747; Tel: +86 431 85262747*
[b]*University of the Chinese Academy of Sciences, No. 19A Yuquanlu, Beijing 100049, China*

overgrowth strategy. For example, concave cubic, trisoctahedral (TOH) or hex-octahedral (HOH) Au, and TOH or octahedral Ag NCs have been synthesized with seed-mediated growth methods.[24-27] Concave tetrahedral/octahedral Pd and trapozehedral Pt can be obtained using solvothermal methods.[28-30] Electrochemical square-wave methods have been applied to the synthesis of HOH Pd and Pt NCs.[31,32] Besides these methods, kinetic controlled overgrowth is emerging as an effective method to produce noble metal NCs with concave surfaces, such as concave cubic Pd, Pt, Rh, Au/Pd alloy, and Pt multipod NCs.[33-37]

Luminol is a sensitive chemiluminescent reagent widely used for testing traces of blood at a crime scene, and detecting proteins, cyanides or metal ions.[38-47] In a basic solution, luminol loses two protons and exists in the form of a dianion, which can be easily oxidized by oxygen in the presence of metal ion catalysts. An unstable, excited state intermediate is formed, which will release photons and return to the ground state. Since the amount of oxygen in ambient conditions is limited, hydrogen peroxide is necessary to promote intense luminescence, as oxygen is formed through the disproportionate reactions of hydrogen peroxide in basic conditions with iron compound catalysts.[48-51]

Electrochemiluminescence (ECL), also called electrogenerated chemiluminescence, is chemiluminescence induced by electrochemical methods. An external light source and catalyst are not required; the electron transfer between the luminophores and co-reactants is induced by the applied potential. ECL has many advantages in biochemical analysis, and has been an important method for immunoassays, DNA analysis, and detection of metal ions or other analytes.[52-59] ECL of luminol-based systems has been investigated and used to construct biosensors with high selectivity and sensitivity.[60-63] Au nanoparticles can catalyze a multitude of chemical transformations, such as oxygen reduction reactions,[64] alkylations,[65] and fluorogenic reductions.[66] Due to the surface-to-volume effect and quantum confinement, Au nanoparticles possess superior catalytic activity compared with their bulk counterparts. The enhancement of the transition dipole moment in nanoparticle antenna by electric currents and the concentration of the optical field in the vicinity of the Au nanoparticle cause a strong fluorescence increase.[67,68] A single Au nanoparticle provides a multitude of active sites and docking sites where the products can stay on the surface before dissociation.[69] The electrocatalysis of Au nanoparticles towards the ECL of luminol in neutral solutions was demonstrated recently.[70] Au nanoparticles can electrochemically catalyze the breaking of chemical bonds and stabilize the radicals through adsorption, showing excellent electrochemical activity. Noble metal NCs with concave surfaces have been synthesized and demonstrated to have superior surface-enhanced properties. However, the synthesis of Au NCs with concave surfaces has been much less frequently reported than that of other noble metal NCs, such as Pd or Pt. Additionally, the electrochemical activity of Au NCs with concave structures has also been investigated to a much lesser extent.

Herein, we report a kinetically controlled synthesis of Pd@Au core–shell NCs, which are similar to cubes with side surfaces concaved into the centres, so the products are termed concave cubic NCs. A suitable ratio of the cubic Pd seeds to Au precursor ($HAuCl_4$) is essential to the formation of the concave cubic nanostructures. At a high ratio of $HAuCl_4$ and Pd seeds, the deposition of Au atoms preferentially occurs on the corners of the cubic Pd seeds, promoting the overgrowth of NCs along the <111> direction and formation of the concave cubic

Pd@Au core–shell NCs. It is believed that CTAC has a weak stabilizing effect and benefits the kinetically controlled growth of metal NCs. Thus, at a low ratio of HAuCl$_4$ and Pd seeds, the surfactant capping agent effect dominates and TOH Pd@Au core–shell NCs are the major product. Meanwhile, Br$^-$ is critical to the kinetically controlled synthesis of the concave cubic Pd@Au core–shell NCs. Without Br$^-$, there is no evolution from TOH to concave cube, even at a high ratio of HAuCl$_4$ and Pd seeds. The growth process of the concave cubic Pd@Au core–shell NCs was monitored by their surface plasmon resonance (SPR) spectra. The extinction peak increased and a broad band from 660 to 950 nm was formed gradually. The electrocatalysis of the ECL of luminol and H$_2$O$_2$ by the concave cubic Pd@Au core–shell NCs was investigated and the ECL intensity increased by 35% compared with the bulk Au electrode. It was suggested that Au shells with concave cubic surfaces electrochemically catalyzed the breaking of HO–OH bonds, producing HO˙ and reactive oxygen radicals. The concave cubic Au shells provided multiple electrochemically active sites, catalyzing the oxidation of luminol anions by HO˙ radicals to luminol radicals. Furthermore, on the electrochemically active sites, luminol radicals were oxidized by reactive oxygen radicals to 3-aminophthalate anions in the excited state, leading to an enhanced ECL signal.

2 Experimental section

2.1 Chemicals and solutions

PdCl$_2$, HAuCl$_4$·4H$_2$O, NaH$_2$PO$_4$·H$_2$O, NaOH and KBr were obtained from Sinopharm Chemical Reagent Co. Ltd. (Shanghai, China). Cetyltrimethylammonium chloride (CTAC, 98.0%) was obtained from Tianjin Guangfu Fine Chemical Research Institute (China). Cetyltrimethylammonium bromide (CTAB) was obtained from Fluka (Switzerland). 3-Aminophthalhydrazide (Luminol, ≥98%) was purchased from Aldrich. L-Ascorbic acid and hydrogen peroxide (30%) were obtained from Beijing Chemical Reagent Company. All the chemicals were of analytical grade and used without further purification. Doubly distilled water was used throughout the experiments. A 10 mM H$_2$PdCl$_4$ stock solution was prepared by dissolving 0.1773 g PdCl$_2$ in 10 mL of 0.2 M HCl solution and further diluting to 100 mL with doubly distilled water. A 10 mM HAuCl$_4$ stock solution was prepared by dissolving 0.1030g HAuCl$_4$·4H$_2$O in 25 mL doubly distilled water. 1 mM luminol stock solution was prepared by dissolving 0.0044 g 3-aminophthalhydrazide into 25 mL of 0.1 M NaOH solution, and the final pH of the luminol solution was about 10. Phosphate buffer (PB, 0.1 M, pH = 7) solutions were prepared by adjusting the pH of standard solutions of NaH$_2$PO$_4$ with NaOH solutions.

2.2 Instruments

Scanning electron microscopy (SEM) images were taken using an FEI XL30 ESEM FEG scanning electron microscope operated at 25 kV. Transmission electron microscopy (TEM) images, high-resolution TEM (HRTEM) images, selected area electron diffraction (SAED) patterns, high-angle annular dark-field scanning TEM (HAAD-STEM) images, and elemental mapping images were all obtained using a FEI Tecnai G2 F20 microscope operated at 200 kV. UV-vis extinction spectra were obtained using a WFZ UV-2802PC spectrometer. Electrochemical measurements were carried out with a 660c electrochemical working station (Chenhua Inc.,

Xi'an, China). ECL intensities were monitored through the bottom of the three-electrode cell with a BPCL Ultra-Weak luminescence analyzer with a photomultiplier tube (PMT).

2.3 Synthesis of Pd nanocubes with an average size of 56 nm

56 nm Pd nanocubes were synthesized using a seed-mediated growth method reported previously.[71] Firstly, Pd nanocubes with an average size of 22 nm were synthesized at higher temperatures. A 500 μL aliquot of 10 mM H_2PdCl_4 solution was added to 9420 μL of 12.5 mM CTAB aqueous solution heated at 95 °C under stirring. After 5 min, an 80 μL aliquot of 100 mM ascorbic acid aqueous solution was added, and the reaction was allowed to proceed for 20 min. The ascorbic acid solution was freshly prepared. Secondly, the 56 nm Pd nanocubes were obtained by the overgrowth of the smaller Pd cubes. An 80 μL aliquot of 22 nm Pd cubes solution was added into a 5 mL aliquot of CTAB solution which was kept at 40 °C. The reaction proceeded for 14 h at 40 °C. Then, the 56 nm Pd nanocubes were collected and washed once with 5 mM CTAC aqueous solution by centrifugation (8000 rpm, 3 min). Finally, the 56 nm cubic Pd seeds were diluted 10-fold and dispersed in 5 mM CTAC aqueous solution, and stored at 30 °C for future use.

2.4 Synthesis of concave cubic Pd@Au core–shell NCs

In a typical synthesis of the concave cubic Pd@Au NCs, a 5 mL aliquot of 5 mM CTAC aqueous solution containing 1 mM KBr was kept at 30 °C for 10 min. Then 56 nm cubic Pd seeds solution, a 65 μL aliquot of 100 mM ascorbic acid and a 100 μL aliquot of diluted 56 nm cubic Pd seeds were added in sequence (mole ratio of $HAuCl_4$ and Pd seeds is 8 : 1). Finally, a 125 μL aliquot of 10 mM $HAuCl_4$ aqueous solution was added, upon which $AuCl_4^-$ ions were rapidly reduced by ascorbic acid, and the yellow colour disappeared instantly. After 2 min, the reaction solution turned light pink and the colour gradually became deeper. The growth process of NCs was monitored by UV-vis absorption spectrometry, and the reaction was complete within 40 min. The resultant NCs solutions were centrifuged at 8000 rpm for 2 min and the precipitate was washed with warm water three times for further characterization.

For the synthesis of the TOH Pd@Au core–shell NCs, the conditions were the same as for the concave cubic Pd@Au core–shell NCs, except that a 400 μL aliquot of the diluted 56 nm cubic Pd seeds was added into the growth solutions (mole ratio of $HAuCl_4$ and Pd seeds is 2 : 1).

2.5 ECL of luminol-based systems catalyzed by concave cubic Pd@Au core–shell NCs

Electrochemical measurements were carried out in a conventional three-electrode cell in PB solutions at pH 7. The working electrodes were a concave cubic Pd@Au core–shell NCs modified glassy carbon grid or a Au bulk electrode, and the auxiliary electrode and the reference electrode were a thin gold grid and a Ag/AgCl electrode (saturated KCl), respectively. The concave cubic Pd@Au core–shell NCs modified glassy carbon electrode was prepared by dropping colloidal NC solution onto a fresh glassy carbon electrode and drying at room temperature. Cyclic voltammetry was used to stimulate ECL of luminol-based systems. The curves of ECL intensity *versus* applied potential and current *versus* applied potential were recorded simultaneously. The electrochemically active surface area (ECSA) was obtained by accumulating the reduction peak of Pd@Au core–shell NCs or bulk Au electrode in 0.1 M NaOH solutions, and the current and ECL intensity were normalized to ECSA.

3 Results and discussion

3.1 Kinetically controlled synthesis of concave cubic Pd@Au core–shell NCs

Firstly, the 56 nm Pd cubes were used as the initial seeds to induce the overgrowth of Au shells on Pd surfaces. Fig. 1A and B show the SEM images of the concave cubic Pd@Au core–shell NCs produced at a high ratio of $HAuCl_4$ and Pd seeds (mole ratio 8 : 1) in the presence of 1 mM Br^-. Eight extrude corners and six concave faces were formed in this case. The average size of the concave cubes was 176 (±13) nm, with a yield of 91%. When the mole ratio of $HAuCl_4$ and Pd seeds was decreased to 2 : 1, the main products were TOH NCs. The average size was 111 (±13) nm with a yield of 81%, as shown by the SEM images in Fig. 1D and E. Only 14% products were concave cubes.

Further TEM characterization of concave cubic Pd@Au core–shell NCs is shown in Fig. 2. A darker contrast in the centre with an "X" shape is shown in Fig. 2A, and is formed due to the concave surfaces of the products. The concave cubic nanostructures were further analyzed by TEM. As shown in Fig. 2B, the eight corners of the concave cube along the <111> directions are projected into four corners, and the measured concave curvity of the side surfaces was about 141°. The corresponding SAED pattern (Fig. 2C) also demonstrates the single-crystalline nature of the concave cubic Pd@Au core–shell NCs. An HRTEM image of a corner of an individual NC is shown in Fig. 2D. The linear atomic arrangements further indicate the single-crystalline nanostructures of the outer Au shell. HAAD-STEM characterization was explored to study the

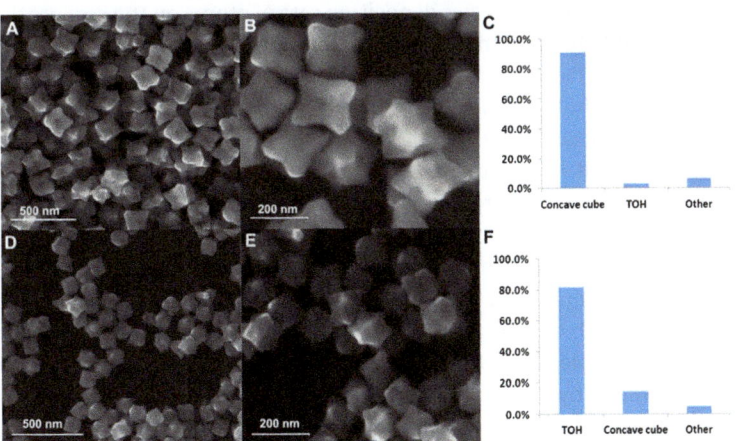

Fig. 1 (A, B) SEM images, at different resolutions, of the concave cubic Pd@Au core–shell NCs produced when the ratio of $HAuCl_4$ and Pd seeds is 8 : 1. (C) Shape distribution of NCs produced when the ratio of $HAuCl_4$ and Pd seeds was 8 : 1. (D, E) SEM images, at different resolutions, of the TOH Pd@Au core–shell NCs produced when the ratio of $HAuCl_4$ and Pd seeds was 2 : 1. (F) Shape distribution of NCs produced when the ratio of $HAuCl_4$ and Pd seeds was 2 : 1.

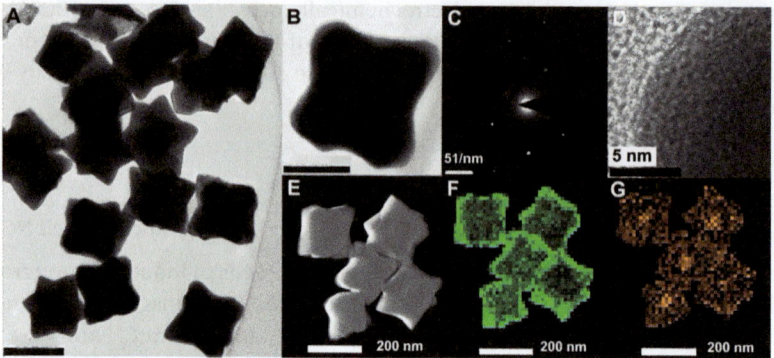

Fig. 2 Further characterization of the concave cubic Pd@Au core–shell NCs produced when the ratio of $HAuCl_4$ and Pd seeds was 8 : 1. (A) TEM image of the concave cubic Pd@Au core–shell NCs (scale bar: 200 nm). (B, C) TEM image of an individual concave cubic NC (scale bar: 100 nm) oriented along the <100> direction and its corresponding SAED pattern, respectively. (E) HAAD-STEM image of the concave cubic core–shell NCs. (F, G) Elemental mapping analysis of Au-M and Pd-L, respectively.

core–shell nanostructures of the products, as shown in Fig. 2E. An enhanced elemental contrast in the centre shows a perfect cubic shape. The elemental mapping analyses also certified the core–shell nanostructures of the products, composed of a concave cubic Au shell and a cubic Pd core in the centre (Fig. 2F and G).

3.2 The growth mechanism of concave cubic Pd@Au core–shell NCs

A fast reduction rate of the precursor can induce anisotropic overgrowth of the NCs. For example, concave cubic and octopod Rh NCs, and concave cubic Pd NCs were formed at a fast reaction rate.[35,37] The reactivity of the corners was higher than that of the edges or side faces for a cubic seed. The fast reduction rate further benefited the faster deposition rate of atoms on the corners than on the edges or side faces. Furthermore, the migration of atoms from the corners to the edges and faces cannot compensate for this disparity of activity on the different sites. Thus, concave cubic Pd or Rh were usually produced at high concentrations of reducing agents or injection speed of precursors. Moreover, Au@Pd core–shell hetero nanostructures with HOH shape were also formed through a kinetically-controlled growth process; a faster reduction rate made Pd atoms deposit preferentially on the corner of the rhombic dodecahedral (RD) Au cores, inducing the formation of HOH shaped pods.[72] On the other hand, CTAC is a common capping agent, which avoids aggregation and directs the crystal growth of metal NCs.[73,74] Since the adsorption of CTAC on the surface of the noble metal NCs is not very strong, the growth direction of NCs can be controlled by kinetic factors.[75,76] For example, the shape evolution of Au NCs among cubic, TOH, and RD has been realized through adjusting the reduction rate when CTAC was used as the capping agent. Meanwhile, low concentrations of Br^- were necessary for the shape transformations.[76]

In this study, the fast deposition rate of Au atoms, the capping agent CTAC, and the selective adsorption of Br^- are three critical factors for the formation of the concave cubic Pd@Au core–shell NCs. At a higher ratio of $HAuCl_4$ and Pd seeds, there are not enough sites for the homogeneous deposition of Au atoms. The deposition of Au atoms preferentially occurs on the corners of the cubic Pd seeds, and the migration rates of the atoms to the edges and faces are slow.[18,35,37] Thus, the NCs grew along the <111> directions of the cubic seeds, and the concave cubic NCs were produced, formed as a result of kinetically controlled overgrowth. At lower ratio of HAuCl$_4$ and Pd seeds, the deposition rate of Au atoms on the different sites of the Pd cube is probably equal, so the crystal morphology was controlled by the surfactant effect. CTAC facilitated the formation of the high-index facets of Au NCs, such as {221} or {321} facets.[24,25,27] In this case, the surfactant effect of CTAC dominated the crystal growth. 81% of the products retained the TOH shape, and only a low proportion of the products had concave cubic shapes, but the concavity was lower than that of the NCs formed at higher ratio of HAuCl$_4$ and Pd seeds. Br$^-$ was critical to the shape evolution from TOH to concave cubes. As shown in Fig. 3, the products were TOH NCs in the absence of Br$^-$ whether the ratio of HAuCl$_4$ and Pd seeds was low or high. On the one hand, Br$^-$ seems to selectively adsorb to the {100} facets of the noble metal NCs, and block the growth of the NCs along the <100> direction, facilitating the formation of the concave cubes.[77-80] When Br$^-$ was added to the CTAC solutions, the [PdBr$_4$]$^{2-}$ complex was formed, and its reduction rate was lower than that of [PdCl$_4$]$^{2-}$.[81] But in this case, a fast reduction rate was beneficial to the formation of the concave cubes, so the formation of [PdBr$_4$]$^{2-}$ did not play an important role in the shape transformations.

The growth process of concave Pd@Au core–shell NCs was monitored by their SPR spectra. As shown in Fig. 4, as the size of the NCs became larger, the extinction intensity gradually became stronger. The extinction cross-section of the SPR spectrum is composed of two sections, scattering and absorption.[82] According to the discrete dipole approximation calculations, the position of the absorption peak did not significantly change with the size of the NCs, and was about 550 nm.[82] For the scattering cross-section, the intensity was weak when the NCs were small, so only one SPR peak was observed when the reaction proceeded for 4 min. As the Au shell of the Pd@Au core–shell NCs became thicker, the scatter intensity became stronger and the scattering peaks became apparent. After the reaction had proceeded for 11 min, a main scattering peak at 630 nm and a shoulder peak at 590 nm were formed. The main scattering peak red shifted and became broader as the reaction continued. The reaction was completed in 39 min,

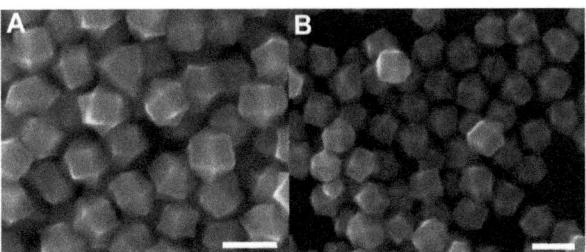

Fig. 3 TOH Pd@Au core–shell NCs produced when Br$^-$ ions were not added into the growth solution for the overgrowth of Au shells. (A, B) SEM images of the TOH Pd@Au core–shell NCs produced when the ratio of HAuCl$_4$ and Pd seeds was 8 : 1 (scale bar: 200 nm) and 2 : 1 (scale bar: 100 nm), respectively.

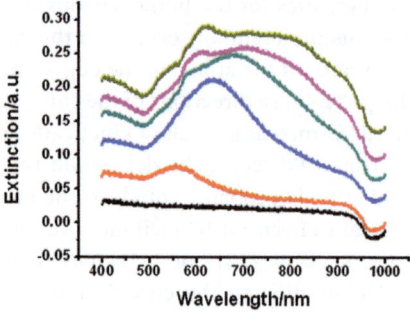

Fig. 4 SPR spectra of the concave cubic Pd@Au core–shell NCs obtained at different reaction times. Bottom line, the CTAC solution containing cubic Pd seeds, HAuCl$_4$, and 1 mM Br$^-$; The other lines, 4, 11, 18, 27, and 39 min after the reaction started (from bottom to top), respectively.

and a broad extinction band from 660 to 950 nm was formed, possibly due to the large size or the sharp extrudes/corners of the concave cubic Pd@Au core–shell NCs.

3.3 ECL of luminol and H$_2$O$_2$ catalyzed by concave cubic Pd@Au core–shell NCs

The cathodic ECL of luminol and H$_2$O$_2$ in a neutral solution catalyzed by concave cubic Pd@Au core–shell NCs was investigated, with a bulk Au electrode as a contrast. The ECL of luminol in the neutral PB buffer is shown in Fig. 5A. The start potential was set at 0 V, and the ECL intensity of the first scan was lower than that of the reverse scan for both the concave cubic Pd@Au core–shell NCs and Au bulk electrode. The ECL intensity at −0.5 V of the concave cubic Pd@Au NCs is about 33% higher compared with the Au bulk electrode. The reactive oxygen species were necessary to oxidize luminol to the 3-aminophthalate dianion in the excited state.[83] In luminol solutions, dissolved oxygen played a role as an oxidant. The general chemiluminescence mechanism of luminol in aqueous solution containing O$_2$ involves one-electron oxidation of luminol, followed by the rapid addition of O$_2$.[84] The initial oxidation was believed to be the rate-determining step, and catalysts such as Fe^{2+}, Cr^{3+}, Co^{2+} or their complexes were essential to induce the production of oxygen-related radicals (HO$^{\cdot}$, O$_2^{\cdot-}$, et al.).[49,50,85–87] These oxygen-related radicals can also be produced at the cathode, inducing the cathodic ECL of luminol.[88,89] On the other hand, Au nanoparticles have been demonstrated to electrochemically catalyze oxygen reduction reactions, and were supposed to stabilize superoxide radicals through Au–O adsorption.[90] A possible ECL mechanism of luminol in the presence of O$_2$ catalyzed by concave cubic Pd@Au core–shell NCs is shown by the following equations:[88]

$$O_2 + e^- \rightarrow O_2^{\cdot-} \tag{1}$$

$$O_2^{\cdot-} + e^- + 2H_2O \rightarrow 2H_2O_2 + 2OH^- \tag{2}$$

$$H_2O_2 + e^- \rightarrow HO^{\cdot} + OH^- \tag{3}$$

$$HO^{\cdot} + LH^- \rightarrow OH^- + LH^{\cdot} \tag{4}$$

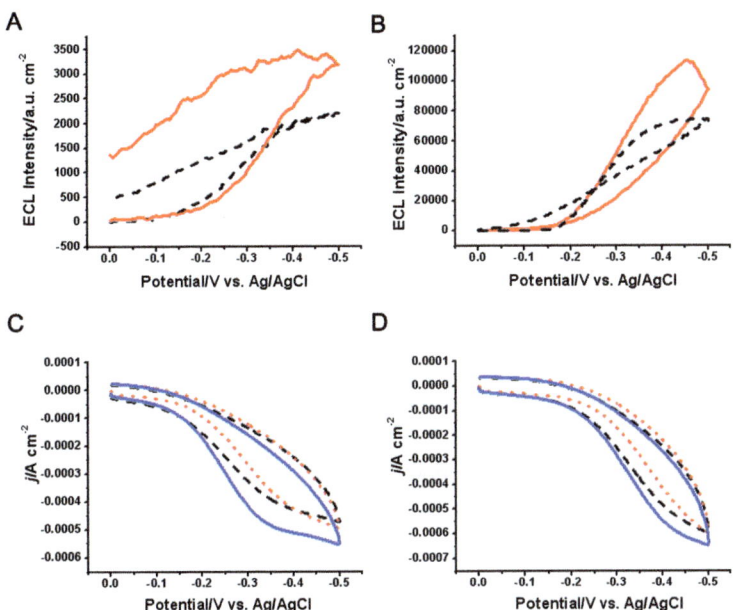

Fig. 5 Electrocatalytic properties of the concave cubic Pd@Au core–shell NCs in the luminol-based systems. (A) Curves of ECL intensity *versus* applied potential on the concave cubic Pd@Au core–shell NCs modified electrode (solid line) and bulk Au electrode (dashed line) in the luminol solutions. Luminol, 100 μM. (B) Curves of ECL intensity *versus* applied potential on the concave cubic Pd@Au core–shell NCs modified electrode (solid line) and bulk Au electrode (dashed line) in the luminol–H$_2$O$_2$ solutions. Luminol, 100 μM. H$_2$O$_2$, 100 μM. (C) Current density *versus* applied potential on bulk Au electrode in the PB buffer (dotted line), luminol (dashed line), and luminol–H$_2$O$_2$ (solid line) solutions. (D) Current density *versus* applied potential on the concave cubic Pd@Au core–shell NCs modified electrode in the PB buffer (dotted line), luminol (dashed line), and luminol–H$_2$O$_2$ (solid line) solutions. PB buffer, 0.1 M, pH 7. Scanning rate, 0.05 V s^{-1}. Start potential, 0 V. Temperature, 25 °C. PMT, −800 V. All solutions were air-saturated.

$$LH^{\cdot} + O_2^{\cdot -} \rightarrow LHO_2^{-} \rightarrow N_2 + AP^{2-*} \qquad (5)$$

$$AP^{2-*} \rightarrow AP^{2-} + h\nu\ (425\ nm) \qquad (6)$$

Pd@Au core–shell NCs provided multiple electrochemically active sites for the reduction of O$_2$ and the oxidation of luminol. First, an O$_2$ molecule received an electron at the cathode, forming a superoxide radical O$_2^{\cdot -}$ (reaction 1, pK_a of O$_2^{\cdot -}$ is 4.8).[91] O$_2^{\cdot -}$ was further reduced to H$_2$O$_2$ and HO$^{\cdot}$ through a one-electron transfer step (reactions 2 and 3). The concentration of O$_2$ in the air-saturated solution approaches 2 × 10^{-4} M.[92] O$_2^{\cdot -}$ is a weak oxidant (formal reduction potential 0.75 V *vs.* SHE)[93] and cannot oxidize luminol anions to luminol radicals (E_0 (LH$^{\cdot}$/LH$^-$) = 0.87 V *vs.* SHE).[94] HO$^{\cdot}$ is a strong oxidant with a reduction potential between 2.2 and 1.8 V *vs.* SHE at pH 7–14.[93] The pK_{a1} and pK_{a2} of luminol are 6.2–6.7 and 15.1, respectively.[95–97] So at pH = 7, luminol is present in the form of its monoanion, LH$^-$. The chemiluminescence started with one electron oxidation of LH$^-$ by HO$^{\cdot}$, forming an LH$^{\cdot}$ radical (pK_a of LH$^{\cdot}$ is 7.7) (reaction 4).[98,99] Further, in the presence of O$_2^{\cdot -}$, an endoperoxide species, LHO$_2^-$, was formed, which can be regarded as an addition reaction of LH˙ and O_2^-. Finally, the endoperoxide species eliminated nitrogen, generating an excited 3-aminophthalate dianion (AP^{2-*}), which returned to the ground state and caused the emission of photons (reactions 5 and 6). The emission wavelength was 425 nm. The structural formula of luminol and its related compounds are shown in Fig. 6.

Fig. 5B shows the curves of ECL intensity *versus* applied potential in the presence of 100 μM H_2O_2. H_2O_2 is a kind of reactive oxygen species, which has a higher reactivity than molecular oxygen.[88,100] At the cathode, H_2O_2 in the solution received one electron from the Pd@Au core–shell NCs on the electrode surface, producing OH˙ directly (reaction 3) which can initiate the first ECL pathway. The superoxide radicals were produced in two ways: by reaction 1, a one electron transfer from the electrode to O_2, or by HO˙ reacting with H_2O_2 or its monoanion, producing O_2^-, shown by the following equation:

$$HO˙ + H_2O_2 \rightarrow O_2^- + H_3O^+ \quad (7)$$

$$HO˙ + HO_2^- \rightarrow O_2^- + H_2O \quad (8)$$

It has been reported that the rate constants for reaction 7 and 8 are 3.7×10^7 M^{-1} s^{-1} and 6.7×10^9 M^{-1} S^{-1}, respectively.[100] The association constant of H_2O_2 is $10^{-11.7}$,[100] thus, reaction 7 was the main source of O_2^- at pH = 7, and this step became the rate determining step. Luminol anions and H_2O_2 competed to react with OH˙ radicals, and luminol radicals and O_2^- were produced, respectively (reactions 4 and 7). Luminol radicals reacted with O_2^- rapidly, forming endoperoxide complexes (reaction 5).

The current density *versus* applied potential in luminol-based systems for the bulk Au electrode and concave cubic Pd@Au core–shell NCs are shown in Fig. 5C and D, respectively. The curves obtained in PB solution show the catalysis current of oxygen reduction from −0.25 V,[101] the current density of Pd@Au core–shell NCs at −0.5 V was about 20% higher than that of the Au bulk electrode. The presence of luminol did not induce the change of reduction current density. When H_2O_2 was added to the solution, the cathodic current densities at −0.5 V were increased by 8% and 12% for the concave cubic Pd@Au core–shell NCs and bulk Au electrode, respectively, corresponding to the reduction current of H_2O_2. The current density of the concave cubic Pd@Au core–shell NCs was 16% higher than that of bulk Au electrode. Compared with ECL intensity, the electrochemical current density changed a little when H_2O_2 was added, indicating the high sensitivity of ECL technology.

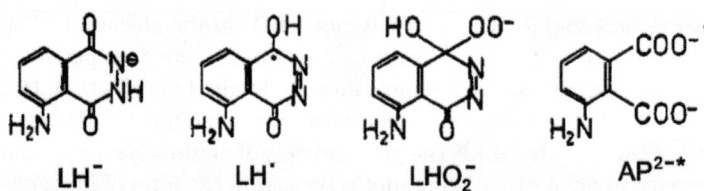

Fig. 6 Structural formula of luminol and its related compounds involved in the ECL pathways. LH^-, luminol anion; LH˙, luminol radical; LHO_2^-, endoperoxide species; AP^{2-*}, excited 3-aminophthalate dianion.

It has been previously reported that Au nanoparticles catalyzed the chemiluminescence of luminol and H_2O_2, and a possible mechanism was also proposed.[90,102] Au nanoparticles were suggested to cleave the O–O bond of H_2O_2 into double HO˙ radicals. The luminol anion formed in the basic solution reacted with the HO˙ radicals, forming luminol radicals. Meanwhile, superoxide radical anions were formed when HO_2^- reacted with the HO˙ radicals. Further, electrons were transferred from the luminol radicals to the superoxide radical anions on the surface of Au nanoparticles, producing the key intermediate endoperoxide complexes, which would decompose with the emission of chemiluminescence. However, the chemiluminescence intensity of Au colloidal nanoparticles was largely affected by their size. 38 nm Au nanoparticles showed the strongest chemiluminescent signal, and 6 nm and 99 nm Au nanoparticles showed relatively weak chemiluminescence signals. The concave cubic Pd@Au core–shell NCs of 176 nm, immobilized on the electrode surface, exhibited catalytic activity for the cathodic ECL of luminol and H_2O_2. The ECL peak of Pd@Au core–shell nanoparticles was set at −0.46 V, and the ECL intensity was about 35% higher than that of the bulk Au electrode. The excellent electrocatalytic properties of concave cubic Pd@Au core–shell NCs may be due to the availability of more low-coordinated Au atoms on the surface.[103] The highly active low-coordinated atoms on the Au shell facilitate electron transfer, resulting in stronger ECL.

4 Conclusion

In this study, we have successfully synthesized a new type of concave cubic Pd@Au core–shell NCs through kinetic control. The weak stabilizing effect of capping agent CTAC benefited the kinetically controlled growth of the Pd@Au core–shell NCs. At a higher ratio of $HAuCl_4$ and Pd seeds, the fast deposition rate of Au atoms on the corner of cubic Pd seeds led to the growth of Pd@Au NCs along the <111> directions, forming the concave cubic Pd@Au core–shell NCs. Br^- was critical to the shape transformations from TOH to concave cube. The reduction process was monitored by SPR spectra, and the extinction band became broader as the Au shell grew thicker. The electrocatalytic properties of the concave cubic Pd@Au NCs were investigated with the cathodic ECL of luminol and H_2O_2 in neutral solutions. Concave cubic Au shells provided electrochemically active sites for the formation of oxygen related radicals and oxidation of luminol, facilitating the electron transfer between superoxide radicals and luminol radicals. This report will benefit the future studies of electrocatalytic properties of noble metal NCs with unusual structures.[104,105] The stronger extinction intensities of the concave cubic Pd@Au NCs in the near-infrared range endow them with potential applications in therapy and cellular detection.

Acknowledgements

This work is kindly supported by the National Natural Science Foundation of China (No. 21175126), Changchun Institute of Applied Chemistry, Chinese Academy of Sciences.

References

1 P. Ghosh, G. Han, M. De, C. K. Kim and V. M. Rotello, *Adv. Drug Delivery Rev.*, 2008, **60**, 1307.

2 P. K. Jain, I. H. El-Sayed and M. A. El-Sayed, *Nano Today*, 2007, **2**, 18.
3 D. Kim, Y. Y. Jeong and S. Jon, *ACS Nano*, 2010, **4**, 3689.
4 R. Popovtzer, A. Agrawal, N. A. Kotov, A. Popovtzer, J. Balter, T. E. Carey and R. Kopelman, *Nano Lett.*, 2008, **8**, 4593.
5 R. A. Alvarez-Puebla, E. R. Zubarev, N. A. Kotov and L. M. Liz-Marzán, *Nano Today*, 2012, **7**, 6.
6 A. Sánchez-Iglesias, P. Aldeanueva-Potel, W. Ni, J. Pérez-Juste, I. Pastoriza-Santos, R. A. Alvarez-Puebla, B. N. Mbenkum and L. M. Liz-Marzán, *Nano Today*, 2010, **5**, 21.
7 Y. Yu, Z. T. Luo, J. Y. Lee and J. P. Xie, *ACS Nano*, 2012, **6**, 7920.
8 X. Yuan, Z. T. Luo, Q. B. Zhang, X. H. Zhang, Y. G. Zheng, J. Y. Lee and J. P. Xie, *ACS Nano*, 2011, **5**, 8800.
9 T. Li, K. Zhu, S. He, X. Xia, S. Liu, Z. Wang and X. Jiang, *Analyst*, 2011, **136**, 2893.
10 S. Q. Liu, Z. Z. Zheng and X. Y. Li, *Anal. Bioanal. Chem.*, 2013, **405**, 63.
11 S. S. Cheong, J. D. Watt and R. D. Tilley, *Nanoscale*, 2010, **2**, 2045.
12 B. Sepulveda, P. C. Angelome, L. M. Lechuga and L. M. Liz-Marzan, *Nano Today*, 2009, **4**, 244.
13 J. Watt, C. Yu, S. L. Y. Chang, S. Cheong and R. D. Tilley, *J. Am. Chem. Soc.*, 2013, **135**, 606.
14 H.-L. Wu, H.-R. Tsai, Y.-T. Hung, K.-U. Lao, C.-W. Liao, P.-J. Chung, J.-S. Huang, I. C. Chen and M. H. Huang, *Inorg. Chem.*, 2011, **50**, 8106.
15 Z. Li, E. Cheng, W. Huang, T. Zhang, Z. Yang, D. Liu and Z. Tang, *J. Am. Chem. Soc.*, 2011, **133**, 15284.
16 J. Wang, J. Gong, Y. Xiong, J. Yang, Y. Gao, Y. Liu, X. Lu and Z. Tang, *Chem. Commun.*, 2011, **47**, 6894.
17 Z. Zhu, H. Meng, W. Liu, X. Liu, J. Gong, X. Qiu, L. Jiang, D. Wang and Z. Tang, *Angew. Chem., Int. Ed.*, 2011, **50**, 1593.
18 H. Zhang, M. Jin and Y. Xia, *Angew. Chem., Int. Ed.*, 2012, **51**, 7656.
19 L. Zhang, W. Niu and G. Xu, *Nano Today*, 2012, **7**, 586.
20 Z. Jiang, Y. Lin and Z. Xie, *Mater. Chem. Phys.*, 2012, **134**, 762.
21 S. Cheong, J. Watt, B. Ingham, M. F. Toney and R. D. Tilley, *J. Am. Chem. Soc.*, 2009, **131**, 14590.
22 X. Han, X. Zhou, Y. Jiang and Z. Xie, *J. Mater. Chem.*, 2012, **22**, 10924.
23 M. J. Mulvihill, X. Y. Ling, J. Henzie and P. D. Yang, *J. Am. Chem. Soc.*, 2010, **132**, 268.
24 J. W. Hong, S.-U. Lee, Y. W. Lee and S. W. Han, *J. Am. Chem. Soc.*, 2012, **134**, 4565.
25 Y. Y. Ma, Q. Kuang, Z. Y. Jiang, Z. X. Xie, R. B. Huang and L. S. Zheng, *Angew. Chem., Int. Ed.*, 2008, **47**, 8901.
26 X. H. Xia, J. Zeng, B. McDearmon, Y. Q. Zheng, Q. G. Li and Y. N. Xia, *Angew. Chem., Int. Ed.*, 2011, **50**, 12542.
27 J. Zhang, M. R. Langille, M. L. Personick, K. Zhang, S. Li and C. A. Mirkin, *J. Am. Chem. Soc.*, 2010, **132**, 14012.
28 X. Q. Huang, S. H. Tang, H. H. Zhang, Z. Y. Zhou and N. F. Zheng, *J. Am. Chem. Soc.*, 2009, **131**, 13916.
29 X. Q. Huang, Z. P. Zhao, J. M. Fan, Y. M. Tan and N. F. Zheng, *J. Am. Chem. Soc.*, 2011, **133**, 4718.
30 M. Chen, B. Wu, J. Yang and N. Zheng, *Adv. Mater.*, 2012, **24**, 862.
31 N. Tian, J. Xiao, Z.-Y. Zhou, H. Liu, Y.-J. Deng, L. Huang, B. Xu and S.-G. Sun, *Faraday Discuss.*, 2013, DOI: 10.1039/C3FD20146E.
32 Z.-Y. Zhou, N. Tian, Z.-Z. Huang, D.-J. Chen and S.-G. Sun, *Faraday Discuss.*, 2009, **140**, 81.
33 C. J. DeSantis, A. A. Peverly, D. G. Peters and S. E. Skrabalak, *Nano Lett.*, 2011, **11**, 2164.
34 T. Herricks, J. Y. Chen and Y. N. Xia, *Nano Lett.*, 2004, **4**, 2367.
35 M. S. Jin, H. Zhang, Z. X. Xie and Y. N. Xia, *Angew. Chem., Int. Ed.*, 2011, **50**, 7850.
36 T. Yu, D. Y. Kim, H. Zhang and Y. N. Xia, *Angew. Chem., Int. Ed.*, 2011, **50**, 2773.
37 H. Zhang, W. Y. Li, M. S. Jin, J. E. Zeng, T. K. Yu, D. R. Yang and Y. N. Xia, *Nano Lett.*, 2011, **11**, 898.
38 D. F. Roswell and E. H. White, *Methods Enzymol.*, **57**, p. 409.
39 A. Castello, F. Frances and F. Verdu, *J. Forensic Sci.*, 2012, **57**, 500.
40 S. Walter, *Angew. Chem.*, 1937, **50**, 155.
41 Y. Chai, D. Tian, W. Wang and H. Cui, *Chem. Commun.*, 2010, **46**, 7560.
42 D. Chen, Z. Wang, Y. Zhang, X. Xiong and Z. Song, *Anal. Methods*, 2012, **4**, 1485.
43 R. S. Chouhan, A. C. Vinayaka and M. S. Thakur, *Anal. Methods*, 2010, **2**, 924.
44 J. Fan, M. Hu, P. Zhan and X. Peng, *Chem. Soc. Rev.*, 2013, **42**, 29.
45 J. A. Murillo Pulgarin, L. F. Garcia Bermejo and A. Carrasquero Duran, *Environ. Monit. Assess.*, 2013, **185**, 573.

46 S. K. Sahoo, D. Sharma, R. K. Bera, G. Crisponi and J. F. Callan, *Chem. Soc. Rev.*, 2012, **41**, 7195.
47 G. Guan, L. Yang, Q. Mei, K. Zhang, Z. Zhang and M.-Y. Han, *Anal. Chem.*, 2012, **84**, 9492.
48 Y.-C. Chen, Y.-L. Jian, K.-H. Chiu and H.-K. Yak, *Anal. Sci.*, 2012, **28**, 795.
49 A. L. Rose and T. D. Waite, *Anal. Chem.*, 2001, **73**, 5909.
50 K. Tsukagoshi, M. Sumiyama, R. Nakajima, M. Nakayama and M. Maeda, *Anal. Sci.*, 1998, **14**, 409.
51 M. Yaqoob, B. F. Biot, A. Nabi and P. J. Worsfold, *Luminescence*, 2012, **27**, 419.
52 M. M. Richter, *Chem. Rev.*, 2004, **104**, 3003.
53 B. A. Gorman, P. S. Francis and N. W. Barnett, *Analyst*, 2006, **131**, 616.
54 W. Miao, *Chem. Rev.*, 2008, **108**, 2506–2553.
55 L. Hu and G. Xu, *Chem. Soc. Rev.*, 2010, **39**, 3275.
56 R. Kurita, K. Arai, K. Nakamoto, D. Kato and O. Niwa, *Anal. Chem.*, 2012, **84**, 1799.
57 M. Sentic, G. Loget, D. Manojlovic, A. Kuhn and N. Sojic, *Angew. Chem., Int. Ed.*, 2012, **51**, 11284.
58 P. Bertoncello and R. J. Forster, *Biosens. Bioelectron.*, 2009, **24**, 3191.
59 T. Hu, T. Li, L. Yuan, S. Liu and Z. Wang, *Nanoscale*, 2012, **4**, 5447.
60 F. Li and H. Cui, *Biosens. Bioelectron.*, 2013, **39**, 261.
61 J. Li, S. Li, X. Wei, H. Tao and H. Pan, *Anal. Chem.*, 2012, **84**, 9951.
62 X. Liu, W. Niu, H. Li, S. Han, L. Hu and G. Xu, *Electrochem. Commun.*, 2008, **10**, 1250.
63 M. Yan, L. Ge, W. Gao, J. Yu, X. Song, S. Ge, Z. Jia and C. Chu, *Adv. Funct. Mater.*, 2012, **22**, 3899.
64 P. Quaino, N. B. Luque, R. Nazmutdinov, E. Santos and W. Schmickler, *Angew. Chem., Int. Ed.*, 2012, **51**, 12997.
65 M. Bandini, A. Bottoni, M. Chiarucci, G. Cera and G. P. Miscione, *J. Am. Chem. Soc.*, 2012, **134**, 20690.
66 P. Chen, W. Xu, X. Zhou, D. Panda and A. Kalininskiy, *Chem. Phys. Lett.*, 2009, **470**, 151.
67 O. L. Muskens, V. Giannini, J. A. Sánchez-Gil and J. Gómez Rivas, *Nano Lett.*, 2007, 7, 2871.
68 H. Yuan, S. Khatua, P. Zijlstra, M. Yorulmaz and M. Orrit, *Angew. Chem., Int. Ed.*, 2013, **52**, 1217.
69 W. Xu, J. S. Kong, Y.-T. E. Yeh and P. Chen, *Nat. Mater.*, 2008, 7, 992.
70 H. Cui, Y. Xu and Z. F. Zhang, *Anal. Chem.*, 2004, **76**, 4002.
71 W. Niu, Z.-Y. Li, L. Shi, X. Liu, H. Li, S. Han, J. Chen and G. Xu, *Cryst. Growth Des.*, 2008, **8**, 4440.
72 D. Kim, Y. W. Lee, S. B. Lee and S. W. Han, *Angew. Chem., Int. Ed.*, 2012, **51**, 159.
73 M. L. Personick, M. R. Langille, J. Zhang and C. A. Mirkin, *Nano Lett.*, 2011, **11**, 3394.
74 J. Zhang, M. R. Langille, M. L. Personick, K. Zhang, S. Li and C. A. Mirkin, *J. Am. Chem. Soc.*, 2010, **132**, 14012.
75 P.-J. Chung, L.-M. Lyu and M. H. Huang, *Chem.–Eur. J.*, 2011, **17**, 9746.
76 H.-L. Wu, C.-H. Kuo and M. H. Huang, *Langmuir*, 2010, **26**, 12307.
77 W. Niu and G. Xu, *Nano Today*, 2011, **6**, 265.
78 W. Niu, L. Zhang and G. Xu, *ACS Nano*, 2010, **4**, 1987.
79 W. Niu, S. Zheng, D. Wang, X. Liu, H. Li, S. Han, J. Chen, Z. Tang and G. Xu, *J. Am. Chem. Soc.*, 2009, **131**, 697.
80 L. Zhang, W. Niu and G. Xu, *Nanoscale*, 2011, **3**, 678.
81 S. C. Srivastava and L. Newman, *Inorg. Chem.*, 1966, **5**, 1506.
82 L. Zhang, W. Niu, Z. Li and G. Xu, *Chem. Commun.*, 2011, **47**, 10353.
83 Y.-P. Dong, *J. Lumin.*, 2010, **130**, 1539.
84 P. B. Shevlin and H. A. Neufeld, *J. Org. Chem.*, 1970, **35**, 2178.
85 T. G. Burdo and W. R. Seitz, *Anal. Chem.*, 1975, **47**, 1639.
86 C. A. Chang and H. H. Patterson, *Anal. Chem.*, 1980, **52**, 653.
87 J. M. Lin, X. Q. Shan, S. Hanaoka and M. Yamada, *Anal. Chem.*, 2001, **73**, 5043.
88 S. Kulmala, T. Ala-Kleme, A. Kulmala, D. Papkovsky and K. Loikas, *Anal. Chem.*, 1998, **70**, 1112.
89 H. Xu, H. Ye, X. Zhu, S. Liang, L. Guo, J. Lin, X. Liu and G. Chen, *Analyst*, 2013, **138**, 234.
90 Z. F. Zhang, H. Cui, C. Z. Lai and L. J. Liu, *Anal. Chem.*, 2005, 77, 3324.
91 J. Rabani and S. O. Nielsen, *J. Phys. Chem.*, 1969, **73**, 3736.
92 W. H. Koppenol and J. Butler, *Adv. Free Radical Biol. Med.*, 1985, **1**, 91.
93 W. Koppenol, *Bioelectrochem. Bioenerg.*, 1987, **18**, 3.
94 P. Neta, *Adv. Phys. Org. Chem.*, 1976, **12**, 223.
95 L. Erdey, I. Buzas and K. Vigh, *Talanta*, 1966, **13**, 463.
96 A. Babko and N. Lukovskaya, *Ukr. Khim. Zh*, 1963, **29**, 479.
97 K. Haapakka, J. Kankare and J. Linke, *Anal. Chim. Acta*, 1982, **139**, 379.

98 J. Lind, G. Merenyi and T. E. Eriksen, *J. Am. Chem. Soc.*, 1983, **105**, 7655.
99 G. Merenyi, J. Lind and T. E. Erikson, *J. Am. Chem. Soc.*, 1986, **108**, 7716.
100 G. Merenyi and J. S. Lind, *J. Am. Chem. Soc.*, 1980, **102**, 5830.
101 Y. Lee, A. Loew and S. Sun, *Chem. Mater.*, 2010, **22**, 755.
102 Y. He and H. Cui, *J. Phys. Chem. C*, 2012, **116**, 12953.
103 N. Lopez, T. V. W. Janssens, B. S. Clausen, Y. Xu, M. Mavrikakis, T. Bligaard and J. K. Nørskov, *J. Catal.*, 2004, **223**, 232.
104 W. Niu, L. Zhang and G. Xu, *Sci. China: Chem.*, 2012, **55**, 2311.
105 W. Niu, L. Zhang and G. Xu, *Nanoscale*, 2013, **5**, 3172.

Faraday Discussions

PAPER

Electrochemical mechanical micromachining based on confined etchant layer technique

Ye Yuan,[a] Lianhuan Han,[a] Jie Zhang,[a] Jingchun Jia,[a] Xuesen Zhao,[b] Yongzhi Cao,[b] Zhenjiang Hu,*[b] Yongda Yan,[b] Shen Dong,[b] Zhong-Qun Tian,[a] Zhao-Wu Tian[a] and Dongping Zhan*[a]

Received 1st February 2013, Accepted 13th February 2013
DOI: 10.1039/c3fd00008g

The confined etchant layer technique (CELT) has been proved an effective electrochemical microfabrication method since its first publication at *Faraday Discussions* in 1992. Recently, we have developed CELT as an electrochemical mechanical micromachining (ECMM) method by replacing the cutting tool used in conventional mechanical machining with an electrode, which can perform lathing, planing and polishing. Through the coupling between the electrochemically induced chemical etching processes and mechanical motion, ECMM can also obtain a regular surface in one step. Taking advantage of CELT, machining tolerance and surface roughness can reach micro- or nano-meter scale.

Introduction

Electrochemical machining is a non-conventional machining method in which metal materials are removed through anodic stripping or formed through cathodic deposition.[1] The metal workpiece acts as the anode while the tool acts as the cathode, or *vice versa*. Since the tool is guided along the desired path close to but without touching the workpiece, there is no tool wear. Essentially, the kinetic rate of metal electrode processes are very fast and a high metal removal rate is possible without thermal or mechanical stresses. Electrochemical machining can produce high aspect ratio or complex microstructures (*e.g.*, LIGA[2-5] and EFAB[6,7]) and also super-smooth surfaces (*e.g.*, electrochemical mechanical polishing).

However, the workpieces in the above-mentioned electrochemical machining methods must be conductive. Distinct from direct metal processes, the confined etchant layer technique (CELT) is actually an electrochemically induced chemical etching method which is proposed by Prof. Tian in 1992.[8] The principle of CELT is described as follows:[9]

---

[a]*College of Chemistry and Chemical Engineering, and State Key Laboratory for Physical Chemistry of Solid Surfaces, Xiamen University, Xiamen 361005, China. E-mail: dpzhan@xmu.edu.cn; Fax: +865922181906; Tel: +865922185797*

[b]*Center for Precision Engineering, Harbin Institute of Technology, P. O. Box 413, Harbin 150001, China. E-mail: lyhoo@163.com; Fax: +8645186415244; Tel: +8645186412924*

1) Generating the etchant at the surface of the tool electrode:

$$R \rightarrow O + ne \qquad (1)$$

Where R is the precursor of etchant and O is the etchant. Actually, the tool electrode used in CELT is both the working electrode and the mold for microfabrication. Due to the diffusion of etchant in the working solution, the shape and thickness are difficult to control. To ensure fabrication tolerance, it is essential to confine the diffusion distance of the etchant in the vicinity of the surface of the tool electrode.

2) Confining the etchant layer to the micro- or nano- meter scale:

$$O + S \rightarrow R + Y \qquad (2)$$

Where S is the scavenger in the working solution, which can react with O and produce R and Y. Due to the scavenging reaction, the etchant is confined at the surface of the tool electrode to form the so-called "confined etchant layer" (CEL). If the concentration of scavenger is significantly higher than the etchant precursor, reaction (2) can be considered as a quasi-first-order reaction. Therefore, the thickness of CEL can be estimated theoretically through the following equation:[9]

$$\mu = (D/K_s)^{1/2} \qquad (3)$$

Where μ is the thickness of CEL, D is the diffusion coefficient of etchant in the working solution and K_s is the quasi-first-order reactive rate constant of the scavenging reaction. If K_s were 10^9 s^{-1}, the CEL would be 1 nm.[10] Actually, the thickness of CEL is tuneable experimentally and determines the fabrication tolerance.

3) Microfabrication through chemical etching:

$$O + M \rightarrow R + P \qquad (4)$$

Where, M is the material of the workpiece, which can react with O and produce R and P. When the tool electrode approaches the workpiece and the CEL contacts the surface of workpiece, chemical etching will occur until the microfabrication process is finished.

CELT has been proved successful in the fabrication of 3D microstructures on metals,[11,12] metal alloys[13] and semiconductors.[10,14–17] In general, a molded tool electrode with certain complementary microstructure is used. The mold material can be a Pt–Ir alloy, silicon or PMMA and so on.[10–20] A thin layer of titanium and then a layer of platinum are deposited on the surface of silicon or PMMA layer through magnetron sputtering to make the mold conductive and stable enough as an electrode. For the metal and metal alloy workpieces, protons are generated as the etchant while sodium hydroxide is used as the scavenger. 3D microstructures have been fabricated on copper,[11,21] nickel,[11,22] aluminum,[12] titanium,[23] nitimol,[13] Ti$_6$Al$_4$V[23] and Mg alloys.[24] For semiconductor workpieces such as silicon and GaAs, the most commonly used etchant is bromine while L-cystine as the scavenger. Recently, we developed CELT as a polishing method to produce super-smooth surface.[25]

Here we present our recent progress, termed electrochemical mechanical machining (ECMM), in which CELT combines well with the motion modes of conventional mechanical machining. The mechanical cutter is replaced by a tool electrode used in CELT. Thus, mechanical machining operations such as lathing, planing and polishing can be performed by electrochemistry. On one hand, the machining precision is improved by employing the chemical principle of CELT. On the other, the mechanical motions enhance the mass transport and balance of the CELT system. Consequently, the machining efficiency is improved.

Experimental

Chemicals and materials

All chemicals (NaBr, H_2SO_4 and L-cystine) were analytical grade or better and provided by Sinopharm Co., China. GaAs wafer was provided by China Crystal Technologies Co. Ltd, China. All aqueous solutions were prepared with deionized water (18.2 MΩ, Milli-Q, Millipore Co.). Both the cylinder and the linear platinum cutters are prepared through precision machining by our multidisciplinary group. The surface roughness of the cutters is lower than 12 nm and sufficient to be employed as the tool electrode for ECMM.

Instrument

The lab-made ECMM equipment has been described elsewhere previously.[26,27] As shown in Fig. 1, the equipment is composed of four parts: a mechanical motion system, an electrochemical system, a monitoring system and an information-processing computer. In the mechanical motion system, a macro–micro dual driven positioning stage (stepper Z1 and piezo motor Z2) moves the tool electrode accurately in the vertical direction. The workpiece is fixed on the working stage,

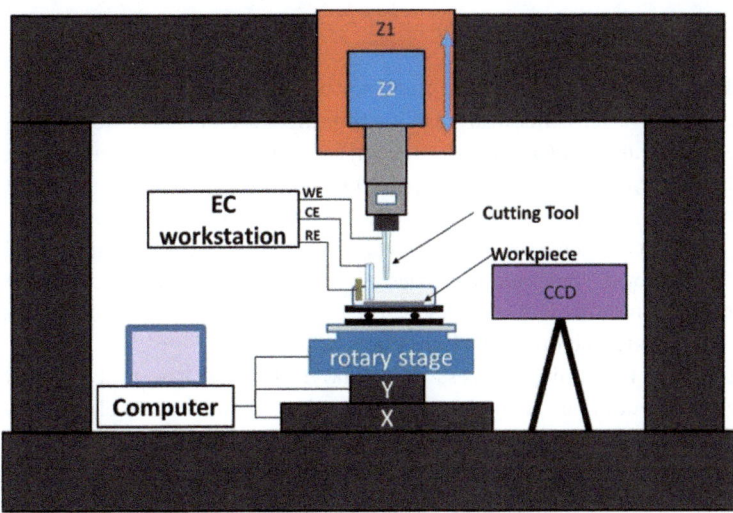

Fig. 1 The schematic diagram of the electrochemical machining instrument, which is composed of a mechanical motion system, an electrochemical system, a monitoring system and an information-processing computer.

which is combined by stepper X, stepper Y and an air-bearing rotary stage (ABRS-150MP-M-X50, Aerotech. Inc., USA). The relative motion between the tool electrode and the workpiece can be programmed in advance. Thus, with the mechanical motion system, almost all mechanical machining processes can be realized. A CHI760 electrochemical workstation (CHI Instrument Co., USA) is used to perform the confined etching processes, including controlling the potential of the tool electrode and detecting the current feedback of the CELT system. The monitoring system includes a force-displacement sensing module and CCD video monitor, which is employed to align the tool electrode with the workpiece, and to control the distance between the tool electrode and the workpiece.[26] Before machining, the working stage is levelled through a SECM current feedback mode as reported previously.[27] Then, the tool electrode is moved to the workpiece by the macro–micro dual driven positioning stage Z with the aid of the CCD video monitor. When the tool electrode touches the workpiece, the force feedback of the force-displacement sensing module changes abruptly, which indicating the zero point of ECMM. After that, the tool electrode is withdrawn for a certain distance, which depends on the expected machining precision. During ECMM, the tool electrode keeps still while the workpiece moves in parallel, vertical or combined manner.

Results and discussion

In the experiments, GaAs wafer was adopted as the workpiece. Considering the mass balance of the ECMM processes, the chemical reactions based on the CELT principle can be formulated as follows:[28]

$$16Br^- \rightarrow 8Br_2 + 16e \quad (5)$$

$$5Br_2 + RSSR + 6H_2O \rightarrow 2RSO_3H + 10Br^- + 10H^+ \quad (6)$$

$$3Br_2 + GaAs + 3H_2O \rightarrow 6Br^- + AsO_3^{3-} + Ga^{3+} + 6H^+ \quad (7)$$

Bromide (Br^-) is used as the precursor to generate the etchant bromine (Br_2) through the electrochemical reaction on the Pt tool electrode. The applied potential is 1.0 V vs. SCE. L-Cystine (RSSR) is employed as the scavenger which can react with Br_2. Through the subsequent homogenous chemical reaction, a CEL is formed on the surface of tool electrode. When the CEL contacts the GaAs workpiece, an heterogeneous etching reaction occurs. Actually, the precision is determined by the reaction properties of the chemical etching system but also the concentration ratio of Br^- over RSSR. The confined etching effect of this system is well investigated before and will not be discussed here.[11,16,28]

The first machining process of ECMM is lathing. The tool electrode (or, machining cutter) used in the experiment is a cylindrical Pt electrode with a diameter of 300 μm. As shown in Fig. 2a, a linear pattern on the GaAs workpiece was produced through ECMM. Fig. 2b shows the lateral profile of the obtained pattern. The average width of the grooves is 322.3 μm and the resulting machining tolerance is about 22.3 μm, which shows the good confining effect of CELT. Beside the parallel motion, the workpiece can also rotate. Fig. 3 gives another lathing example in the case of a rotating workpiece, in which a group of

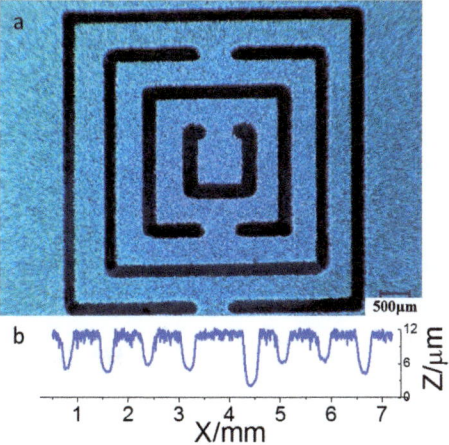

Fig. 2 (a) Optical image of the lathing pattern; (b) profilometric result of lathing pattern in the lateral direction. The working electrolyte solution is an aqueous solution containing 0.3 M NaBr + 0.1 M L-cystine + 2 M H_2SO_4. The distance between the Pt tool electrode and the GaAs workpiece is 25 μm. The moving speed of Pt tool electrode is 60 μm s^{-1}.

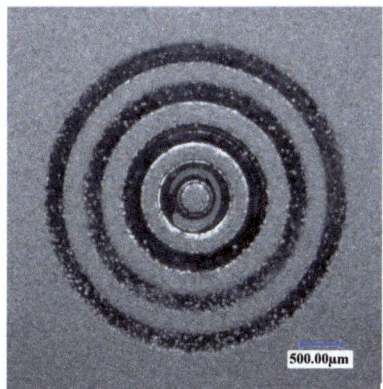

Fig. 3 Optical image of the lathing pattern on a rotating GaAs workpiece. The rotating speed is 2.09 rad min^{-1}; the distance between the GaAs workpiece and the Pt cutting tool electrode is 25 μm.

concentric circles were made through an electrochemical "cutting" process. It should be noted that the product of both the scavenging reaction (6) and the chemical etching traction (7) is bromide, which forms a recycling of Br^-/Br_2 in the narrow space between the tool electrode and the GaAs workpiece. Furthermore, the relative motion between the tool electrode and the GaAs workpiece enhances the mass transport and the balance of the reactant of the CELT system. Thus, the ECMM efficiency and quality can be promoted by optimizing the technical parameters.

ECMM can also be employed to perform planing and polishing processes. Fig. 4 shows the planing effect when the workpiece is moving laterally under a Pt cutting tool electrode with a linear blade shape. The linear platinum blade is 5 mm long, 0.5 mm high and 0.5 mm wide. The moving distance of the Pt cutting

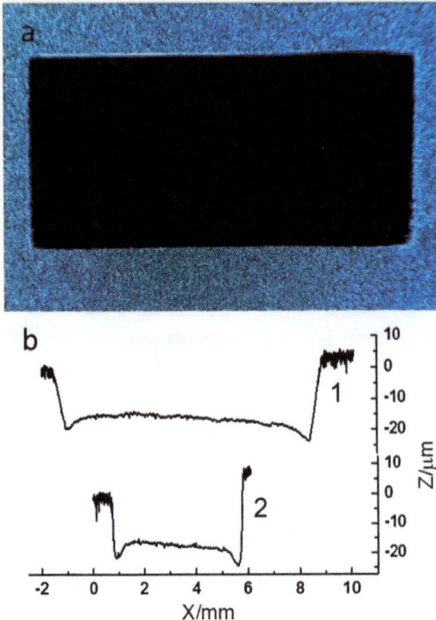

Fig. 4 (a) Optical image of the planing area on a GaAs workpiece; (b) profilometric results of the planing area in the X (Curve 1) and Y (Curve 2) directions. The working electrolyte solution is an aqueous solution containing 0.1 M NaBr + 0.1 M L-cystine + 2 M H_2SO_4. The distance between the Pt tool electrode and the GaAs workpiece is 25 μm. The moving speed of the Pt tool electrode is 60 μm s^{-1}.

tool electrode is 10 mm. The planing depth is about 16.23 μm while the surface roughness of the planing area is 23.01 nm. From these technological parameters it can be concluded that CELT has a competitive potential to be a planing and polishing method for large-scale supersmooth surface machining.

In general, conventional mechanical machining is a point-by-point cutting operation. However, ECMM provides a possibility to work in a more intensified way. This means that an irregular surface can be formed in one step, resulting in a higher machining efficiency. Fig. 5a shows a case where the GaAs workpiece rotates under a platinum tool electrode with a linear blade shape. The etching depth is bigger in the central area but smaller in the outer area due to the difference of radial-linear velocity. An edge effect can also be observed. This irregular surface is formed through the coupling effect of the CELT etching system and the mechanical motion.

If the chemical etching system of CELT is in steady-state during the ECMM process, as in the case of Fig. 5, a simplified relationship can be derived based on Faraday's law and the mass balance of the CELT system:

$$\int_0^l v \frac{\rho \cdot h \cdot 2\pi \cdot R \mathrm{d}R}{M} = \int_0^l \eta \frac{i \cdot a \cdot \mathrm{d}R \cdot t}{nF} \qquad (8)$$

Where v is the stoichiometric ratio of the etching reaction (7), ρ the density of GaAs, h the etching depth, R the radical distance from the central point, M the molecular weight of GaAs, η the current efficiency, a the width of the tool

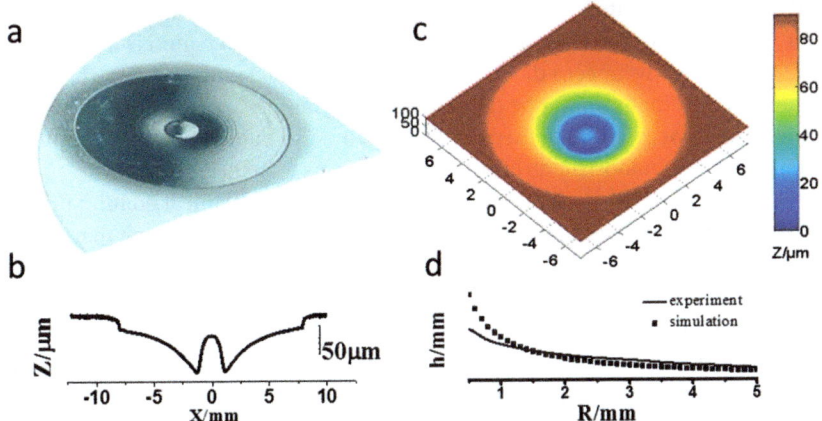

Fig. 5 (a) Optical image of an irregular surface obtained on a GaAs workpiece through the coupling effect of the CELT etching system and the mechanical motion; the rotating speed is 2.09 rad min^{-1} and the distance between the Pt tool electrode and the GaAs workpiece is 10 μm; (b) profilometric results of the irregular surface; (c) a preliminary 3D simulation image of the obtained irregular surface; (d) simulation result of the lateral profile of the irregular surface.

electrode and t the etching time. From eqn (8), the following relationship between etching depth and radical distance can be derived:

$$h = \frac{\eta i a t M}{\upsilon \rho \pi n F} \cdot \frac{1}{R} \qquad (9)$$

The preliminary simulation results are shown in Fig. 5c and 5d, which indicates the etching depth is in proportion to $1/R$.

The above examples show the capacity of ECCM to perform large-scale machining with micro- or nano-precision. The last case will show how low in dimensions ECMM can reach. As shown in Fig. 6, the gap between the two grooves (width: 200 μm) is decreased deliberately down to about 10 μm. Then, silver is

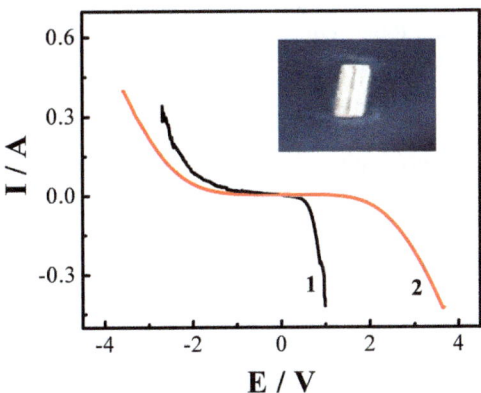

Fig. 6 The I–V behaviours of the diodes fabricated by ECMM with different gaps: Curve 1 (500 μm) and Curve 2 (10 μm). The insert is the optic microscopic image of the diode with a gap of 10 μm.

electrolessly deposited into the grooves to construct a diode. Curve 2 shows the *I–V* behaviour of this micro-diode. The threshold potential region is doubled that of the comparative macro-diode with a GaAs gap of 500 μm. However, the results seem to show that it might be possible to make simple structures with a size of a micrometer or even nanometers. At present, the fabrication of complex or continuous 3D nanostructures by ECMM remains a challenge.

Since CELT is an *in situ* electrochemically-induced chemical etching process, ECMM can work on conductive, semiconductive and even insulating materials. It can also work on flexible, fragile or fissile materials, even materials harder than the tool. This depends on whether the etching reaction can occur or not, and how fast the etching reaction occurs. Since the tool electrode doesn't contact directly with the workpiece, there is no tool wear and no residual stress the on workpiece. Once the coupling between the chemical process and mechanical motion is optimized, ECMM is highly efficient for both regular and irregular surface machining. In brief, ECMM is a prospective multi-scale machining method with micro & nano precision.

Conclusions

The confined etchant layer technique (CELT) has been developed as a large-scale electrochemical mechanical micromachining (ECMM) method with micro & nano precision. Through the coupling of confined chemical etching and mechanical motion, ECCM can perform conventional lathing, planing and polishing in an electrochemical way. Compared with traditional mechanical machining, ECMM is free of tool-wear, thermal and mechanical stresses due to its confined chemical etching characteristics. Thus, ECMM has prospective applications in micro and nano manufacture.

Acknowledgements

The financial support by the National Science Foundation of China (No. 91023006), the Natural Science Foundation of Fujian Province of China (No. 2012J06004), the National Science Foundation of China (No. 91023047, 91023043, 21061120456 and 21021002), the Fundamental Research Funds for the Central Universities (No. 2010121022), and the Scientific Research Foundation for the Returned Overseas Chinese Scholars (State Education Ministry) are appreciated. Y. Yuan and L. Han have made an equal contribution to this paper. The authors declare no competing financial interest.

Notes and references

1 J. A. McGeough, J. McGeough and J. McGeough, *Principles of Electrochemical Machining*, Chapman and Hall, London, UK, 1974.
2 E. Becker, W. Ehrfeld, P. Hagmann, A. Maner and D. Münchmeyer, *Microelectron. Eng.*, 1986, **4**, 35–56.
3 L. T. Romankiw, *Electrochim. Acta*, 1997, **42**, 2985–3005.
4 Y. Yang, B. Imasogie, S. Allameh, B. Boyce, K. Lian, J. Lou and W. Soboyejo, *Mater. Sci. Eng., A*, 2007, **444**, 39–50.
5 L. Singleton, *J. Photopolym. Sci. Technol.*, 2003, **16**, 413–421.
6 A. Cohen, G. Zhang, F. G. Tseng, F. Mansfield, U. Frodis and P. Will, *EFAB: Batch Production of Functional, Fully-Dense Metal Parts with Micron-Scale Features*, The University of Texas at Austin, 1998.

7 S. E. Alper, I. E. Ocak and T. Akin, *J. Microelectromech. Syst.*, 2007, **16**, 1025–1035.
8 Z. Tian, Z. Fen, Z. Tian, X. Zhuo, J. Mu, C. Li, H. Lin, B. Ren, Z. Xie and W. Hu, *Faraday Discuss.*, 1992, **94**, 37–44.
9 A. J. Bard and L. R. Faulkner, *Electrochemical Methods: Fundamentals and Applications*, Wiley India Pvt. Ltd, 2001.
10 J. J. Sun, H. G. Huang, Z. Q. Tian, L. Xie, J. Luo, X. Y. Ye, Z. Y. Zhou, S. H. Xia and Z. W. Tian, *Electrochim. Acta*, 2001, **47**, 95–101.
11 L. M. Jiang, Z. F. Liu, J. Tang, L. Zhang, K. Shi, Z. Q. Tian, P. K. Liu, L. N. Sun and Z. W. Tian, *J. Electroanal. Chem.*, 2005, **581**, 153–158.
12 L. M. Jiang, X. M. Huang, Z. Q. Tian and Z. W. Tian, *Chem. J. Chin. Univ.*, 2006, **8**, 1540–1544.
13 X.-Z. Ma, L. Zhang, G.-H. Cao, Y. Lin and J. Tang, *Electrochim. Acta*, 2007, **52**, 4191–4196.
14 Y. B. Zu, L. Xie, B. W. Mao, J. Q. Mu, Z. X. Xie and Z. W. Tian, *Electrochemistry*, 1997, **3**, 11–14.
15 Y. Zu, L. Xie, B. Mao and Z. Tian, *Electrochim. Acta*, 1998, **43**, 1683–1690.
16 L. Zhang, X. Ma, J. Tang, D. Qu, Q. Ding and L. Sun, *Electrochim. Acta*, 2006, **52**, 630–635.
17 T. Jing, W. Wen-Hua, Z. Jin-Liang and C. Chen, *Acta Phys.- Chim. Sinica*, 2009, **25**, 1671–1677.
18 L. Xie, J. Luo, B. W. Mao and Z. W. Tian, *Chin. J. Sci. Instrum.*, 1996, **17**, 193–198.
19 Y. Zu, L. Xie, B. Mao and Z. Tian, *Electrochim. Acta*, 1998, **43**, 1683–1690.
20 T. Jing, Z. Li, L. M. Jiang, X. Lei, Y. B. Zu and Z. W. Tian, *Three-Dimensional Electrochemical Micromachining on Metal and Semiconductor by Confined Etchant Layer Technique (CELT)*, 2007.
21 Z. F. Liu, L. M. Jiang, J. Tang, P. K. Liu, L. N. Sun, Z. Q. Tian and Z. W. Tian, *Chin. J. Appl. Chem.*, 2004, **21**, 227–230.
22 Z. F. Liu, L. M. Jiang, J. Tang, L. Zhang, Z. Q. Tian, Z. W. Tian, P. K. Liu and L. M. Sun, *Electrochemistry*, 2004, **4**, 249–253.
23 L. Jiang, W. Li, A. Attia, Z. Cheng, J. Tang and Z. Tian, *J. Appl. Electrochem.*, 2008, **38**, 785–791.
24 L. M. Jiang, Z. Y. Cheng, N. Du, W. Li, Z. Q. Tian and Z. W. Tian, *Acta Phys.- Chim. Sinica*, 2008, **24**, 1307–1312.
25 *CN Pat.*, 201010219037.5, 2010.
26 L. J. Lai, H. Zhou, Y. J. Du, J. Zhang, J. C. Jia, L. M. Jiang, L. M. Zhu, Z. W. Tian, Z. Q. Tian and D. P. Zhan, *Electrochem. Commun.*, 2013, **28**, 135–138.
27 L. Han, Y. Yuan, J. Zhang, X. Zhao, Y. Cao, Z. Hu, Y. Yan, S. Dong, Z.-Q. Tian, Z.-W. Tian and D. Zhan, *Anal. Chem.*, 2013, **85**, 1322–1326.
28 L. Zhang, X. Z. Ma, J. L. Zhuang, C. K. Qiu, C. L. Du, J. Tang and Z. W. Tian, *Adv. Mater.*, 2007, **19**, 3912–3918.

Faraday Discussions

PAPER

Decoration of active sites to create bimetallic surfaces and its implication for electrochemical processes†

Blake J. Plowman, Ilija Najdovski, Andrew Pearson and Anthony P. O'Mullane*

Received 14th February 2013, Accepted 5th March 2013
DOI: 10.1039/c3fd00015j

The creation of electrocatalysts based on noble metals has received a significant amount of research interest due to their extensive use as fuel cell catalysts and electrochemical sensors. There have been many attempts to improve the activity of these metals through creating nanostructures, as well as post-synthesis treatments based on chemical, electrochemical, sonochemical and thermal approaches. In many instances these methods result in a material with active surface states, which can be considered to be adatoms or clusters of atoms on the surface that have a low lattice co-ordination number making them more prone to electrochemical oxidation at a wide range of potentials that are significantly less positive than those of their bulk metal counterparts. This phenomenon has been termed pre-monolayer oxidation and has been reported to occur on a range of metallic surfaces. In this work we present findings on the presence of active sites on Pd that has been: evaporated as a thin film; electrodeposited as nanostructures; as well as commercially available Pd nanoparticles supported on carbon. Significantly, advantage is taken of the low oxidation potential of these active sites whereby bimetallic surfaces are created by the spontaneous deposition of Ag from $AgNO_3$ to generate Pd/Ag surfaces. Interestingly this approach does not increase the surface area of the original metal but has significant implications for its further use as an electrode material. It results in the inhibition or promotion of electrocatalytic activity which is highly dependent on the reaction of interest. As a general approach the decoration of active catalytic materials with less active metals for a particular reaction also opens up the possibility of investigating the role of the initially present active sites on the surface and identifying the degree to which they are responsible for electrocatalytic activity.

School of Applied Sciences, RMIT University, GPO Box 2476V, Melbourne, VIC 3001, Australia. E-mail: anthony.omullane@rmit.edu.au

† Electronic supplementary information (ESI) available. See DOI: 10.1039/c3fd00015j

Introduction

The electrochemical behaviour of noble metals has received significant attention given their use in green energy generation and conversion applications. In particular, for fuel cells there have been numerous reports on the synthesis, characterisation and application of Pt, Au, Ag and Pd nanomaterials for proton exchange membrane, direct methanol, formic acid and alkaline fuel cells[1–9] where advantage is taken of the electrocatalytic activity of the relevant metal under appropriate pH conditions. For instance Pt and Pd are mainly investigated and commercially used for proton exchange membrane, formic acid and direct methanol fuel cells under acidic conditions[3–5,7] whereas both Au and Ag demonstrate remarkable activity under alkaline conditions.[8–13] In many electrocatalytic studies the synthesis of novel nanostructured materials has been addressed and the role that shape and size plays on electrocatalysis has been systematically investigated by numerous groups.[3,5,7,14–21] Likewise there have been important fundamental insights into the role that exposed crystal facets play on electrocatalytic processes which are undertaken at well-defined extended single crystal surfaces.[13,18,21–25] From these studies it has been found that the low index planes of (111), (100) and (110) show different activity for a specific electrocatalytic reaction but noteworthy is that the trend in activity is not uniform across different electrocatalytic reactions.[10,21] Electrocatalytic studies at these surfaces is critical to fundamental understanding, however they are highly unlikely to be utilised in a functional device. Furthermore there is debate whether very small nanoparticles possess the crystal structure of the bulk metal,[21] which would be expected to alter the electrochemical behaviour of the material. Therefore, experimental investigation of technologically relevant nanomaterials and surfaces also needs to be pursued even though a significant level of complexity is added to the system under study.

The presence of active sites on nanomaterials is also an important parameter and the presence of kinks, steps, ledges and other crystallographic defects that are present on arguably all surfaces are expected to play a significant role in improving the electrocatalytic performance of a material. In recent years a somewhat controversial model of electrocatalysis has been proposed, namely the incipient hydrous oxide adatom mediator (IHOAM) model.[26–30] In this approach a metal is assumed to consist of: (i) bulk atoms that are fully lattice stabilised and are oxidised at the thermodynamically predicted potential by Pourbaix; and (ii) a surface containing defects, adatoms or clusters of adatoms. The latter have low lattice co-ordination numbers and are more readily oxidised than their bulk counterparts leading to electrochemical oxidation processes at potentials that are significantly lower than bulk metal oxide formation. This phenomenon has been termed "pre-monolayer oxidation" and demonstrated on a range of metals including Au, Pt, Pd, Cu and Ag.[30–39] The evidence for these pre-monolayer oxidation processes is mounting from reports that have studied this effect using contact electroresistance,[40,41] electroreflectance techniques,[42] electrochemical quartz crystal microbalance,[43] large amplitude Fourier transformed ac (FT-ac) voltammetry[44,45] and the surface interrogation mode of scanning electrochemical microscopy (SECM).[46] For the latter approach Bard demonstrated that the coverage of gold with such incipient oxides can be as high as 0.2 of a monolayer.

However, the characterisation of such active sites has proven extremely difficult due to low surface coverage and their somewhat transient nature, yet significantly a recent study using aberration-corrected scanning transmission electron microscopy in the high-angle annular dark-field imaging mode revealed the presence of low coordination Au adatoms on truncated octahedral gold clusters (Au_{923}).[47] Even so new approaches for studying such active site behaviour need to be explored, particularly under aqueous conditions as encountered in electrochemical studies. A recent investigation by Scholz[48] showed an elegant method of investigating active site behaviour on gold whereby Fenton's reagent was used to chemically dissolve asperities on the surface. This resulted in "knocking out" the active sites thereby shutting down reactions such as oxygen reduction and hydrogen evolution. In this work an analogous approach is taken in which it is described how advantage can be taken of pre-monolayer oxidation to drive the electroless deposition of metals such as Ag onto Pd surfaces, which is normally a thermodynamically forbidden process. Pd was chosen as it has been identified as a suitable catalyst material for fuel cells; it has comparable activity to Pt but is significantly cheaper. This approach is undertaken at a variety of Pd surfaces including evaporated films, electrodeposited nanostructures and commercially available Pd/C where it is shown that active sites across all materials are involved in the decorating process which significantly impacts on many electrocatalytic reactions.

Experimental

Chemicals

$AgNO_3$ (Sigma), $Pd(NO_3)_2$ (BDH), NaOH (BDH), H_2SO_4 (Sigma), ferrocenemethanol (Sigma), formaldeyde (BDH) were used as received. All aqueous solutions were made up with deionized water (resistivity of 18.2 MΩ cm) purified by use of a Milli-Q reagent deioniser (Millipore).

Electrode materials

Glassy carbon plates (HTW) were used for the electrodeposited Pd nanostructures and porous Pd and the Pd thin films were deposited by a Balzers™ electron beam evaporator. The layer composed of 1500 Å of Pd and 10 Å of Ti. The films were deposited sequentially by an electron evaporation process onto the bare Si substrates (Supplier). Pd/C was drop-cast onto 3 mm diameter GC electrodes (BAS). A 3 mm diameter Pd electrode (BAS) was also used.

Modification of evaporated Pd films and nanostructured Pd with silver

The relevant substrate was immersed in an aqueous 1 mM $AgNO_3$ solution under dark conditions after which it was thoroughly rinsed with Milli-Q water and dried with a gentle flow of nitrogen gas.

Modification of Pd/C with silver

To 1 mL of an aqueous 1 mM $AgNO_3$ solution, 0.5 mg of a commercial Pd/C powder (Sigma-Aldrich) was added and allowed to react for a period of 5 min under dark conditions. After 5 min of reaction, the material was centrifuged at 5000 rpm for 10 min, the supernatant was removed and the powder washed with

1 mL ethanol. The material was then centrifuged and washed an additional two times. After washing the powder was re-suspended in 1 mL of 0.2% Nafion in ethanol. Materials for electrochemical reactions were prepared by drop-casting 10 μL of well dispersed particles in solution onto a GC electrode and allowing the solvent to evaporate.

Electrochemical measurements

All experiments were conducted at $(20 \pm 2)°C$ with either a CH Instruments CHI 760C or CHI 760D electrochemical analyser in an electrochemical cell that allowed reproducible positioning of the working, reference, and counter electrodes. Working electrodes consisting of glassy carbon (GC) plates (0.159 cm^2) or Pd (BAS) were polished with an aqueous 0.3 μm alumina slurry on a polishing cloth (Microcloth, Buehler), sonicated in acetone and then deionized water for 5 min, and dried with a flow of nitrogen gas prior to use and Pd films were washed with acetone and ethanol and then blown dry with nitrogen. The exposed geometric area of these films was controlled by using Kapton tape with a circular cutout (4.5 mm diameter). The reference electrode was Ag/AgCl (aqueous 3 M KCl). A graphite rod (6 mm diameter, Johnson Matthey Ultra "F" purity grade) was used as the counter electrode. All electrochemical experiments were commenced after degassing the electrolyte solutions with nitrogen for at least 10 min prior to any measurement except for the oxygen reduction reactions. For the latter an aqueous solution of 1 M NaOH was purged with high purity oxygen (5.0 coregas) for at least 20 min. Open circuit potential (OCP) *versus* time experiments were carried out in a 3 electrode arrangement under stirring conditions where the Pd electrode was first immersed in Milli-Q water followed by injection of a solution AgNO$_3$ to give a final concentration of 1 mM.

Large amplitude FT-ac measurements

A description of the FT voltammetric instrumentation used in this study is available elsewhere.[49] Sine waves of frequencies $f = 21.46$ Hz and amplitudes of $\Delta E = 100$ mV were employed as the ac perturbation. DC voltammetric experiments were also carried out with this instrumentation by using a zero amplitude perturbation to compare with results obtained using the CH Instruments potentiostat.

Materials characterisation

Scanning electron microscopy (SEM) measurements were performed on a FEI Nova NanoSEM instrument (Nova 200). X-Ray diffraction (XRD) measurements were carried out on a Bruker AXS X-ray diffraction system operated at a voltage of 40 kV and current of 40 mA with Cu-Kα radiation. X-Ray photoelectron spectroscopy (XPS) measurements were obtained with a Thermo K-Alpha XPS instrument at a pressure better than 1×10^{-9} Torr with core levels aligned with C 1s binding energy of 285 eV. Samples for transmission electron microscopy (TEM) were drop cast onto a carbon coated copper grid and performed using a JEOL 1010 TEM instrument operated at an accelerating voltage of 100 kV.

Results and discussion

Illustrated in Fig. 1 is the conventional voltammetry of a commercial Pd electrode (BAS) in 1 M H_2SO_4. The adsorption/desorption of hydrogen is present over the potential range of −0.20 to 0.30 V but is rather sluggish in nature without any indication of distinct peaks. The oxide formation region commences at *ca.* 0.73 V until the end of the positive sweep at 1.20 V. On the reverse sweep the oxide layer is removed in a sharp process centred at 0.50 V. This is the expected behaviour for a relatively smooth Pd surface in acidic electrolyte.[30,50]

There have been reports on activating Pd surfaces *via* thermal pretreatment[51] or by an electrochemical procedure whereby a thick hydrous oxide is grown on the Pd surface followed by slow reduction to produce a disrupted or active surface.[30] This usually leads to additional features in the hydrogen adsorption/desorption region and pre-monolayer oxide processes. In this work similar effects can be seen for electrodeposited Pd films that have a nanostructured morphology. Illustrated in Fig. 1b is the CV of nanostructured Pd in 1 M H_2SO_4 that was achieved through a simple electrodeposition procedure at a GC electrode held at −0.30 V for 300 s in an electrolyte containing 20 mM $Pd(NO_3)_2$. It can be seen that the CV of nanostructured Pd is distinctly different to that of the bulk Pd electrode. The region from −0.20 to 0.30 V contains an additional pair of well-defined conjugate peaks with a midpoint potential of 0.045 V with a peak-to-peak separation of 60 mV. The appearance of these peaks at this potential region has been observed previously for both activated Pd electrodes and Pd nanoparticles and nanowires.[30,51,52] The interpretation of these peaks has been disputed, with Burke and others attributing this behaviour to quasi-reversible pre-monolayer oxide formation process at highly active Pd adatoms or clusters of adatoms which is accompanied by field assisted specific adsorption of sulphate and/or bisulphate anions,[30,51,52] whereas other groups have ascribed this to better-defined H UPD processes at different exposed crystal facets.[25,50] At more positive potentials it can be seen that there is clearly an earlier onset in the formation of the monolayer oxide region which has also been attributed to pre-monolayer oxidation processes at activated Pd sites.[30,51,53]

This can be related to the formation of a discontinuous layer of Pd dendrites distributed over the entire surface (Fig. 2a). A higher magnification image is shown in Fig. 2b: the branched nature of the individual dendrites can be seen

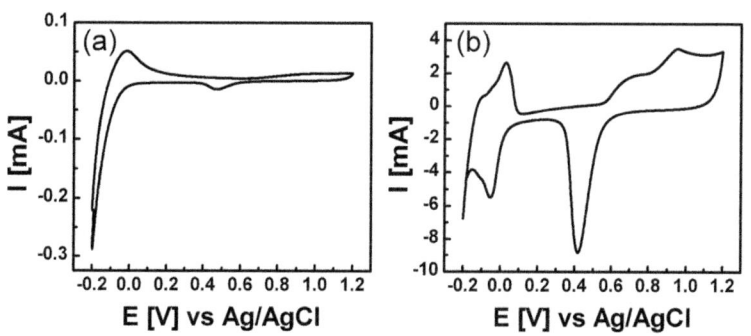

Fig. 1 CVs recorded in 1 M H_2SO_4 at a sweep rate of 50 mV s^{-1} at (a) bulk Pd electrode and (b) electrodeposited nanostructured Pd surface on a GC electrode.

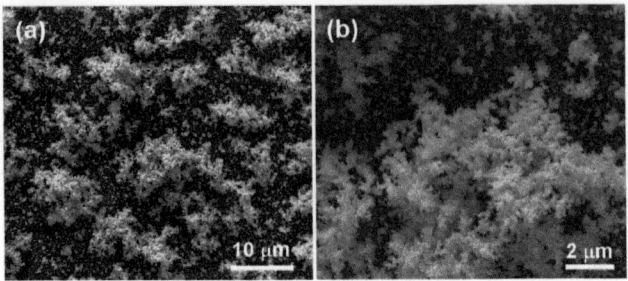

Fig. 2 SEM images of Pd electrodeposited at −0.30 V for 300 s on a GC electrode from an electrolyte containing 20 mM Pd(NO$_3$)$_2$.

where the branches taper to a typical tip size of *ca.* < 100 nm in diameter. Previous studies have shown that highly anisotropic nanostructures of metals such as Pd, Pt and Au are quite active electrocatalytic materials[1,14,54–56] and are prone to oxidation at low potentials.

The electrochemical behaviour of a bulk Pd electrode in 1 M NaOH (Fig. 3a) was also investigated which showed analogous behaviour to that observed in 1 M H$_2$SO$_4$ consisting of a broad hydrogen desorption region, double layer and monolayer oxide formation on the positive sweep and an oxide reduction process at −0.24 V followed by hydrogen adsorption on the negative sweep. For the nanostructured Pd electrode there was no evidence of the pair of conjugate peaks in the hydrogen adsorption/desorption region (Fig. 3b) as was the case for this electrode in acidic electrolyte. The nanostructured Pd surface however showed an additional feature in the double layer region with a broad process occurring prior to the onset of monolayer oxide formation. To probe this effect further a highly porous Pd surface was created by the relatively new method of hydrogen bubble templating which involves metal deposition around hydrogen bubbles that are evolved from the surface. A SEM image of this type of structure is shown in the ESI Fig. S1.† A typical honeycomb-like structure is observed with a highly branched internal wall structure which is consistent with previous studies.[57] However, the CV response for this sample, shown in Fig. 3c, in 1 M NaOH shows a distinctly different behaviour to the bulk Pd electrode and electrodeposited Pd nanostructures. There are numerous oxidation process that occur at significantly lower potential than bulk oxide formation at −0.56, −0.46 and −0.30 V. This is consistent with previous studies by Burke and also Pletcher *et al.* who also observed similar behaviour at activated and electrodeposited mesoporous Pd, respectively[30,51,58] and attributed it to the oxidation of the Pd surface at particularly favourable sites. It is interesting to note that in all cases in Fig. 3 there is an absence of the sharp conjugate peaks that were observed in Fig. 1b. This suggests that the appearance of these peaks in acidic electrolyte is not related to the adsorption/desorption of hydrogen. This was also concluded recently from electrochemical impedance studies where the data suggested that it was the formation of incipient oxides on the surface of palladium that was the origin of this type of response.[52] The lack of a reduction component below −0.50 V in Fig. 3b may be related to the recalcitrant nature of the oxidation product of active Pd[51,53] which has been suggested to be in the form of $[Pd(OH)_6]^{2-}_{ads}$.

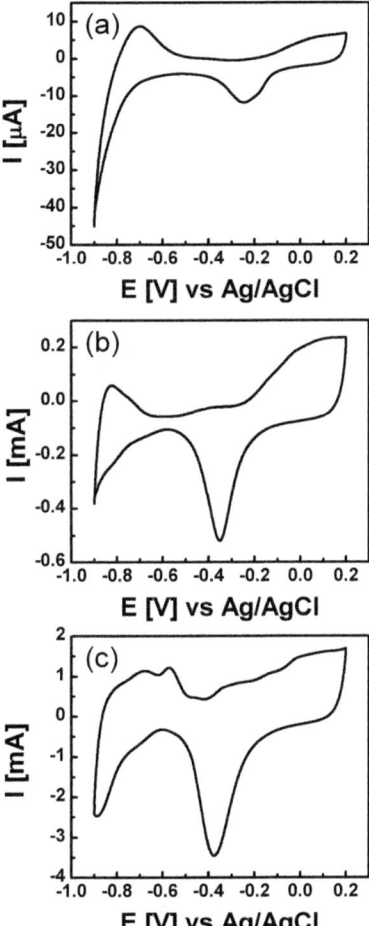

Fig. 3 CVs recorded at 50 mV s^{-1} in 1 M NaOH at: (a) bulk Pd electrode; (b) electrodeposited Pd surface (as in Fig. 2); and (c) highly porous Pd surface generated *via* hydrogen bubble templating.

It has recently been shown that FT-ac voltammetry is highly effective in detecting electrochemical processes at metal surfaces such as Au and Cu that are not clearly observable *via* dc voltammetry due to the capacitive nature of the double layer region. The FT-ac technique allows the capacitive and Faradaic components to be separated and has been used to demonstrate that significant pre-monolayer oxidation responses occur on both gold and copper in the double layer region.[44,45] This approach was applied to the bulk Pd electrode in acidic solution and the results are shown in Fig. 4. The dc component is also shown to demonstrate the consistency with the response recorded under regular dc voltammetric conditions. The forward and reverse sweeps are represented separately for clarity of presentation. In the 1st and 2nd ac harmonic responses the current is still quite significant in the −0.20 to 0.30 V region and still has a contribution from capacitive current. In the 3rd and 4th ac harmonic responses there is a distinct process centred at 0.10 V which is Faradaic in nature and coincides with the pair of conjugate peaks seen in the case of nanostructured Pd (Fig. 1b). This

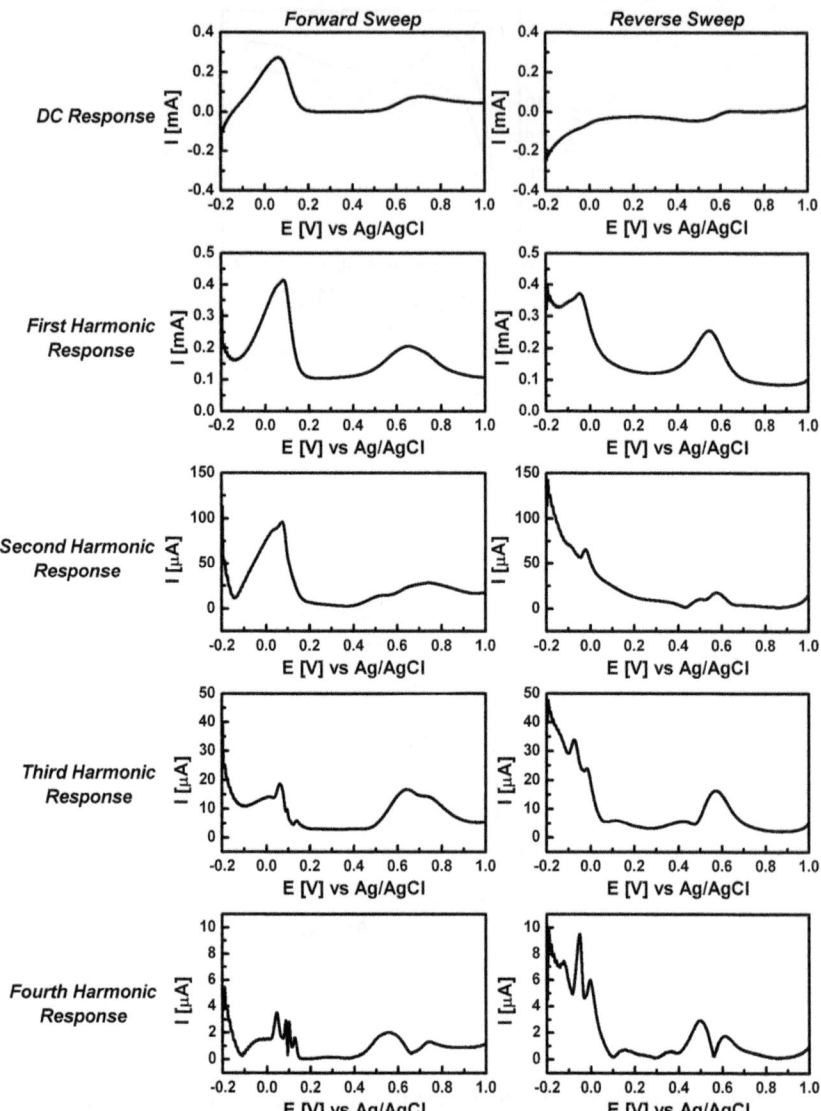

Fig. 4 Large amplitude FT-ac voltammetry at a Pd electrode in 1 M H_2SO_4 for the dc and 1st to 4th ac harmonic responses recorded at a sweep rate of 44.70 mV s^{-1} with an amplitude of 100 mV at a frequency of 21.46 Hz.

illustrates the effectiveness of the technique whereby processes that are not readily observed in dc voltammetry due to capacitance can be identified in the higher harmonics. This implies that the active site response at 0.10 V is also present on the bulk Pd electrode. The response is not as well defined for what is expected for a surface confined electron transfer process (*e.g.* a quartet of 1 : 3 : 3 : 1 intensity ratio for a fully reversible process on both the forward and reverse sweeps) which implies that electron transfer for this reaction is not highly reversible at the bulk Pd electrode in acidic conditions.[44,59] The lack of a clear response on the reverse sweep at this potential suggests an EC type mechanism which is not unexpected if the oxidation product is a species like $[Pd(OH)_6]^{2-}_{ads}$ which may be unstable under acidic conditions. There is also evidence for a process at ca. 0.55 V which is just prior to the onset of monolayer oxide formation and could be an additional active site response.

It is clear from this data that both nanostructured and bulk Pd electrodes can be oxidised at potentials well below that recorded for the oxidation of bulk Pd containing more stabilised atoms with a higher lattice co-ordination than the described metastable surface state of adatoms or clusters of adatoms. Such large shifts in oxidation potential have also been shown for extremely small gold nanoparticles of <4 nm in diameter that were oxidised in a halide containing media at potentials ca. 850 mV negative of the bulk Au oxidation peak.[60] A recent discussion by Kucernak[61] on Pt suggested that the formation of a hydrogen bonded water–OH_{ads} network on polycrystalline surfaces occurs which dictates the rate of CO_{ads} oxidation. The potential at which this occurs is highly dependent on the nature of the surface and the presence of strongly adsorbed ions, however it was suggested that this can occur in the double layer region of Pt.

Therefore, if Pd can be oxidised at potentials that are significantly less positive than the formation of bulk oxide then advantage can be taken of this as a means of identifying active site behaviour *via* the spontaneous deposition of a less noble metal onto the active surface that is thermodynamically forbidden at the bulk material. Taking Ag as an example the standard reduction potential for Ag^+/Ag is 0.799 *vs.* SHE compared to 0.915 V *vs.* SHE for the Pd^{2+}/Pd couple. Therefore, it is not expected that Ag can be reduced on to the Pd surface. However, given that oxidation processes are observed on Pd at potentials as low as 0.100 V *vs.* Ag/AgCl (or 0.315 V *vs.* SHE) it should provide enough driving force for the spontaneous deposition of silver to occur at these active sites. As an initial study a highly reproducible surface was employed, namely a Pd film of 150 nm thickness that was evaporated onto a silicon substrate. All experiments in this work were conducted using samples cut from the same evaporated film. The sample was immersed for 5 min in 1 mM $AgNO_3$ in the dark and then washed thoroughly with Milli-Q water and dried with nitrogen. From XPS analysis it can be clearly seen that Ag deposits on to the surface (Fig. 5) where the surface concentration was calculated to be 15 atomic %. The Ag $3d_{5/2}$ and Ag $3d_{3/2}$ peaks of the decorated surfaces at 368.4 and 374.5 eV clearly indicate the formation of metallic Ag.[62] The

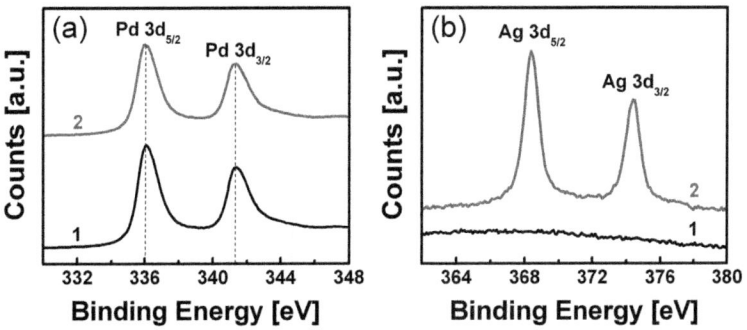

Fig. 5 XPS spectra of (a) Pd 3d and (b) Ag 3d recorded at an evaporated Pd film before (1) and after (2) decoration with silver.

Pd $3d_{5/2}$ and $3d_{3/2}$ peaks did not shift in the presence of Ag (Fig. 5a) which indicates that surface alloy formation is unlikely.

CV experiments performed at the evaporated film and the Ag decorated film are shown in Fig. 6. In the absence of Ag the voltammetry recorded at the Pd substrate is particularly sluggish in the hydrogen adsorption/desorption region. Significantly, there is a distinct oxidation process at *ca.* 0.80 V prior to the onset of monolayer oxide formation. After decoration with Ag the potential region from 0.60 to 1.20 V is significantly affected. The magnitude of the peak at 0.80 V on the Pd substrate is diminished and replaced with a distinct peak at 0.87 V which is attributed to the oxidation of Ag. It must be emphasised that in this decoration process the Pd film is not electrochemically treated prior to immersion in $AgNO_3$ to avoid any possibility of forming adsorbed hydrogen that would drive the deposition process *via* its oxidation in the presence of the metal salt. Upon cycling both electrodes, the conventional Pd voltammetry in acid is observed (as in Fig. 1a) but both the hydrogen adsorption/desorption and monolayer oxide region for the Pd/Ag sample is quite different. This is consistent with the voltammetry obtained for PdAg nanocrystals immobilised on graphene formed *via* a galvanic replacement reaction.[63] It is interesting in both cases that electrochemical cycling is required to attain more defined voltammetry. This suggests that the initially deposited Pd film is not particularly active but can still readily drive the spontaneous deposition of silver on to the surface.

The decoration process was monitored by performing open circuit potential *versus* time experiments. This is a particularly useful approach as it gives information on the accumulation or dissipation of surface charge. Generally it is assumed that an increase in the OCP value above the point of zero charge reflects an increase in positive charge on the surface.[64] This has been used to investigate the formation of self-assembled monolayers,[64] monitor catalytic reactions[65] and galvanic replacement reactions.[66] It can be seen from Fig. 7 that there is a very sharp increase in the OCP value up to 0.53 V when $AgNO_3$ is introduced into solution which is consistent with oxidation of the surface, in this case at the active sites, followed by a decrease which could indicate the formation of an oxide on Pd that dissipates the accumulated surface charge. The spontaneous decoration of noble metals such as Pt with Ru or *vice versa* has been reported[67,68] but there remains contention over the mechanism by which this occurs. It has been

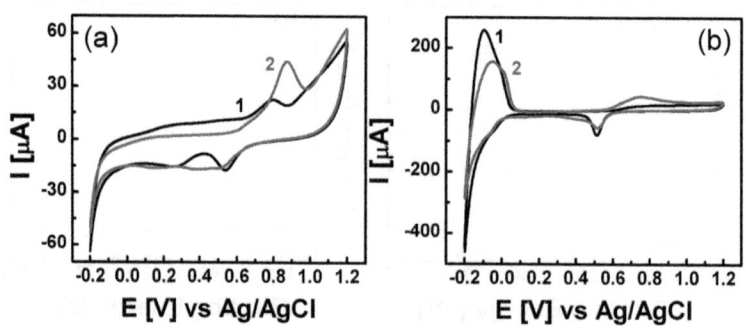

Fig. 6 CVs recorded in 1 M H_2SO_4 at 50 mV s^{-1} at an evaporated Pd film before (1) and after (2) Ag decoration where (a) shows the 1st cycle and (b) shows the 10th cycle.

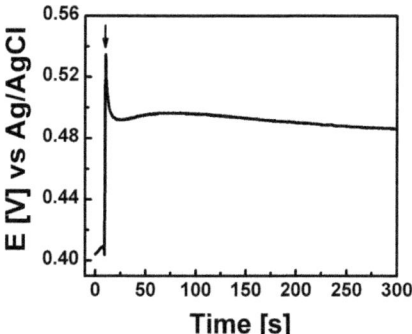

Fig. 7 OCP *versus* time for an evaporated Pd film immersed in stirred Milli-Q water with the addition of AgNO$_3$ (as indicated by the arrow) to a total concentration of 1 mM.

suggested that: (i) the oxidative dissolution of metal occurs in the presence of metal cations, however such a large driving force to dissolve Pd is not present in this system when using AgNO$_3$; (ii) a chemical reaction with adsorbed hydrogen, this has been carefully avoided as discussed above; and (iii) in the case of Pt a local cell mechanism was proposed where Pt is oxidised to Pt-OH that facilitates metal deposition.[68] The latter appears to be the most viable under the conditions of this study and supports the oxidation of active sites facilitating the reduction of silver cations to metallic silver on the surface.

The recent work by Scholz[48] demonstrated that gold active sites could be "knocked out" using Fenton's reagent where it was demonstrated that gold asperities were chemically dissolved, which resulted in severe inhibition of inner sphere reactions while outer sphere reactions remained unperturbed. From that rationale the location of active sites on those asperities was inferred. In this work a similar approach is taken using the oxidation of ferrocenemethanol as an outer sphere electron transfer reaction. It can be seen in Fig. 8a that the reversible oxidation of ferrocenemethanol is not affected by the presence of Ag on Pd as expected. However, taking two inner sphere reactions such as the oxygen reduction reaction (ORR) and the hydrogen evolution reaction (HER) shows that the presence of Ag on Pd has a profound effect. It is immediately apparent that the activity for the ORR is severely suppressed (Fig. 8b) with Ag on the surface and is reflected by a shift in the onset potential of 180 mV to less positive potentials from *ca.* 0.40 to 0.22 V. This is highly significant given that 85 atomic % of the surface is Pd. This suggests that the active sites on Pd that are responsible for the reduction of oxygen are covered with silver *via* this decoration process, *i.e.* in effect "knocked out". The HER produced quite a surprising result in that the reaction was slightly enhanced in the presence of Ag (Fig. 8c) which is unexpected given that Ag is such a poor electrocatalyst for this reaction under acidic conditions. This infers that the active sites that are responsible for the HER may not be involved in the ORR. This is consistent with the IHOAM model of electrocatalysis where distinct M*/hydrous oxide transitions are observed on most metals that mediate a variety of electrocatalytic reactions. For instance, in the case of Au in an acidic electrolyte four distinct potentials were identified for the catalytic mediator, which were postulated to play a major role in the onset/termination potential of numerous electrocatalytic reactions. In highly alkaline solutions five distinct catalytic

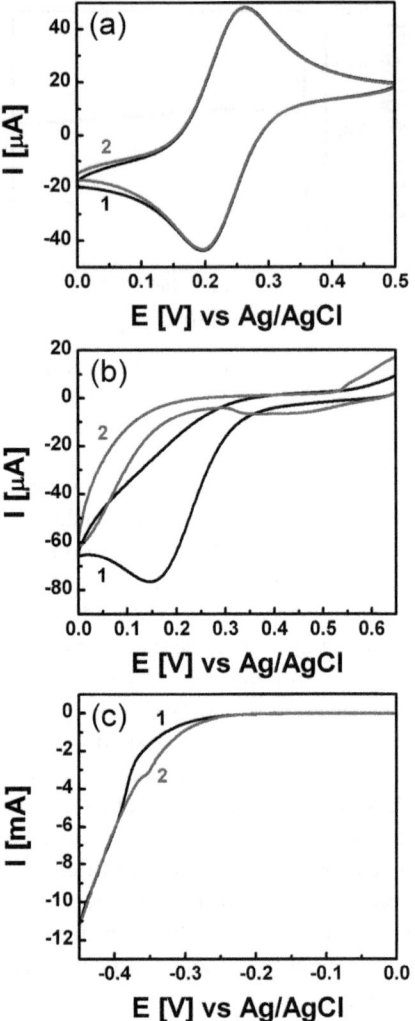

Fig. 8 CVs recorded at (1) evaporated Pd film and (2) Ag decorated Pd film in (a) 1 mM ferrocene-methanol in 0.1 M KNO$_3$, $v = 100$ mV s^{-1}, (b) 1 M H$_2$SO$_4$ saturated with oxygen gas, $v = 50$ mV s^{-1} and (c) 1 M H$_2$SO$_4$, $v = 5$ mV s^{-1}.

mediator potential values were identified.[69] Also Norskov's work illustrated, by density functional theory calculations, that overlayers of silver on palladium increase the free energy of hydrogen adsorption (ΔG_H) from 0–0.1 eV to >0.5 eV, which is a reliable indicator of activity for HER where a ΔG_H value of 0.0 eV leads to optimal activity.[70,71] This supports the postulation that sites with different activity are responsible for this reaction than those involved in the ORR.

Recently Scholz also investigated the effect of OH$^\bullet$ radical attack on Pd surfaces where a dramatic decrease in surface area and a smoothing of the morphology was observed yet the inner sphere reaction quinone/hydroquinone was not affected.[72] The reason suggested was that active sites were present over the entire Pd surface and not specifically located at asperities. It should be noted that using Fenton's reagent results in severe chemical dissolution of the surface, whereas in this work such a substantial change in the surface morphology does not occur due to the lower oxidising power of Ag⁺ which provides a less intrusive method for studying active site behaviour. This was confirmed by SEM imaging of the evaporated Pd film before and after decoration (Fig. 9). It can be seen that the morphology of the surface remains intact and that there is no obvious dissolution of the surface as the grain sizes are comparable in both size and shape. It is surprising however that there is no evidence of Ag formed on the surface in the form of nanoparticles.

This lends support to Scholz's argument that active sites are located homogeneously across the Pd surface. In an effort to possibly increase the loading of the Pd surface with Ag, the nanostructured Pd samples shown in Fig. 2 were also decorated with Ag as confirmed by XPS with a surface loading of 10.2 atomic % (Fig. S2†) and cyclic voltammetry. It can be clearly seen in Fig. 10 that the response at *ca.* 0.05 is diminished after decoration with Ag, which also supports that this process is associated with the oxidation of surface active sites that are utilised in the metal decoration process and not hydrogen adsorption/desorption peaks, given that the substrate was not used prior to the decoration process.

SEM imaging however did not reveal the obvious presence of Ag as nanoparticles on the surface. It should be noted that this decoration process is not

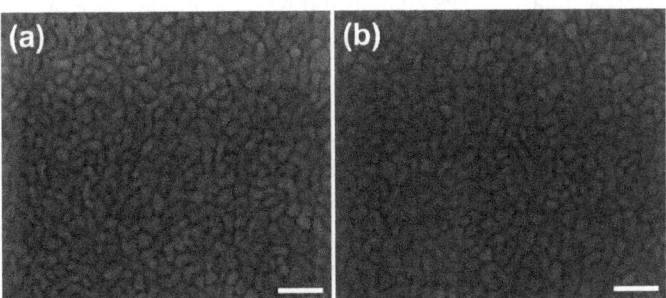

Fig. 9 SEM images of an evaporated Pd film before (a) and after (b) decoration with silver. Scale bar is 100 nm in each case.

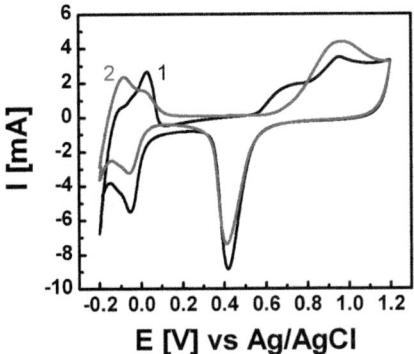

Fig. 10 CVs recorded in 1 M H_2SO_4 at 50 mV s⁻¹ at an evaporated Pd film before (1) and after (2) decoration with Ag.

reported for the highly porous Pd sample due to hydrogen being evolved during the electrodeposition procedure. The possible presence of adsorbed hydrogen on the surface post synthesis could contribute to the deposition of Ag and is therefore avoided in this discussion. Finally a commercial catalyst was investigated (10% Pd loaded on carbon) and decorated with Ag. The presence of Ag was confirmed by XPS (Fig. S3†) which gave a surface loading of 14.5 atomic %. TEM imaging before and after decoration is shown in Fig. 11 where there are a number of new features in the decorated sample. The morphology of the Pd nanoparticles transforms from a relatively smooth oblong shape (Fig. 11b) into a rough porous particle that appears to be an agglomeration of much smaller nanoparticles (Fig. 11d) while there is also the appearance of numerous particles on the carbon support which are assumed to be Ag. This was further confirmed by powder X-ray diffraction where the presence of peaks at $2\theta = 44.3$ and 64.2 and a shoulder at 38.2 are indicative of the (200), (220) and (111) crystal planes of fcc Ag, respectively (Fig. 12).[12] Interestingly the (111) and (220) planes of Pd become quite broad after the decoration procedure and this is consistent with a porous morphology that consists of smaller particles as seen by TEM imaging. However the peak positions do not change, implying that alloy formation does not occur as supported by the previous XPS results (Fig. 5). To exclude the possibility of surface functional groups on carbon driving the metal reduction process, the same experiment was performed using activated carbon where no evidence of Ag nanoparticles was found *via* TEM imaging. XPS analysis did reveal a minute trace of Ag (Fig. S4†) which is assumed to be due to the oxidation of organic functional groups which

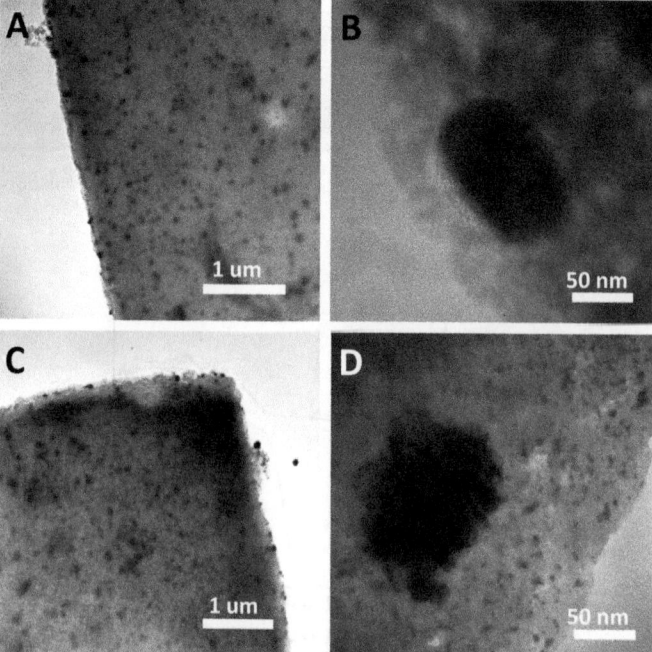

Fig. 11 TEM images of commercial Pd on carbon catalyst before (A, B) and after (C, D) decoration with Ag.

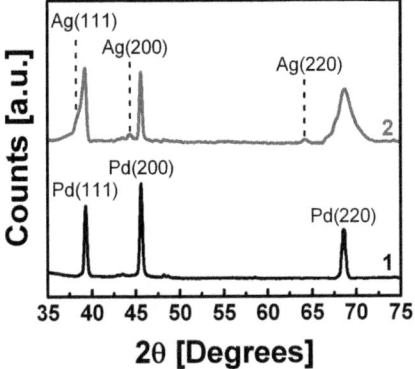

Fig. 12 XRD pattern of Pd supported on carbon particles before (1) and after (2) modification with Ag.

facilitates Ag metal ion reduction.[73] Therefore, the significant number of additional nanoparticles on the support must be related to the presence of Pd nanoparticles on the surface. Interestingly some of these nanoparticles have quite anisotropic shapes; several prisms are evident in Fig. 11. Ag is well known to form such shapes under controlled chemical synthesis conditions using capping agents.[74] It appears that the oxidation of active sites on Pd results in electrons being laterally propagated into the underlying support material that can facilitate the reduction of Ag^+ into Ag^0. This type of effect has been seen previously for a galvanic replacement process where controlling the rate of charge propagation on the surface of Cu in the presence of Au ions impacted on the morphology and porosity of the resultant film.[75] Given the resolution of field emission SEM, nanoparticles of the size observed in Fig. 11 should be detectable and, this therefore suggests that Ag is deposited as a submonolayer on both the evaporated Pd film and nanostructured Pd surface.

Under acidic conditions it was found that the presence of Ag on evaporated Pd does not significantly enhance its electrocatalytic performance but rather inhibits it. This was also confirmed at nanostructured Pd where the oxidation of formic acid was investigated. This reaction has received a significant amount of attention as it proceeds *via* a direct pathway to CO_2 formation involving dehydrogenation; $HCOOH = CO_2 + 2 H^+ + 2e^-$ rather than a dehydration step that produces CO_{ads}; $HCOOH = CO_{ads} + H_2O = CO_2 + 2H^+ + 2e^-$, which is highly detrimental to performance, as seen for Pt catalysts. Fig. 13 shows the CV behaviour for a nanostructured Pd electrode and one decorated with Ag for the oxidation of formic acid. The onset potential coincides with the pair of conjugate peaks seen in the absence of formic acid (Fig. 1b) which is consistent with a Pd*/hydrous oxide couple mediating the reaction. It is apparent that the presence of Ag inhibits the reaction over the entire potential range. On the reverse sweep the sharp increase in current at 0.45 V is due to the monolayer oxide being removed from the surface, allowing a fresh surface to participate in the oxidation of formic acid. For the decorated sample it can be seen that the re-oxidation of formic acid is significantly more sluggish. There is a rapid decay in current from 0.20 V to the end of the sweep due to the transition through the active site response and also the adsorption of hydrogen on the surface that competes with formic acid oxidation.

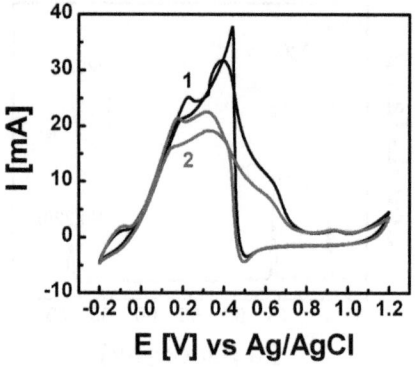

Fig. 13 CVs recorded in 1 M formic acid in 1 M H_2SO_4 at a sweep rate of 50 mV s^{-1} at a nanostructured Pd (as in Fig. 2) before (1) and after (2) Ag decoration.

However, this approach does offer an alternative route for the investigation of active site behaviour and its role in electrocatalytic reactions. It is known that Ag is a much more effective electrocatalyst in alkaline conditions and therefore the possibility of an effective Pd/Ag bimetallic catalyst being formed through this process was investigated for the ORR and formaldehyde oxidation. The latter reaction has received attention from a fuel cell point of view as an alternative fuel source to methanol, ethanol or hydrogen[76,77] and also in the sensing area, as formaldehyde is classified as a probable human carcinogen by the EPA.[78] Fig. 14a shows the CV recorded at a Pd substrate and Pd/Ag substrate in 1 M NaOH saturated with oxygen. For Pd an irreversible reduction process occurs with a peak potential of −0.19 V followed by a gradual decay in current until the end of the sweep at −0.40 V. For Pd/Ag the onset potential is shifted to slightly more negative potentials as well as the peak maximum by *ca.* 30 mV. However the rate of oxygen reduction is maintained at a near steady state value until the end of the sweep. This indicates that the ORR is facilitated to a greater extent over a larger potential range. The prolonged activity of both surfaces for ORR was determined by chronoamperometric experiments where the potential was held at −0.40 V (Fig. 14b). It can be seen that the current recorded at Pd/Ag is higher than that for Pd which indicates that the presence of Ag has a beneficial effect on the stability of the catalyst under alkaline conditions. Illustrated in Fig. 14c are CVs recorded in formaldehyde under alkaline conditions which demonstrate that the presence of Ag on the surface promoted the reaction which is reflected in a shift in onset potential to more negative potentials for the Pd/Ag surface compared to Pd. This enhancement has been reported previously for Ag/Pd nanoalloys where it was suggested that Ag can activate water at lower potentials than Pd, which in turn oxidises adsorbed CO thereby liberating Pd sites to participate in the reaction.[78] In this case there is no direct evidence of alloy formation, but the presence of Ag on the surface of Pd appears to have a similar effect. Given that improvements were seen in the case of evaporated Pd films decorated with Ag, the nanostructured Pd surface was also investigated for the oxidation of formaldehyde (Fig. 14d). A similar effect is seen in that the onset potential is shifted to more negative potentials as well as an increase in current magnitude when Ag is present on the surface.

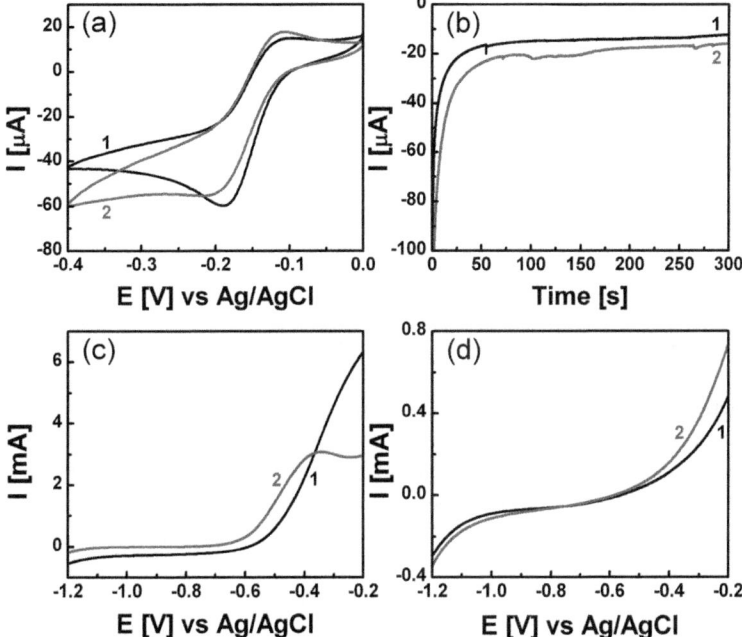

Fig. 14 CVs recorded at an evaporated Pd film before (1) and after (2) silver decoration in (a) 1 M NaOH saturated with oxygen gas, $\nu = 50$ mV s^{-1}; (b) chronoamperometric curves obtained at -0.40 V in 1 M NaOH saturated with oxygen; (c) CVs in 0.1 M formaldehyde in 1 M NaOH, $\nu = 50$ mV s^{-1}; and (d) same conditions as (c) at a nanostructured Pd electrode.

Electrocatalysis was then investigated at the Pd catalyst supported on carbon to see whether this approach may be a simple way of improving the performance of a commercial catalyst. The influence of Ag can again be seen by CV experiments undertaken in 1 M NaOH only where the peak at *ca.* 0.05 V is diminished after decoration with Ag along with the significant change in the monolayer oxide formation region (Fig. 15a). The overall capacitance is also decreased, as evident from the response in the double layer region, and may be due to the carbon

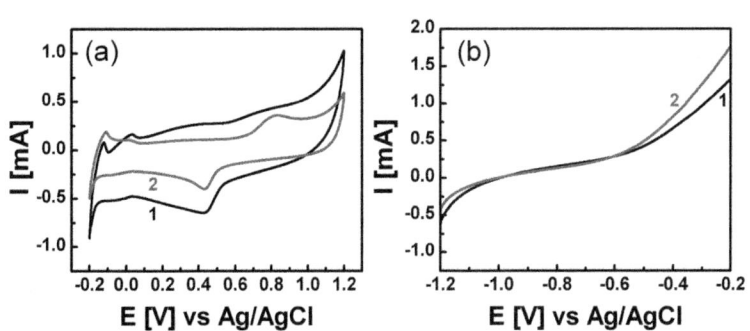

Fig. 15 CVs recorded at 50 mV s^{-1} at a commercial Pd catalyst before (1) and after (2) modification with Ag in (a) 1 M H$_2$SO$_4$ and (b) 1 M NaOH with 1 M formaldehyde.

support being covered with Ag nanoparticles. As expected the performance for formic acid oxidation under acidic conditions was inhibited by the presence of Ag (data not shown) but in 1 M NaOH the oxidation of formaldehyde (Fig. 15b) showed an improvement as seen above for the decorated Pd film (Fig. 14c).

Conclusions

In this work the electrochemical behaviour of Pd was investigated at a range of materials that would typically be used in research studies as well as in industrial applications. It was found that oxidation responses occur at potentials well below that for bulk metal oxidation and are attributed to the oxidation of surface active sites. This allows Pd surfaces to be spontaneously decorated with Ag which is thermodynamically forbidden at the bulk metal. The driving force for the reduction of the Ag salt at the surface is the facile oxidation of active sites. The Ag coverage was found to be no higher than 15 atomic % in all cases and had a significant impact on the electrocatalytic performance of the Pd/Ag material. Outer sphere reactions were unperturbed, but the HER, ORR and formaldehyde oxidation reactions were highly dependent on the presence of Ag. In acidic conditions Ag had an inhibitory effect allowing active site behaviour of the Pd surface to be investigated while the electrocatalytic properties of Ag in alkaline conditions could be harnessed to give improved performance for formaldehyde oxidation. It is envisaged that this approach is not just confined to the Pd/Ag system but could be applied to a variety of combinations where active site behaviour of a metal is present. The choice of a less active metal for decoration offers a new insight in to the behaviour of active sites on the substrate metal and what role they play in electrocatalytic reactions. By careful choice of reaction, pH conditions and combination of metals, improved electrocatalytic performance could be achieved using this straightforward approach. The possibility of increasing the number of defects on the substrate material *via* electrochemical, chemical or thermal pre-treatments prior to decoration is another possibility and needs further exploration.

Acknowledgements

The authors acknowledge the facilities, and the scientific and technical assistance, of the Australian Microscopy and Microanalysis Research Facility at the RMIT Microscopy & Microanalysis Facility. The provision of FT-ac voltammetry instrumentation by Prof. Alan M. Bond (Monash University) is gratefully acknowledged. AOM also acknowledges the ARC for a Future Fellowship (FT110100760).

References

1 X. Han, D. Wang, J. Huang, D. Liu and T. You, *J. Colloid Interface Sci.*, 2011, **354**, 577.
2 A. Rabis, P. Rodriguez and T. J. Schmidt, *ACS Catal.*, 2012, **2**, 864.
3 J. Solla-Gullon, F. J. Vidal-Iglesias and J. M. Feliu, *Annu. Rep. Sect. C*, 2011, **107**, 263.
4 A. Chen and P. Holt-Hindle, *Chem. Rev.*, 2010, **110**, 3767.
5 C. K. Rhee, B.-J. Kim, C. Ham, Y.-J. Kim, K. Song and K. Kwon, *Langmuir*, 2009, **25**, 7140.
6 J. H. Shim, J. Kim, C. Lee and Y. Lee, *J. Phys. Chem. C*, 2011, **115**, 305.
7 N. Tian, Z.-Y. Zhou, S.-G. Sun, Y. Ding and Z. L. Wang, *Science*, 2007, **316**, 732.
8 N. Wang, X. Cao, Q. Chen and G. Lin, *Chem.–Eur. J.*, 2012, **18**, 6049.
9 S. Maheswari, P. Sridhar and S. Pitchumani, *Electrocatalysis*, 2011, **3**, 13.

10 Y. Chen, W. Schuhmann and A. W. Hassel, *Electrochem. Commun.*, 2009, **11**, 2036.
11 W. Chen and S. Chen, *Angew. Chem., Int. Ed.*, 2009, **48**, 4386.
12 V. Bansal, V. Li, A. P. O'Mullane and S. K. Bhargava, *CrystEngComm*, 2010, **12**, 4280.
13 B. B. Blizanac, P. N. Ross and N. M. Markovic, *J. Phys. Chem. B*, 2006, **110**, 4735.
14 F. Ye, L. Chen, J. Li, J. Li and X. Wang, *Electrochem. Commun.*, 2008, **10**, 476.
15 C. Wang, H. Daimon, T. Onodera, T. Koda and S. Sun, *Angew. Chem., Int. Ed.*, 2008, **47**, 3588.
16 M. Subhramannia and V. K. Pillai, *J. Mater. Chem.*, 2008, **18**, 5858.
17 J. Hernandez, J. Solla-Gullon, E. Herrero, A. Aldaz and J. M. Feliu, *J. Phys. Chem. C*, 2007, **111**, 14078.
18 F. J. Vidal-Iglesias, J. Solla-Gullón, P. RodrIguez, E. Herrero, V. Montiel, J. M. Feliu and A. Aldaz, *Electrochem. Commun.*, 2004, **6**, 1080.
19 M. J. Rodriguez-Vazquez, M. C. Blanco, R. Lourido, C. Vazquez-Vazquez, Carlos, E. Pastor, G. A. Planes, J. Rivas and M. A. Lopez-Quintela, *Langmuir*, 2008, **24**, 12690.
20 F. J. Perez-Alonso, D. N. McCarthy, A. Nierhoff, P. Hernandez-Fernandez, C. Strebel, I. E. L. Stephens, J. H. Nielsen and I. Chorkendorff, *Angew. Chem., Int. Ed.*, 2012, **51**, 4641.
21 R. Gilliam, D. Kirk and S. Thorpe, *Electrocatalysis*, 2011, **2**, 1.
22 S. C. S. Lai and M. T. M. Koper, *Phys. Chem. Chem. Phys.*, 2009, **11**, 10446.
23 S. C. S. Lai and M. T. M. Koper, *J. Phys. Chem. Lett.*, 2010, **1**, 1122.
24 T. H. M. Housmans, A. H. Wonders and M. T. M. Koper, *J. Phys. Chem. B*, 2006, **110**, 10021.
25 H.-X. Zhang, H. Wang, Y.-S. Re and W.-B. Cai, *Chem. Commun.*, 2012, **48**, 8362.
26 L. D. Burke, J. K. Casey, J. A. Morrissey and M. M. Murphy, *Bull. Electrochem.*, 1991, 7, 506.
27 L. D. Burke and P. F. Nugent, *Gold Bull.*, 1997, **30**, 43.
28 L. D. Burke and P. F. Nugent, *Gold Bull.*, 1998, **31**, 39.
29 L. D. Burke, L. M. Kinsella and A. M. O'Connell, *Russ. J. Electrochem.*, 2004, **40**, 1105.
30 S. Garbarino and L. D. Burke, *Int. J. Electrochem. Sci.*, 2010, **5**, 828.
31 A. P. O'Mullane, S. J. Ippolito, Y. M. Sabri, V. Bansal and S. K. Bhargava, *Langmuir*, 2009, **25**, 3845.
32 L. D. Burke and L. M. Hurley, *J. Solid State Electrochem.*, 2000, **4**, 353.
33 L. D. Burke and A. P. O'Mullane, *J. Solid State Electrochem.*, 2000, **4**, 285.
34 L. C. Nagle, A. J. Ahern and L. D. Burke, *J. Solid State Electrochem.*, 2001, **6**, 320.
35 L. D. Burke, A. M. O'Connell, R. Sharna and C. A. Buckley, *J. Appl. Electrochem.*, 2006, **36**, 919.
36 V. Díaz, S. Real, E. Téliz, C. F. Zinola and M. E. Martins, *Int. J. Hydrogen Energy*, 2009, **34**, 3519.
37 V. Diaz and C. F. Zinola, *J. Colloid Interface Sci.*, 2007, **313**, 232.
38 L. M. Doubova, S. Daolio, C. Pagura, A. De Battisti and S. Trasatti, *Russ. J. Electrochem.*, 2002, **38**, 20.
39 B. K. Jena and C. R. Raj, *Langmuir*, 2007, **23**, 4064.
40 V. A. Marichev, *Russ. J. Electrochem.*, 1999, **35**, 434.
41 V. A. Marichev, *Surf. Sci. Rep.*, 2001, **44**, 51.
42 C. Nguyen Van Huong, C. Hinnen and J. Lecoeur, *J. Electroanal. Chem.*, 1980, **106**, 185.
43 Ž. Petrović, M. Metikoš-Huković, R. Babić, J. Katić and M. Milun, *J. Electroanal. Chem.*, 2009, **629**, 43.
44 B. Lertanantawong, A. P. O'Mullane, W. Surareungchai, M. Somasundrum, L. D. Burke and A. M. Bond, *Langmuir*, 2008, **24**, 2856.
45 M. J. A. Shiddiky, A. P. O'Mullane, J. Zhang, L. D. Burke and A. M. Bond, *Langmuir*, 2011, **27**, 10302.
46 J. Rodriguez-Lopez, M. A. Alpuche-Aviles and A. J. Bard, *J. Am. Chem. Soc.*, 2008, **130**, 16985.
47 Z. W. Wang and R. E. Palmer, *Nano Lett.*, 2012, **12**, 91.
48 A. M. Nowicka, U. Hasse, G. Sievers, M. Donten, Z. Stojek, S. Fletcher and F. Scholz, *Angew. Chem., Int. Ed.*, 2010, **49**, 3006.
49 A. M. Bond, N. W. Duffy, S.-X. Guo, J. Zhang and D. Elton, *Anal. Chem.*, 2005, **77**, 186A.
50 J. Sun, C. Zhang, T. Kou, J. Xu and Z. Zhang, *ACS Appl. Mater. Interfaces*, 2012, **4**, 6038.
51 D. Burke and L. Hurley, *J. Solid State Electrochem.*, 2003, **7**, 529–538.
52 V. Diculescu, A.-M. Chiorcea-Paquim, O. Corduneanu and A. Oliveira-Brett, *J. Solid State Electrochem.*, 2007, **11**, 887.
53 L. D. Burke and L. C. Nagle, *J. Electroanal. Chem.*, 1999, **461**, 52.
54 W. Ye, J. Yan, Q. Ye and F. Zhou, *J. Phys. Chem. C*, 2010, **114**, 15617.
55 A. Ott, L. A. Jones and S. K. Bhargava, *Electrochem. Commun.*, 2011, **13**, 1248.
56 B. J. Plowman, A. P. O'Mullane, P. R. Selvakannan and S. K. Bhargava, *Chem. Commun.*, 2010, **46**, 9182.

57 G.-M. Yang, X. Chen, J. Li, Z. Guo, J.-H. Liu and X.-J. Huang, *Electrochim. Acta*, 2011, **56**, 6771.
58 G. Denuault, C. Milhano and D. Pletcher, *Phys. Chem. Chem. Phys.*, 2005, **7**, 3545.
59 J. Zhang and A. M. Bond, *J. Electroanal. Chem.*, 2007, **600**, 23.
60 R. A. Masitas and F. P. Zamborini, *J. Am. Chem. Soc.*, 2012, **134**, 5014.
61 A. R. Kucernak and G. J. Offer, *Phys. Chem. Chem. Phys.*, 2008, **10**, 3699.
62 K. Seong-Eun, H. Young-Hwan, L. Byung Cheol and L. Jong-Chan, *Nanotechnology*, 2010, **21**, 075302.
63 M. Liu, Y. Lu and W. Chen, *Adv. Funct. Mater.*, 2013, **23**, 1289–1296, DOI: 10.1002/adfm.201202225.
64 C.-J. Zhong, N. T. Woods, G. B. Dawson and M. D. Porter, *Electrochem. Commun.*, 1999, **1**, 17.
65 M. Mahajan, S. K. Bhargava and A. P. O'Mullane, *RSC Adv.*, 2013, **3**, 4440, DOI: 10.1039/c3ra22936j.
66 J.-B. Raoof, R. Ojani, A. Kiani and S. Rashid-Nadimi, *Int. J. Hydrogen Energy*, 2010, **35**, 452.
67 F. Maillard, F. Gloaguen and J. M. Leger, *J. Appl. Electrochem.*, 2003, **33**, 1.
68 S. R. Brankovic, J. McBreen and R. R. Adžić, *J. Electroanal. Chem.*, 2001, **503**, 99.
69 L. D. Burke, *Gold Bull.*, 2004, **37**, 125.
70 J. Greeley and J. K. Norskov, *Surf. Sci.*, 2007, **601**, 1590.
71 J. Greeley, T. F. Jaramillo, J. Bonde, I. Chorkendorff and J. K. Norskov, *Nat. Mater.*, 2006, **5**, 909.
72 A. Nowicka, U. Hasse, M. Donten, M. Hermes, Z. Stojek and F. Scholz, *J. Solid State Electrochem.*, 2011, **15**, 2141.
73 A. Dekanski, J. Stevanovic, R. Stevanovic and V. M. Jovanovic, *Carbon*, 2001, **39**, 1207.
74 C. M. Cobley, S. E. Skrabalak, D. J. Campbell and Y. Xia, *Plasmonics*, 2009, **4**, 171.
75 A. Ott, S. K. Bhargava and A. P. O'Mullane, *Surf. Sci.*, 2012, **606**, L5.
76 L. D. Burke and W. A. O. Leary, *J. Electrochem. Soc.*, 1988, **135**, 1965.
77 J. Geng, Y. Bi and G. Lu, *Electrochem. Commun.*, 2009, **11**, 1255.
78 A. Safavi, S. Momeni and M. Tohidi, *Electroanalysis*, 2012, **24**, 1981.

Faraday Discussions

DISCUSSIONS

General discussion

DOI: 10.1039/C3FD90029K

Dr Tschulik opened the discussion of the paper by Professor Schmuki: In your paper you discuss the influence of different organic electrolytes on the growth of porous TiO_2 from titanium in the presence of fluoride ions. Fluoride ions remaining in the formed TiO_2 pores after the anodization of titanium are assumed to cause delamination of the pores from the substrate. How can these ions be removed from the pores, avoiding delamination during this removal?

Professor Schmuki responded: Indeed, during tube formation, as the migration of fluoride ions during anodization is competitive with that of oxygen ions, the ions are accumulated as a F-rich layer at the metal–oxide interface. This fluoride layer is the cause of the mechanical instability. The longer an anodization process takes, the more fluoride is accumulated and the less mechanically stable is the layer; *i.e.*, all processes, for example recently published lactic acid electrolytes,[1] where fluoride accumulation is hampered lead to significantly more stable tube layers.

Additionally, typically during annealing treatments, most of the fluoride is lost, *i.e.*, this also makes the layer more stable (annealing is used for most applications of TiO_2 nanotubes to convert the amorphous – as formed – tubes into crystalline anatase).

1 S. So *et al.*, *Chem.–Eur. J.*, **19**, 2966.

Professor Bund opened the discussion of the paper by Professor Kornyshev: For the estimation of the Bjerrum length you assume a dielectric constant of $\varepsilon \approx 2$. The value of ε should reflect the electronic polarisability of the ions. How strongly will ε vary with the nature of the ions?

Professor Kornyshev replied: The dielectric constant due to electronic polarisation for optically transparent materials is assessed through the square of the refractive index. For typical ionic liquids such dielectric constants lie between approximately 1.8–2.5 (for a review of refractive indices of ionic liquids see *e.g.*: S. Seki *et al.*, *J. Chem. Eng. Data*, 2012, **57**, 2211). The effective value of the electronic polarisability dielectric constant to be taken in the Ising model depends, however, not only on the liquid, but on the density of ions in the pore. Smaller values of this dielectric constant will increase interaction between ions, *i.e.* reduce the coupling parameter in the Ising Hamiltonian, thereby reducing the capacitance. The effect will be noticeable, but not huge. However it will have a different character in a generalisation of the Ising model onto three states, one of which is the empty state. Indeed, if a non-polarised pore is not 'super-ionophilic', it will gradually become filled with counterions with the increase of the electrode polarisation replacing most of the voids. The overall density of ions in the pore will then be increased. Hence, in this case the effective dielectric constant will gradually increase with the voltage. We are currently investigating the consequences of this effect.

Professor Barbero said: I have two questions. 1) In real supercapacitor applications, the capacitance should be invariant with potential. However, dispersion of the pore sizes, which makes the capacitance less dependent on potential, strongly decreases the maximum capacitance density (uF cm^{-2}). Could that problem be solved by building capacitors using a mixture of materials with various pore sizes and an electrolyte having a mixture of ions of different sizes?

2) Most real capacitors, including those with high capacitance density made by Gogotsi's group, are made of some form of carbon, not metals. Kotz and coworkers[1] show a similar dependence of electronic conductance of activated carbon and measured capacitance. Could the poor conducting nature of the material coupled with thin walls have an effect on the model you are proposing?

1 M. Hahn *et al.*, *Electrochem. Solid-State Lett.*, 2004, **7**, A33.

Professor Kornyshev responded: It is a matter of optimization, whether the capacitance should be voltage independent or not; this depends on the application. If the capacitance may be substantially larger at some voltages, this fact can be used to benefit higher energy storage, depending on the regime of operation voltage. This question is studied in detail in Kondrat *et al.*,[1] where it is shown that, on the contrary to your suggestion, a monodispersed pore size distribution may be optimal for maximal energy storage. But generally, yes, the pore size distribution makes the voltage dependence milder, as shown in the above reference, as well as in my paper in this issue.

Now let us move to your second question. The non-ideal metal character of the electrode in carbon materials is mainly due to a lower concentration of charge carriers. For equilibrium properties it will not matter how low the electronic conductivity is because of that. Neither will it matter for kinetic properties, as long as the electronic conductivity of the electrode is still much higher than the ionic conductivity of the electrolyte. But what will matter is the finite electronic screening length in these materials, which will depend on doping: the higher the concentration of charge carriers, the shorter it is, approaching the limit of an ideal metal. For a reasonably good electronic conductance of electrodes, the screening length is just a couple of Angstroms. Keeping these circumstances in mind it is relatively simple to answer your question, when pore walls are wide. The finite screening length will renormalize the effective thickness of the pore in the law for the exponentially screened interionic interactions inside the pore (eqn (2) and (3) of my paper), making the pore effectively wider by roughly two screening lengths. This conjecture was used by Shklovskii,[2] and it has been verified by a detailed calculation by Rochester *et al.*[3] Hence, the interionic interactions will become slightly less screened, *i.e.*, stronger, although still exponentially decaying along the axis of the pore. Because of the exponential character of the decay, it will still be possible to use the Ising Hamiltonian model with the nearest neighbour interactions only, but with a somewhat larger value of the coupling constant α. Then, as in Fig. 5 of our paper, the capacitance curve will run a bit lower with its maximum shifted to higher voltages. There will also be another effect of finite electronic screening: weaker image attraction of ions to the walls in a more general spin model,[4] in which the pores are not exclusively ionophilic; the effect will contribute to the lower propensity of ions to enter the pore. However, if the walls between the pores are very thin, the mode of screening will be changed. This question will require further investigation, although there are first results in this direction: see the paper by the Shklovskii group[2] and the somewhat different results of the work of Rochester et al.[3]

1 S. Kondrat et al., *Energy Environ. Sci.*, 2012, **5**, 6474.
2 M. Loth et al., *Phys. Rev. E*, 2011, **83**, 056102.
3 C. C. Rochester et al., *ChemPhysChem*, DOI: 10.1002/cphc.201300834.
4 A. A. Lee et al., resubmitted to *Phys. Rev. Lett.*, 2013.

Professor Barbero enquired: You show that dispersion of the pore size decreases the maximum specific capacitance. However, it also seems to make the range of potential with high capacitance wider. In real applications, a capacitor should have a nearly constant capacitance over a wide potential range, otherwise it works like a battery. Do you think that your model predicts that a large dispersion of pore sizes will produce a nearly constant capacitance over a wide potential range?

Professor Kornyshev replied: As I have answered to another question, a wide pore size distribution will smear the 'resonances' in the voltage dependence of the capacitance, making the latter almost constant. But it is an open question regarding what is the best for maximal energy storage; the paper by Kondrat et al.[1] shows that mono-pore-size structures can be optimal in some situations.

1 S. Kondrat et al., *Energy Environ. Sci.*, 2012, **5**, 6474.

Professor Schuhmann opened the discussion of the paper by Professor Chen: An irreversible oxidation wave is seen in the cyclic voltammogram which is shifted to lower potentials upon modification of the electrode with CNTs. What is the mechanism leading to the observed potential shift? A catalytic effect? And if so, how could one assume the reaction mechanism?

Professor Chen responded: This is a very good question. The shift in the peak of the electrochemical oxidation of valacyclovir to a lower potential may be attributed not only to a catalytic effect, but also to a change in the diffusional regime upon the modification of the glassy carbon electrode with CNTs. On one hand, there has been a broad debate relating to the enhancement of the electrocatalytic activity of CNTs. It has been recognized that in most cases, the electrochemical activity of carbon nanotubes is due to edge plane-like site defects and metallic impurities within the CNTs. On the other hand, the immobilized CNTs form a porous layer on the surface of the glassy carbon electrode. As demonstrated by Compton and co-workers,[1] the porous layer may serve as a thin layer cell, which results in the shift of the oxidation peak.

1 I. Streeter et al., Sens. Actuators, B, 2008, **133**, 462.

Professor Williams continued the discussion of the paper by Professor Kornyshev: I'd like to ask about the effect of coupling between pores. If the pore walls are very thin, and the material is carbon (hence not necessarily highly conductive) then the charges in the pore would not be fully screened and may couple into adjacent pores, dependent on the conductance of the pore walls. Could you comment on what such effects might be and whether there are any implications for the design of supercapacitors and for the choice of materials for the electrodes?

Professor Kornyshev answered: This is a difficult question which is currently under investigation. When the pore walls are very thin there will still be screening of Coulomb interactions between the ions in such pores in the direction along the pore. However, the decay of the interaction will not be exponential, although will still remain short range.[1] As for the interaction of ions across the pores, this effect and how it will influence the capacitance are yet to be fully understood; part of the paper by Shklovskii's group[2] is devoted to exactly this subject.

1 C. C. Rochester et al., ChemPhysChem, DOI: 10.1002/cphc.201300834.
2 B. Skinner et al., Phys. Rev. E, 2011, **83**, 056102.

Professor Bond continued the discussion of the paper by Professor Chen: The limit of detection in a voltammetric experiment is usually governed by the Faradaic to background current ratio. In Fig. 2 and 3 of your paper, which contain cyclic and differential pulse voltammograms respectively, the Faradaic peak current to background ratio looks to be similar at both the SWCNT-modified and unmodified glassy carbon electrodes. That is, even though the magnitude of the peak current increases significantly because of the increase in electrode surface area, and for other reasons when using the carbon nanotube modified electrode configuration, the background current also increases substantially, leaving the important ratio term almost unaltered. Thus, is anything analytically useful actually achieved by using the more complicated chemically modified carbon nanotube version of a glassy carbon electrode, noting that measurement of smaller currents at a bare glassy carbon electrode does not really introduce an electronic noise issue?

Professor Chen responded: Although the Faradaic peak current to background ratio is similar at both the SWCNT-modified and unmodified glassy carbon electrodes (GCEs), there are some notable differences in the cyclic and differential pulse voltammograms for the electrochemical oxidation of valacyclovir. A small and broad peak centred at 0.98 V was observed with the bare GCE; in contrast, a large and well-defined peak appeared with a peak potential that was shifted to 0.91 V when a SWCNT-modified GCE was employed. More significantly, the SWCNTs formed a porous layer on the GCE. A recent study carried out by Compton and co-workers[1] showed that conducting porous layers on the surface of electrodes can modify the mass transport regime from linear (planar) diffusion to one of an approximately 'thin layer' character, and that this alteration may lead to improved selectivity in amperometric discrimination. In our study, the recovery tests performed in human serum plasma confirmed that the fabricated SWCNT-modified GCE exhibited high selectivity and sensitivity in the detection of valacyclovir, demonstrating a marked improvement over the bare glassy carbon electrode.

1 M. C. Henstridge et al., Sens. Actuators, B, 2010, **145**, 417.

Dr Kataky continued the discussion of the paper by Professor Kornyshev: Apart from size effects on the storage of charge in the nanopore, is there a factor that accounts for the relative hydrophilicity of the wall surfaces on charge confinement?

Professor Kornyshev replied: Yes. In my paper I have considered a special case of ionophilic pores, when they are filled both with cations and anions in equal or unequal proportions, but there are no voids between them. When the electrode is polarized, counterions tend to replace the co-ions in the pore. But there could be a more general situation when the free energies of transfer of cations and anions from the bulk to the interior of the pore are not overwhelmingly negative or can even be positive. There is then a balance between the presence of ions or voids inside the pore which shifts with the polarisation of the electrode. At positive free energies of transfer much greater than the thermal energy, the ions do not fill an unpolarised pore: the pore is initially empty; with the polarisation of the pore, predominantly counterions get into it. Such a situation cannot be described by a two state Ising model, but requires a more general 3 state model, where a third state is due to a void (or a solvent molecule). Such a model has been developed[1] and it does show a dramatic effect of ionophilicity-ionophobicity on the pore capacitance. A special question, to be addressed to experimentalists in the first place, are pores in real electrodes ionophobic or ionophilic? This is a subject of systematic studies by John Griffin et al. in Clare Grey's group at Cambridge, using originally developed techniques based on NMR spectroscopy.

1 A. A. Lee et al., resubmitted to Phys. Rev. Lett, 2013.

Professor Compton enquired: Leaving aside the issue of the composition of nanopores, can I ask, if we considered a planar electrode in an ionic liquid, is there the possibility of significant water accumulation or depletion at the electrode–solvent interface, remembering that even well purified ionic liquids often contain significant levels of trace water (see for example A. M. O'Mahony et al., J. Chem. Eng. Data, 2008, **53**, 2884; E. I. Rogers et al., J. Chem. Eng. Data, 2009, **54**, 2049)?

Professor Kornyshev responded: Professor Compton raised a question that I am currently asking myself. As we are in the process of investigating this issue, I will not go into detail, but just tell you my general thoughts about it. The answer may appear to be different in different voltage regimes. At small voltages the electric field in the double layer of a pure ionic liquid will undergo oscillations related to the overscreening effect; at very large voltages, in the so-called regime of lattice saturation, the field will be more homogeneous – this is all well described in our papers[1,2] and a number of follow up atomistic simulations. Based on that knowledge, and recalling that any dipole is known to be drawn into the region of maximal inhomogeneity of an external electric field, one may naively expect that we will have more water dipoles in the double layer than on average in the bulk of the liquid, particularly at small voltages. But this is most likely to be wrong. Let us not forget that these dipoles would in turn affect the interactions and the structure of the double layer: they will dilute the concentration of ions in it. Common sense advises, if counterions 'wish' to condense at the electrode to form the double layer, why they should let dipoles mess around? Thus water molecules should rather be expelled from the double layer.

A responsible answer to this question can, however, be given only after a detailed computer simulation is performed or theory developed. At the same time, I understand the importance of this question, because if the double layer becomes richer in water than the bulk, we could expect the onset of water electrolysis at voltages much lower than those at which electrochemical reactions in pure ionic liquids would take place and this may cause a discharge and degradation of supercapacitors. Has this been seen experimentally?

1 M. V. Fedorov and A. A. Kornyshev, *Electrochim. Acta*, 2008, **53**, 6835.
2 M. Z. Bazant *et al.*, *Phys. Rev. Lett.*, 2011, **106**, 046102.

Professor Compton remarked: Carbon dioxide shows a very high solubility in many room temperature ionic liquids (see for example Barosse-Antle *et al.*, *Chem. Commun.*, 2009, 3744) where the concentrations may reach the molar (M) level. Under these conditions would you expect enrichment or depletion of the dissolved gas at interfaces or pores relative to bulk solution?

Professor Kornyshev replied: Since carbon dioxide molecules have no dipole, but only quadruple moments, why should they replace ions in the double layer that like to form the latter when the electrode is polarised? These neutral molecules should rather be expelled from the double layer. In any case this is something that will be easy to test by molecular dynamic simulation.

Professor Compton continued the discussion of the paper by Professor Chen: Is your voltammetric signal diffusive or adsorptive?

Please note that when considering nanotube modified electrodes there can be a change of diffusive mass transport between the unmodified electrode (where semi-infinite diffusion would operate) and the modified electrode where solution occluded within the porous modifying layer can give rise to 'thin layer' diffusion. The effect of the latter is to give larger current and to shift the voltammetric wave to lower overpotentials thus giving the illusion of 'electrocatalysis'. We have described these effects in I. Streeter *et al.*, *Sens. Actuators, B*, 2008, **133**, 462. Further I believe that this effect – simply one of porosity leading to a movement of the observed voltammetric wave towards the formal potential of the redox couple of interest – is the primary cause of the selectivity shown by many modified electrodes covered where porous layers of nanomaterials, polymers *etc.* are used as modifiers. We have simulated this behaviour in M. C. Henstridge *et al.*, *Sens. Actuators, B*, 2010, **145**, 417.

Professor Chen responded: Thank you for the insightful information. In our case, it is likely that both thin layer diffusion and adsorption effects are present. Our pre-accumulation tests revealed that the peak current increased in association with the duration of the accumulation time, indicating the presence of the adsorption effect. The shift of the oxidation peak is in excellent agreement with the simulation and experimental results that you have mentioned, showing that the thin layer effect had indeed played a role. This is further supported by the porous layer structure of the CNTs that were formed on the surface of the GCE (Fig. 1A in our paper), and a recent study of the electrochemical oxidation of nicotinamide adenine dinucleotide at a functionalized porous graphene electrode.[1]

1 C. Punckt et al., J. Phys. Chem. C., 2013, **117**, 16076.

Professor Schuhmann commented: Following Professor Compton's question concerning mass-transfer control in the CNT network, it is suggested that the peak shift with the scan rate in the voltammogram is evaluated, however, keeping in mind that this has to be done outside of the mass-transfer limited region, to obtain further insight into the mechanism.

Professor Chen replied: Thank you for the further information.

Professor Amatore noted: When adsorption–desorption dynamics are fast, all reactions can take place exclusively in between adsorbed moieties although the overall electrochemical behavior may appear as occurring entirely in solution. This was pointed out theoretically by Laviron a few decades ago in a seminal paper and has been generalized recently by us.[1] So would that not be a way to rationalize your observations?

1 O. V. Klymenko et al., J. Electroanal. Chem., 2013, **688**, 320.

Professor Chen replied: I greatly appreciate your perceptive information. Yes, it would indeed serve as an additional viable approach to explain the experimental results that were observed in the present study.

Professor Amatore commented: Even if the exact mechanism followed by valacyclovir oxidation is irrelevant if you use your method as an electroanalytical tool with calibration curves under each condition, it would be interesting to know it. This is a complex system which may well depend on many interfering species (*e.g.*, some controlling the acido–basic strength of the analyte).

Professor Chen answered: Thank you very much for your comments, and I agree with you.

Professor Williams remarked: Your electrode worked well in human plasma, which is impressive given that the protein content of this material can act efficiently to adsorb on electrodes and block electrochemical activity. Is it possible that the nanotube mesh acted like a thin layer cell that confined the analytical electrochemistry, but also acted like a filter to keep the proteins at the outside?

Professor Chen responded: Yes, you have made an excellent and valid point. In the present study, treatment with acetonitrile was utilized to precipitate proteins, which were then removed *via* centrifugation. As can be seen in the SEM image (Fig. 1A in our paper), the carbon nanotubes formed a porous layer on the glass carbon electrode surface, which might indeed have served as a thin layer cell, as well as a filter, to externally segregate any remaining protein residues.

Ms Horwood continued the discussion of the paper by Professor Schmuki: According to current models of nanotube formation, nanotube growth rate slows as diffusion limitations (of reactants in and products out of the nanotubes) increase, as shown by a current decrease in I–t curves. Is there a different mechanism or explanation as to why the current increases during nanotube formation in DMSO at 60 °C? Could this be due to further heating of the solution, or unstable nanotube films delaminating the surface?

Professor Schmuki replied: In the present case we believe this is mainly due to a conductivity increase.

For HF/DMSO electrolytes the initial conductivity of the solution is usually considerably lower than for EG based electrolytes, *i.e.*, in the initial phase the slow formation (low currents) of reaction products leads to a slow but considerable increase in conductivity and thus to a lower IR drop.

Professor Amatore commented: Thank you very much for this excellent contribution. However, I am puzzled by your view about the involvement of $(O_2)^{2-}$ ions. Indeed, this is an extremely basic species and I doubt that it may survive enough to travel sufficiently far to account for your mechanistic views, unless it may somewhat 'creep' on the solid oxide surface and thus be sufficiently stabilized. Could you comment on this point?

Professor Schmuki answered: Thanks for the kind comment. I believe the question is mainly based on a typo in the slides (not $(O_2)^{2-}$ ions but O^{2-} ions). The common model in anodic oxide growth assumes deprotonation of H_2O or OH^- at the oxide–electrolyte interface and the insertion of O^{2-} ions into the oxide lattice. This is then followed by high field transport of O^{2-} through the oxide – possibly *via* vacancy hopping – and reaction at the oxide–metal interface to form new oxide.

Professor Kanoufi enquired: You nicely demonstrate how you can control the mechanism of thinning of nanotubes by using DMSO as a trap of HO radicals. Do you think you could use other radical traps? For example, would monomers help generate hybrid materials or allow the polymeric coverage of the inside nanotube walls?

Professor Schmuki replied: It is a very interesting idea, at present we did not really try to trigger a radical polymerization reaction by anodic radical generation (valence band-ionization), but if we are successful we certainly need to acknowledge you…

Professor Bond continued the discussion of the paper by Professor Kornyshev: I have seen many cartoons of the structure of an ionic liquid that are often based on analysis of NMR data. These usually imply aggregates of cation and anions. What is the physical reality of the cartoons you provide in your paper for charge storage, which in some scenarios seem to imply the presence of a single file of cations and anions in a channel?

Professor Kornyshev answered: Complex aggregates of cations and anions may of course take place in ionic liquids also at electrified interfaces (see *e.g.* discussion in A. A. Kornyshev, *J. Phys. Chem. B*, 2007, **111**, 5545). They may form in voluminous pores and even in one-ionic-layer-thick slit nanopores – there they will be quasi-two-dimensional. But in the pores such as those considered in this paper there are no clusters as they are single-file. Generally such pores are not practical for supercapacitors, because it is difficult to rapidly charge or discharge them. But they are interesting in the first place as tutorial models for which exact solutions are possible.

Professor Compton asked: Do you believe that the layering effect you have described for ionic liquids at the planar electrode–solution interface will measurably influence the electron transfer kinetics of simple redox probes? We have found that the Butler–Volmer formulations of electron transfer are quantitatively applicable; see for example E. I. Rogers *et al.*, *J. Phys. Chem. C*, 2008, **112**, 10976–10981; 2008, **112**, 2729 and 2007, **111**, 13957).

Professor Kornyshev responded: Generally, I do. One may expect a strong effect of layering in electrode kinetics due to the so-called Frumkin Correction. To remind the audience what it is, I recall that the driving force of an elementary electrochemical electron transfer reaction is the difference of the potential drop between the electrode and the point of location of the reacting species (donor or acceptor of the electron) minus the value of the same quantity at equilibrium. People use concentrated background electrolyte to localise the potential drop between the electrode and the reactant plane, to achieve the maximal driving force for the target reaction. One can never achieve a 100% localisation, and in diluted electrolytes only some part of the voltage drop is utilized for driving the reaction. The means of accounting for this effect is called Frumkin Correction, and the whole effect following Frumkin is called, historically, the Ψ'-effect. In ionic liquids the potential at low voltages oscillates, due to the overscreening effect. Thus one might envisage situations with an 'over-drop' rather than 'under-drop'. I mean that the driving force may get eventually amplified, becoming larger than the one determined by the overall potential drop across the double layer, let us call it the 'over-Frumkin' effect. But this is a pure speculation. We discuss it in detail in our review.[1] However, the reality may be different as the reactant (although in minority) may somehow disturb the structure of the double layer (*cf.* Lynden-Bell *et al.*[2]) or sit in a place between the overscreening layers. Anyway, if experiments do not show this, theorists must find the reason why this is so. The computer simulations performed so far show no signatures of over-Frumkin. In a paper by S. K. Reed *et al.*,[3] which studied heterogenous electron transfer parameters at metal–model ionic liquid interface, the authors write that "pronounced oscillations in the mean electrical potential seen in molten salt systems in the double-layer region are not reflected in the reaction free energy for the electron transfer event". But they stress that "The reorganization energy depends markedly on the distance of the redox ion from the electrode surface because of image charge effects."

1 M. V. Fedorov et al., Chem. Rev., 2013, in press.
2 R. M. Lynden-Bell et al., Phys. Chem. Chem. Phys., 2012, **14**, 2693.
3 S. K. Reed et al., J. Chem. Phys., 2008, **128**, 124701.

Professor Kanoufi commented: In the proposed model you consider a unique pore of infinite wall thickness. I guess considering an array of pores is more difficult to model, or at least if an analytical solution is sought. One could suspect cooperative effects will be encountered in this configuration of an array of nanopores. Such a cooperative effect in the diffusion field towards the nanopore array is well known.[1,2] Are such cooperative effects expected in nanopore arrays? It will likely be important when interpore distances are small. Is it possible to anticipate this effect on the voltage dependence of capacitance?

1 C. Amatore et al., J. Electroanal. Chem., 1983, **147**, 39.
2 N. Godino et al., J. Phys. Chem. C, 2009, **113**, 11119.

Professor Kornyshev answered: This is a very important and just question! My model is not by chance called the "Simplest model". It cannot be directly applied to the case of 1–3 atomic layer thin walls between the pores. For very thin walls, the screening of interionic interactions inside each pore will be weaker, although "nearest neighbour interaction only" models can still be used (cf. the quoted paper by Rochester et al.[1]). But, more importantly, the interaction of ions between the pores may become non-negligible, which will lead to novel cooperative effects. A rather bold analysis of this question is presented in a very interesting paper from the Shklovskii group,[2] and a crude estimate of the effect on capacitance was suggested there for slit pore systems. However, already for 5–6 atomic layer thick walls, ions will not be able to communicate across the walls.

1 C. C. Rochester et al., ChemPhysChem, DOI: 10.1002/cphc.201300834.
2 M. Loth et al., Phys. Rev. E, 2011, **83**, 056102.

Professor Hillman queried: Your model attempted to introduce greater chemical reality in terms of molecular geometry by considering bead-like molecular structures. You then went on to propose that these bead-like structures might re-orient at the interface in response to changes in the applied potential. This is a physically reasonable proposal, but can you identify experimental evidence of a structural nature that provides direct support for this idea?

Professor Kornyshev responded: You seemingly mean our 2010 Electrochem. Commun. and J. Electroanal. Chem. papers.[1,2] We have indirect evidence through systematic observation of capacitance shapes in dense ionic liquids correlated with the length of the neutral tail of cations, as well as data for such correlations coming from 'in silico microscopy' – simulations with fully atomistic force-fields (papers to appear soon). Data for the structure near electrified interfaces, obtained from synchrotron X-ray diffraction, of the kind reported for a spontaneously charged sapphire surface,[3] distinguishing cations and anions, have not been, to my knowledge, performed for electrodes with controllable voltage. In the quoted review article[4] we write more about it. But with a reasonable estimate of charge on the sapphire surface the data qualitatively agree with our one bead model simulation.[5]

1 M. V. Fedorov et al., Electrochem. Commun., 2010, **12**, 296.
2 N. Georgi et al., J. Electroanal. Chem., 2010, **649**, 261.
3 M. Mezger et al., Science, 2008, **322**, 424.
4 M. V. Fedorov et al., Chem. Rev., 2013, in press.
5 M. V. Fedorov and A. A. Kornyshev, Electrochim. Acta, 2008, **53**, 6835.

Mr Je Hyun Bae asked: Does the suggested model for charge storage apply to organic electrolytes?

Professor Kornyshev replied: Literally this, a two state Ising model cannot. But its extension to three states, with competitive adsorption of ions and solvent molecules will do the job. As I have mentioned, such a model has been already developed.[1] In the discussion of the results this reference focussed more on the interpretation of the third state as the one corresponding to a void (vacancy). But it can equally stand for a neutral solvent molecule (as indicated in that paper). In that case the model will have fewer problems with a likely variation of the effective dielectric constant ε with the population of the vacancies in the pore. Unlike a vacancy, a solvent molecule entering the pore will bring its own dielectric polarisability there, so that the overall effective ε in the pore may be considered approximately independent of the voltage.

1 A. A. Lee et al., resubmitted to Phys. Rev. Lett., 2013.

Dr Batchelor-McAuley continued the discussion of the paper by Professor Chen: The drug you have electrochemically investigated is a derivative of guanine. On the basis of work by Goyal and Dryhurst,[1] and more recently with the work of Oliveira-Brett et al.,[2] the oxidation mechanism of guanine on carbon is commonly taken to be overall a four-electron, four-proton oxidation. In this mechanism a two-proton, two-electron oxidation occurs to form 8-oxoguanine with the addition of water. The formed 8-oxoguanine may then undergo a further reversible two-electron, two-proton oxidation. The formal potential of this second reversible step is below that of the initial chemically irreversible oxidation by around $\sim$300 mV.

In your paper you state that the electrochemical response of the valacyclovir corresponds to a two-proton, two-electron oxidation, resulting in the formation of an 8-oxoguanine derivative. My first question is what evidence do you have for the oxidation mechanism, and secondly, why do you believe the process differs from that of the parent guanine molecule?

1 R. N. Goyal and G. Dryhurst, J. Electroanal. Chem., 1982, **135**, 75.
2 A. M. Oliveira-Brett et al., Bioelectrochemistry, 2002, **55**, 61.

Professor Chen replied: Thank you for your question. The mechanism mentioned in our paper was based on a previous study of the electrochemical oxidation of valacyclovir and the electrochemical behaviour of guanine at a carbon-based electrode; the oxidation process initially occurred at a guanine moiety within the molecule. We have not, as yet, undertaken a study of the oxidation mechanism in detail. Based on the literature,[1-4] the oxidation peak that appeared in the cyclic and differential pulse voltammograms might be attributed to the initial electrochemical oxidation of valacyclovir to form the intermediate 8-oxoguanine, which involves a two-electron and two-proton transfer process, as illustrated in Scheme 2 in our paper. We agree, the formed 8-oxoguanine may subsequently undergo a further reversible two-electron, two-proton oxidation.

1 B. Uslu et al., Anal. Chim. Acta, 2006, **555**, 341.
2 R. N. Goyal and A. Dhawan, J. Electroanal. Chem., 2006, **591**, 159.
3 A. M. Oliveira-Brett et al., Bioelectrochemistry, 2002, **55**, 61.
4 H. S. Wang et al., Anal. Chim. Acta, 2002, **461**, 243.

Professor Mount commented: Granero et al.[1] state that in the pH range studied, at pH 7.5 valacyclovir "exists as a mixture of neutral and cationic forms". This is attributed to the protonation–deprotonation reaction of the N-terminal amine in the valine unit, which is said to have a pK_a of 7.47. How does this affect the stoichiometry of the reaction with respect to m (the number of protons) compared to n (the number of electrons, equal to 2) in Scheme 2 of your paper in the pH range studied? In well buffered solution, the value of redox potential, half wave potential (and/or peak potential for a reversible system) should vary with pH, giving a magnitude of the gradient of $2.303mRT/nF$, from which m can be extracted. Has this been measured and what does it give for m across the range studied? Finally, concerning Fig. 4 in your paper, what is the chemical reason for the maximum in peak current observed near pH 7.5? Is the peak sharper at this pH and/or larger, and why is this?

1 G. E. Granero et al., Int. J. Pharm., 2006, **317**, 14.

Professor Chen replied: I certainly appreciate your question and the information. To investigate the effect of pH on the electrochemical oxidation of valacyclovir, differential pulse voltammograms were recorded at the SWCNTs-modified GCE under various pH conditions, ranging from pH 5 to 9 at a valacyclovir concentration of 15 nM. The peak potential shifted to lower electrode potentials with the increase of pH of the electrolyte. At pH 7.4, a larger and slightly sharper oxidation peak was observed in the differential pulse voltammograms. Two linear segments were observed under the studied pH range. The initial segment was located between pH 5.0 and 7.0 with a slope of −31.0 mV/pH ($R^2 = 0.998$), which indicated that the ratio of the number of protons (m) to the number of electrons (n) was equal to 0.5. The second linear segment was located between pH 7.0 and 9.0 with a slope −59.2 mV/pH ($R^2 = 0.987$), revealing that the ratio of m to $n = 1$. According to Scheme 2 in our paper, the number of protons involved in the oxidation reaction changed from 1 to 2 with the increase of pH of the electrolyte, which might be explained by alterations in the protonation of the acid–base functionalities within the molecule.

Professor Hillman opened the discussion of the paper by Professor Barbero: Your paper includes probe beam deflection (PBD) measurements of the interfacial compositional changes in response to redox switching under potentiodynamic (cyclic voltammetric) and potentiostatic (following a potential step) conditions. Since the voltammetric control function involves a combination of potential- and time-dependent factors, while the potentiostatic control function isolates the temporal response, it is interesting to compare the two outcomes. In the voltammetric experiment, I note that the minima in the responses during the anodic and cathodic half cycles are separated by $ca.$ 200 mV, $i.e.$ $ca.$ 10 s in time. Superficially, this seems consistent with the transit time of species between the interface and the probe beam, but in fact the minimum in the anodic direction is at a *less* positive potential than the minimum in the cathodic direction. This is the wrong way round, $i.e.$, opposite to the expectation for conventional hysteresis, $i.e.$, optical response lagging interfacial transfer. Can you explain this? Additionally, the short time 'dip' in the optical response following the potential step seems rather smaller and faster than one would expect based on the diffusion time. Again, can you explain this?

Professor Barbero responded: I agree that conventional hysteresis, that is optical response lagging interfacial transfer, would produce the opposite signal. However, this is the case with immobilized redox couples in the surface where a defined peak potential can be seen in the cyclic voltammogram for the interfacial transfer. The CV in our case is featureless because it involves the charging of a double layer. The minima in the cyclic deflectogram is related to the potential of minimum charge of the material. Therefore, the observed hysteresis means that the potential of minimum charge (PMC) measured potentiodynamically is less anodic in the forward than in the backward scan. This is related to the fact that cyclic deflectometry measures in a non equilibrium condition where previous conditions affect the data. Such behaviour has been observed before[1] and makes sampled deflection voltammetry[2] a better technique for the purpose of measuring PMC. In this case the effect was so clear that scanning at slower scan rates, which will make the conventional hysteresis potential difference smaller, makes the potential difference larger, suggesting a different phenomena.

The prepeak observed in Fig. 9 in our paper is not a complete negative peak due to ion expulsion, which should occur at a time determined by the diffusional delay, but the combination of two larger peaks with opposite sign. If both processes occur at the same time inside the electrode only the dominant peak (in this case anion intake) could be observed. However, when the two discontinuous processes inside the electrode occur with some delay, the double peak structure is observed ($e.g.$ Grumelli et $al.$[3]) and can be simulated, as it is shown in Fig. 9 (broken line) in our paper. The prepeak reveals a ion flux in the opposite direction which we assign to proton expulsion related to water dissociation and hydroxide ion used inside the pores of the carbon to charge the double layer.

1 G. A. Planes et $al.$, $Chem.$ $Commun.$, 2005, 2146.
2 G. Lang and C. A. Barbero, $Laser$ $Techniques$ for the $Study$ of $Electrode$ $Processes$, Springer, Berlin, 2012.
3 D. E. Grumelli et $al.$, $Chem.$ $Commun.$, 2003, 3014.

Dr Tschulik asked: In your probe beam deflection measurements you determine the local change in ion concentration in the electrolyte by measuring the refractive index of the electrolyte in the direct vicinity of the electrode surface. The refractive index of a solution is also strongly temperature dependent, which makes quantitative analysis of concentration differences very difficult even in the case of small temperature changes. For your experiments you use a 5 mW laser and you focus the beam in a small area (ca. 60 μm diameter) very close to the surface of your carbon sample. Do you expect or see any heating effects, considering that the carbon is likely to absorb the laser light in this area and that heat dissipation will be limited due to the relatively small thermal conductivity coefficient of the substrate and the electrolyte? Does this limit your ability to quantify concentration changes during the measurement?

Professor Barbero replied: The photothermal effect due to laser light absorption by the carbon will produce a constant refractive index gradient which is usually detected as a signal drift. The use of pulse techniques is one of the best ways, due to the relatively fast measurement, to eliminate such interference and in fact we have not observed a noticeable effect of drift in our measurement. It should be borne in mind that the magnitude of the PBD signal is directly related to the amount of ions exchanged and porous materials exchange a significant amount of ions per geometric area. Therefore, it is likely that the photothermal signal is negligible in comparison with the signal due to ion fluxes.

Dr Kataky opened the discussion of the paper by Professor Xu: You are working above the CMC of the surfactants so you have micelles. Is it likely that the crystals are seeded within the micelles?[1,2]

1 J. Lui et al., *Langmuir*, 2007, **23**, 7286.
2 S. J. Cooper et al., *J. Chem. Phys.*, 2008, **129**, 124715.

Professor Xu answered: Yes. As you pointed out, some crystals are seeded within the micelles.

Professor Kanoufi commented: Fig. 5 in your paper shows an enhancement of both the current and light intensity during the ECL of luminol at electrodes coated with Pd@Au nanoparticles. However this enhancement may be small; isn't it only due to differences in the electroactive surface area? Could you provide more information on the estimation of this electroactive surface area?

Professor Xu responded: We have made a comparison using ECL density and current density, not intensity. The electrochemical active surface area (EASA) of Au catalysts were calculated from the reduction peak area of the surface $Au(OH)_3$. The characteristic reduction peak in the cyclic voltammograms standards the reduction of $Au(OH)_3$ on the Au surface, reflecting the number of active Pd sites. The EASA value is calculated from the following equation: EASA= $Q/(C)$, where Q is the Coulombic amount of the reduction peak area of $Au(OH)_3$, C is the reduction charge of $Au(OH)_3$ monolayer on the Au surface (440 $\mu C\ cm^{-2}$).

Professor Chung commented: The nanoparticles that were presented have interesting and novel shapes. But the electrocatalytic activity of those is not directly explained by the apparent shape and size of the given nanoparticles. One of the widely considered parameters is the crystal facets that they have. I am wondering if you have any evidence or observation that indicates the contribution of the specific crystal facets to the electrocatalytic power you have shown. Or are there any other factors you would like to suggest?

Professor Xu answered: We have tested electrochemiluminescence at metal nanocrystals with different specific crystal facets. The results show that electrocatalytic power depends significantly on specific crystal facets.[1–4] Electrocatalytic studies at single crystal electrodes by many other groups have also shown that electrocatalytic power depend on specific crystal facets.

1 L. Zhang et al., Chem. Commun., 2013, **49**, 8836.
2 L. Zhang et al., Faraday Discuss., DOI: 10.1039/C3FD00016H.
3 L. Zhang et al., Nano Today, 2012, **7**, 586.
4 L. Zhang et al., Chem. Commun., 2011, **47**, 10353.

Professor Chung said: The electrocatalytic activity that was presented might have come from the interstitial voids among the nanoparticles. Many papers which report new nanoparticles explain the observed electrocatalysis on the basis of surface area, plasmonics or crystalline facets. However, electrochemistry in practice, such as what you saw, was often from a stack of nanoparticles that have numerous interstices. So we need to separate the contributions of multiple catalytic factors to determine what is new in terms of electrocatalysis. Considering the geometric effects of nanopores, pore size information would help with unraveling the underlying causes.

Professor Xu replied: As you have mentioned, we have also found that the electrocatalytic activity might have come from the interstitial voids among the nanoparticles. For example, we have observed that electrochemistry of uric acid became reversible at a single-walled carbon nanohorn modified electrode (Zhu et al., Biosens. Bioelectron., 2009, **24**, 940). It seems that the interstitial voids may play an important role in the electrocatalytic activity when reactants, intermediates and/or products interact strongly with nanoparticles in the confined interstitial voids. Your interesting research also shows that nanopore structure is a cause of electrocatalysis under certain conditions.

Professor Barbero asked: The increased ECL signal is assigned to an increased catalytic activity. However, it has been shown that local electric fields in the surface of gold nanoparticles increase the fluorescence efficiency of fluorophores (the so-called surface enhanced fluorescence), could this effect also increase the luminescence of luminol?

Professor Xu replied: Your question may lead to the development of a new way to improve the ECL signal. More experiments are needed to check the effect on the luminescence of luminol.

Dr O'Mullane asked: Did you investigate the influence of the gold shell thickness on the electrocatalytic reaction, as the presence of the underlying Pd core may significantly influence the electronic properties of the gold shell?

Professor Xu responded: Currently, we have not investigated the influence of the gold shell thickness on the electrocatalytic reaction. However, as you have mentioned, the presence of the underlying Pd core may significantly influence the electronic properties of the gold shell when the gold shell thickness is very small. By taking advantage of the interaction between core and shell, it may provide an effective way to develop new electrocatalytic materials.

Professor Williams opened the discussion of the paper by Dr O'Mullane: Your deposition method relies on the mixed potential developed at the surface, which is sensitive both to the nature of the surface sites (both those that are driving the reductions and those on which the metal is deposited). This then raises the question whether it is possible to learn more about the nature of the sites involved by manipulating the depositing metal and the solution in order to explore a range of mixed potential – for example using a range of different metals to deposit, altering complexants for the depositing metal, and also by manipulating the oxygen partial pressure (for example) in the deposition solution. Have you been able to explore any such ideas?

Dr O'Mullane replied: The presented paper utilises only one approach, however you are correct in stating that it would be useful to use different depositing metals in order to observe which sites are involved in driving the reduction reaction at the electrode surface. For instance using Cu^{2+} would be a useful probe to see if the extent of deposition decreases compared to silver given the lower standard reduction potential for $Cu^{2+/0}$ compared to $Ag^{+/0}$. It is then believed that only the active site responses observed at the lower potentials of ca. 0.05 V (Fig. 10 in our paper) would be involved and not the active site responses at the higher potential of 0.70 V. Also changing the solvent environment and complexing ligands on the metal cations could fine tune the redox potential to give greater control for this approach, as well as having the added benefit of being able to probe the oxidation product on the substrate metal surface. We have not attempted this yet, but have looked at other metal combinations such as the spontaneous deposition of Pd and Ag on Au,[1] as well as Au deposition on electrochemically activated Pt electrodes.[2] In the latter the extent of gold coverage increases as a function of activation time of the Pt electrodes. We have also conducted experiments for Ag deposition on Au where the presence of oxygen in the solution was not found to have an influence on the reaction.

1 B. J. Plowman et al., Faraday Discuss., 2011, **152**, 43.
2 A. P. O'Mullane and S. K. Bhargava, Electrochem. Commun., 2011, **13**, 852.

Dr Wain asked: Do you always observe evidence of silver stripping in your voltammetry after silver decoration, and can the silver metal be removed completely from the surface?

Dr O'Mullane replied: Under acidic conditions, as shown in Fig. 6a in our paper, there is a peak at 0.87 V which is attributed to silver stripping. XPS analysis of the evaporated Pd surface that was decorated with Ag and cycled over a potential range of −0.20 to 1.20 V for 10 cycles (Fig. 6b) showed no evidence of silver on the surface. However, under alkaline conditions, Ag is not stripped from the surface but forms silver oxide when an appropriate potential is applied.

Professor Xu continued the discussion of the paper by Professor Barbero: Did the pyrolysis of resorcinol/formaldehyde resin at 850 degrees result in much decrease in weight? Have you pyrolyzed resorcinol/formaldehyde resin at temperatures higher than 850 degrees? What would happen if resorcinol/formaldehyde resin is pyrolyzed at temperatures higher than 850 degrees?

Professor Barbero answered: The pyrolysis of the resin results in a weight decrease of 20–30%, due to the loss of hydrogen and oxygen. We have pyrolyzed the resin up to 1000 °C. There is a small increase of density, with some weight loss. Additionally, the conductivity increases somewhat even though no sign of graphitization is observed.

Professor Compton continued the discussion of the paper by Professor Xu: In the ECL experiments you discuss in your paper what is the quantum yield for the luminescence? For each electron transferred electrochemcially, how many photons are emitted and what are the loss processes? If the photochemical emission is a minor reaction pathway, does that not make the approach analytically vulnerable to interference?

Professor Xu answered: It is quite difficult to measure ECL quantum yields, and most ECL papers have not reported ECL quantum yields. We have also not measured the ECL quantum yields in our papers. The selectivity of ECL detection depends on the ECL systems used. Some ECL systems show better selectivity.

Professor Compton asked: In the ECL experiments you describe, to what extent is the luminescence subject to possible quenching, for example by dissolved oxygen or other species? To what extent does this limit the analytical usefulness of the approach?

Professor Xu replied: Sometimes the quenching phenomena in ECL are similar to those in fluorescence. Sometimes the quenching phenomena in ECL are quite different from those in fluorescence since some compounds are electrochemically transformed to other compounds during ECL reactions. ECL reactions also involve some very unstable intermediates; it is difficult to predict to what extent the luminescence is subject to possible quenching. However, by combining with molecular recognition techniques or separation techniques, ECL can be a very useful analytical technique. For example, ECL immunoassays have been widely used in clinical analysis.

Professor Amatore continued the discussion of the paper by Dr O'Mullane: In our experience with platinum detection of oxidative stress species we observed that electrogenerated 'platinum black' yields more stable and electroactive electrodes compared to polished platinum. Bartlett and Denuault observed the same with another nanoporous Pt material. Our rationale is to say that nano platinum clusters provide numerous reactive electrocatalytic centers. What is your opinion based on your work?

Also, Simonet showed that in the presence of reduction of difficult to reduce organic halides, noble metal surfaces may be corroded leading to the formation of negative metal nanoclusters embedded in the electrode material and compensated by countercations. Would such a mechanism be of any importance in your work?

Dr O'Mullane answered: Yes, a similar reasoning is appropriate for this work. We believe that the electrodeposited Pd which has a nanostructured morphology contains a significant number of electrocatalytically active centres compared to the smooth evaporated Pd film, as evidenced by comparing the voltammetry shown in Fig. 10 and Fig. 6a of our paper. So this is certainly analogous to the case of 'platinum black'. However, the hypothesis here is that essentially all surfaces contain such active centres, albeit at different coverage which facilitates the reduction of Ag cations to metallic Ag at such sites. The second comment would certainly be applicable to electrochemically treated electrode surfaces as reported extensively by Burke *et al.* who cathodically treated noble metal electrodes to induce active site responses *via* hydrogen embrittlement where such negatively charge metal clusters may indeed be formed. In this work, care was taken to avoid this type of surface disruption, *i.e.*, electrodes that had been used extensively and also the Pd surface created *via* hydrogen bubble templating that applies a significantly low potential were not used in the decoration process to avoid complicating the study. However, an investigation into the role that such a corroded surface as in Simonet's work might play in this decoration process would be interesting to pursue.

Professor Kornyshev continued the discussion of the paper by Professor Barbero: The dynamic of charging nanopores is not a simple issue. If the pore size is large, the classical transmission line approach could well describe the data; for a small voltage jump there is a relatively simple formula that describes such characteristics as, *e.g.* the total charge in a pore as a function of time. For large voltage jumps the same approach gives numerical solutions. However, when the effective resistance of the pore cannot be 'spatially' separated from the interfacial capacitance, which is the case in narrow nanopores of, say, 1 nm diameter, one will need a different theory.[1]

1 S. Kondrat and A. Kornyshev, *J. Phys. Chem. C*, 2013, **117**, 12399.

Dr Batchelor-McAuley opened a discussion of Professor Xu's and Dr O'Mullane's papers: Many of the electrochemical process that we are interested in are inner-sphere reactions, such as oxygen reduction. Consequently, these processes are likely to be sensitive to both surface defects and the local surface environment *i.e.*, the presence of capping agents. In the pursuit of new and more electrochemically active materials, what criteria can we use to find them or are we limited to a purely empirical process?

Professor Xu responded: Many interesting electrochemically active materials and some relevant theories have been reported. Based on these studies, it is possible to design some new and more electrochemically active materials. Since inner-sphere reactions are quite sensitive to both surface defects and the local surface environment, it is quite difficult to theoretically predict it, particularly in complex situations. Empirical processes will play an important role in the search for new and more electrochemically active materials.

Dr O'Mullane answered: For many electrocatalytic reactions the most highly active metallic and bimetallic catalysts are known and have been studied in detail. The influence of size, shape and exposed crystal facets for single metallic nanoparticles and extended single crystal surfaces has been the focus of extensive studies. The benefit of bimetallic systems, whether as alloys or core/shell materials, has gained significant momentum. A significant breakthrough was the identification of the $Pt_3Ni(111)$ alloy from single crystal studies which shows extremely high activity for the oxygen reduction reaction due to an unusual electronic structure and arrangement of surface atoms.[1] However, to a large extent it does remain quite empirical as to what is the optimum composition, size and shape for chemically synthesised bimetallic electrocatalysts that would give comparable behaviour to well defined single crystal surfaces. An example of high throughput screening advances for the identification of the optimum composition of nanoparticles involves probing bimetallic catalyst inks immobilised on an insulating surface using SECM.[2] However, there are also significant efforts to predict the electrocatalytic activity of metallic and bimetallic surfaces using DFT calculations and an excellent illustration of this is the work by Greeley and Nørskov who predicted the activity of a wide variety of bimetallic catalysts with different compositions for the hydrogen evolution reaction.[3] Indeed the work on the $Pt_3Ni(111)$ system also employed DFT calculations to quantitatively describe the effects promoting the oxygen reduction reaction. The application of such an approach to more complicated inner sphere reactions such as the oxidation of organic molecules as well as accounting for the presence of defects on nanomaterials may be more difficult. Regarding the presence of capping agents, for the majority of electrocatalytic reactions these are detrimental to performance and need to be removed prior to any electrochemical study.

1 V. R. Stamenkovic *et al.*, *Science*, 2007, **315**, 493.
2 J. L. Fernández *et al.*, *J. Am. Chem. Soc.*, 2005, **127**, 357.
3 J. Greeley and J. K. Nørskov, *Surf. Sci.*, 2007, **601**, 1590.

Dr Batchelor-McAuley addressed Professor Xu and Dr O'Mullane: It would seem likely that the higher Miller index crystal facets are expressed in the synthesis due to the presence of the organic capping agent. The organic layer will serve to stabilise the surface. However, the presence of these capping agents may be problematic and even if we attempt to remove them we can not be certain that the surface remains unaltered. So just because one can fabricate a high order surface does this necessarily lead to the production of a more electroactive material?

Professor Xu replied: The higher Miller index crystal facets are beneficial to the electrochemical reactions of some compounds, not all compounds, even if the surface of higher Miller index crystal facets is clean. Moreover, similar to the modification of electrodes, the modification of crystal facets can change the catalytic properties of nanocrystals.

Dr O'Mullane responded: As mentioned in the previous answer, the presence of a capping agent is often detrimental to electrocatalytic performance and can be difficult to remove given the strong binding affinity of many of these capping agents to the metal surface. However, for the realisation of a practically applicable material this is a requirement. The surface treatment of chemically synthesised nanoparticles is highly dependent on the nature of the capping agent and there are a range of possibilities including exposure to UV light, heat treatment, chemical treatment with organic solvents, acids, bases *etc.* However, as you mentioned, there is always the possibility of changing the shape of the material or agglomeration effects. Therefore it is important that for any chemical synthesis procedure nanoparticles are produced with capping agents that can be readily and completely removed by simple methods. A good example is the removal of KBr and PVP from the surface of Pd nanocubes by using *tert*-butylamine which preserves the shape of the catalyst and results in high activity for the oxygen reduction reaction (N. Naresh *et al.*, *J. Mater. Chem. A*, 2013, **1**, 8553).

Dr Batchelor-McAuley addressed Professor Xu and Dr O'Mullane: Just to return to one of your previous points, I agree that theoretical calculations such as DFT can give physical insights into the origins of observed electrocatalytic effects. However, to heavily paraphrase a recent review article on DFT,[1] 'the results are only as good as their approximations.' To what extent do you believe that the current state-of-the-art DFT calculations are truly predictive in nature?

1 K. Burke, *J. Chem. Phys.*, 2012, **136**, 150901.

Dr O'Mullane replied: The paper of Burke is an excellent description of the current state of the DFT field. At present it is a useful tool to combine with experimental observation to strengthen arguments regarding mechanistic pathways. However, with regards to it being a predictive tool I believe new developments will be required to be able to predict the often complex behaviour observed in electrocatalytic reactions. As stated by Burke new approximations need to be developed to keep the DFT research field moving forward.

Professor Xu remarked: For ideal systems, DFT calculations may predict well. Since catalysts generally do not have perfect structures and may have some unknown impurities that may affect catalytic properties, I think it is still challenging to predict electrocatalytic systems which are sensitive to local environments by DFT calculations.

Professor Williams continued the discussion of the paper by Professor Barbero: I'd like to raise two points. Firstly, the charging of your electrodes is certainly analogous to the charging of a transmission line and the hierarchical structure can be considered as a truncated transmission line. You showed a characteristic timescale for the system. Could that be explored simply by varying the conductivity of the electrolyte (to change the resistance per unit length of the transmission line) and hence explore the length scale in the system? Secondly, your system should act as a highly confined thin-layer electrochemical cell, characterised by a confinement time. Hence in principle it should be possible to significantly modify the course of reactions, such as many electro-organic reactions, that are sensitive to the timescale of confinement in the vicinity of the electrode, maybe even to tailor an electrode structure to be appropriate for a particular course of reaction. Have you been able to explore any such effects?

Professor Barbero replied: I agree that changing the conductivity of the electrolyte should change the characteristic time for the transmission line to be truncated and we will try such an experiment. We would try to model the system more precisely to be able to quantitatively correlate the characteristic time with the conductivity. We have been able to show confinement of redox reactions inside the confined space pores of porous carbon[1] and we are interested in controlling the mechanism of reaction of methanol oxidation. This was the main reason for studying the immobilization of PtRu nanoparticles inside the hierarchical porous carbon, as described in our paper. The use of the system for more complex electro-organic reactions is worth considering.

1 Balach *et al.*, *J. Power Sources*, 2012, **199**, 386.

Professor Mount remarked: We note the MNEE array also has an effective confinement time near the electrode due to the nanodimension and enhanced mass transport. We plan to investigate the effect on electro-organic reactions.

Professor Chung asked: When a nanoporous material has pores with a variety of diameters and it is hard to directly see or measure the pore size and its contribution to the electrochemical behavior, you can consider electrochemical impedance analysis. Once you obtain a conventional Nyquist plot under appropriate conditions, *e.g.*, dc offset potential for the electrochemical reaction of your own interest, and you can find a linear region with 45 degree tilt in the high frequency regime. At a frequency, such a region begins to deviate from linearity. That is the characteristic frequency that can give you an implication on the mass transport in the porous structure. Plus, you can also extract valuable information of pore resistance from the Bode plot, which offer how ion moves in response to the electric field inside the pores.

Professor Barbero responded: I agree that electrochemical impedance analysis could be quite helpful to study the electrochemical response of porous carbon materials. As can be seen in Fig. 7 in our paper, the Nyquist plot of hierarchical carbon does not show the 'Warburg' region with a 45 degree slope, probably because the pores are too short to present a transmission line behavior. However, in porous carbon based on the same RF but non-hierarchical (having mesopores but no macropores) we have observed a clear 'Warburg' region. In those cases, the analysis suggested in the comment could be performed.

Professor Mount said: In previous work in the early 1990s, John Albery and I published a series of papers on the transmission line (TL) analysis of the electrochemical oxidation and reduction of conducting polymer modified electrodes. In this work,[1-3] both distributed capacitance (non-Faradaic charging) and redox (Faradaic) charging of the conducting polymer was modelled using a single transmission line model, along with charge transport through ion diffusion and migration due to fields in the solvent and ion filled pores. The combined overall capacitance of the film in the TL, C_Σ was shown to be given by $1/C_\Sigma = 1/C + 1/C_N + 1/C_X$, i.e., the smaller the distributed (non-Faradaic double layer charging) capacitance, C, the redox (Faradaic) capacitance, C_N, and the counterion capacitance due to the counterions in the pores, C_X. To what extent does this general TL treatment also apply to these systems, with the conducting polymer of huge internal surface area being replaced by the porous graphite and the ion and solvent filled pores being present in both systems? Does this not explain both Warburg ac and Cottrell (current versus $\sqrt{t}$) behaviour in both the stated mechanisms (as ion transport is explicitly linked to charge transport in this single model)? Is an assumption of this work and equations (4) and (5) in your paper that ion transport from solution is never rate limiting? Finally, as there are different pore sizes, what would be the effect of this and any pore size distribution on equation (4)?

1 W. J. Albery et al., *J. Electroanal. Chem.*, 1990, **288**, 15.
2 W. J. Albery and A. R. Mount, *J. Electroanal. Chem.*, 1991, **305**, 3.
3 W. J. Albery and A. R. Mount, *J. Electrochem. Soc.*, 1991, **138**, 440.

Professor Barbero responded: The transmission line analysis developed by Mount and Albery for conducting polymers should also apply to porous carbon electrodes, in the special case where Faradaic capacitance is negligible. In that way, as in the simpler TL model described in our paper, there is a Warburg region in Nyquist plots of AC impedance and a Cottrell region in chronoamperometry. Additionally, the complete model should be applied to porous carbon electrodes having significant Faradaic pseudocapacitance, for example due to the presence of redox species adsorbed on the pore surface (e.g. J. Balach et al., *J. Power Sources*, 2012, **199**, 386). A modelling of such experimental data using the complete model would be most interesting. Probe beam deflection measurements show that the amount of ions inside the pores is not sufficient for double layer charging since a clear concentration gradient develops outside the porous surface. However, in the case of a solid having relatively long pores, where the current scales with the square root of t (i.e. Cottrell behaviour), the ion transport seems to be controlled by the mass transport inside the porous layer since the effective diffusion coefficient outside the layer is 5–8 orders of magnitude larger than the ones inside the layer. The effect of pore size distribution will not affect the 'Cottrellian' part of the current–time plot, because the slope of the current versus $\sqrt{t}$ plot (eqn 6), is not related to pore size. However, a pore size distribution will produce a distribution of characteristic times (eqn 5) and, in that way, affect eqn 4. Instead of having a single time value where the plot departs from the straight line in a current versus $\sqrt{t}$ plot, a distribution of pore sizes will cause a range of times where the plot departs from the straight line. The current–time plot for times in the order of the characteristic time will be also affected and could be calculated by integrating the pore size distribution into eqn 5.

Faraday Discussions

RSC Publishing

PAPER

Mapping fluxes of radicals from the combination of electrochemical activation and optical microscopy

Sorin Munteanu,[a] Jean Paul Roger,[b] Yasmina Fedala,[ab] Fabien Amiot,[c] Catherine Combellas,[a] Gilles Tessier[b] and Frédéric Kanoufi*[a]

Received 25th February 2013, Accepted 28th March 2013
DOI: 10.1039/c3fd00024a

The coating of gold (Au) electrode surfaces with nitrophenyl (NP) layers is studied by combination of electrochemical actuation and optical detection. The electrochemical actuation of the reduction of the nitrobenzenediazonium (NBD) precursor is used to generate NP radicals and therefore initiate the electrografting. The electrografting process is followed *in situ* and in real time by light reflectivity microscopy imaging, allowing for spatio-temporal imaging with sub-micrometer lateral resolution and sub-nanometer thickness sensitivity of the local growth of a transparent organic coating onto a reflecting Au electrode. The interest of the electrochemical actuation resides in its ability to finely control the grafting rate of the NP layer through the electrode potential. Coupling the electrochemical actuation with microscopic imaging of the electrode surface allows quantitative estimates of the local grafting rates and subsequently a real time and *in situ* mapping of the reacting fluxes of NP radicals on the surface. Over the 2 orders of magnitude range of grafting rates (from 0.04 to 4 nm s^{-1}), it is demonstrated that the edge of Au electrodes are grafted $\sim$1.3 times more quickly than their centre, illustrating the manifestation of edge-effects on flux distribution at an electrode. A model is proposed to explain the observed edge-effect, it relies on the short lifetime of the intermediate NP radical species.

Introduction

Monitoring the chemical transformation of surfaces is of considerable importance in many fields and related applications including biology, catalysis, electronics or energy. The most promising approaches to monitor the chemical reactivity of surfaces rely on high-resolution spatio-temporal inspection. Mapping the local chemical reactivity of surfaces is possible using real "chemical"

---

[a]CNRS UMR7195, ESPCI ParisTech, 10 rue Vauquelin, 75231 Paris Cedex 05, France. E-mail: frederic. kanoufi@espci.fr; Tel: +33 14079 4526
[b]Institut Langevin CNRS UMR 7587, ESPCI ParisTech, 1 rue Jussieu, 75238 Paris Cedex 05, France
[c]FEMTO-ST Institute, CNRS-UMR 6174/UFC/ENSMM/UTBM, 24 chemin de l'Épitaphe, 25030 Besançon, France

microscopes, allowing for the micrometric observation of local chemical reactions. As such, the scanning electrochemical microscope (SECM) has been proposed since the early 1990s as a true scanning "chemical" probe microscope allowing for *in situ* imaging of surface chemical processes. More recently, imaging optical microscopes have emerged in the chemical sciences, allowing one to optically image surface transformation processes.[1] If they provide a full-field view of surface transformations, they usually require dedicated labelling species, as in the numerous fluorescence microscopies,[2–4] or surface enhanced Raman spectroscopy.[5–8] Alternatively, label-free approaches, such as those provided by surface plasmon resonance[9,10] or ellipsometry[11] microscopies, provide imaging of the local change in the refractive index associated with the chemical transformation of an interface.

To map the chemical transformation of a surface, the next requisite is to develop strategies to direct such a transformation during the local inspection of the surface. In this respect, electrochemistry or electrochemical activation offer promising potential. Indeed, the electrochemical activation of an electrode surface allows for the minute generation of gradients of chemical reagent concentrations.

Combining electrochemical activation of chemical reagent gradients with chemical microscopy surface imaging then provides a promising strategy to modulate and visualize the spatio-temporal repartition of the chemical reactivity of surfaces. It has been illustrated in many aspects by SECM for the controlled local etching of surfaces[12–14] or in more recent examples for local surface reactivity mappings.[15,16] The observation of electrochemically generated chemical gradients by optical microscopies has also been described.[17–19] In seminal works, electrochemical activation was used to activate either gradients of concentrated highly absorbing species[17] or a fluorescent probe by electrochemiluminescence,[4,20–23] allowing one to map the electrochemically-driven fluxes of the electrogenerated species at the electrode. More recently, label-free approaches have been proposed and SPR microscopy was used to analyse local traces of explosives on fingerprint-decorated surfaces.[9] Our group has developed a versatile alternative approach using ellipsometric microscopy to monitor *in situ* and in real-time the chemical transformation of surfaces by organic layers.[11]

Here, we exploit the combination of such opto-electrochemical microscopies to visualize and map the reaction of electrogenerated gradients of radicals with an electrode surface. Indeed, the electrochemical reductive activation of an electrode in an aryldiazonium solution allows for the generation of phenyl radical species in the vicinity of the electrode. These reactive species are then prone to react with their generating surface and the deposition of a thin organic aryl multilayer (several nanometres thick) on the electrode ensues. Even though phenyl radicals are highly reactive, their reaction with a surface is rather slow, as attested by the slow growth rate of the organic layer. The rate of deposition of the organic layer on the electrode is then an indirect measure of the flux of radicals reacting with the surface. Visualizing and mapping the spatio-temporal growth of an aryl multilayer from an electrode surface then provides access to the mapping of the reactivity of phenyl radicals with the surface.

The present report aims at inspecting and mapping the reactivity of nitrophenyl (NP) radicals at surfaces. As illustrated in Fig. 1, NP radicals are generated at different gold (Au) microelectrodes from reduction of nitrobenzenediazonium (NBD), while the extent of the local reaction of NP radicals with the Au surface is

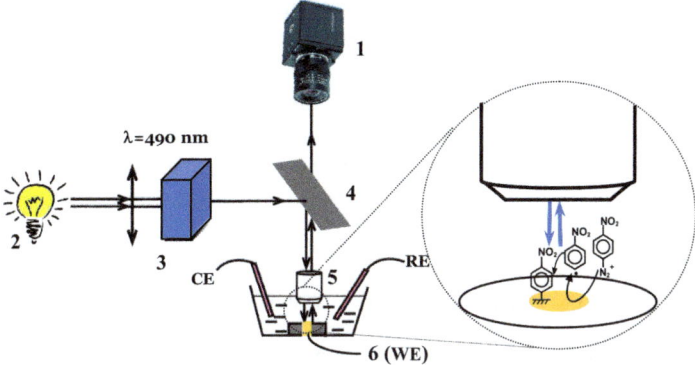

Fig. 1 Schematic representation of light reflectivity mapping of the grafting of nitrophenyl (NP) radicals onto Au microelectrodes under electroreduction of nitrobenzenediazonium (NBD) solution.

monitored by light reflectivity microscopy.[24] After a brief description of the optical method, its performances are tested through its coupling with the electrochemical activation of the surface for the inspection of local fluxes of electrogenerated reactive radicals.

Experimental

Nitrobenzenediazonium (NBD), acetonitrile (ACN), tetrabutylammonium tetrafluoroborate (NBu$_4$BF$_4$) from Sigma-Aldrich were used as received. Macroscopic gold (Au) electrodes were plates cut in Au coated silicon wafers obtained from Sigma-Aldrich (100 nm coating). Gold microelectrodes were home-made from Au wires (25 or 250 μm diameter, Goodfellow) sealed in borosilicate glass capillaries (Sutter).

The electrografting of the different Au surfaces was performed at room temperature, 20 °C, in a 5 mM NBD + 0.1 M NBu$_4$BF$_4$ ACN solution by chronoamperometric experiments in a three-electrode configuration (counter electrode, CE: 1 mm diameter Pt wire, and quasi-reference, RE: 1 mm diameter anodized Ag wire) using a CH 660C potentiostat (CH Instruments).

The light reflectivity microscopic observation schematized in Fig. 1 was obtained using a standard microscope (Olympus), equipped with a camera (1, Dalsa 1M30). The light of a halogen white lamp (2), spectrally filtered (3) to obtain a blue illumination ($\lambda = 490$ nm, $\Delta\lambda = 20$ nm), is directed towards the sample by an optical separator (4). The substrate (6) is therefore illuminated from the top by a light beam with a known numerical aperture at zero mean incidence *via* a microscope objective (5). The reflected light is collected by the same objective and sent on the CCD camera (1), which allows real time imaging of the light flux reflected by the analyzed surface. The incident light intensity is chosen so that the reflected light reaching the camera detector is close to but smaller than the saturation level. It corresponds to an incident light beam of measured radiant power of 170 μW, which illuminates a 3 mm diameter spot of the studied substrate (light flux density of 2.4 mW cm^{-2}). The light intensity is so low that the illumination cannot heat the substrate/solution by more than 10^{-6} °C.

The first step of the observation procedure consists of the precise adjustment of the illumination incidence on the surface. This is achieved with an accuracy of

10^{-2} °, in air by using a Mirau interferential objective (Nikon). The electrolytic solution is then poured into the electrochemical cell in which *in situ* observation is obtained with a water-immersion objective (10×, NA = 0.3, WD = 3.3 mm, or 40×, NA = 0.80, WD = 3.3 mm). Images of the reflecting surface are captured with a CCD camera interfaced with home-programmed Labview software. Each acquired image is a stack of 6 snapshots, each integrated over 50 ms, for a total duration of 300 ms. Image analysis is achieved using Matlab routines.

A. Principle of light reflectivity microscopy

The principle of the method is based on monitoring the surface reflectivity variations by simply measuring the intensity of light that has been reflected by the illuminated reflective surface. The light reflectivity, R, is defined as the ratio of the intensity of the reflected beam, I_{refl}, to the intensity of the incident beam, I_{inc}: $R = I_{\text{refl}}/I_{\text{inc}}$. Under normal incidence, as obtained in our set-up, the expression of the light reflectivity of a beam propagating in a transparent ambient medium (medium A, real refractive index n_A) reflecting on an opaque substrate (medium S), is given by:

$$R_{AS} = |\tilde{r}_{AS}|^2 = \left|\frac{n_A - \tilde{n}_S}{n_A + \tilde{n}_S}\right|^2 \tag{1}$$

The substrate is optically characterized by a complex refractive index $\tilde{n}_S = n_S + i k_S$, where n_S represents the real part of the refractive index, and k_S is the imaginary part, related to the extinction coefficient.[25]

The light reflectivity is equal to the square of the modulus of the reflection coefficient of an electromagnetic wave, $\tilde{r}_{AS}$, which is a complex number obtained through Fresnel equations.

When the light passes through an intermediate organic layer (medium L with refractive index $n_L \approx 1.5$, and thickness th) between the ambient medium and the reflecting substrate, the reflection coefficients at each interface (ambient/layer interface, A/L and layer/substrate interface, L/S) can be expressed by relations similar to (1). The overall light reflectivity takes into consideration the multiple reflections at each interface (A/L described by $\tilde{r}_{AL}$ and L/S described by $\tilde{r}_{LS}$), along with the phase shift induced in the deposited layer. The expression of light reflectivity extends to:[26]

$$R = |\tilde{r}|^2 = \left|\frac{\tilde{r}_{AL} + \tilde{r}_{LS} e^{2i\frac{2\pi}{\lambda} n_L th}}{1 + \tilde{r}_{AL}\tilde{r}_{LS} e^{2i\frac{2\pi}{\lambda} n_L th}}\right|^2 \tag{2}$$

If the incident light beam has a constant intensity, one can measure, from recorded snapshots, the evolution of the reflected light beam intensity collected at time t, compared with the one recorded at time $t = 0$, which is related to the relative variation of reflectivity:

$$\frac{I_{\text{refl}}(t)}{I_{\text{refl}}(0)} = 1 + \frac{\Delta R}{R} \tag{3}$$

Thus, a continuous acquisition of images allows one to determine the local evolution of a thin layer deposition on a reflecting surface.

The sensitivity of the technique is directly dependent on the optical properties of the studied surface (optical indices of the substrate, and of the deposited thin layer to be analyzed). If the refractive indices are known, the thickness, th, can be deduced from optical measurements for low thicknesses. For thick films, the full knowledge of the evolution of the reflectivity is necessary in order to resolve ambiguities introduced by the periodicity of (2).

Gold is a particularly appealing substrate for light reflectivity measurements and a high sensitivity is obtained for blue light illumination (λ < 500 nm). The theoretical relative variations of light reflectivity on an Au surface ($\tilde{n}_S = 1.02 + 1.83\ i$ for $\lambda = 490$ nm),[27] covered by an homogeneous organic layer (thickness th and refractive index $n_L = 1.5$), for a light beam ($\lambda = 490$ nm) propagating in ACN ($n_A = 1.34$) can be obtained from (2). The relative variations of light reflectivity that are presented in Fig. 2 show periodic oscillations with the layer thickness with a period in thickness of $\Delta th = \lambda/2n_L$ ($\Delta th = 163$ nm for $\lambda = 490$ nm and $n_L = 1.5$). For a thin layer deposition, th < 30 nm, the relative light reflectivity change decreases linearly with the deposited layer thickness, th, according to:

$$\frac{\Delta R}{R} = -\sigma\ th \quad (\text{for } th < 30 \text{ nm}) \quad (4)$$

where σ is the sensitivity of the optical method, $\sigma = 0.0043$ nm^{-1} for $\lambda = 490$ nm and $n_L = 1.5$, and for measurements performed in a liquid ambient medium ($n_A = 1.34$). This sensitivity is much higher than that expected for electroreflectance, the change of light reflectivity associated with the change of refractive index due to potential activation of the double layer at the electrode–electrolyte interface.[24,28] Owing to the good stability of the light source, the noise level or light drift induces a limit of detection for the variations of $1 + \Delta R/R$ < 0.002.

This work is devoted to the mapping of fluxes of NP radical species from the mapping of the local deposition of thin NP multilayers. Indeed, the growth rate of the thin layer is related to the flux of incoming and reacting NP radicals. It could then be easier to express the relative light reflectivity variation in terms of change in surface concentration of the deposited material, Γ (in mol cm^{-2}), which can be obtained from the molecular mass of the deposited moiety, M, and the density of the formed thin layer, ρ:

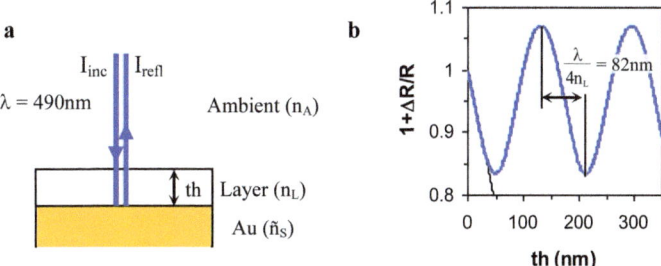

Fig. 2 Variation of reflectivity of light propagated in an ambient medium at an Au surface coated with a transparent layer of thickness th and refractive index n_L. a) Optical model. b) Theoretical relative variation of the light reflectivity with the layer thickness for a liquid ambient ($n_A = 1.34$), $\tilde{n}_S = 1.02 + 1.83i$ for $\lambda = 490$ nm, and $n_L = 1.5$.

$$th = \frac{M}{\rho}\Gamma \tag{5}$$

In the approximation of a thin layer deposition, the relative light reflectivity change also decreases linearly with the deposited material surface concentration and (4) becomes:

$$\frac{\Delta R}{R} = -\sigma \frac{M}{\rho} th \quad (\text{for } th < 30 \text{ nm}) \tag{6}$$

The latter expression allows one to express more easily layer growth rates in units of molecular fluxes in mol cm^{-2} s^{-1}.

B. Results and discussion

1. Inspecting nitrophenyl radical reactions on macroscopic electrodes

First, we illustrate briefly the potential of light reflectivity microscopy for the real time and *in situ* monitoring of the electrografting of a macroscopic Au wafer surface by the NBD salt. Fig. 3a presents an example of a snapshot of a 850 × 850 μm^2 area of an Au surface (total area 0.7 × 0.7 cm^2) in a 5 mM NBD solution before electrochemical activation. After 10 s, the electrode grafting is initiated by potential activation at a constant $E_{Au} = -0.4$ V *vs.* Ag/AgCl, while snapshots are continuously acquired every 2 s. The last snapshot taken at the end of the electrografting is then transcribed from eqn (3) into a local variation of light reflectivity at each CCD camera pixel of coordinate (*x*,*y*); for that, the reflected light

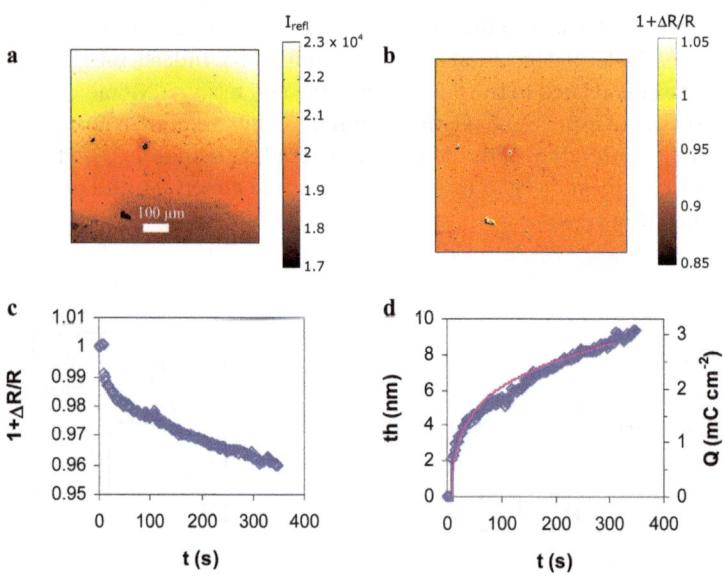

Fig. 3 Light reflectivity observation of NBD electrografting on an Au macroelectrode. a) Image of the intensity of light reflected by the Au macroelectrode before the electrografting. b) Image of the variation of light reflectivity at the end of the grafting (electrochemical actuation at $E_{el} = -0.4$ V *vs.* Ag/AgCl for 300 s). c) variation of light reflectivity (symbols) averaged on a 10 × 10 μm^2 region during the grafting, transcribed d) into NP layer thickness by eqn (4) and compared with the measured charge density (line).

intensity at each pixel $I_{\text{refl}}(x,y,t)$ is compared with that obtained on the same pixel in the initial image, $I_{\text{refl}}(x,y,t = 0)$. The local variation of light reflectivity on the inspected region of the Au macroelectrode is given in Fig. 3b. As predicted from the optical model presented in section A, the electrografting procedure results in the decrease of the intensity of the light reflected by the surface.

The time evolution of the overall reflectivity change (Fig. 3c), $1 + \Delta R/R$, averaged over a 10×10 µm² area of the imaged macroelectrode is transcribed, using eqn (4), into the time evolution of the thickness, $th(t)$, of the NP layer. It is presented in Fig. 3d and paralleled with the evolution of the charge, $Q(t)$ passed in the whole macroscopic electrode during the electrografting process. The growth of the coating layer can be described by two characteristic timescales. During a first short induction period, <4–6 s, the layer grows quickly while a large amount of charge is quickly exchanged. After this induction period, both the layer thickness and the injected charge steadily increase. The steady layer thickness growth is characterized by a growth rate,

$$v = \frac{d\,th}{dt} \tag{7}$$

estimated from the reflectivity measurement and eqn (4) as $v = 0.016$ nm s^{-1}. Using eqn (5), a grafted NP moiety rate of 13 pmol cm^{-2} s^{-1} is obtained taking $\rho = 1$ g cm^{-3} as the grafted layer density[29] and $M = 122$ g mol^{-1} as the molecular mass of the NP moiety.

As demonstrated with other related coupled mass deposition detection methods, such as quartz crystal microbalance (QCM)[29,30] or local ellipsometric imaging,[11] the growth of the aryl multilayer is sustained by the flow of charge in the electrode and therefore the generation of the nitrophenyl (NP) radicals at the coated electrode. The good correlation between both the reflectivity variation and the charge, $Q(t)$, for different times of the experiment shows that the overall grafting process requires the exchange of a constant amount of electrons per NBD molecule, n_e. The latter can be estimated from eqn (8) derived from Faraday's law:

$$n_e = \frac{M}{F\rho A} \frac{Q(t)}{th(t)} \tag{8}$$

where A is the electrode surface area and F the Faraday constant.

An average value of $n_e = 3.7$ is obtained for the macroscopic Au surface. If one assumes that one electron is required for the generation of one radical by the reductive cleavage of the diazonium ion, $n_e = 3.7$ indicates that approximately three out of four generated radicals are not involved in the grafting process and are lost in competing chemical paths (*e.g.* radical reactions in solution). This is actually in agreement with diazonium electrografting processes monitored by other electrochemically-coupled techniques.[11,29,30]

2. Visualizing nitrophenyl radical reactions on microelectrode surfaces

The versatility of light reflectivity microscopy relies on its simplicity of operation. Indeed, as for Au electrode surfaces, different electrode geometries and configurations can be monitored by this optical detection mode. Particularly, it can be applied to the monitoring of electrografting processes at microelectrodes that can be either microfabricated or made by standard glass-embedding techniques. Fig. 4a presents snapshots of the surface of a 25 µm diameter Au microelectrode

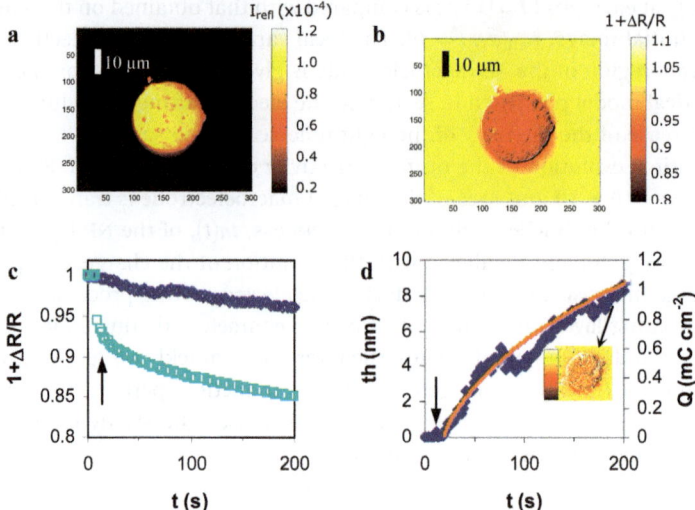

Fig. 4 Light reflectivity monitoring of the electrografting of NP layers on a 25 μm diameter Au microelectrode in the low flux regime. a) Light intensity image taken at $t = 0$ and b) light reflectivity image deduced from snapshot taken after 200 s of grafting actuated at $E_{el} = -0.5$ V vs. Ag/AgCl in a 5 mM NBD + 0.1 M NBu$_4$BF$_4$ ACN solution. c) Change of light reflectivity averaged over 10 × 10 μm^2 during electrografting performed from top to bottom at $E_{el} = -0.3$ and -0.5 V vs. Ag/AgCl. d) $E_{el} = -0.3$ V, NP layer thickness variation (symbols) during the electrografting compared with measured charge (line); the inset is the light reflectivity image taken at $t = 200$ s; on the associated scale bar 1 + ΔR/R ranges from 1.02 to 0.9. Vertical arrows indicate the initiation of the electrografting.

embedded in glass, before the application of a constant $E_{Au} = -0.5$ V vs. Ag/AgCl in a 5 mM NBD solution. The snapshot taken after 200 s of electrografting and converted into a light reflectivity variation image is given in Fig. 4b. As on macroscopic electrodes, the Au microelectrode electrografting is associated with an overall decrease of the Au surface reflectivity. Moreover, the heterogeneity of the reflected light intensity detected on the last image of the Au microelectrode grafting procedure (Fig. 4b) is indicative of the heterogeneity of the thickness of the organic layer coating over the whole microelectrode. Even though these heterogeneities can be correlated to defects of the microelectrode (scratches), as was also detected on the macroscopic Au electrode in Fig. 3b, it is also obvious that the edge of the microelectrode is notably darker than its inner part. This qualitatively indicates the occurrence of higher grafting rates at electrode edges.

We then evaluated quantitatively the heterogeneity of reaction rates on microelectrodes submitted to NBD electrografting. This is easily amenable by coupling the electrochemical initiation of the grafting at the microelectrode surface with its *in situ* and real time optical monitoring. We then used light reflectivity imaging microscopy to quantitatively monitor the electrografting operated at different overall rates on Au microelectrodes. Indeed, the overall rate of the deposition of a thin NP layer can be tuned by adjustment of the microelectrode potential, E_{el}. Particularly in the case of the electroactive NP moieties, thick multilayers can be obtained when the electrografting is performed at E_{el} close to $E^0{}_{NP}$, the standard reduction potential of nitrobenzene moieties. The NP moieties of the grafted layer then act as electron acceptor relays (through generation of the NP radical anions) to transport charges from the electrode to the NBD solution. A catalytic growth of the NP multilayer is then observed.[31,32] Under such conditions as $E_{el} \sim E^0{}_{NP}$, the electrografting is performed in a regime of high driving force for the electrogeneration of NP radicals, while when $E_{el} \gg E^0{}_{NP}$ or E_{el} close to $E_{p,NBD}$, the peak potential associated with NBD reduction, the grafting is performed in a low driving force regime.

A. **Electrografting at low driving force ($E^0{}_{NP} \ll E_{el} \sim E_{p,NBD}$).** In this regime, electrografting experiments have been performed at two different potentials, $E = -0.3$ and -0.5 V vs. Ag/AgCl, and therefore at different overall rates. The overall rate can be evaluated from the time evolution of $1 + \Delta R/R$ averaged over a large domain ($>10 \times 10$ μm²) of the microelectrode as presented in Fig. 4c. The more negative the electrode potential, the more the reflectivity decreases (for $th < 50$ nm as seen in Fig. 2b) and therefore the thicker the multilayer grows. The overall growth of the NP layer is also faster for both the induction period and the coarsening period when a more negative E_{el} is used. The experiment performed at $E_{el} = -0.5$ V vs. Ag/AgCl give information on the reactivity of the electrogenerated NP radicals towards the NP multilayer, and more generally can be used to assess the reactivity of the NP radical on surface-immobilized organic moieties.

The activation at $E_{el} = -0.3$ V vs. Ag/AgCl presents a much slower grafting. The transcription of the reflectivity variation into the organic layer thickness is shown in Fig. 4c and 4d and compared with the charge density variation detected at the microelectrode. As for the macroscopic electrode, the correlation between both quantities is good and an apparent number of electrons transferred per NBD molecule, $n_e \sim 1.6$, is obtained from eqn (8) for the grafting of one NP radical on the Au microelectrode surface. During the early stage ($t < 30$ s) of this grafting procedure, the monitoring of the NP monolayer formation reveals the reactivity of the NP radicals on the Au surface, which is characterized by an average layer growth rate of $v \sim 0.1$ nm s^{-1} or an equivalent flux of reacting NP radicals of 80 pmol cm^{-2} s^{-1}. At longer times ($t > 30$ s), the layer growth is slower and ranges between $v = 0.035$ to 0.06 nm s^{-1} (29 to 50 pmol cm^{-2} s^{-1}).

The grafting at the Au microelectrode was operated under E_{el} conditions comparable with the one presented on the macroscopic Au surface. Even though grafting rates on both surfaces are of the same order of magnitude, the NP layer deposition proceeds at a faster rate on the microelectrode than on the macroscopic Au surface.

We next addressed the local grafting of the microelectrodes under both E_{el} conditions. As shown in Fig. 3b and 4b, the coating thickness on the microelectrode at long grafting times is heterogeneous and reveals the heterogeneity of the grafting reactivity. This clearly shows the importance of surface defect sites and, more importantly, the higher coating along the edges of the microelectrode. The same fast grafting produced at $E_{el} = -0.5$ V is analyzed at shorter times, during the induction period. Then, to reveal the heterogeneity of the grafting, Fig. 5a and b present distribution images of the time derivative of local relative reflectivity changes, $d(\Delta R/R)/dt$, for $t = 8$ and 12 s (the grafting starts at $t = 10$ s). Even though data differentiation usually introduces noise, the microelectrode activity is clearly revealed as regions of $d(\Delta R/R)/dt < -0.02$ s^{-1}, or by layer thickness growth rates: $v > 4$ nm s^{-1}.

This image also reveals heterogeneous regions as lighter micrometric areas of slower coating rates. Except from these individual defects, a cross section of the

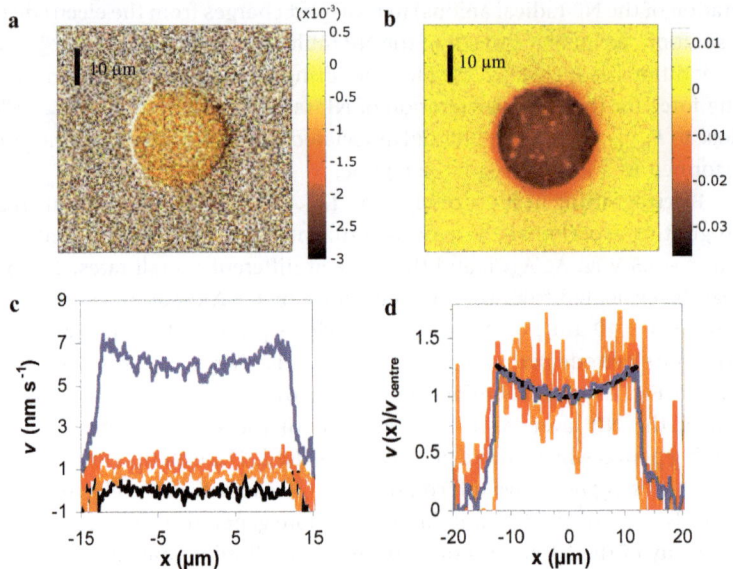

Fig. 5 Distribution of electrografting rates over a 25 μm diameter Au microelectrode, $E_{el} = -0.5$ V vs. Ag/AgCl (the grafting starts at $t = 10$ s). a, b) Image of reflectivity variation d($\Delta R/R$)/dt at 8 s (a) and 12 s (b). The color bars correspond to values of d($\Delta R/R$)/dt in s^{-1}. c, d) Distribution of the grafting rate (c) over a microelectrode diameter at $t = 12, 14, 16, 8$ s (from top to bottom) or (d) normalized by the rate at the microelectrode center (black line: simulated distribution rate with $k = 5 \times 10^{-4}$ cm s^{-1} and $k_c > 100$ s^{-1}).

growth rate, $v(x)$, along a microelectrode diameter is presented in Fig. 5c for $t = 8$, 12, 14 and 16 s. For $t > 10$ s, spatio-temporal variations of the reactivity of NP radicals on the surface are clearly seen. Indeed, the grafting rate $v(x)$ depends on the radial position, x, on the microelectrode. As observed at longer times in Fig. 4b, microelectrode regions closer to the edge are regions of faster NP radicals grafting. The ratio of edge-to-centre grafting rates is however rather small: $v_{edge}/v_{centre} = 1.27$. As observed for $t = 14$ and 16 s, the grafting rate greatly decreases with time, but the radial evolution of the grafting rate is still observed. At longer times, $t > 30$ s, the layer growth stabilizes to a constant rate of 0.09 nm s^{-1} (74 pmol cm^{-2} s^{-1}) with still an apparent lower reflectivity around the edge of the microelectrode than on its centre (Fig. 4b). Even if this experiment actually characterizes the reactivity on an already coated surface and therefore on surface-immobilized NP moieties, it suggests that the spatio-temporal reactivity of the NP radical is maintained along the whole grafting process.

Similar analysis during the early stage of Au microelectrode monolayer coverage is more delicate, as the signal detected when $E_{el} = -0.3$V is close to the limit of detection of the optical method (~0.002 in term of 1 + $\Delta R/R$). However, the grafting rate is not significantly different along the different positions of the microelectrode, and again the v_{edge}/v_{centre} ratio is likely close to 1.

B. Electrografting at high driving force ($E^0_{NP} \sim E_{el} \ll E_{p,NBD}$). A similar analysis has been undertaken at much higher NP radical flux regimes. This is obtained by performing the electrografting at E_{el} where the NP radical anion can be formed (typically $E_{el} = -0.9$ V vs. Ag/AgCl). Here, thick NP multilayers are obtained and their generation is sustained by the NP immobilized units. The latter act as redox catalytic relays for electron transfer to propagate electrons from the electrode to the NBD solution for further NP radical generation. A typical example of reflectivity variation during such thick NP multilayer generation is presented in Fig. 6a and compared with the evolution of the charge transferred electrochemically for the NBD reduction.

The reflectivity variations clearly show oscillations, as predicted optically (see eqn (5)), even though the average variation decreases and deviates from the theoretical prediction for a transparent layer. It is not the aim of this work to explain the origin of such deviation and this will be described elsewhere. Even without theoretical agreement, approximate layer thicknesses may be obtained at the maximum and minimum of $1 + \Delta R/R$. These characteristic occurrences should be met for destructive or constructive interferences and therefore they correspond to layer thicknesses similar to those observed in Fig. 2b. For an illumination at $\lambda = 490$ nm and a layer of refractive index $n_L = 1.5$, the first minimum corresponds to the deposition of a 51 nm thick layer, and the layer grows by $\Delta th \sim \lambda/4n_L = 82$ nm between two consecutive minima and maxima. As observed in Fig. 6b, the grafting performed at $E_{el} = -0.9$ V vs. Ag/AgCl yields a multilayer of thickness $th > 135$ nm, in agreement with the possible generation of very thick multilayers[30,31] under such high driving force regimes for NP radical generation.

Owing to the coarse optical model used here, it is difficult to correlate the light reflectivity change with the electrochemical charge transferred during the whole experiment. However, for an optical observation made at the microelectrode centre (red symbols in Fig. 6a), the extrema of $1 + \Delta R/R$ correspond to a constant layer growth rate (red symbols in Fig. 6b) of $v = 0.36$ nm s^{-1}, or equivalently 300 pmol cm^{-2} s^{-1}. Such a growth rate is about 4 to 6 times faster than the one observed on microelectrodes at lower driving force ($E_{el} = -0.5$ and -0.3 V respectively). This shows the impact of the driving force on the generation of the NP radicals, that is the impact of the electrode potential, E_{el}, on the grafting rate. Moreover, as observed for NP grafting at lower driving force, the constant v at long times is also consistent with the linear variation of the charge with the NBD reduction time by 0.035 mC cm^{-2} s^{-1}. From eqn (8), an average number of electrons $n_e \sim 1.23$ is transferred per NBD molecule on this 250 μm diameter Au electrode.

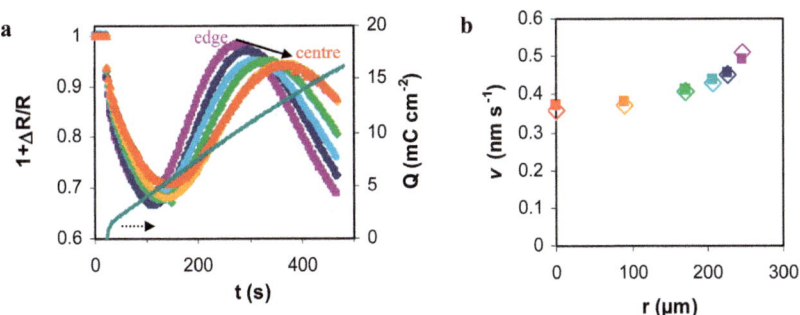

Fig. 6 Growth of a NP multilayer of thickness $th > 130$ nm on a 500 μm diameter disk Au electrode ($E_{el} = -0.9$ V vs. Ag/AgCl). a) Light reflectivity measured at different locations along a radius of the electrode going from the edge to the centre, compared with electrochemical charge density variation (grey-blue line). b) Evolution along the electrode radius ($r = 0$: centre, $r = 250$ μm: edge) of the local grafting rate estimated assuming resp. 51 (■) and 133 (◇) nm are grafted at the minimum and maximum of $1 + \Delta R/R$.

The characteristic reflectivity variation also allows the interpretation of the spatio-temporal distribution of the grafting on the electrode in these very fast coating regimes. Fig. 6a presents the time evolution of the reflectivity changes estimated at different locations on the 250 μm radius electrode, as indicated by the arrow in Fig. 6a and the positions represented in Fig. 6b. The spatial distribution of the grafting is deduced from the apparent arrival of the reflectivity wavefront at different positions on the electrode. Clearly, the wavefront first reaches the edge of the electrode. It propagates and is delayed when going to its centre. Quantitative spatio-temporal analysis is obtained from the estimate of the grafting rate at the time of arrival of the minimum and maximum reflectivity along the different positions on the microelectrode. The variations of this grafting rate are presented in Fig. 6b. Again, faster rates are observed in the microelectrode edge region than at its centre, with an edge-to-centre ratio $v_{edge}/v_{centre} = 1.40 \pm 0.04$, closely related to that observed at lower NP driving force.

C. Tentative mechanistic interpretation

1. Summary of the results. The electrochemical actuation of the NBD reductive electrografting allows one to tune the overall grafting of NP layers on Au electrode surfaces with growth rates, v, spanning over ~2 orders of magnitude: from 0.035 nm s^{-1} (long time and low driving force, $E_{el} = -0.3$ V vs. Ag/AgCl) to 0.36 nm s^{-1} (long time and high driving force, $E_{el} = -0.9$ V vs. Ag/AgCl) to >2 nm s^{-1} during the fast induction period (short time, $E_{el} = -0.5$ V vs. Ag/AgCl). Coupling such electrochemical actuation of thin NP organic layer coatings on a reflecting microelectrode surface to its optical detection then allows one to monitor the local grafting rates or the local fluxes of reacting NP radicals. It provides an indirect detection of the local electrochemical current densities associated with the NBD reduction.

Over the whole range of fluxes/growth rates investigated, the NP layer deposition rate is slightly higher at the microelectrode edges (and defects) than at its centre. At first sight, this is in line with edge effects such as the generation of higher diffusion fluxes of electrogenerated reactive species at the edge of a microelectrode rather than at its centre. If this has been theoretically predicted, it has scarcely been demonstrated experimentally.[17,19–22,33] We believe that the observed grafting rate distribution provides another qualitative illustration of the manifestation of such edge effects.

2. Simplified kinetic model for NBD electrografting. To obtain a more quantitative insight into the phenomenon, the electrochemical process leading to the NP layer deposition should be modelled. Even though the electrografting of surfaces by reduction of aryldiazonium salts is largely employed and different mechanistic aspects have been discussed,[34–36] a mechanism that satisfactorily encompasses the whole aspects of the grafting has not yet been proposed. This objective is particularly difficult to reach as the aryldiazonium reduction yields reactive species that are prone to react with a wide range of materials, such as their generating electrode surface but also the deposited film or the close environment of the electrode. Moreover, if the reaction of the electrogenerated radical with the surface leads to the modification (partial attenuation) of its electrochemical properties, it hardly blocks it completely, allowing for a continuous radical generation and sustained grafting process. It is therefore not reasonable to propose a mechanism based on these results, and we will only concentrate on a tentative interpretation of the observed phenomenon following a simplified approach proposed by Savéant.[37]

The grafting of a surface by aryldiazonium salts, ArN_2^+, results from the initial dediazonation of the aryldiazonium, leading to the generation of a reactive species. The mechanism of such a preliminary step and the nature of the reactive step is controversial. Depending on the experimental conditions, mostly on the solvent, actuation and temperature, either heterolytic or homolytic dediazonation routes have been proposed and reviewed several times over the years.[38-40] As described by Zollinger,[38,41] the heterolytic dediazonation (9) yields a reactive aryl cation, Ar^+, through intermediate caged fragments.

$$ArN_2^+ \rightleftarrows [Ar^+, N_2] \rightleftarrows Ar^+ + N_2 \quad (9)$$

Such an S_N1 mechanism and carbocation intermediate can be observed in solvents of low nucleophilicity, such as water or acetonitrile, and is favoured at elevated temperature[42] or during the spontaneous chemical grafting of surfaces.[40] Further reduction of the aryl carbocation into an aryl radical may be invoked owing to its highly positive potential. However, the characteristic time for the aryl carbocation generation is of the order of several hours, >3 orders of magnitude higher than that imposed by the electrochemical activation proposed here.

Homolytic dediazoniation prevails in more nucleophilic solvents or conditions,[43] and more particularly when an electron is transferred to the aryldiazonium salt as in the case of electrografting processes. In the following, we consider, as generally admitted,[44] that NP radicals, denoted $Ar^{\bullet}$, are generated from the irreversible electrochemical reduction of NBD according to a concerted electron transfer bond-breaking process following Savéant's formalism:[45]

$$NBD + e \xrightarrow{k} Ar^{\bullet} + N_2 \quad (10)$$

The NBD consumption at the electrode is described either by mass-transfer limitation or by kinetic limitation from the charge transfer process associated with the rate constant k (in cm s^{-1}). The flux of consumed NBD is then given, on the one hand, from a Butler–Volmer-like relationship in the case of kinetic limitation:

$$D \frac{\partial [NBD]}{\partial z}\bigg|_{z=0} = k\,[NBD]_{z=0} \quad (11)$$

where z is the direction normal to the electrode surface, $[NBD]_{z=0}$ the concentration of NBD at the electrode surface and D the NBD diffusion coefficient. On the other hand, for fast charge transfer, the boundary condition at the electrode surface, eqn (12), implies that the concentration of NBD at the electrode surface, $[NBD]_{z=0}$ is constant for any radial position, $r \leq a$ (a: electrode radius), and its value is dictated by the electrode potential:

$$[NBD]_{z=0,\,r \leq a} = \text{cst} \quad (12)$$

The flux of NBD is then obtained from the resolution of the diffusion equation, eqn (13), in both directions: radial, r, and normal, z, to the electrode surface.

$$\frac{\partial [NBD]}{\partial t} = D\left(\frac{\partial^2 [NBD]}{\partial r^2} + \frac{1}{r}\frac{\partial [NBD]}{\partial r} + \frac{\partial^2 [NBD]}{\partial z^2}\right) \quad (13)$$

The flux of NBD consumption is also the rate of production of NP radicals, which can diffuse away in solution from their generating electrode surface and also react on the electrode surface to grow the NP layer. Comparison with the experimental observation indicates that at long times, both the overall grafting rate and the electrochemical current are constant (steady increase of thickness and charge). It then allows a rough estimate of the average flux of NBD consumption at the electrode surface:

$$D\frac{d[NBD]}{dz}\bigg|_{z=0} = \frac{i}{FS} \approx \frac{\Delta Q}{FS\Delta t} \quad (14)$$

From the long time behaviours observed in Fig. 3, 4 and 6, this flux is $\sim 10^{-10}$ mol cm^{-2} s^{-1}. This is considerably small compared with the steady-state flux that can be sustained by a 25 μm diameter microelectrode for complete reduction of a 5 mM NBD solution ($4DC^0/(\pi a) = 5 \times 10^{-8}$ mol cm^{-2} s^{-1}). It suggests that the long time grafting is performed under slow charge transfer rate for the NBD reduction (boundary (11) holds instead of (12)), and that the NBD concentration at the electrode is close to the bulk concentration, $[NBD]^0$. An estimate of the charge transfer rate, k, is then given by:

$$\frac{\Delta Q}{FS\Delta t} \approx k[NBD]^0 \quad (15)$$

which yields a lower boundary of the apparent NBD consumption rate of $k \sim 2 \times 10^{-5}$ cm s^{-1}, considering that the whole electrode surface is active.

In a simplified model, the NP radical grafting on the electrode surface assumes that NP radicals react with any part of the surface (bare Au surface or NP multilayer):

$$Ar^\bullet + \text{Surface} \xrightarrow{k_{gr}} \text{surface} - NP \quad (16)$$

The evolution of Γ, the surface concentration of the grafted NP, Surface-NP in eqn (16), is then described by a first-order law (eqn (17)) respective to the concentration of the reactive NP radical, Ar$^\bullet$, at the solution/electrode interface:

$$\frac{d\Gamma}{dt} = k_{gr}[Ar^\bullet]_{z=0} \quad (17)$$

In eqn (17), k_{gr} characterizes the grafting reaction rate constant (in cm s^{-1}), which contains the overall diversity of the grafting process. For example, it may include the difference of reaction rates of the NP radical with the bare electrode surface or with the covering NP multilayer. It may also include the number of active sites of the surface, or their surface concentration, which are prone to grafting and to sustain the layer growth. In a simplified version, we assume the reaction rate k_{gr} is constant over the whole grafting process.

Eqn (10) and (16) indicate that the rate of production of Ar$^\bullet$ is balanced by two competing routes from the electrode surface: their escape by diffusion and their grafting. This gives:

$$-D\frac{\partial[Ar^\bullet]}{\partial z}\bigg|_{z=0} = k \, [NBD]_{z=0} - k_{gr}[Ar^\bullet]_{z=0} \quad (18)$$

Based on eqn (17), the grafting rate, $v = d\Gamma/dt$, and also the amount of grafted material, Γ, should reflect the concentration profile of the NP radical, Ar$^\bullet$, above the electrogenerating electrode surface. The problem then relies on the resolution of the diffusion equations for NBD and Ar$^\bullet$ (eqn (13)), taking into account the boundary conditions given by (11) and (18).

3. **Simulated distribution of NP radical over the electrode surface.** As observed experimentally for long grafting rates, the NBD consumption (or Ar˙ generation) at the electrode is slow compared with the radical attack on the surface ($k \ll k_{gr}$); this is consistent with the observed values of k_{gr} of the order of 1 cm s^{-1} for the growth of polyphenyl multilayers.[36] The electrode coating is then dictated by the generation/diffusion of the NP radical from the electrode and the electrode coating follows the diffusion-controlled concentration profile of Ar˙ above the electrode surface acting as a source of NP radicals. These concentration profiles, $[Ar˙]_{z=0}$, at the electrode surface can be obtained from 2D-axisymmetric finite-elements simulation (COMSOL®, as sketched in Fig. 7a). An example is shown in Fig. 7a and 7b, which present respectively the spatial distribution of the NBD concentration in solution or the evolution of the Ar˙ concentration along the radial coordinate, r, generated above a source electrode, of radius $a = 250$ µm (mimicking the experiment of Fig. 6), using $k = 2 \times 10^{-5}$ cm s^{-1} and two extreme values for the grafting rate k_{gr} of 2×10^{-3} (Fig. 7a and 7b) and 2 cm s^{-1} (Fig. 7b) respectively.

Both situations yield a flux of diffusing Ar˙, which diverges at the electrode edge (not shown), and this illustrates the known electrode edge effect. Obviously from examination of the boundaries (11) and (12) for Ar˙ generation at the electrode, only the kinetically-limited charge transfer situation can introduce some radial evolution of the electrogenerated Ar˙ radical and therefore some spatial evolution of the grafting rate along the electrode. This is indeed confirmed from the simulated concentration profiles of Ar˙ given in Fig. 7b. However, the charge transfer kinetic limitation at the microelectrode rather predicts a higher [Ar˙] and therefore a higher grafting rate at the microelectrode centre than at the edge. Increasing k_{gr} to values >0.2 cm s^{-1} allows to inverse the Ar˙ spatial distribution over the electrode surface, but the concentration is only 1% higher at the electrode edge than at its centre. For both limiting k_{gr} values, the predicted grafting distributions are in disagreement with the experimental observations.

The experimentally observed inverted grafting distribution then suggests, based on the simple grafting model proposed, that higher Ar˙ concentrations are found at the microelectrode edge, while they are also regions of higher NBD concentrations. This can be met if considering that either (i) NBD and/or Ar˙ have

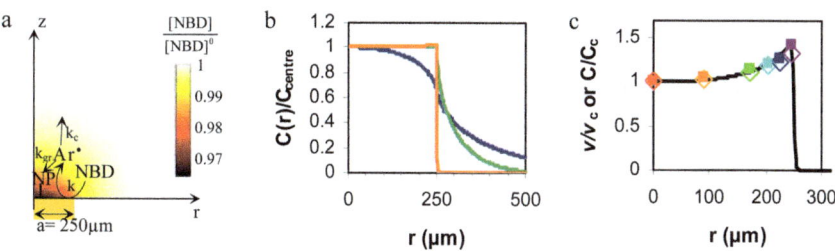

Fig. 7 Simulation of NP radical, Ar˙, generation at a 250 µm radius electrode mimicking the experimental grafting of Fig. 6. a) Schematic description of the problem. b) Simulated normalized concentration profile of $[Ar˙]_{z=0}$ at 100 s, along the electrode radial coordinate, r, for kinetically-limited situations for $k = 2 \times 10^{-5}$ cm s^{-1}: low-consumption regime (blue line), $k_c = 0$, $k_{gr} = 2 \times 10^{-3}$ cm s^{-1}, high-consumption regime for (green) $k_c = 0$ and $k_{gr} = 2$ cm s^{-1} or (orange) $k_c > 10^2$ s^{-1} and $k_{gr} = 2 \times 10^{-3}$ cm s^{-1}. c) Comparison of the simulated concentration profile for $k = 10^{-3}$ cm s^{-1}, $k_{gr} = 2 \times 10^{-3}$ cm s^{-1} $k_c > 10^2$ s^{-1} with the experimental radial distribution of the grafting rate (from Fig. 6d).

different intrinsic reactivities at the electrode centre *vs.* edge, or more simply if (ii) Ar˙ presents some chemical instability in the solution phase.

4. *Influence of the NP radical chemical stability.* As a tentative first hint, we consider the simplest second possibility as phenyl radicals are indeed highly reactive species and are scavenged in solution at a rate k_c according to:

$$\text{Ar}^\bullet \xrightarrow{k_c} \text{product} \qquad (19)$$

which is accounted for by the introduction in the right-hand part of the diffusion equation of Ar˙ (eqn (13) with Ar˙ instead of NBD) of a consumption term – $k_c[\text{Ar}^\bullet]$. If the chemical scavenging of Ar˙, the NP radical, is likely not the only regulating matter, it is also a reasonable explanation for the low electrochemical efficiency of the grafting ($n_e > 1$). In view of this simple model, the grafting efficiency is obtained from comparison of both competing routes for Ar˙ consumption: its grafting on the electrode at a rate k_{gr} *vs.* its solution diffusion/ scavenging at a rate $(Dk_c)^{1/2}$. The apparent number of exchanged electrons is then given by:[37,46,47]

$$\frac{1}{n_e} = \frac{k_{gr}}{k_{gr} + \sqrt{Dk_c}} \qquad (20)$$

Indeed, phenyl radicals are known to be scavenged in various radical reactions, and particularly *via* H-atom abstraction in most common organic solvents. This reaction actually has a rate constant estimated as $k_c > 10^6$ s^{-1} in ACN,[48] that is fast enough to impose the heterogeneity of [Ar˙] above the eletrogenerating electrode surface. For a value of the diffusion coefficient of 10^{-5} cm^2 s^{-1}, the observed values of n_e (from 1.2 to 4) ensure values of $(Dk_c)^{1/2}/k_{gr}$ of the order of 0.2 to 3, yielding values of k_{gr} in the 1 to 15 cm s^{-1} range, in agreement with values described for the growth of polyphenyl multilayers.[36]

This homogeneous scavenging can be implemented[13,49,50] in the problem of diffusion/solution reaction/grafting of Ar˙, as presented in Fig. 7b when considering as negligible the contribution from the grafting reaction, $k_{gr} = 2 \times 10^{-3}$ cm s^{-1}, and observing the effect of $k_c = 10^3$ s^{-1}. As a fast grafting rate, the solution scavenging of Ar˙ results in the observation of a higher [Ar˙] at the electrode edge than at its centre, and the same limiting concentration profile is observed (1% higher concentration at the electrode edge).

Therefore, whatever the value of the Ar˙ generation rate k, adjusting its surface grafting rate, k_{gr}, and/or its homogeneous scavenging rate, k_c, allows one to obtain a limiting situation for which the edge of the electrode becomes a region of higher [Ar˙]. The radial distribution of the simulated grafting rate in the limited situation is then related to the value of k (the higher k, the higher the possible the edge effect). To reproduce satisfactorily the experimentally observed evolution at both the 500 and 25 μm diameter electrodes, a minimum value of $k = 10^{-3}$ and 5×10^{-4} cm s^{-1} respectively is necessary as shown in Fig. 7c for the 500 μm diameter electrode and in Fig. 5d (black line) for the 25 μm one. Despite its crudeness, the proposed mechanism predicts a distribution of grafting rates (higher rate at the electrode edge) in good agreement with those detected experimentally. It is not reasonable to provide a deeper analysis (determination of k_{gr}, k and k_c) or a more complex mechanistic approach as it requires too many unknown parameters.

Conclusion

The coating of gold (Au) electrode surfaces with nitrophenyl (NP) layers was studied by combination of electrochemical actuation and optical detection of the grafting process. The electrochemical actuation of the reduction of the nitrobenzenediazonium (NBD) precursor is used to generate NP radicals and therefore initiate the electrografting. This electrografting process is followed *in situ* and in real time by light reflectivity microscopy imaging, allowing for spatio-temporal observation with sub-micrometric lateral resolution and sub-nm thickness sensitivity in the case of a reflecting Au electrode coated by a transparent organic layer. The interest of an electrochemical actuation lies in its ability to finely control the electrode potential and therefore to tune the overall grafting of NP layers onto Au electrode surfaces with growth rates, v, spanning over almost 2 orders of magnitude: from 0.035 nm s^{-1} (long time and low driving force, $E_{el} = -0.3$ V vs. Ag/AgCl) to 0.36 nm s^{-1} (long time and high driving force, $E_{el} = -0.9$ V vs. Ag/AgCl) to >2 nm s^{-1} during the fast induction period (short time $E_{el} = -0.5$ V vs. Ag/AgCl). Coupling such an electrochemical actuation of a thin NP organic layer coating on a reflecting microelectrode surface to its optical detection then allowed monitoring the local grafting rates or the local fluxes of reacting NP radicals.

Over the whole range of fluxes/growth rates investigated, the NP layer deposition rate is slightly higher at the microelectrode edges (and defects) than at its centre. At first sight, this is in line with edge effects such as the generation of higher diffusion fluxes of electrogenerated reactive species at edges of a microelectrode rather than at its centre. If this has been theoretically predicted, it has been scarcely demonstrated experimentally. We believe that the observed grafting rate distribution provides another qualitative illustration of the manifestation of such edge effect phenomena. Despite the general complexity of the mechanism involved in diazonium electrografting processes, a simplified model showed how the chemical instability of the electrogenerated radical (by fast scavenging in solution or fast grafting at the electrode) is a possible source of the observed distribution of grafting rates over the electrode surface.

More generally, this work demonstrates the potentiality of combined optoelectrochemical microscopy to image surface reaction kinetics from the quantification (or indirect detection) of the local electrochemical current densities associated with these surface reactions. This is demonstrated here in the case of NP layer grafting from NBD reduction and it can be generalized to a wide range of surface transformation processes.

Acknowledgements

The "Agence Nationale de la Recherche", ANR, is gratefully acknowledged for its financial support *via* the ANR-08-JCJC0088 µECOLIERS project.

References

1. B. M. Weckhuysen, *Angew. Chem., Int. Ed.*, 2009, **48**, 4910.
2. X. Zhou, N. M. Andoy, G. Liu, E. Choudhary, K.-S. Han, H. Shen and P. Chen, *Nat. Nanotechnol.*, 2012, **7**, 237.
3. A. Meunier, O. Jouannot, R. Fulcrand, I. Fanget, M. Bretou, E. Karatekin, S. Arbault, M. Guille, F. Darchen, F. Lemaître and C. Amatore, *Angew. Chem., Int. Ed.*, 2011, **50**, 5081.
4. C. Amatore, C. Pebay, L. Servant, N. Sojic, S. Szunerits and L. Thouin, *ChemPhysChem*, 2006, **7**, 1322.

5   C. Amatore, F. Bonhomme, J. L. Bruneel, L. Servant and L. Thouin, *Electrochem. Commun.*, 2000, **2**, 235.
6   F. Deiss, N. Sojic, D. J. White and P. R. Stoddart, *Anal. Bioanal. Chem.*, 2010, **396**, 53.
7   C. L. Brosseau, F. Casadio and R. P. Van Duyne, *J. Raman Spectrosc.*, 2011, **42**, 1305.
8   D. P. dos Santos, M. L. A. Temperini and A. G. Brolo, *J. Am. Chem. Soc.*, 2012, **134**, 13492.
9   X. Shan, U. Patel, S. Wang, R. Iglesias and N. Tao, *Science*, 2010, **327**, 1363.
10 S. Szunerits, N. Knorr, R. Calemczuk and T. Livache, *Langmuir*, 2004, **20**, 9236.
11 S. Munteanu, N. Garraud, J. P. Roger, F. Amiot, J. Shi, Y. Chen, C. Combellas and F. Kanoufi, *Anal. Chem.*, 2013, **85**, 1965.
12 G. Wittstock, M. Burchardt, S. E. Pust, Y. Shen and C. Zhao, *Angew. Chem.*, 2007, **119**, 1604.
13 S. Nunige, H. Hazimeh, R. Cornut, C. Lefrou, C. Combellas and F. Kanoufi, *Angew. Chem., Int. Ed.*, 2012, **51**, 5208.
14 D. Mandler, in *Scanning Electrochemical Microscopy*, ed. A. J. Bard and M. V. Mirkin, CRC Press: Boca Raton, 2012, ch. 15, pp. 489–524.
15 H. V. Patten, K. E. Meadows, L. A. Hutton, J. G. Iacobini, D. Battistel, K. McKelvey, A. W. Colburn, M. E. Newton, J. V. Macpherson and P. R. Unwin, *Angew. Chem., Int. Ed.*, 2012, **51**, 7002–7006.
16 A. N. Patel, K. McKelvey and P. R. Unwin, *J. Am. Chem. Soc.*, 2012, **134**, 20246.
17 Q. G. Li and H. S. White, *Anal. Chem.*, 1995, **67**, 561.
18 O. Andersson, C. Ulrich, F. Björefors and B. Liedberg, *Sens. Actuators, B*, 2008, **134**, 545.
19 A. R. Perry, M. Peruffo and P. R. Unwin, *Cryst. Growth Des.*, 2013, **13**, 614.
20 R. C. Engstrom, C. M. Pharr and M. D. Koppang, *J. Electroanal. Chem.*, 1984, **221**, 251.
21 C. Amatore, A. Chovin, P. Garrigue, L. Servant, N. Sojic, S. Szunerits and L. Thouin, *Anal. Chem.*, 2004, **76**, 7202.
22 R. G. Maus, E. M. McDonald and R. M. Wightman, *Anal. Chem.*, 1999, **71**, 4944.
23 S. Szunerits, J. M. Tam, L. Thouin, C. Amatore and D. R. Walt, *Anal. Chem.*, 2003, **75**, 4382.
24 N. Garraud, Y. Fedala, F. Kanoufi, G. Tessier, J. P. Roger and F. Amiot, *Opt. Lett.*, 2011, **4**, 594.
25 M. Born and E. Wolf, in *Principles of Optics*, Pergamon Press: London, 1980.
26 P. Drude, *Ann. Phys. Chem.*, 1889, **36**, 865.
27 Edward D. Palik, *Handbook of Optical Constants of Solids*, Academic Press: Boston, 1985.
28 M. Stedmann, *Symp. Faraday Soc.*, 1970, **4**, 64.
29 A. Laforgue, T. Addou and D. Bélanger, *Langmuir*, 2005, **21**, 6855.
30 S. Chernyy, A. Bousquet, K. Torbensen, J. Iruthayaraj, M. Ceccato, S. U. Pedersen and K. Daasbjerg, *Langmuir*, 2012, **28**, 9573.
31 M. Ceccato, A. Bousquet, M. Hinge, S. U. Pedersen and K. Daasbjerg, *Chem. Mater.*, 2011, **23**, 1551.
32 A. Bousquet, M. Ceccato, M. Hinge, S. U. Pedersen and K. Daasbjerg, *Langmuir*, 2012, **28**, 1267.
33 C. Combellas, F. Kanoufi and S. Nunige, *Chem. Mater.*, 2007, **19**, 3830.
34 J. Pinson and D. Bélanger, *Chem. Soc. Rev.*, 2011, **40**, 3995.
35 C. Combellas, F. Kanoufi, J. Pinson and F. I. Podvorica, *Langmuir*, 2005, **21**, 280.
36 A. Adenier, C. Combellas, F. Kanoufi, J. Pinson and F. I. Podvorica, *Chem. Mater.*, 2006, **18**, 2021.
37 I. Bhugun and J.-M. Savéant, *J. Electroanal. Chem.*, 1995, **395**, 127.
38 H. Zollinger, *Angew. Chem., Int. Ed. Engl.*, 1978, **17**, 141 and references therein.
39 C. Galli, *Chem. Rev.*, 1988, **88**, 765.
40 J.-M. Seinberg, M. Kullapere, U. Mäeorg, F. C. Maschion, G. Maia, D. J. Schiffrin and K. Tammeveski, *J. Electroanal. Chem.*, 2008, **624**, 151.
41 I. Szele and H. Zollinger, *Helv. Chim. Acta*, 1978, **61**, 1721.
42 K. Ishida, N. Kobori, M. Kobayashi and H. Minato, *Bull. Chem. Soc. Jpn.*, 1970, **43**, 285.
43 R. Pazo-Llorente, C. Bravo-Díaz and E. Gonzalez-Romero, *Eur. J. Org. Chem.*, 2004, 3221.
44 M. Delamar, R. Hitmi, J. Pinson and J.-M. Savéant, *J. Am. Chem. Soc.*, 1992, **114**, 5883.
45 J.-M. Savéant, *Adv. Phys. Org. Chem.*, 2000, **35**, 117.
46 H. Hazimeh, F. Kanoufi, C. Combellas, J.-M. Mattalia, C. Marchi-Delapierre and M. Chanon, *J. Phys. Chem. C*, 2008, **112**, 2545.
47 C. Amatore, in *Organic Electrochemistry: An Introduction and a Guide*, ed. M. M. Baizer and H. Lund, M. Dekker: New York, 3rd edn, 1991, pp. 207–232.
48 C. P. Andrieux and J. Pinson, *J. Am. Chem. Soc.*, 2003, **125**, 14801.
49 H. Hazimeh, S. Nunige, R. Cornut, C. Lefrou, C. Combellas and F. Kanoufi, *Anal. Chem.*, 2011, **83**, 6106.
50 K. Torbensen, K. Malmos, F. Kanoufi, C. Combellas, S. U. Pedersen and K. Daasbjerg, *ChemPhysChem*, 2012, **13**, 3303.

Faraday Discussions

PAPER

Electrochemically assisted self-assembly of ordered and functionalized mesoporous silica films: impact of the electrode geometry and size on film formation and properties

Grégoire Herzog, Emilie Sibottier, Mathieu Etienne and Alain Walcarius*

Received 21st February 2013, Accepted 23rd April 2013
DOI: 10.1039/c3fd00021d

Surfactant-templated mesoporous silica thin films can be deposited onto solid electrode surfaces by electrochemically assisted self-assembly (EASA). The method involves a cathodically triggered self-assembly of cationic surfactants (cetyltrimethyl ammonium bromide, CTAB) and local pH increase leading to the polycondensation of silica precursors (i.e., tetraethoxysilane, alone or in the presence of (3-mercaptopropyl) trimethoxysilane (MPTMS)) and concomitant growth of the ordered mesoporous silica or organosilica film. The present work shows that the EASA method can be applied to film deposition on electrode supports of various morphologies, geometries and sizes (large and flat discs or non-flat streaked supports, i.e., gold CD-trodes, as well as several kinds of ultramicroelectrodes, including carbon fibers, platinum wires, and platinum microdiscs). Galvanostatic conditions were mainly preferred to potentiostatic conditions to avoid problems related to various overpotentials and surface areas experienced with the various working electrodes used here. The results indicate that film deposition was possible on each electrode support but also that both the film formation and properties were dependent on the experimental conditions for EASA. For example, passing from large electrodes to ultramicroelectrodes required the application of larger current densities to ensure film deposition, which can be due to faster loss of the hydroxyl species in solution in the case of radial or spherical diffusion, in comparison to the linear. Highly porous deposits were obtained after template removal, as ascertained by cyclic voltammetry using $Ru(NH_3)_6^{3+}$ as a redox probe. The advantage of better signal-to-background current ratios for ultramicroelectrodes relative to the macroscopic ones was maintained after film deposition, also resulting in higher sensitivity when used in conditions of preconcentration electroanalysis (using silver(I) or mercury(II) as a probe being accumulated by complexation to MPTMS-based films).

Laboratoire de Chimie Physique et Microbiologie pour l'Environnement (LCPME), UMR 7564, CNRS – Université de Lorraine, 405 rue de Vandoeuvre, Villers-lès-Nancy, F-54600, France. E-mail: alain. walcarius@univ-lorraine.fr; Fax: (+33) 3 83 27 54 44

1. Introduction

Nanostructuration of electrode surfaces *via* a bottom-up approach has become a well-established area of modern electrochemistry. Recently, the use of templates such as supramolecular assemblies, packed colloidal crystals or hard porous materials, appear to be increasingly attractive for the generation of nanosystems with ordered pore structure at the meso- or macro-scale.[1] Among the more versatile templates are the packed colloidal crystal assemblies (to prepare periodic macroporous solids by the so-called sphere templating method[2–4]) and the molecular or supramolecular aggregates (that can be used as soft templates for the generation of ordered mesoporous materials, such as silicates[5] or other metal oxides and organic–inorganic hybrids,[6,7] non oxide inorganic mesostructures,[8] or mesoporous polymers and carbons[9]). Both approaches can be also applied to the generation of ordered macro-[10] or mesoporous[11] metallic structures. Template-based ordered materials, especially those deposited as thin films on electrode surfaces, are highly promising for applications, notably in the fields of electroanalytical chemistry and sensors,[12–18] and energy conversion and storage.[18–22]

Besides the various existing chemical methods to generate such nanostructured thin films,[2–11] electrochemistry has become an attractive, and sometimes unique, means to synthesize ordered mesoporous (and also macroporous) deposits on electrode surfaces. This can be basically achieved by direct electrodeposition or indirect electro-assisted deposition of selected precursor compounds (to give metals, semiconductors, metal oxides, polymers, or sol–gel-derived materials) through soft and/or hard templates.[14,16,18,23] The soft templates are typically lyotropic liquid crystalline phases[24] (made from highly concentrated surfactants or block copolymers) or more diluted surfactant solutions containing precursors, which are likely to undergo self-assembly co-electrodeposition in the form of a nanostructured thin film.[11] The hard templates are mainly preformed mesoporous films, whose mesopores are filled with the precursors, which are then subjected to electrochemical deposition in the void volumes of the template material (formation of a structural replica).[25] An alternative approach is electrodeposition through packed colloidal crystal assemblies,[14] which was mainly applied to the generation of ordered macroporous deposits rather than the mesoporous ones.[24] One can distinguish three mechanisms that have been exploited to electrogenerate mesoporous thin films around supramolecular assemblies or through hard templates: (1) the direct electrodeposition of metals, which is usually performed by electro-reduction of metal cations in solution; (2) the indirect precipitation of metal hydroxides or oxides by electrogenerated hydroxide ions; (3) sol–gel film deposition by polycondensation of hydrolyzed precursors (*e.g.*, $Si(OH)_4$) catalyzed by electrogenerated hydroxide ions.[18]

An interesting breakthrough in the field is the discovery that ordered mesoporous metal, metal oxide or metal hydroxide thin films can be generated by electrodeposition from dilute surfactant solutions containing the appropriate inorganic/metal precursors according to a cooperative templating mechanism.[26] It takes advantage of the possible assembly of ionic surfactants onto electrode surfaces under potential control (*i.e.*, electrochemical interfacial surfactant templating).[27] The driving force to incorporate a mesostructure into the inorganic films is determined by cooperative interactions between the surfactant, inorganic ions, and the working electrode during electrodeposition. The first examples were mesostructured ZnO films[28] and 2D hexagonal ordered platinum mesostructures,[29] as successfully obtained using sodium dodecyl sulfate as a template.

More recently, the electrochemical interfacial surfactant templating approach was combined with the electro-assisted deposition of sol–gel materials, to give the first example of electrogenerated highly ordered mesoposous silica films.[30] In this case, the electrogenerated species (e.g., OH^-) do not serve to precipitate a metal hydroxide but they act as catalysts to gelify a sol onto the electrode surface. Indeed, applying a cathodic potential to an electrode immersed in a hydrolyzed sol solution is likely to generate locally OH^- species at the electrode/solution interface, inducing thereby the polycondensation of the silane precursors and growth of silica films onto the electrode surface.[31–34] If operating in the presence of a cationic surfactant (i.e., cetyltrimethylammonium bromide, CTAB), one can obtain hexagonally packed mesoporous silica channels growing perpendicularly to the electrode surface, as a result of electrochemically driven cooperative self-assembly of surfactant micelles and concomitant silica formation. This new method has been called "Electro-Assisted Self-Assembly" (EASA).[35] Such vertically aligned mesopores are very difficult to obtain by other sol–gel synthesis methods (such as evaporation-induced self-assembly),[36] and often require substrate patterning,[37] magnetically induced orientation,[38] epitaxial growth,[39] or a Stobër-solution growth approach.[40] Mesoporous silica films obtained by EASA generally show granular domains, each of which is composed of hexagonally packed one-dimensional channels oriented uniquely perpendicular to the film surface,[41] with uniform thicknesses ranging typically between 50–200 nm.[35] Aminopropyl-, mercaptopropyl- or methyl-functionalized mesoporous silica films can be produced by EASA, keeping a high degree of mesostructural order up to a maximum functionalization level depending on the nature of the organic group (i.e., 10% for amino groups[42] and 50–60% for the methyl ones[43]). The films can be prepared on various conducting supports (carbon, platinum, gold, indium–tin oxide),[30,35] and even on insulating supports using higher electric fields.[44] The EASA approach was also exploited for the preparation of bimodal macro–mesoporous films[45] and the deposition of micron sized mesoporous silica spots using a scanning electrochemical microscope[46] (in a parallel way to that applied to the generation of non-templated silica deposits at a local scale[47]).

Following these pioneering approaches, the present investigation aims at showing that EASA can be basically applied to the generation of templated and ordered silica and organically modified silica films of electrode supports exhibiting non-planar topography (e.g., gold CD-trodes made from recordable CDs[48]) or variable shape and size (e.g., from macro- to ultramicroelectrodes with tubular or disk geometry) and different natures (carbon, platinum). The effect of the electrode size and shape on both the deposition process (influence of the electrochemical deposition conditions on the film formation) and the permeability properties of the resulted coatings will be discussed. Using thiol-functionalized films and silver(I) or mercury(II) as model redox probes, attempts will be made to define the most promising systems for electroanalytical applications.

2. Experimental

2.1 Chemicals, reagents and electrodes preparation

Tetraethoxysilane (TEOS, 98%, Alfa Aesar), (3-mercaptopropyl)trimethoxysilane (MPTMS, 95%, Alfa Aesar), ethanol (95–96% Merck), $NaNO_3$ (99%, Fluka), HCl

(Riedel de Haan, 1 M solution) and cetyltrimethyl ammonium bromide (CTAB, 99%, Acros) were used for film synthesis. Ruthenium hexamine chloride (Ru(NH$_3$)$_6^{3+}$, 98%, Aldrich) was used for the film permeability characterisation. Silver(I) stock solutions were prepared from silver nitrate powder (99.8%, Prolabo) while mercury(II) stock solution was a standard solution (1000 ppm in 2% HNO$_3$, VWR International). All solutions were prepared with high purity water (18.2 MΩ cm^{-1}) obtained from a Purelab Option-Q from ELGA.

Four types of electrodes (macroscopic, μwire, μdisc and Au-CD-trode) were used. Macroscopic electrodes were discs made of either Pt (Ø = 2 mm) or glassy carbon (Ø = 3 mm). μwire electrodes were made of carbon fiber (Ø = 7 μm) or Pt wire (Ø = 25 μm), which were attached to a copper wire for connection purposes using a silver epoxy (EPOTEK H2SO from Epoxy Technology USA). The epoxy was then cured at 80 °C for 16 h. The wire was inserted in a borosilicate capillary and then elongated using a home-made pipette puller. The aperture between the capillary walls and the electrode is sealed with glue (Araldite, Bostik, France). The Pt μdiscs were prepared as follows: a 1 cm long Pt wire (Ø = 25 μm) was inserted in a borosilicate capillary, which had an end sealed in a Bunsen burner flame. The capillary was melted around the Pt wire over a distance of 5 mm. A copper wire was connected to the Pt wire end using the silver epoxy. The borosilicate capillary was then polished using sandpaper of two granularities (starting with 2500 and then 4000) until a Pt microdisc was apparent. Gold CD-trodes were prepared by attaching the gold film taken from a recordable compact disc (CD) onto a polyvinyl chloride slide (PVC).[48] Each CD-trode was connected using a copper wire and a plastic slide with a disc opening of 9 mm of diameter. The gold surface was then cleaned in 65% nitric acid. The geometry of the gold electrode prepared this way was a succession of troughs (1 μm wide and 100 nm high) separated by 500 nm.

2.2 Preparation of surfactant templated thin films

The mesoporous silica films were prepared by the electrochemically assisted self-assembly (EASA) method described elsewhere.[30,35] A typical sol was prepared as follows: ethanol and 0.1 M NaNO$_3$ were mixed in 1 : 1 v/v ratio. Silanes (TEOS, or TEOS:MPTMS 9 : 1 molar ratio) and CTAB were added to the mixture. Total silane concentrations varied between 25 and 340 mM. The molar ratio of CTAB : silanes was maintained at 0.32 as it was reported to form regular mesoporous structures.[35] The pH of the sol was adjusted to 3 by addition of HCl and the sol was hydrolysed under stirring at room temperature for 2.5 h prior to use as electrodeposition medium. The sol was prepared on the day of its use for electrode modification. The electrodes were modified in the hydrolysed sol either by a potentiostatic or a galvanostatic method. A current density, j, of −0.74 mA cm^{-2} (unless stated otherwise) or a potential of −1.3 V for Au, −2.2 V for carbon and −0.9 V for Pt was applied at the working electrode with respect to a silver wire acting as pseudo-reference electrode and a gold disc as a counter electrode (note that on Pt μdiscs, a galvanostatic conditioning step (i = 30 nA, t = 5 s) was applied prior to successful deposition of the silica film). After film formation, the electrodes were treated overnight at 130 °C (to ensure good cross-linking of the silica network). Template removal was achieved using 0.1 M HCl in ethanol, as previously described.[42]

2.3 Electrochemical methods

All electrochemical experiments were done with a PGSTAT 12 or a μAutolab from Ecochemie (Metrohm, Switzerland). The bare and modified electrodes were characterised by cyclic voltammetry (CV) at 10 mV s^{-1} in a de-aerated solution of 5 mM Ru(NH$_3$)$_6^{3+}$ in 0.1 M NaNO$_3$, in order to obtain information on the film permeability properties. In these experiments, a stainless steel rod was used as a counter electrode and the reference electrode was Ag|AgCl|1 M KCl (purchased from Metrohm, Switzerland). Ag$^+$ and Hg^{II} ions were also used as probes to assess the electroanalytical properties of the modified electrodes. Ag$^+$ and Hg^{II} were detected as described in previous publications.[49,50] Briefly, the modified electrodes were immersed at open-circuit potential for 2 min (unless stated otherwise) in a solution containing a known concentration of Ag$^+$ (or Hg^{II}) in 0.1 M HNO$_3$. Detection was performed in a metal-free 0.5 M HCl solution for Ag$^+$ or in a 3 M HCl solution for (Hg^{II}). The potential was held at −0.3 V for 60 s to allow deposition of metal species on the electrode surface prior to anodic stripping by differential pulse voltammetry (experimental conditions: step potential: 2 mV, modulation amplitude: 25 mV, modulation time: 50 ms; interval time: 100 ms) from −0.3 to 0.3 V. For Ag$^+$ detection, a silver wire was used as a pseudo-reference electrode and a stainless steel rod as a counter electrode. For Hg^{II} detection, a Ag wire electrode was used as a pseudo-reference electrode and a platinum mesh as a counter electrode.

2.4 Microscopy

Scanning electron microscopy (SEM) images were obtained using a Philips XL30 microscope, while transmission electron microscopy (TEM) images were achieved with a Philips CM20 microscope at an acceleration voltage of 200 kV. Atomic force microscopy measurements were carried out at room temperature using a commercial microscope (Thermomicroscope Explorer Ecu+, Veeco Instruments SAS), using V-shaped silicon nitride tips (MLCT-EXMT-BF, Veeco Instruments) with a spring constant of 0.1 N m^{-1} (manufacturer specifications). The images were collected in contact mode with a scan size ranging from 10 to 50 μm.

3. Results and discussion

3.1 Electrochemically assisted self-assembly on non-planar electrodes

As illustrated at the top of Fig. 1, the EASA method involves the immersion of an electrode into a surfactant-containing hydrolysed sol (stable at pH 3) and the application of a cathodic potential likely to generate hydroxyl ions at the electrode/solution interface, causing thereby a local pH increase resulting in the precursors' condensation and concomitant growing of a surfactant-templated mesoporous silica film. Such an electrochemically triggered pH change inducing the sol–gel transition is now a well-established process.[31–34,51–54] In the presence of ionic surfactants, however, the applied potential not only plays the role of generating the basic catalysts (*i.e.*, hydroxyl groups), but also contributes to the self-assembly of the surfactants onto the electrode surface.[30,46] Such assemblies were claimed to induce an unique orientation of silica mesopore channels perpendicularly to the underlying support (see bottom of Fig. 1), a configuration ideal for many applications but difficult to get by other methods.[23]

A first objective of this work was to check if EASA allows the achievement of uniform silica thin films, which are also mesoporous and organised, on heterogeneous non-flat surfaces. Fig. 2 shows the modification by EASA of 3D Au structures obtained from compact discs (*i.e.*, gold CD-trodes[48]). SEM images, respectively obtained before (Fig. 2A) and after (Fig. 2B) modification of the gold surface, reveal that although the thin film is not visible, the presence of small beads of silica can be noticed, indicating that there is a modification of the surface. The 3D structure of the electrodes is still visible despite the modification process. This is even more noticeable on the AFM picture (Fig. 2C) and the cross-sectional SEM view (Fig. 2H), demonstrating that the streaked microstructure was not significantly affected after film deposition by EASA, whereas the same support covered with a similar mesoporous silica film prepared by spin-coating did not exhibit the pristine topography (it is obvious from the image depicted in Fig. 2D that the troughs are filled with silica, confirming the limitation of traditional sol–gel techniques (*e.g.* drop- and dip-coating methods) to be applied to non-flat surfaces). More overwhelming is the fact that a uniform film (80 nm thick) was deposited by EASA over the whole non-flat surface exhibiting a step height between moulds and hills of the same magnitude as the film thickness (*i.e.*, 100 nm step height between 0.5 μm wide moulds and 1.0 μm wide hills). The AFM profiles of cross-sections recorded before and after film deposition indicate that the vertical sidewalls of the CD-trode are indeed uniformly coated with a 80-nm thick film (Fig. 2E). The modified surface obtained by EASA was also analysed by TEM (Fig. 2F and 2G). On Fig. 2G, a dark area is visible, which corresponds to the sidewalls of the 3D structure. Fig. 2G shows that the film formed by EASA is highly ordered throughout the film, and the electron diffraction pattern confirms the existence of a hexagonal packing of mesopore channels over the whole surface.

3.2 Electrochemically assisted self-assembly on microdisc and microwire electrodes

The above results demonstrate that EASA is likely to deposit surfactant-templated silica films at the micron scale, suggesting that the method could be also

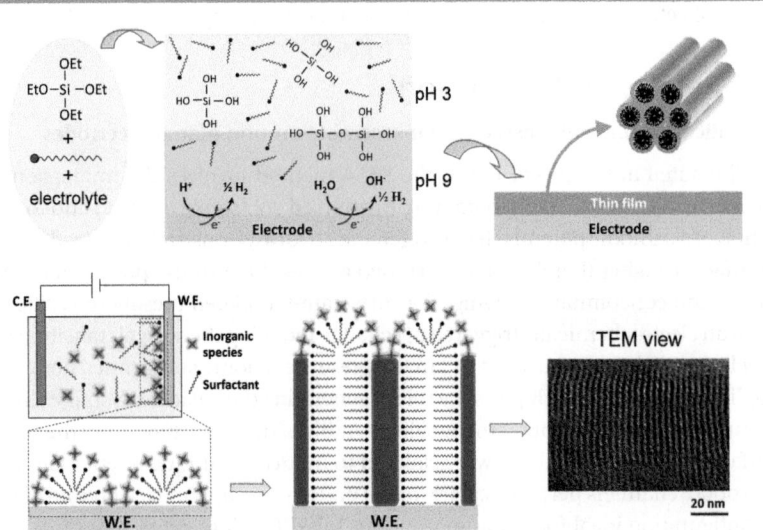

Fig. 1 Illustration of the EASA method to generate ordered and oriented mesoporous silica films onto an electrode surface.

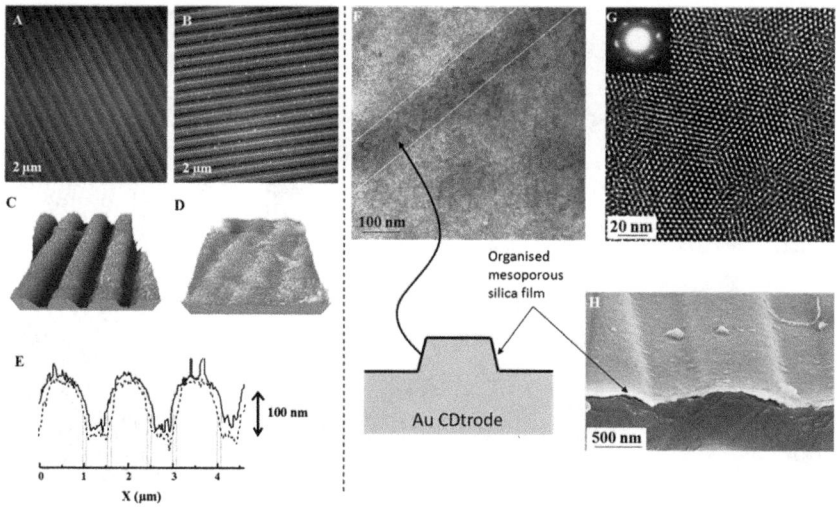

Fig. 2 Modification of Au CD-trodes by EASA. SEM images of a bare Au CD-trode (A) and the same electrode modified by EASA (−1.3 V for 5 s) (B). AFM images of Au CD-trodes coated with a mesoporous silica film prepared by EASA (C) or by spin-coating (D). AFM profiles (E) of the cross-section of Au CD-trode prior to (dotted line) or after (plain line) film deposition by EASA. TEM images (F–G) of mesoporous silica films obtained by EASA (−1.3 V for 10 s) at two different enlargements; insert of G is the corresponding electron diffraction pattern. Field emission SEM image (H) of the cross section of the same thin mesoporous silica film as in (C).

applicable to the modification of ultramicroelectrodes. The first attempts were made using microwire electrodes based on either carbon fiber (Ø = 7 μm) or Pt wire (Ø = 50 μm). Typical SEM and TEM micrographs of the modified microwires are shown in Fig. 3. The modification of the microwires is clearly visible by comparing the SEM images obtained before (Fig. 3A and 3D) and after (Fig. 3B and 3E) film deposition. It should be noted that the formation of H_2 gas bubbles was not observed during the film electrogeneration, otherwise significant macropores would have formed in the film, as reported elsewhere.[55] As shown, an important amount of silica beads and aggregates are formed on the surface of both carbon fiber and platinum wire electrodes. Nevertheless, underneath the beads/aggregates, an ordered mesoporous film is formed as is revealed by TEM (Fig. 3C and 3F). These micrographs indicate indeed the existence of hexagonally packed silica mesochannels. The sample shown in Fig. 3F was obtained after ultramicrotomy of the film and the resulting cross-sectional view demonstrates that mesopores were oriented perpendicular to the surface of the wire. Due to the different hydrogen evolution on carbon and platinum (distinct potentials for reduction of protons/water), the applied potential likely to induce film formation on these two electrodes was also different (*i.e.*, −0.9 V for Pt and −2.2 V for C). To overcome this problem, and to offer a generic way to generate mesoporous silica films on electrodes of different nature and size, galvanostatic conditions have been used on the basis of the application of an adequate cathodic current density (values determined from the previous potentiostatic experiments and corrected by the real electroactive surface area of the electrode). Note that the electroactive surface area of the microfiber electrodes was determined from CV experiments, and determined using the following equation:[56]

Fig. 3 Modification of Pt wire (A–C) and carbon fiber (D–F) with mesoporous silica thin films generated by EASA. (A, B) SEM images of a bare (A) and a modified (B) Pt wire (−0.9 V for 1 s); (C) TEM image of the corresponding film (top view). (D, E) SEM images of a bare (D) and a modified (E) carbon fiber (−2.2 V for 5 s); (F) Cross-section TEM image of the corresponding film (sample prepared by ultramicrotomy).

$$i = \frac{2nFADC}{r\ln\tau}$$

C is the concentration (in mol cm^{-3}), D the diffusion coefficient (in mol cm^{-2}), A the surface area of the electrode (in cm^2), F the Faraday constant, n the number of electrons transferred in the reaction, r the radius of the fibre (in cm). Finally, τ is defined by $\tau = 4Dt/r^2$ where t is the duration of the experiments (calculated from potential difference from the foot of the wave up to the reverse potential divided by the scan rate). Once the surface area was determined, microfiber electrodes were then modified by applying a current for a duration, which varied typically between 5 and 15 s. For instance, a current density of −3.7 × 10^{-3} A cm^{-2} gave rise to the generation of similar films as in Fig. 3 for the carbon fiber and Pt wire. The next step is the evaluation of the quality and permeability properties of the films, which can be made using CV of a redox probe in solution, with an expected absence of signal prior to surfactant removal in case of good quality films covering the entire electrode surface and significant CV responses observed after template removal due to the porous nature of the mesoporous film.[35]

Microwire electrodes made of carbon fibers modified by a mesoporous silica film generated by the galvanostatic method were first characterised by CV in 5 mM Ru(NH$_3$)$_6$$^{3+}$ (in 0.1 M NaNO$_3$, see typical curves in Fig. 4). CV curves were recorded before and after surfactant extraction from the pores. The first observation is that prior to surfactant removal, the signal was never totally/efficiently blocked by the presence of the electrodeposited film onto the electrode surface, contrary to what was observed on larger electrodes.[30,35] This suggests either an incomplete coverage of the underlying electrode with the mesoporous film (some heterogeneous growing?) or the generation of some defects/damages during the curing step (stress/strain or lattice contraction, which would be more constraining for the curved surface of a microfiber in comparison to the flat electrodes?). This second hypothesis is probably more plausible as SEM images (Fig. 3) seem to indicate full coverage of the fiber by the mesoporous film (the aggregates are not present everywhere but the thin film seems to be). The gain/increase in CV currents recorded after template removal (which was expected as a result of the mesoporosity of the film) was not so obvious (yet clearly noticeable in some cases, see right part of Fig. 4), pointing out some restricted diffusion of the probe through the film to the electrode surface. Similar results were obtained from pure silica films (*i.e.*, prepared with TEOS alone) and for the mercaptopropyl-functionalised ones (*i.e.*, prepared from a 9 : 1 TEOS–MPTMS mixture). A most intricate issue is the lack of reproducibility for both film formation and properties. Indeed, Fig. 4A and 4B represent the CV characterisation for two carbon fibers using the same film formation conditions. It is clear that the same experimental conditions lead to different electrochemical behaviours, which is basically unacceptable for practical applications. These film imperfections could be partly due to the rather poor adhesion of the silica material to the carbon support and/or to the curved shape of the fiber surface which could induce additional stress during post-treatment heating. In attempting to circumvent some of these limitations, platinum electrodes (macro- and ultramicroelectrodes) were used in the following.

Three Pt electrodes with three distinct geometries and sizes (macrodisc, microwire and microdisc) were used as electrode substrates for mesoporous silica film deposition and CV was then applied to characterise their permeability properties using $Ru(NH_3)_6^{3+}$ as redox probe (Fig. 5). It should be first reminded that the bare electrodes already behave differently as they are characterised by the distinct mass transport behaviours that are observed at these electrodes: linear diffusion of species was achieved at a Pt macrodisc electrode, whereas spherical diffusion was operating at a Pt microdisc electrode and cylindrical diffusion at a Pt microwire. As expected, the highest current density was observed for the CV of $Ru(NH_3)_6^{3+}$ at a microdisc, its limiting current density being 15 times greater than

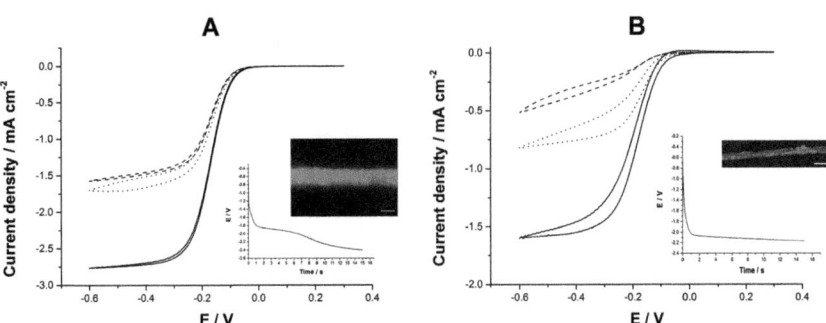

Fig. 4 Electrochemical characterisation of two carbon fibers (A–B) coated with a mercaptopropyl-functionalized mesoporous silica film obtained by galvanostatic deposition ($j = -3.69 \times 10^{-3}$ A cm^{-2}, $t = 15$ s; [TEOS + MPTMS] = 340 mM). CV of 5 mM $Ru(NH_3)_6^{3+}$ recorded in a 0.1 M NaNO$_3$ solution at a bare carbon fiber (solid line), and at the modified carbon fiber before (dash line) and after (dotted line) extraction of the surfactant. $\nu = 10$ mV s^{-1}. Inset: variation of the electrode potential during the galvanostatic modification of the carbon fibers and optical microscope image of the fibers after modification. Scale bars represent 10 μm (A) or 50 μm (B).

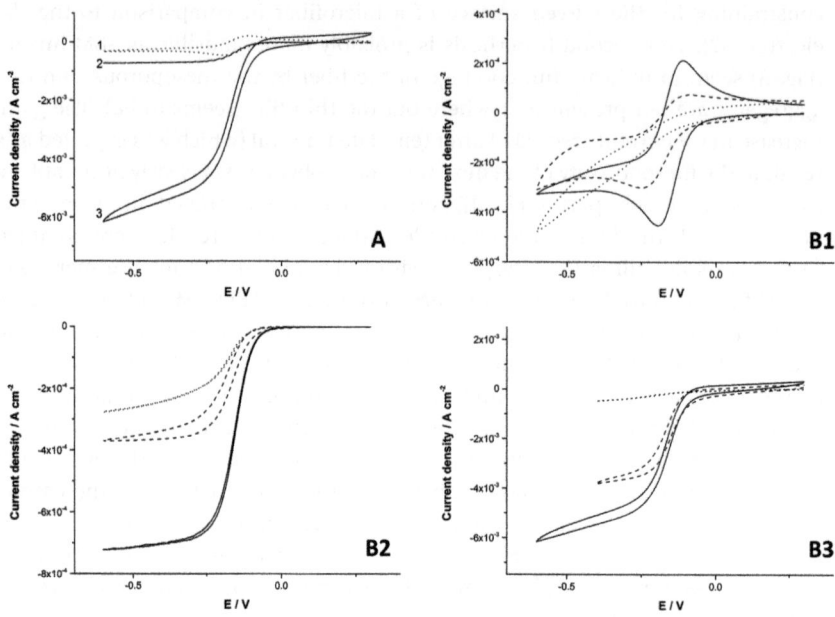

Fig. 5 (A) Comparison of the current densities obtained at (1) a Pt macrodisc (Ø = 0.2 cm), (2) a Pt wire (Ø = 25 × 10^{-4} cm) and (3) a Pt microdisc (Ø = 25 × 10^{-4} cm) by CV in 5 mM Ru(NH$_3$)$_6$$^{3+}$, 0.1 M NaNO$_3$. All electrodes are bare (unmodified). v = 10 mV s^{-1}. (B) CV of 5 mM Ru(NH$_3$)$_6$$^{3+}$ (in a 0.1 M NaNO$_3$ solution) before modification (solid line), after modification with a mercaptopropyl-functionalized mesoporous silica film obtained by EASA, respectively before (dash line) and after extraction of the surfactant (dotted line), at (B1) a Pt macrodisc (j = −0.74 × 10^{-3} A cm^{-2}, t = 15 s; [TEOS + MPTMS] = 340 mM), (B2) a Pt microwire (j = −1.5 × 10^{-3} A cm^{-2}, t = 15 s; [TEOS + MPTMS] = 110 mM) and (B3) a Pt microdisc (j = −5.0 × 10^{-3} A cm^{-2}, t = 15 s; [TEOS + MPTMS] = 110 mM). v = 10 mV s^{-1}.

the peak current density recorded at the macrodisc and 7.5 times greater than the limiting current density at the Pt microwire (Fig. 5A). All three kinds of electrodes were modified by EASA and characterised by CV. As above (*i.e.*, for carbon microfibers), CV curves for Ru(NH$_3$)$_6$$^{3+}$ were recorded at the bare electrodes and after film deposition, before and after extraction of the surfactant (Fig. 5B). At both the macro- and microdisc, the presence of the film before surfactant removal results in an almost complete blocking of the electrode surface by inhibiting drastically any electron transfer between the species in solution and the electrode (Fig. 5B1 and 5B3), confirming complete coverage of the electrodes surface with a defect-free (or few defects) templated mesoporous silica film. By comparison, and similarly to what was observed for carbon fibers (Fig. 4), the film deposited onto the Pt microwire was not characterised by such a blocking behaviour (Fig. 5B2), highlighting again the difficulty in completely covering a microelectrode with such geometry with a mesoporous silica film while maintaining a homogeneous and adherent crack-free coating. Extraction of the surfactant from the film re-establishes the possibility for the probe to cross the mesoporous film and the signal for Ru(NH$_3$)$_6$$^{3+}$ almost reaches the value that was achieved at the bare electrode. Again, the electrochemical response was more sensitive when using the microdisc electrode (larger current densities), suggesting a possible interest for electroanalytical purposes with enhanced performance (see preliminary observations in section 3.3).

Another point that should be mentioned and discussed is the fact that larger current densities must be applied when passing from the Pt macrodisc to a microwire and to a microdisc (see caption in Fig. 5) in order to observe the formation of the mesoporous silica film onto the electrode surface. Indeed, for instance, applying a current density of -0.74×10^{-3} A cm^{-2} (which is a typical value leading to good electro-assisted deposition on the macroscopic electrode) did not result in any film deposition on the Pt microdisc. Keeping in mind that the applied current induces the generation of the sol–gel polycondensation catalyst (OH$^-$), it seems that the diffusion regime (linear, cylindrical or spherical), which controls the OH$^-$ gradient at the electrode/solution interface, also affects the polycondensation rates of the sol–gel precursors. A plausible explanation is a faster "loss" of OH$^-$ species in the bulk of the solution in thecase of spherical diffusion with respect to the linear one, requiring thereby the application of higher current densities (*i.e.*, to generate more OH$^-$ species per unit area) or increasing the deposition time (to counterbalance the slower polycondensation kinetics) to ensure effective polycondensation catalysis and concomitant film deposition.

A more detailed investigation of the factors affecting the deposition process and film permeability properties has thus been performed on the Pt microdisc. As for classical electro-assisted deposition of sol–gel materials,[31,34,57,58] the EASA process is expected to be mainly affected by three parameters: the deposition time, the applied potential (for potentiostatic mode) or current (for galvanostatic mode), and the concentration of silane precursors.[35] Fig. 6 illustrates the effect of deposition time (A) and deposition current density (B) on the mesoporous silica film quality, *via* the variations of limiting currents observed before template removal (significant/high current values would indicate poor quality of the film because the probe (Ru(NH$_3$)$_6^{3+}$) can reach the underlying electrode surface through defects/holes, whereas low current values would indicate good coverage of the electrode surface by a film covering the whole surface). The results clearly point out that sufficiently long deposition times ($t_{dep} > 14$–15 s) and current densities large enough ($j_{dep} \geq 5 \times 10^{-3}$ A cm^{-2}) are required to get thin films likely to block the access of the redox probe to the electrode surface. Actually, one can distinguish three limiting cases (Fig. 6C): (C1) too short deposition time and/or too low current density leading to an important CV signal before surfactant extraction (film incompletely covering the Pt surface) and no real enhancement of the limiting current after template removal; (C2) good film quality at intermediate t_{dep} and j_{dep} values resulting in complete suppression of the CV signal before surfactant extraction and good permeability (high limiting current) after template removal; (C3) longer t_{dep} and/or larger j_{dep} values giving rise to rather good electrode coverage with the mesoporous film which is however much thicker, inducing thereby restricted diffusion of Ru(NH$_3$)$_6^{3+}$ species across the film to the underlying electrode surface (see lower currents on the dotted curve in part C3 in comparison to C2 in Fig. 6). The optimal ranges (leading to films covering completely the electrode surface and remaining highly open after template removal) for t_{dep} and j_{dep} values are respectively between 14 and 18 s and between -5 and -8 mA cm^{-2}, for electro-assisted self-assembly of mesoporous silica films on a Pt microdisc. Finally, the influence of the precursors (90 : 10 TEOS–MPTMS) concentration was studied in the range between 25 and 340 mM (using $t_{dep} = 15$ s and $j_{dep} = -6$ mA cm^{-2}). The results (not shown) indicate that mesoporous films covering well the whole electrode surface area (*i.e.*, resulting

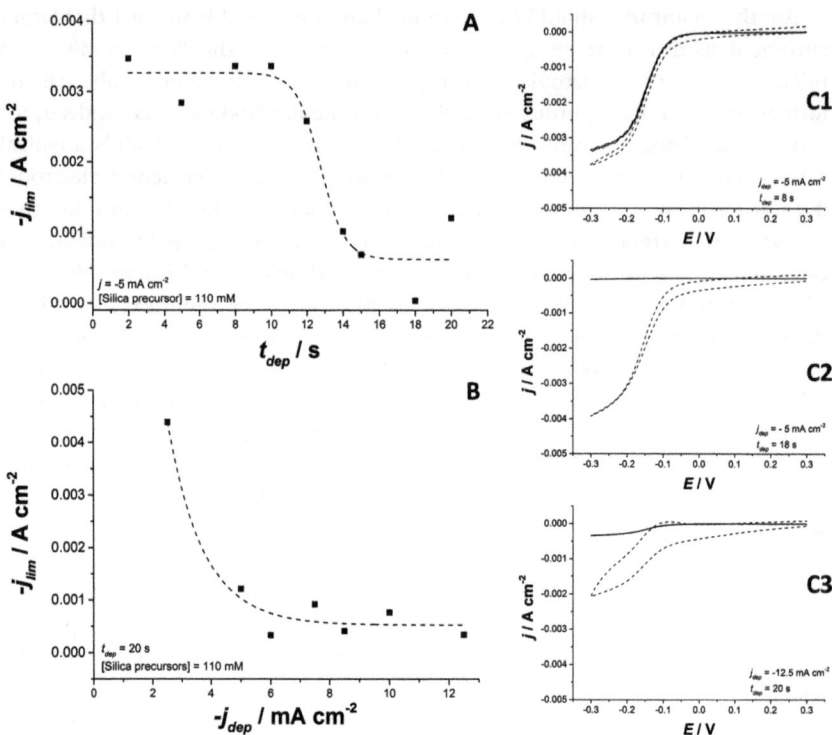

Fig. 6 Influence of (A) the deposition time (t_{dep}) and (B) the deposition current density (j_{dep}) on the limiting current density (j_{lim}) measured by CV (in 5 mM Ru(NH$_3$)$_6^{3+}$, 0.1 M NaNO$_3$; $v = 10$ mV s^{-1}) before extraction of the surfactant from mercaptopropyl-functionalized mesoporous silica film ([TEOS + MPTMS] = 110 mM) generated by EASA on Pt microdisc electrodes. (C1–C3) Typical CV curves obtained before (solid line) and after (dashed line) extraction of the surfactant for three characteristic films prepared in three distinctive experimental conditions (see j_{dep} and t_{dep} as inserts).

in mostly (>90%) blocked CV signals prior to template removal) can be obtained in the 50 to 200 mM silane precursor concentration range. Below 50 mM, the amount of deposited material was not enough to cover the whole electrode surface area, while too high precursor concentrations (>200 mM) resulted in much thicker deposits which are known to induce significant cracks upon drying, as already observed for mesoporous silica films generated by EASA on macroscopic electrodes.[30] In conclusion, the optimal conditions for film preparation by EASA on Pt microdiscs are: deposition times in the range 14–18 s, applied current densities between −5 and −8 mA cm^{-2}, and precursor concentrations ranging between 50 and 200 mM.

3.3 Functionalised films for accumulation – detection by stripping voltammetry

In order to check the potential interest of depositing mesoporous silica films on ultramicroelectrodes instead of macroelectrodes for electroanalysis purposes, some preliminary experiments have been performed using model analytes (silver or mercury ions) that have been accumulated at mercaptopropyl-functionalized

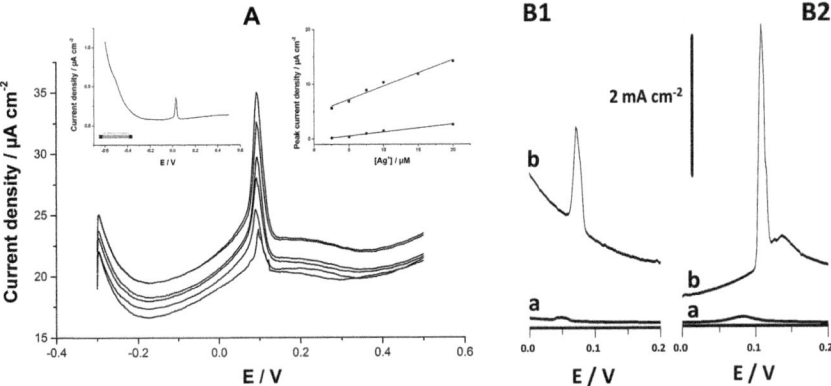

Fig. 7 (A) Differential pulse anodic stripping voltammetry (DPASV) of Ag^+ ([Ag^+] ranging from 2.5 to 20 μM) after open-circuit accumulation (for 2 min) at a modified carbon fiber modified with a mercaptopropyl-functionalized mesoporous silica film obtained by EASA ($j = -3.69 \times 10^{-3}$ A cm^{-2}, $t = 15$ s). Left inset: DPASV of 20 μM Ag^+ at a modified glassy carbon macrodisc (Ø = 0.3 cm); Right inset: calibration curves for Ag^+ obtained by DPASV at a modified carbon fiber (■) and glassy carbon macrodisc (●). (B) DPASV of Ag^+ (B1) or Hg^{II} (B2) at a modified (EASA conditions as in Fig. 5) Pt macrodisc (a) or Pt microdisc (b). [Ag^+] 15 μM or [Hg^{II}] = 10 μM; $t_{acc} = 7$ min (Ag^+) or 2 min (Hg^{II}); detection medium: 0.5 M HCl for Ag^+ and 3 M HCl for Hg^{II}; other conditions as in part (A) of the Figure.

mesoporous silica film obtained by EASA prior to their voltammetric detection. The first attempts were made using modified carbon fiber electrodes for the detection of Ag^+ after open-circuit accumulation. The presence of 10% MPTMS in the mesoporous silica thin film was responsible for the accumulation of Ag^+ ions by complexation to the thiol binding sites.[49] The modified electrode was then transferred to a silver-free detection solution (0.5 M HCl) and silver species were detected by anodic stripping voltammetry in the differential pulse mode (DPASV). A cathodic potential was applied to reduce the Ag^+ ions to form metallic Ag which was then stripped from the surface of the electrode. Typical DPASV curves are shown in Fig. 7A for increasing concentrations of Ag^+ in the accumulation medium. The stripping peak currents increased linearly with concentrations of Ag^+ between 2.5 and 20 μM at both modified macroscopic glassy carbon and carbon fiber electrodes (inset of Fig. 7A). The carbon fiber electrode is clearly more sensitive than the macroscopic one as the slope of the calibration curve is 3.4 times steeper than the one obtained at the modified glassy carbon electrode. It should be noted that the accumulation step was done under stirring for the macroscopic electrode, whereas the solution was still for the accumulation step using modified carbon fiber electrodes. Under cylindrical diffusion at the carbon fiber electrodes, accumulation of Ag^+ ions is more efficient, even in the absence of stirring. The second example concerns the preconcentration electroanalysis of either Ag^+ or Hg^{II} at Pt macro- or microelectrodes covered with a similar mercaptopropyl-functionalized mesoporous silica film obtained by EASA (Fig. 7B). Again, in both cases, a higher stripping peak current density was obtained at the microelectrodes in comparison to the macroelectrodes. This advantage is attributed to the spherical diffusion regime at the ultramicroelectrodes, which is maintained after modification of the electrode surface thanks to the fast mass transport ensured by the highly porous organosilica film.

4. Conclusions

In this work, we have shown that the EASA method can be basically applied to the generation of mesoporous silica and organosilica films on electrodes of various topographies, shapes and sizes. The experimental conditions (deposition time and applied potential/current) have to be carefully adapted to ensure the formation of defect-free films covering the whole electrode surface as well as to maintain good permeability properties after template removal. Preliminary experiments in preconcentration electroanalysis suggest a significant increase in sensitivity when passing from macroelectrodes to ultramicroelectrodes coated with functionalized mesoporous silica thin films.

Acknowledgements

The authors are grateful to Mr Sébastien Gourhand (LCPME, CNRS – Université de Lorraine) for the preparation of carbon fiber and Pt wire electrodes, to Ms Marie-Anne François (Université de Lorraine) for her help with some preliminary experimental works and to Dr Jaafar Ghanbaja (Institut Jean Lamour – Université de Lorraine) for the SEM and TEM images.

Notes and references

1. M. Etienne and A. Walcarius, in *Electrochemistry Volume 11 – Nanosystems Electrochemistry*, Edited by R. Compton and J. Wadhawan, The Royal Society of Chemistry, RSC Publishing, 2013, pp. 124–197.
2. A. Stein, *Microporous Mesoporous Mater.*, 2001, **44–45**, 227.
3. J. C. Lytle and A. Stein, *Annu. Rev. Nano Res.*, 2006, **1**, 1.
4. A. Stein, F. Li and N. R. Denny, *Chem. Mater.*, 2008, **20**, 649.
5. Y. Wan and D. Zhao, *Chem. Rev.*, 2007, **107**, 2821.
6. F. Hoffmann, M. Cornelius, J. Morell and M. Froeba, *Angew. Chem., Int. Ed.*, 2006, **45**, 3216.
7. C. Sanchez, C. Boissiere, D. Grosso, C. Laberty and L. Nicole, *Chem. Mater.*, 2008, **20**, 682.
8. Y. Shi, Y. Wan and D. Zhao, *Chem. Soc. Rev.*, 2011, **40**, 3854.
9. Y. Wan, Y. Shi and D. Zhao, *Chem. Mater.*, 2008, **20**, 932.
10. L. Lu and A. Eychmueller, *Acc. Chem. Res.*, 2008, **41**, 244.
11. Y. Yamauchi and K. Kuroda, *Chem.–Asian J.*, 2008, **3**, 664.
12. A. Walcarius, D. Mandler, J. Cox, M. M. Collinson and O. Lev, *J. Mater. Chem.*, 2005, **15**, 3663.
13. A. Walcarius, *Electroanalysis*, 2008, **20**, 711.
14. A. Walcarius and A. Kuhn, *TrAC, Trends Anal. Chem.*, 2008, **27**, 593.
15. A. Walcarius and M. M. Collinson, *Annu. Rev. Anal. Chem.*, 2009, **2**, 121.
16. A. Walcarius, *Anal. Bioanal. Chem.*, 2009, **396**, 261.
17. A. Walcarius, *TrAC, Trends Anal. Chem.*, 2012, **38**, 79.
18. A. Walcarius, *Chem. Soc. Rev.*, 2013, **42**, 4098.
19. L. M. Peter, *Phys. Chem. Chem. Phys.*, 2007, **9**, 2630.
20. H. Chang, S. H. Joo and C. Pak, *J. Mater. Chem.*, 2007, **17**, 3078.
21. A. Reynal and E. Palomares, *Eur. J. Inorg. Chem.*, 2011, 4509.
22. H. Jiang, J. Ma and C. Li, *Adv. Mater.*, 2012, **24**, 4197.
23. M. Etienne, Y. Guillemin, D. Grosso and A. Walcarius, *Anal. Bioanal. Chem.*, 2012, **405**, 1497.
24. P. N. Bartlett, *Electrochem. Soc. Interface*, 2004, **13**, 28.
25. C. W. Wu, Y. Yamauchi, T. Ohsuna and K. Kuroda, *J. Mater. Chem.*, 2006, **16**, 3091.
26. K.-S. Choi and E. M. P. Steinmiller, *Electrochim. Acta*, 2008, **53**, 6953.
27. M. Chen, I. Burgess and J. Lipkowski, *Surf. Sci.*, 2009, **603**, 1878.
28. K.-S. Choi, H. C. Lichtenegger and G. D. Stucky, *J. Am. Chem. Soc.*, 2002, **124**, 12402.
29. K. S. Choi, E. W. McFarland and G. D. Stucky, *Adv. Mater.*, 2003, **15**, 2018.
30. A. Walcarius, E. Sibottier, M. Etienne and J. Ghanbaja, *Nat. Mater.*, 2007, **6**, 602.
31. R. Shacham, D. Avnir and D. Mandler, *Adv. Mater.*, 1999, **11**, 384.

32 M. M. Collinson, N. Moore, P. N. Deepa and M. Kanungo, *Langmuir*, 2003, **19**, 7669.
33 S. Sayen and A. Walcarius, *Electrochem. Commun.*, 2003, **5**, 341.
34 E. Sibottier, S. Sayen, F. Gaboriaud and A. Walcarius, *Langmuir*, 2006, **22**, 8366.
35 A. Goux, M. Etienne, E. Aubert, C. Lecomte, J. Ghanbaja and A. Walcarius, *Chem. Mater.*, 2009, **21**, 731.
36 C. J. Brinker and D. R. Dunphy, *Curr. Opin. Colloid Interface Sci.*, 2006, **11**, 126.
37 M. Dutreilh-Colas, M. Yan, P. Labrot, N. Delorme, A. Gibaud and J.-F. Bardeau, *Surf. Sci.*, 2008, **602**, 829.
38 Y. Yamauchi, M. Sawada, M. Komatsu, A. Sugiyama, T. Osaka, N. Hirota, Y. Sakka and K. Kuroda, *Chem.-Asian J.*, 2007, **2**, 1505.
39 E. K. Richman, T. Brezesinski and S. H. Tolbert, *Nat. Mater.*, 2008, **7**, 712.
40 Z. Teng, G. Zheng, Y. Dou, W. Li, C.-Y. Mou, X. Zhang, A. M. Asiri and D. Zhao, *Angew. Chem., Int. Ed.*, 2012, **51**, 2173.
41 F.-F. Xu, F.-M. Cui, M.-L. Ruan, L.-L. Zhang and J.-L. Shi, *Langmuir*, 2010, **26**, 7535.
42 M. Etienne, A. Goux, E. Sibottier and A. Walcarius, *J. Nanosci. Nanotechnol.*, 2009, **9**, 2398.
43 Y. Guillemin, M. Etienne, E. Aubert and A. Walcarius, *J. Mater. Chem.*, 2010, **20**, 6799.
44 X. Wang, R. Xiong and G. Wei, *Surf. Coat. Technol.*, 2010, **204**, 2187.
45 M. Etienne, S. Sallard, M. Schröder, Y. Guillemin, S. Mascotto, B. M. Smarsly and A. Walcarius, *Chem. Mater.*, 2010, **22**, 3426.
46 Y. Guillemin, M. Etienne, E. Sibottier and A. Walcarius, *Chem. Mater.*, 2011, **23**, 5313.
47 L. Liu, R. Toledano, T. Danieli, J.-Q. Zhang, J.-M. Hu and D. Mandler, *Chem. Commun.*, 2011, **47**, 6909.
48 L. Angnes, E. M. Richter, M. A. Augelli and G. H. Kume, *Anal. Chem.*, 2000, **72**, 5503.
49 M. Etienne, J. Cortot and A. Walcarius, *Electroanalysis*, 2007, **19**, 129.
50 M. Etienne and A. Walcarius, *Electrochem. Commun.*, 2005, **7**, 1449.
51 P. N. Deepa, M. Kanungo, G. Claycomb, P. M. A. Sherwood, M. M. Collinson and N. Moore, *Anal. Chem.*, 2003, **75**, 5399.
52 M. Sheffer, A. Groysman, D. Starosvetsky, N. Savchenko and D. Mandler, *Corros. Sci.*, 2004, **46**, 2975.
53 R. Shacham, D. Mandler and D. Avnir, *Chem.-Eur. J.*, 2004, **10**, 1936.
54 R. Shacham, D. Mandler and D. Avnir, *J. Sol-Gel Sci. Technol.*, 2004, **31**, 329.
55 S. Yang, W.-Z. Jia, Q.-Y. Qian, Y.-G. Zhou and X.-H. Xia, *Anal. Chem.*, 2009, **81**, 3478.
56 A. J. Bard and L. R. Faulkner, *Electrochemical Methods: Fundamentals and Applications*, John Wiley and Sons, 2001.
57 L. Shapiro, S. Marx and D. Mandler, *Thin Solid Films*, 2007, **515**, 4624.
58 F. Qu, R. Nasraoui, M. Etienne, Y. Bon Saint Côme, A. Kuhn, J. Lenz, J. Gajdzik, R. Hempelmann and A. Walcarius, *Electrochem. Commun.*, 2011, **13**, 138.

Faraday Discussions

Metallic impurities availability in reduced graphene is greatly enhanced by its ultrasonication

Rou Jun Toh and Martin Pumera*

Received 17th January 2013, Accepted 8th March 2013
DOI: 10.1039/c3fd00005b

Ultrasonication is an inherent part of the major routes for preparation of reduced graphene. It is used to exfoliate graphite oxide to graphene oxide with consequent reduction to reduced graphenes. Metallic impurities in graphenes, originating from the starting material, graphite, have a profound influence on many properties of graphene, such as the electrochemical, catalytic and electronic properties. We show here that ultrasonication greatly enhances the redox availability of metallic impurities within reduced graphenes. Such findings will have a dramatic influence on future graphene processing methodology and applications of graphene.

Introduction

Graphene is an aromatic two-dimensional sheet of sp^2 bonded carbon atoms. It belongs to a class of materials exhibiting high electrical conductivity, structural flexibility, high mechanical strength, and large surface area.[1-3] Due to the exceptional features they possess, graphenes have generated a market of high demand, as observed in sensing,[4,5] catalysis,[6,7] and energy storage applications.[8]

Production of graphene follows two routes: (i) a bottom-up approach, where the graphene sheets are prepared from hydrocarbon gas; and (ii) a top-down approach, using graphite or carbon nanotubes as starting materials. The decomposition of these higher structures forms graphene nanostructured materials. In either route, the graphene sheets synthesized may restack due to the reformation of highly cohesive intersheet van der Waals' interactions.[9,10] Restacking of graphene sheets hampers us from fully exploiting the desirable properties of graphene. Hence, dispersion of these restacked graphene in solution to obtain single/few-layered graphene is crucial, and is commonly performed by the technique of ultrasonication.

The presence of metallic impurities in graphite has been known for a long time.[11-13] Such contamination persists in samples of graphene even after the

---

Division of Chemical & Biological Chemistry, School of Physical and Mathematical Sciences, Nanyang Technological University, 21 Nanyang Link, Singapore 637371, Singapore. E-mail: pumera@ntu.edu.sg; Fax: +65 6791-1961

synthesis process.[14,15] In particular, it is noteworthy that nickel and iron are the most abundant elements present.[12] Metallic impurities may be trapped within the graphene layers as restacking occurs, or be incorporated into graphene sheets covalently.[16,17] The availability of metallic impurities may alter dramatically the electrochemical behaviour[12,13,18] as well as the toxicity of graphenes. This is similar to carbon nanotubes (CNT), where metallic impurities were found to be responsible for most of the observed "electrocatalytic properties" as well as for their toxicity.[19-26] We have recently shown that the redox availability of metallic impurities from CNTs is greatly enhanced by ultrasonication.[27]

The ultrasonication of graphite oxide is one of the major routes to prepare graphene oxide.[28,29] It is surprising to learn that little is known and studied about how ultrasonication treatment, in the process of dispersing graphene sheets, affects the availability of these metallic impurities.

In this paper, we would like to address the following basic question: is the redox activity of metallic impurities influenced by their ultrasonication treatment? For these purposes, reduced graphene oxide (RGO) derived from natural graphite was selected as a model, with nickel and iron-based metallic impurities as the focal points of our investigation.

Results and discussion

The effects of ultrasonication on the availability of metallic impurities were monitored over a range between 0 and 180 min treatment. Specifically, we investigated if the ultrasonication process influences two major parameters: (1) redox activity of metallic impurities towards the oxidation of L-glutathione; (2) redox activity of metallic impurities towards the reduction of cumene hydroperoxide. The same parameters were considered for the mechanical shaking treatment as a control experiment. It was found that there was a major influence of ultrasonication on the redox availability of metallic impurities.

For the first point, cyclic voltammetry (CV) was used to investigate the oxidation of L-glutathione in the presence of graphene. Present in all living cells at a concentration of approximately 5 mM, and in lower concentrations in the blood, glutathione functions as a vital regulatory molecule to protect cells from oxidative stress, acts as a cofactor for a variety of enzymes, and governs higher-order cell systems, such as redox regulatory proteins.[30-33] The redox properties of L-glutathione are affected by the presence of nickel-oxide impurities, which catalyse its oxidation.[34] The extent to which the redox chemistry of L-glutathione is affected by the presence of nickel-oxide impurities in graphenes may represent the toxicological consequences of graphene materials.

L-Glutathione exhibited an oxidation onset potential of +0.35 V on a reduced graphene oxide (RGO) electrode and showed an oxidation peak at about +0.5 V (Fig. 1A). This oxidation peak potential is similar to that of +0.6 V obtained on a NiO_x electrode (Fig. 2). We may hence relate the oxidation peaks obtained to the oxidation of L-glutathione on nickel-oxide impurities in graphenes. The oxidation peak height of L-glutathione was studied in comparison to the different durations of ultrasonication treatment of the RGO dispersion. Fig. 1B shows the trend line of the average oxidation peak height obtained at different durations of ultrasonication or mechanical shaking treatment. It is observed that the oxidation peak height increases steeply from 0 to 45 min of ultrasonication treatment, to a

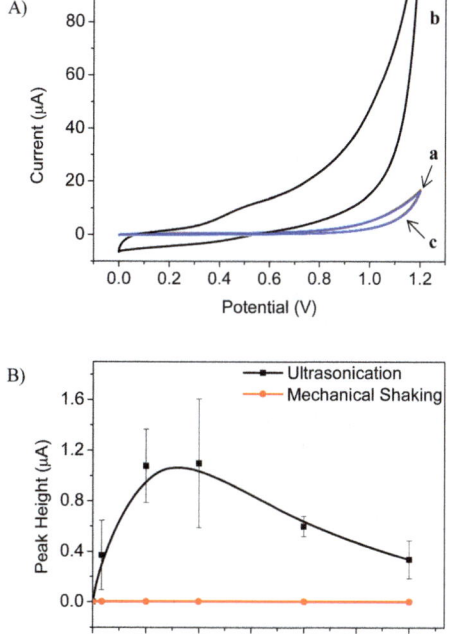

Fig. 1 (A) Cyclic voltammograms resulting from electrochemical oxidation of L-glutathione (5 mM) at GC electrodes modified with RGO after 0 min treatment (a, red line), 120 min ultrasonication (b, black line), and mechanical shaking (c, blue line). (B) Oxidation peak height of cyclic voltammograms at different ultrasonication (black line) and mechanical shaking (red line) time. Error bars represent RSD ($n = 3$; 95% confidence interval).

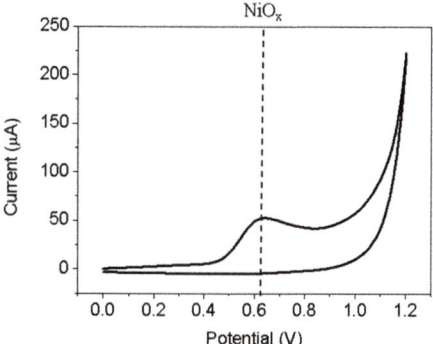

Fig. 2 Cyclic voltammogram resulting from electrochemical oxidation of L-glutathione (5 mM) at GC electrodes modified with NiO_x nanoparticles.

current value of 1.05 µA. It is interesting to note that after this, the oxidation peak height follows a gradual decrease as the duration of ultrasonication treatment increases. From the trend line, it can be observed that such an effect was not evident with mechanical shaking treatment. An inference may be made that ultrasonication has a significant effect on the redox availability of nickel-oxide impurities in reduced graphene oxides.

In order to verify the relationship between the redox availability of metallic impurities within graphene and duration of ultrasonication treatment, we further examined the electrochemical reduction of cumene hydroperoxide in the presence of graphene. It is important to note that cumene hydroperoxide is electrocatalytically reduced only on an iron-based electrocatalyst.[35]

It can be observed from Fig. 3A that 0 min treatment of the RGO dispersion gave a reduction peak starting at −0.6 V and peaks at −0.75 V. This reduction peak of cumene hydroperoxide exhibits a potential similar to that of −0.85 V on bare glassy carbon electrode (Fig. 4A). Hence, it is likely that the reductive peak obtained with 0 min treatment is associated with reduction of cumene hydroperoxide on a bare glassy carbon electrode. The distinct reduction peak of cumene hydroperoxide at −0.75 V was also observed for both 120 min ultrasonication and mechanical shaking treatment. Moreover, an additional reduction peak with an onset potential of ∼0.0 V, and peak potential of about −0.4 V, was produced with an 120 min ultrasonication treatment. This peak may be attributed to the reduction of cumene hydroperoxide on Fe_3O_4 which could be observed at a comparable peak potential of −0.55 V (Fig. 4B).

The reductive peak height at −0.4 V, attributed to iron-based impurities, was monitored in relation to different durations of ultrasonication and mechanical

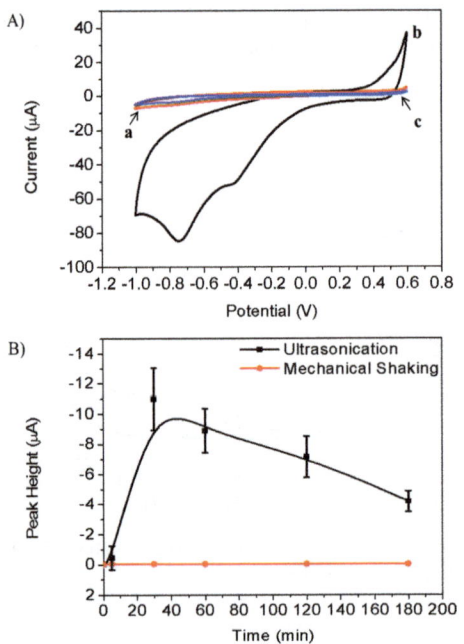

Fig. 3 (A) Cyclic voltammograms resulting from electrochemical reduction of cumene hydroperoxide (10 mM) at GC electrodes modified with RGO after 0 min treatment (a, red line), 120 min ultrasonication (b, black line), and mechanical shaking (c, blue line). (B) Reduction peak height, attributed to iron-based impurities (−0.4 V), of cyclic voltammograms at different ultrasonication (black line) and mechanical shaking (red line) time. Error bars represent RSD ($n = 3$; 95% confidence interval).

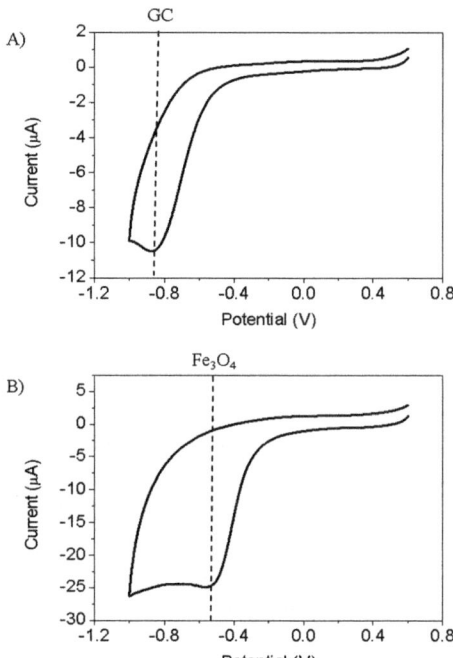

Fig. 4 (A) Cyclic voltammograms resulting from electrochemical oxidation of cumene hydroperoxide (10 mM) at bare GC electrodes. (B) Cyclic voltammograms resulting from electrochemical oxidation of cumene hydroperoxide (10 mM) at GC electrodes modified with Fe_3O_4 nanoparticles.

shaking treatments of the RGO dispersion, and a trend line was constructed (Fig. 3B). A steep increase in reduction peak height to a maximum of -9.5 μA is observed after 0 to 40 min of ultrasonication treatment. This is followed by a gradual decline in reduction peak height as the duration of ultrasonication treatment increases. Once again, mechanical shaking treatment did not produce such an effect, giving zero current values. The trend line produced with the reduction of cumene hydroperoxide was consistent with that for the oxidation of L-glutathione. Hence, we can say that ultrasonication treatment has a significant effect on the redox availability of metallic impurities in RGO, in particular nickel and iron-based metallic impurities. ICP-MS measurements were carried out to find how much metal leached out of graphene. The concentration of Fe in the solution was 0.1 ppb with 0 min ultrasonication treatment. This concentration showed a twentyfold increase when the solution was subjected to an 180 min ultrasonication treatment.

Conclusion

We demonstrated that ultrasonication significantly affected the redox availability of metallic impurities within graphenes. Such is reflected in altered redox behaviours of L-glutathione and cumene hydroperoxide. A consistent trend line was observed where the redox availability of metallic impurities within graphenes peaked at 40–45 min ultrasonication treatment, followed by a gradual decrease with longer durations. These findings are of high importance for the processing of graphene with a wide impact on electrochemical, electronic and catalytic applications of this material.

Experimental

Materials

N,N-Dimethylformamide (DMF) was purchased from Merck, cumene hydroperoxide was obtained from Alfa Aesar, phosphate buffer powder (0.1 M) and L-glutathione were purchased from Sigma-Aldrich. Glassy-carbon (GC), Ag/AgCl reference and Pt counterelectrodes were obtained from Metroohm.

Apparatus

Ultrasonication was carried out at a frequency of 37 kHz, with a FB 11203 ultrasonic unit (Fisherbrand). Mechanical shaking was carried out with a TS-100 thermoshaker (BioSan), at 500rpm with rotation diameter of 2 mm. All voltammetric experiments were performed by employing a μAutolab type III electrochemical analyser (Eco Chemie, The Netherlands) connected to a computer, and controlled by General Purpose Electrochemical Systems, Version 4.9 software (Eco Chemie). Electrochemical experiments were conducted in an electrochemical cell (5 mL) at r.t. (25 °C) by employing a three-electrode setup. A platinum electrode performed as an auxiliary electrode, and an Ag/AgCl electrode served as a reference electrode.

Procedures

Graphite oxide from natural graphite (GO) was synthesized according to the modified Hummers method.[36] Graphite (0.5 g), $NaNO_3$ (0.5 g), and H_2SO_4 (23 mL) were stirred together in an ice bath. Subsequently, $KMnO_4$ (3 g) was added gradually to the mixture at 0 °C. Once mixed, the solution was heated to 35 °C and stirred for approximately 1 h. A thick paste was produced. After which, water (40 mL) was added, and the solution was stirred for 30 min while the temperature was increased to 90 °C. The temperature was kept at a constant of 90 °C for 15 min, and 100 mL of water was then added. This was followed by the slow addition of H_2O_2 (30%, 3 mL), turning the dark brown solution to a yellow colour, until no further gas evolved. The warm solution was then filtered and washed with warm water (100 mL). Via mechanical agitation, the filter cake created was re-dispersed in water. Centrifugation was done at about 6600 × g for 15 min in water until a neutral pH of the supernatant was acquired. Finally, GO powder was dried in the oven at 60 °C for 2 days before usage. The GO powder contained 529.6 ppm of Fe and 5.2 ppm of Ni, as determined by ICP-MS analysis.

Thermally reduced graphite oxide (TRGO) was obtained through thermal-reduction/exfoliation at a high temperature of 1000 degrees Celsius. TRGO was placed in a porous quartz-glass capsule which was connected to a magnetic manipulator inside a vacuum-tight tube furnace under a controlled atmosphere. The application of a magnetic manipulator allowed a temperature gradient of over 1000 degree Celsius min^{-1} to be created. By repeated evacuation of the tube furnace, the sample was flushed with inert nitrogen to remove any traces of oxygen. The sample was then swiftly inserted into a preheated furnace using a magnetic manipulator, and held in the furnace for 12 min. The flow of nitrogen

(99.9999% purity) during the exfoliation procedure was set at 1000 mL min^{-1} to remove the by-products of the exfoliation procedure.

Influence of ultrasonication was studied using a range of ultrasonication timings between 0 and 180 min. Control experiments were carried out in a similar fashion by using a mechanical shaker. GC electrode surfaces were renewed by polishing with 0.05 μm alumina particles on a polishing pad. TRGO (2 mg mL^{-1}) was first dispersed in DMF. The suspension was placed into an ultrasound bath (operating at 150 W) for various timings, and the suspension (1 μL) was then pipetted onto the GC electrode surface. The suspension was left to evaporate at RT for 30 min to produce a randomly distributed TRGO film on the GC electrode surface. Amperometry was then performed at a potential of -1.2 V for 900 s using phosphate buffer solution (50 mM, pH 7.2) to further reduce any oxygen containing functional groups which may be present on TRGO, producing highly reduced graphene oxide (RGO). CV experiments were performed at a scan rate of 100 mV s^{-1}, with phosphate buffer solution (50 mM, pH 7.2) as the supporting electrolyte. For the study of the amount of leached Fe into the solution, the SWCNTs were exposed to ultrasound for 0 or 180 min, followed by filtration to remove the solid materials. The filtrate was then analyzed by using ICP-MS.

Acknowledgements

M. P. thanks to NAP fund (NTU) and JSPS-NTU fund for financial support. We also thank R. D. Webster and B. Khezri for ICP-MS measurements. We would like to acknowledge the funding support for the project from Nanyang Technological University under the Undergraduate Research Experience on Campus (URECA) program.

References

1 M. Pumera, *Chem. Rec.*, 2012, **12**, 201.
2 K. S. Novoselov, *Angew. Chem., Int. Ed.*, 2011, **50**, 6986.
3 D. R. Dreyer, R. S. Ruoff and C. W. Bielawski, *Angew. Chem., Int. Ed.*, 2010, **49**, 9336.
4 A. Bonanni, A. H. Loo and M. Pumera, *TrAC, Trends Anal. Chem.*, 2012, **37**, 12.
5 S. Alwarappan, A. Erdem, C. Liu and C.-Z. Li, *J. Phys. Chem. C*, 2009, **113**, 8853.
6 J. Pyun, *Angew. Chem., Int. Ed.*, 2011, **50**, 46.
7 C. Huang, H. Bai, C. Li and G. Shi, *Chem. Commun.*, 2011, **47**, 4962.
8 S. Guo and S. Dong, *Chem. Soc. Rev.*, 2011, **40**, 2644.
9 R. Zacharia, H. Ulbricht and T. Hertel, *Phys. Rev. B: Condens. Matter Mater. Phys.*, 2004, **69**, 155406.
10 I. V. Lebedeva, A. A. Knizhnik, A. M. Popov, Y. E. Lozovik and B. V. Potapkin, *Phys. Chem. Chem. Phys.*, 2011, **13**, 5687.
11 E. A. Heintz and W. E. Parker, *Carbon*, 1966, **4**, 473.
12 D. W. Mckee, *Carbon*, 1974, **12**, 453.
13 Y. Koshino and A. Narukawa, *Analyst*, 1993, **118**, 827.
14 A. Ambrosi, S. Y. Chee, B. Khezri, R. D. Webster, Z. Sofer and M. Pumera, *Angew. Chem., Int. Ed.*, 2012, **51**, 500.
15 A. Ambrosi, C. K. Chua, B. Khezri, Z. Sofer, R. D. Webster and M. Pumera, *Proc. Natl. Acad. Sci. U. S. A.*, 2012, **109**, 12899.
16 M. F. Chisholm, G. Duscher and W. Windl, *Nano Lett.*, 2012, **12**, 4651.
17 W. Zhou, M. D. Kapetanakis, M. P. Prange, S. T. Pantelides, S. J. Pennycook and J.-C. Idrobo, *Phys. Rev. Lett.*, 2012, **109**, 206803.
18 S. Y. Chee and M. Pumera, *Analyst*, 2012, **137**, 2039.
19 C. E. Banks, A. Crossley, C. Salter, S. J. Wilkins and R. G. Compton, *Angew. Chem., Int. Ed.*, 2006, **45**, 2533.
20 B. Sljukic, C. E. Banks and R. G. Compton, *Nano Lett.*, 2006, **6**, 1556.

21 C. Batchelor-McAuley, G. G. Wildgoose, R. G. Compton, L. Shao and M. L. H. Green, *Sens. Actuators, B*, 2008, **132**, 356.
22 X. Dai, G. G. Wildgoose and R. G. Compton, *Analyst*, 2006, **131**, 901.
23 L. Guo, D. G. Morris, X. Liu, C. Vaslet, R. H. Hurt and A. B. Kane, *Chem. Mater.*, 2007, **19**, 3472.
24 X. Liu, V. Gurel, D. Morris, D. W. Murray, A. Zhitkovich, A. B. Kane and R. H. Hurt, *Adv. Mater.*, 2007, **19**, 2790.
25 M. Pumera, A. Ambrosi and E. L. K. Chng, *Chem. Sci.*, 2012, **3**, 3347.
26 S. Koyama, Y. A. Kim, T. Hayashi, K. Takeuchi, C. Fujii, N. Kuroiwa, H. Koyama, T. Tsukahara and M. Endo, *Carbon*, 2009, **47**, 1365.
27 R. J. Toh, A. Ambrosi and M. Pumera, *Chem.-Eur. J.*, 2012, **18**, 11593.
28 Y. Zhu, S. Murali, W. Cai, X. Li, J. W. Suk, J. R. Potts and R. S. Ruoff, *Adv. Mater.*, 2010, **22**, 3906.
29 A. Ambrosi, A. Bonanni, Z. Sofer, J. S. Cross and M. Pumera, *Chem.-Eur. J.*, 2011, **17**, 10763.
30 *Glutathione: Chemical, Biochemical, and Medical Aspects*, ed. D. Dolphin, R. Poulson and O. Avramovic, Wiley, New York, 1989.
31 A. Meister, *J. Biol. Chem.*, 1988, **263**, 17205.
32 J. Vina, *Glutathione: Metabolism and Physiological Functions*, CRC, Boca Raton, 2000.
33 O. N. Oktyabrsky, G. V. Smirnovam and N. G. Muzika, *Free Radical Biol. Med.*, 2001, **31**, 250.
34 A. Ambrosi and M. Pumera, *Chem.-Eur. J.*, 2010, **16**, 1786.
35 E. J. E. Stuart and M. Pumera, *J. Phys. Chem. C*, 2010, **114**, 21296.
36 W. S. Hummers and R. E. Offeman, *J. Am. Chem. Soc.*, 1958, **31**, 1481.

Faraday Discussions

PAPER

Highly sensitive detection of nitroaromatic explosives at discrete nanowire arrays

Sean Barry,[a] Karen Dawson,[a] Elon Correa,[b] Royston Goodacre[b] and Alan O'Riordan*[a]

Received 4th March 2013, Accepted 18th April 2013
DOI: 10.1039/c3fd00027c

We show a photolithography technique that permits gold nanowire array electrodes to be routinely fabricated at reasonable cost. Nanowire electrode arrays offer the potential for enhancements in electroanalysis such as increased signal-to-noise ratio and increased sensitivity while also allowing quantitative detection at much lower concentrations. We explore application of nanowire array electrodes to the detection of different nitroaromatic species. Characteristic reduction peaks of nitro groups are not observed at nanowire array electrodes using sweep voltammetric methods. By contrast, clear and well-defined reduction peaks are resolved using potential step square wave voltammetry. A Principle Component Analysis technique is employed to discriminate between nitroaromatic species including structural isomers of DNT. The analysis indicates that all compounds are successfully discriminated by unsupervised cluster analysis. Finally, the magnitude of the reduction peak at −671 mV for different concentrations of TNT exhibited excellent linearity with increasing concentrations enabling sub-150 ng mL^{-1} limits of detection.

1 Introduction

The dual-use molecule, 2,4,6-trinitrotoluene (TNT), and its analogues, is used within the bulk chemical industry for the manufacture of dyes, plasticisers, herbicides, *etc.* while it is also employed as ammunition and an explosive for military applications.[1] TNT is highly soluble in water (∼150 mg L^{-1})[2] and its presence in waste and ground water is an on-going concern for environmental, health and security reasons. TNT is highly toxic and has been linked with anaemia, skin irritation, and abnormal liver function in humans.[3,4] Based on animal studies, the US Environmental Protection Agency (EPA) has recently identified TNT as a potential human carcinogen as both TNT and its metabolites are toxic and genotoxic at relatively low concentrations, *i.e.*, at

[a]*Tyndall National Institute – University College Cork, Lee Maltings, Dyke Parade, Cork, Ireland. E-mail: alan. oriordan@tyndall.ie; Tel: +353 21 2346403*
[b]*School of Chemistry, Manchester Institute of Biotechnology, University of Manchester, 131 Princess Street, Manchester, M1 7DN, UK*

levels above 2 μg mL^{-1}.[5] As a result, the EPA has issued health advisories concerning its maximum allowable levels in potable and ambient waters.

Field analyses for environmental monitoring have shown the need for portable, easy-to-use, low maintenance and sensitive methods to enable rapid detection and quantification of these contaminants of interest at reasonable economic cost. A further impetus in this regard is on-going security concerns and the need to develop approaches for remote and decentralised security screening. A variety of techniques including enhanced Raman,[6] UV absorption,[7] surface plasmon resonance[8] and fluorescence[9] spectroscopies, enzyme linked immunosorbent assays,[10] solid/liquid phase extraction (pre-concentration) followed with GC- or HPLC- mass spectrometry[11-13] are commonly employed for the detection of TNT and its analogues. While all of these approaches can quantitatively detect nitroaromatic species within the detection range of 0.01–5000 ng ml^{-1}, the associated instruments are complex, have a high cost of ownership and are not ideally suited to field analysis.

Recent advances in nanofabrication have enabled the development of a new range of highly sensitive nanosensors that are compact, low power, highly sensitive and have extremely small sample volume requirements. One-dimensional nanostructure based field effect transistors (FET) employing silicon,[14] carbon nanotubes,[15] III–V semiconductors,[16] conducting polymers[17] and metal oxide nanowires[18] are an emerging class of nanosensor finding application in label-free sensing. However, the sensing mechanisms of FET devices arise from perturbations of their local electrostatic environment by charged analyte species. As such, they are limited to the detection of large and highly charged species, such as biomolecules, and are mostly insensitive to small or uncharged molecules important in environmental/security monitoring.

Unlike FETs, noble metal nanowires such as gold, platinum and silver, are important because of their excellent electrical and optical properties. Gold nanowires are of particular interest due to their high chemical and thermal stability, and excellent electrical conductivity and have tremendous potential in electroanalysis. In electrochemistry, nanoelectrodes offer a number of enhancements compared to macroelectrodes due to their many advantageous properties: low background charging; high current density due to enhanced mass transport; low depletion of target molecules; low supporting electrolyte concentrations; and faster response times.[19,20,21] Recently, highly sensitive gold nanowire-based sensors for the electrochemical detection of small molecules including hydrogen peroxide,[22,23] dopamine,[24] and glucose[25] have been reported.

Nanoscale electroanalytical approaches are particularly suited for detection of nitroaromatic species due to the inherent redox activity of these molecules. The electroreduction pathways of the nitro groups in nitroaromatics have been well documented in the literature.[26] Firstly, a nitroso intermediate followed by a hydroxylamine intermediate are formed via a 4e$^-$/4H$^+$ electron and proton transfer system. The hydroxylamine group is further reduced to an amine via a 2e$^-$/2H$^+$ electron and proton transfer system, see Scheme 1. In acidic media, each nitro group in TNT is sequentially reduced to amino groups producing three peaks at increasing negative potentials at glassy carbon electrodes. As tri-nitro compounds, including TNT, may be more readily reduced at metallic electrodes compared to mono- or dinitro-compounds,[27] in this work we develop gold

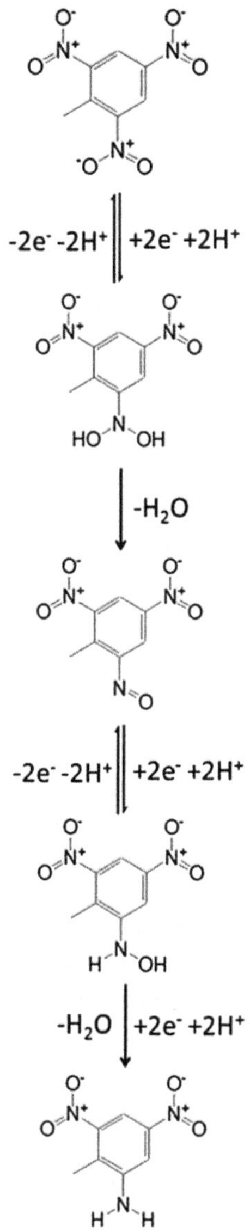

Scheme 1 Electrochemical reduction pathway of nitroaromatic species – 2,4,6-trinitrotoluene.

nanowire-based devices and explore their application to the sensitive detection of TNT and its analogues including nitrotoluene (NT), 2,4-dinitrotoluene (2,4-DNT), 2,6- dinitrotoluene (2,6-DNT) and 1,3-dinitrobenzene (1,3-DNB).

2 Experimental

Chemicals and materials

Sodium phosphate monobasic, sodium phosphate dibasic, 3-nitrotoluene (3-NT), 2,4-dinitrotoluene (2,4-DNT), 2,6-dinitrotoluene (2,6-DNT), 1,3-dinitrobenzene (1,3-DNB), 2,4,6-trinitrotoluene (TNT) and acetonitrile are purchased from

Sigma-Aldrich and used as received. A sample of military grade 2,4,6-trinitrotoluene is provided by the Ordnance School, Defence Forces Ireland. Deionized water (18.2 MΩ cm) from an ELGA Pure Lab Ultra system is used for the preparation of samples. All electrochemical measurements are undertaken in a 50 mM phosphate buffer solution (pH 6.5).

Electrode fabrication

Gold nanowire array electrodes, interconnection tracks, peripheral contact pads and two central micron scale half-disc microelectrodes are fabricated using a photolithography process on four inch wafer silicon substrates bearing a ~300 nm layer of thermally grown silicon dioxide (Si/SiO_2). In this process, metallic structures are patterned in a photoresist layer (Microposit LOR 10A, 500 nm) using an optical mask. Following resist development, gold layers (Ti/Au, 10/100 nm) are blanket deposited by metal evaporation (Temescal FC-2000 E-beam evaporator) and removed from un-patterned areas using standard lift-off techniques. A further metal deposition (Ti/Pt 10/90 nm) step is then performed onto one half-disk electrode. In this manner, one central half-disk electrode may be employed as a gold counter and the other as a platinum pseudo-reference electrode, respectively.

To prevent unwanted electrochemical reactions occurring between metal interconnection tracks and electrochemically active species, a silicon nitride passivation layer (~500 nm) is then deposited by plasma-enhanced chemical vapour deposition (PECVD) onto the wafer surface. Photolithography and dry etching are then employed to selectively open windows (~45 × 100 μm) in the passivation layer directly above the gold nanowire working electrodes to allow exclusive contact between them and an electrolytic solution. Openings are also maintained above the counter and reference electrodes, along with peripheral contact pads. Following fabrication, wafers are diced into 16 × 16 mm chips, such that each chip contains twelve individually contacted gold nanowire working electrode arrays, an integrated gold counter electrode and a platinum pseudo-reference electrode.

Optical and electrical characterisation

Optical micrographs of nanowire electrode arrays are acquired using a calibrated microscope (Axioskop II, Carl Zeiss Ltd.) equipped with a charge-coupled detector camera (CCD; DEI-750, Optronics). As a quality control check, to confirm electrical functionality, two-point electrical measurements are performed using a probe station (Model 6200, Micromanipulator Probe Station) in combination with a source meter (Keithly 2400) and a dedicated LabVIEW™ V8.0 program. In these current–voltage (I–V) measurements, the source electrode is grounded, a bias sweep up to ±10 mV is applied to the drain electrode, and the current through the nanowire is measured.

Electrochemical analysis

All electrochemical studies are performed using a CHI660a Electrochemical Analyzer and a Faraday Cage CHI200b (CH Instruments) connected to a PC. Experiments employ a standard three-electrode cell configuration using a gold nanowire arrays as the working electrodes *versus* the on-chip gold counter electrode and on-chip platinum pseudo-reference electrode. A custom holder is employed comprising a 180 μL sample volume well and spring loaded gold pins to permit electrical contact to electrodes. Cyclic voltammetry experiments (CV) are conducted in 10 mM phosphate buffer saline (PBS, pH 7.4) containing 1 mM ferrocenemonocarboxylic acid (FcCOOH, Sigma Aldrich) in the voltage range of −0.15 V to 0.45, *versus* on-chip platinum pseudo-reference to first confirm electrochemical functionality. Stock solutions of 1000 μg mL^{-1} of 3-NT, 2,4-DNT, 2,6-DNT, 1,3-DNB and TNT (military sourced) are freshly prepared in acetonitrile prior to use. Analytical TNT standards (Spectroscopic grade) are also purchased pre-prepared in ampules as 1000 μg mL^{-1} in acetonitrile (Aldrich). Stock solutions are diluted using phosphate buffer as required. Typical measurements are performed using 100 μL of solution (covering all three electrodes) without stirring or purging of dissolved oxygen with nitrogen. Concerning calibration plots, serial additions are performed by adding 5 or 10 μL aliquots (of stock solution) as appropriate and mixed by agitating the solution using within the volume ranges of 95 μL to 180 μL. All experiments are performed at room temperature.

Cyclic voltammetry (CV) is conducted in 50 mM phosphate buffer solution (PB, pH 6.5) containing 50 μg mL^{-1} of the compound of interest in the voltage range of 0 V to −1.0 V, *versus* on-chip platinum pseudo-reference using a scan rate of 40 mV s^{-1}. A blank measurement (not shown) of PB only is also recorded. Square Wave Voltammetry (SWV) is performed by sweeping the potential from −0.4 V to −1.02 V for individual species. A scan range of −0.4 V to −0.97 V is employed for serial additions of TNT. The slightly longer scan range for the individual compounds is necessitated by the cathodic potential of the 3-NT reduction to ensure a complete peak is observed. All scans are performed using a frequency of 10 Hz, amplitude of 50 mV and a potential step of 4 mV. Blank SWV are also obtained for the purpose of background subtraction.

3 Results and discussion

Gold nanowire arrays are fabricated using a photo-lithography process at Si/SiO$_2$ substrates as described in the Experimental section. Each chip contains twelve different nanowire electrode arrays along with integrated gold counter and

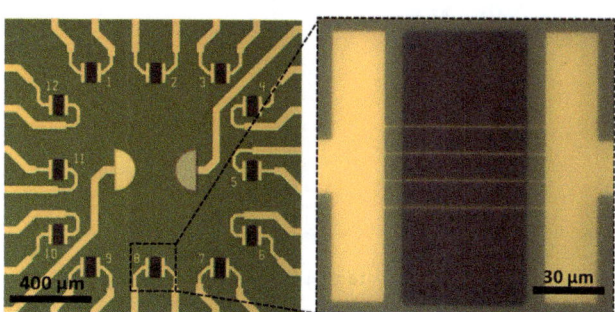

Fig. 1 (a) plan view of the central chip region which contains 12 individually addressable nanowire electrodes, a gold on-chip counter electrode (left) and a platinum on-chip pseudo reference electrode (right). (b) Higher magnification image of a nanowire array electrode comprising four nanowires.

platinum pseudo-reference electrodes and are structurally characterised using optical microscopy following fabrication. Fig. 1(a) shows an optical micrograph of the central chip region which contains the 12 individually addressable nanowire electrodes, along with the half-disc gold on-chip counter electrode (left) and platinum on-chip pseudo reference electrode (right). Fig. 1(b) optical image of a fully fabricated, integrated and passivated nanowire array device comprising four nanowires. The width of the passivation window (central dark rectangle) defines the exposed nanowire length at 45 μm. Standard two-point I–V measurements in air are undertaken as a quality control check to ensure that all nanowire electrodes are fully functional prior to electroanalysis. A voltage bias of ±10 mV is selected to confirm Ohmic behaviour. The low voltage range is chosen to avoid undesired electro-migration effects which may damage the nanowires. The electrical behaviour observed at fully fabricated single nanowires is very reproducible with all functioning nanowire array devices displaying linear Ohmic responses with low resistances (62 ± 4 Ω), see Fig. 2(a). This low variation (~6.4%) in electrical performance for discrete nanowire arrays is excellent and is consistent with that observed for nanowire dimensions. Control electrical measurements were obtained in the absence of nanowire arrays, yielding very high resistances (~10 GΩ) typical of an open circuit. This confirms that the underlying silicon oxide functioned as an effective insulating layer preventing electrical coupling with the silicon beneath and that the observed electrical characteristics were exclusively generated by the nanowire arrays. Nanowire array devices that exhibit high resistances or open circuits are discarded and not used for electrochemical analysis.

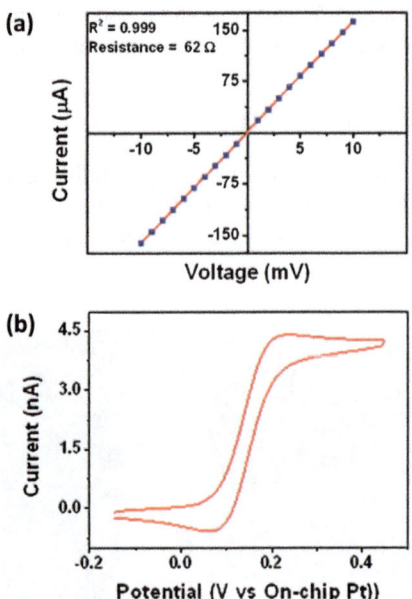

Fig. 2 Typical two point I–V characteristics measured for nanowire arrays comprising of four nanowires exhibit Ohmic behaviour and low contact resistances. Average resistance is 62 ± 4 Ω. (b) Typical voltammogram obtained for a nanowire array and in 1 mM FcCOOH in 10 mM PBS in the voltage range of −0.15 V to 0.45, pH 7.4.

Cyclic voltammetry (CV) is conducted in 10 mM phosphate buffer saline (PBS, pH 7.4) containing 1 mM FcCOOH in the voltage range of −0.15 V to 0.45, *versus* the on-chip platinum pseudo-reference. Fig. 2(b) shows a typical CV for FcCOOH recorded at 100 mV s^{-1}. CVs exhibit semi-infinite diffusive behaviour and demonstrate that the silicon nitride passivation layer has been successfully removed to expose the underlying nanowire arrays. The magnitude of the current, typical of nanowire arrays, confirms that electrochemistry only occurs at the nanowire electrodes and the passivation layer successfully prevents unwanted electrochemistry occurring at on-chip metallisation. Nanowire devices that exhibited lower or no electrochemical current were discarded and not used for further experiments.

The electrochemical reduction and oxidation of the nitroaromatics, shown in Scheme 1, are explored using cyclic voltammetry. Cyclic voltammograms of nitroaromatics are undertaken for 50 μg mL^{-1} concentrations of 3-NT, 2,4-DNT, 2,6-DNT, 1,3-DNB and TNT by scanning the voltage from 0 V to −1.0 V (reduction) followed by sweeping back to 0 V (oxidation) see Fig. 3. The measured peak from −0.6 V onwards may be attributed to the reduction of the phosphate buffer. No cathodic faradaic peaks are observed for any of the nitroaromatic molecules. A very small shoulder is evident on the TNT peak at −0.81 V *versus* the on-chip platinum pseudo-reference electrode. Very little hysteresis is evident on the CVs due to the inherent low capacitance of these nanowire electrode arrays. These results are in contrast to measurements reported using glassy carbon electrodes

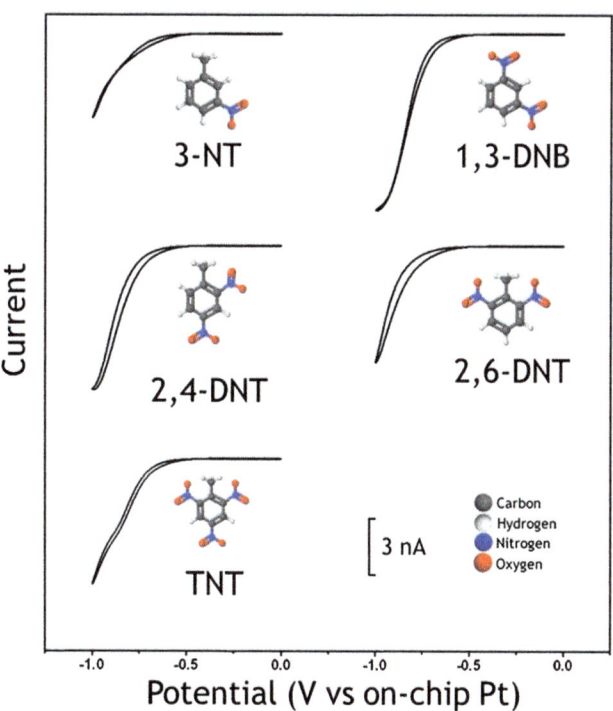

Fig. 3 Cyclic voltammograms recorded for five nitro aromatic species. The reductive peak above −0.6 V may be attributed to the reduction of the phosphate buffer.

which typically show clear reduction peaks and may be attributed to the low activity of gold towards the reduction of nitro groups.[27] Consequently CV is not a sensitive enough technique to maximise signal output.

By contrast, square wave voltammetry is a very sensitive electrochemical method, compared to cyclic voltammetry, which permits fast scan rates and is suitable for remote electroanalysis. Fig. 4 shows background subtracted square wave voltammograms for 50 μg mL^{-1} concentrations of 3-NT, 2,4-DNT, 2,6-DNT, 1,3-DNB and TNT in phosphate buffer undertaken using a nanowire array electrode comprising four nanowires and the following conditions: $E_{initial}$: −0.4V; E_{Final}: −1.02 V; amplitude: 50 mV; frequency 10 Hz; potential step: 4 mV. The reductive signals for the cathodic SWV show well defined peak potentials and are equalised to the 3-NT peak. The reduction of the nitro groups of TNT is characterised by a pair of reduction peaks at −671 mV and −882 mV, respectively. These peaks represent the reduction of the *ortho* nitro groups which are reduced at a lower potential due to their proximity to the electron donating methyl group at the 1 position.[26] The reduction of the *para* positioned nitro group is obscured possibly due to the sensitivity of our electrodes to reduction of the buffer at higher cathodic potentials. A single reduction peak is observed for 2,6-dinitrotoluene (−856 mV), 2,4-dinitrotoluene (−799 mV) and dinitrobenzene (−760 mV). It is again suspected that further expected peaks are obscured by the reduction of the phosphate buffer. A single peak is seen for 3-nitrotoluene at −900 mV. The higher cathodic potential needed for the reduction of 3-NT may be due to the position of the nitro group in relation to the methyl group. A methyl group in aromatic compounds is *ortho* and *para* directing due to resonance allowing the delocalisation of electrons around the aromatic ring, while reduction of the *meta* position nitro group in 3-NT is less favourable. From Fig. 4, it is evident that, due to the high sensitivity and low capacitance of the nanowires, the positions of reduction peaks are clearly resolved for the different molecules. This is exemplified by the ability to resolve the *ortho* reduction peaks of the two isomers 2,4-DNT and 2,6-DNT.

The ability to resolve clearly the position of the SWV cathodic potential peaks lends the data suitability for principle component analysis (PCA). PCA is an

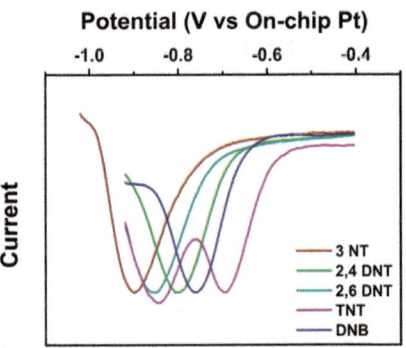

Fig. 4 Background subtracted square wave voltammograms for 50 μg mL^{-1} concentrations of 3-NT, 2,4-DNT, 2,6-DNT, 1,3-DNB and TNT in phosphate buffer pH 6.5, $E_{initial}$: −0.4V; E_{Final}: −1.02 V; amplitude: 50 mV; frequency 10 Hz; amplitude: 4 mV.

unsupervised method with no *prior* knowledge of experimental structure and is used to reduce the dimensionality of the data and explain the variance–covariance structure of a set of variables through a few linear combinations of these variables.[28] In PCA the linear combinations (PCs) of the original variables are orthogonal (uncorrelated) to each other and much of the original data variability can be accounted for by a small number of PCs which are then used for data dimensionality reduction and visual data interpretation. Before applying PCA, data needs to be adequately pre-processed and scaled. In this work, data is autoscaled so that the variable measured (peak potential) had a mean equal to zero and a standard deviation equal to one. Rows on the data matrix used for PCA represent nitro compounds tested using a unique column representing their respective measured peak potential values. The software package used for the PCA analysis was R version 2.9.2 (R: A Language and Environment for Statistical Computing, Vienna, Austria, 2009).

The first principal component scores (PC1) against a sample index to visualize the discrimination between compounds; see Fig. 5. As this data set is composed by only one variable, only one PC can be computed and PC1 explains 100% of the data variance. Each compound forms a clearly distinguishable cluster and all compounds are successfully discriminated by unsupervised cluster analysis; see Fig. 5. In addition, two different samples of the TNT compound were tested (one military grade and one analytical standard, represented by a diamond and a down triangle, respectively). The results show that the two different TNT samples also form unique clusters. As PCA is an unsupervised technique and does not know the identity (nitro compound class) of the samples, these results suggest that the devices successfully detected consistent similarities between readings of the same compound and consistent dissimilarities between readings of different compounds which, when translated into peak potential signals, enable accurate discrimination of nitro compounds.

Calibration curve for TNT

To demonstrate the suitability of the nanowire arrays as sensors for nitroaromatics, calibration experiments were undertaken to examine the effects of

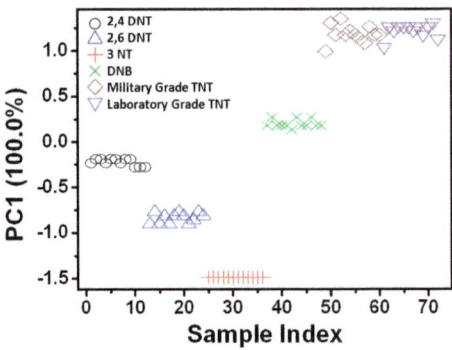

Fig. 5 Cluster analysis: scores plot from PCA applied to the data collected by the "8 (4 × 700 nm Band)" device. The data were autoscaled to previous PCA and as the dataset is composed by only 1 variable, the peak potential, PC1 explains 100% of the data variance.

increasing concentration on the SWV signals using a serial addition approach. Once recorded, the measurements are background-subtracted to give the distinct TNT peaks shown in Fig. 4.

Fig. 6 shows the calibration plot for military grade TNT. Each measurement was undertaken four times to monitor the stability and reproducibility of the electrodes and the error bars represent the standard deviation from the mean value. The limit of detection was sub-150 ng mL^{-1} and a coefficient of $R^2 = 0.996$ showing good linearity with increasing concentration in this concentration range. The reproducibility and selectivity of cathodic SWV is a useful analytical tool and clearly demonstrates the viability of this approach to the selective detection of nitroaromatic species. Work is now on-going to develop this approach further: to examine different mixtures of nitroaromatic species; to increase sensitivity further by increasing the number of nanowires within an array; and by development of an ionic liquid overlayer, selectively permeable to nitroaromatics, to prevent unwanted interference from other reductive species.

4 Conclusions

We show a photolithography technique that permits gold nanowire array electrodes to be fabricated routinely at reasonable cost. Nanowire electrode arrays offer the potential for enhancements in electroanalysis including: increased signal-to-noise ratio and increased sensitivity while also allowing quantitative detection at much lower concentrations. Characteristic reduction peaks of nitro groups were not observed at nanowire array electrodes using sweep voltammetric methods. By contrast, clear and well defined reduction peaks were resolved using potential step square wave voltammetry. Principle component analysis was employed to discriminate between nitroaromatic species, including structural isomers of DNT. Finally, the magnitude of the reduction peak at −671 mV for different concentrations of TNT exhibited very good linearity enabling limits of detection at concentrations sub-150 ng mL^{-1}.

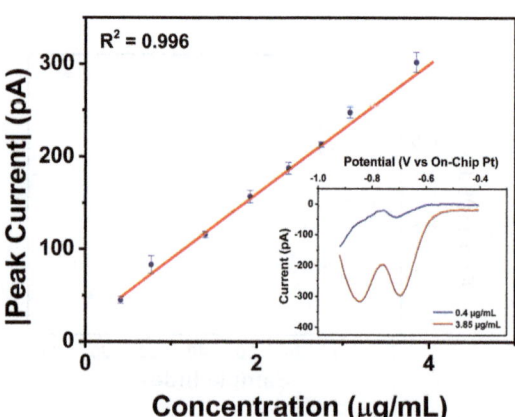

Fig. 6 Calibration plot of TNT obtained using peak currents measured at −671 mV. Inset: SWV voltammograms obtained for 0.40 and 3.8 µg mL^{-1}, respectively.

Acknowledgements

The authors gratefully appreciate assistance provided by the Ordnance School of Defence Forces, Ireland. This work was supported by Science Foundation Ireland under the Research Frontiers Programme (SFI/09/RFP/CAP2455), by the European Commission under the FP7 Security Project CommonSense (261809) and FP7 ICT project "Nanofunction" (257375) and the Irish Higher Education Authority PRTLI programs (Cycle 3 "Nanoscience" and Cycle 4 "INSPIRE").

Notes and references

1 M. Park, L. N. Cella, W. Chen, N. V. Myung and A. Mulchandani, *Biosens. Bioelectron.*, 2010, **26**, 1297–1301.
2 S. S. Talmage, D. M. Opresko, C. J. Maxwell, C. J. Welsh, F. M. Cretella, P. H. Reno and F. B. Daniel, *Rev. Environ. Contamination Toxicol.*, 1999, **161**, 1–156.
3 D. L. Kaplan and A. M. Kaplan, *Environ. Sci. Technol.*, 1982, **16**, 566–571.
4 W. D. Won, L. H. Disalvo and J. Ng, *Appl. Environ. Microbiol.*, 1976, **31**, 576–580.
5 W. C. C. Roberts, B. J. Commons, H. T. Bausum, C. O. Abernathy and J. J. Murphy, *U. S. Environmental Protection Agency*, 1993, EPA 822-R-93-022.
6 A. M. C. J. I. Jerez Rozo, S. L. Pena and S. P. Hernandez-Rivera, *Proc. SPIE-Int. Soc. Opt. Eng.*, 2007, **6538**, 653824.
7 C. A. Weisberg and M. L. Ellickson, *Am. Lab.*, 1998, **30**, 32N.
8 D. R. Shankaran, K. V. Gobi, T. Sakai, K. Matsumoto, T. Imato, K. Toko and N. Miura, *IEEE Sens. J.*, 2005, **5**, 616–621.
9 C. A. Heller, R. R. McBride and M. A. Ronning, *Anal. Chem.*, 1977, **49**, 2251–2253.
10 C. Keuchel, L. Weil and R. Niessner, *Anal. Sci.*, 1992, **8**, 9–12.
11 Gaurav, A. K. Malik and P. K. Rai, *J. Hazard. Mater.*, 2009, **172**, 1652–1658.
12 S. Babaee and A. Beiraghi, *Anal. Chim. Acta*, 2010, **662**, 9–13.
13 M. Agah, G. R. Lambertus, R. Sacks and K. Wise, *J. Microelectromech. Syst.*, 2006, **15**, 1371–1378.
14 Y. Cui, Z. H. Zhong, D. L. Wang, W. U. Wang and C. M. Lieber, *Nano Lett.*, 2003, **3**, 149–152.
15 C. Staii and A. T. Johnson, *Nano Lett.*, 2005, **5**, 1774–1778.
16 C. P. Chen, A. Ganguly, C. Y. Lu, T. Y. Chen, C. C. Kuo, R. S. Chen, W. H. Tu, W. B. Fischer, K. H. Chen and L. C. Chen, *Anal. Chem.*, 2011, **83**, 1938–1943.
17 C. M. Hangarter, M. Bangar, A. Mulchandani and N. V. Myung, *J. Mater. Chem.*, 2010, **20**, 3131–3140.
18 G. M. Cohen, M. J. Rooks, J. O. Chu, S. E. Laux, P. M. Solomon, J. A. Ott, R. J. Miller and W. Haensch, *Appl. Phys. Lett.*, 2007, **90**, 233110.
19 D. W. M. Arrigan, *Analyst*, 2004, **129**, 1157–1165.
20 K. Dawson, A. Wahl, R. Murphy and A. O'Riordan, *J. Phys. Chem. C*, 2012, **116**, 14665–14673.
21 K. Dawson, M. Baudequid, N. Sassiat, A. J. Quinn and A. O'Riordan, *Electrochim. Acta*, 2013, **101**, 169–176.
22 S. Guo, D. Wen, S. Dong and E. Wang, *Talanta*, 2009, 77, 1510–1517.
23 K. Dawson, J. Strutwolf, K. P. Rodgers, G. Herzog, D. W. M Arrigan, A. J. Quinn and A. O'Riordan, *Anal. Chem.*, 2011, **83**, 5535–5540.
24 P. Tyagi, D. Postetter, D. L. Saragnese, C. L. Randall, M. A. Mirski and D. H. Gracias, *Anal. Chem.*, 2009, **81**, 9979–9984.
25 K. Dawson, M. Baudequin and A. O'Riordan, *Analyst*, 2011, **136**, 4507–4513.
26 C. K. Chua, M. Pumera and L. Rulisek, *J. Phys. Chem. C*, 2012, **116**, 4243–4251.
27 M. Galik, A. M. O'Mahony and J. Wang, *Electroanalysis*, 2011, **23**, 1193–1204.
28 R. A. Johnson and D. W. Wichern, *Applied Multivariate Statistical Analysis*, Prentice Hall, 2007.

Faraday Discussions

RSC Publishing

PAPER

A systematic study of the influence of nanoelectrode dimensions on electrode performance and the implications for electroanalysis and sensing

Ilka Schmueser,[ab] Anthony J. Walton,[b] Jonathan G. Terry,[b] Helena L. Woodvine,[a] Neville J. Freeman[cd] and Andrew R. Mount*[a]

Received 11th March 2013, Accepted 20th March 2013
DOI: 10.1039/c3fd00038a

Micron resolution photolithography has been employed to make microsquare nanoband edge electrode (MNEE) arrays with reproducible and systematic control of the crucial dimensional parameters, including array element size and spacing and nanoelectrode thickness. The response of these arrays, which can be reproducibly fabricated on a commercial scale, is first established. The resulting characteristics (including high signal and signal-to-noise, low limit of detection, insensitivity to external convection and fast, steady-state, reproducible and quantitative response) make such nanoband electrode arrays of real interest as enhanced electroanalytical devices. In particular, the nanoelectrode response is presented and analysed as a function of nanometre scale electrode dimension, to assess the impact and relative contributions of previously postulated nanodimensional effects on the resulting response. This work suggests a significant contribution of migration at the band edges to mass transfer, which affects the resulting electroanalytical response even at ionic strengths as large as 0.7 mol dm^{-3} and for electrodes as wide as 50 nm. For 5 nm nanobands, additional nanoeffects, which are thought to arise from the fact that the size of the redox species is comparable to the band width, are also observed to attenuate the observed current. The fundamental insight this gives into electrode performance is discussed along with the consequent impact on using such electrodes of nanometre dimension.

1 Introduction

There has been considerable interest in the fabrication and deployment of miniaturized electrodes for over twenty years. The potential benefits of these

---

[a]*EaStCHEM, School of Chemistry, The University of Edinburgh, King's Buildings, Edinburgh, EH9 3JJ, UK. E-mail: A.Mount@ed.ac.uk*
[b]*School of Engineering, The University of Edinburgh, King's Buildings, Edinburgh, EH9 3JF, UK*
[c]*School of Physics & Astronomy, Manchester University, Oxford Road, Manchester M13 9PL, UK*
[d]*NanoFlex Ltd, Daresbury Innovation Centre, Keckwick Lane, Daresbury, WA4 4FS, UK*

electrodes have been widely discussed in the literature.[1-7] The enhanced mass transport due to the hemispherical or radial diffusion that occurs on the micron scale gives high current densities and steady-state or near steady-state currents, reducing the effects of convection. Additionally, shorter RC time constants allow the investigation of faster electrode processes, while smaller iR drops enable the electrodes to work in a wide range of resistive electrolytes, with little or no added ions. As these effects are enhanced with decreasing electrode size, there has been considerable recent interest in scaling electrodes from the micron[8] to the nanoscale.[9-20] Such electrode structures also offer the opportunity to exploit nanostructural effects such as enhanced surface catalytic activity,[21-23] and nanoelectrode systems have been constructed for applications as diverse as enhanced thermal energy[24] and light harvesting.[25,26] However, there are two main issues with current nanoelectrode technology. The first is the lack of a suitable technology for reproducibly manufacturing electrodes with accurately controlled dimensions at a reasonable cost and volume to enable commercialisation. The second is the low current levels measured with nanoelectrodes (often in the pA range),[15,27] which require precision low current, low noise potentiostats, with efficient electromagnetic shielding. These are the issues addressed by this work.

One way by which this latter issue can be addressed is by multiplying the observed current through the simultaneous measurement of large numbers of arrayed nanoelectrodes.[9,12] The simplest methods involve no specific control over the inter-electrode spacing of array devices, which are often referred to as ensembles.[12] A range of fabrication techniques have been adopted to produce ensemble devices, typically employing either a method to open pores in insulating layers deposited over conductive materials, thereby exposing an ensemble of active electrode areas (*e.g.* ultrasound[28] or neutron track etching[29]), templated growth in the nanopores of membranes[30,31] or the imprinting of structures.[32] This is at present an area of considerable research interest,[33] but the challenge with ensemble technologies is that it is currently not possible to produce large nanoelectrode arrays with systematic control of the size and separation of each electrode. Without this, there is heterogeneity of individual electrode response, with individual electrodes being affected by diffusional layer overlap from neighbouring electrodes to variable extents as a function of time and frequency. The average spacing is also often hard to control and optimise.

The alternative is to produce ordered nanoscale arrays. Some ensemble methods (X-rays[34] and structure imprinting) show potential for control of spacing, (indeed recent work has demonstrated production of nanopore arrays with embedded ring electrodes in each pore[35]) but these involve considerable expense and/or complexity. Conventional photolithography, such as is used in the semiconductor industry, typically requires wafer stepper based technology, which is clearly capable of delivering well-controlled and reproducible nanoarrays. However, the cost of ownership of a modern state-of-the-art, nanometre capable wafer stepper is extremely high. Another option is e-beam lithography.[36] However, this suffers from being an essentially serial writing process and hence is both time consuming and costly. Nano imprint lithography[37] potentially addresses the throughput issue, but has the high cost of producing the nanoscale mask required for making the mould. As a result of the expense and complication of the above approaches most of the nanoelectrodes reported have employed alternative design and production strategies.

It is also possible to use lower cost, micron resolution photolithography and reactive ion etch techniques to fabricate arrays of individually addressable submicron band electrodes, (e.g. 2 mm long with widths as small as 37 nm[38,39]). Similarly, groups have used (optical) lithographically patterned nanowire electrodeposition (LPNE) to prepare a sacrificial nickel nanoband electrode, which acts as a seed layer for the subsequent electrodeposition of a noble metal. The nickel electrode is recessed in a trench, such that the electrodeposition forms a nanowire of rectangular cross-section, after subsequent removal of the sacrificial layers. This results in closely spaced, millimetre long gold nanowires in the form of bands and grids.[40,41] Both of the above process architectures produce long nanobands or nanowires with significant resistance, which are ideally suited for applications such as resistive gas sensing.[42] However, long nanowires/bands are potentially susceptible to line breakages occurring during the fabrication process and/or operation, and as with all band electrodes, their mass transport limited electroanalytical response is not steady-state.

In addition to photolithographic methods, other approaches have been reported for patterning arrays[43] to give electrode elements arranged in an ordered manner with well-defined inter-electrode spacing, including controlled polymer deposition, orientation and selective dissolution,[44] as well as carbon nanotube (CNT)[45,46] and boron-doped diamond[47] nanowire assemblage. However, the challenge with polymer deposition is to control the electrode size and spacing, while CNT assemblies still require standard lithographic patterning (e.g. e-beam)

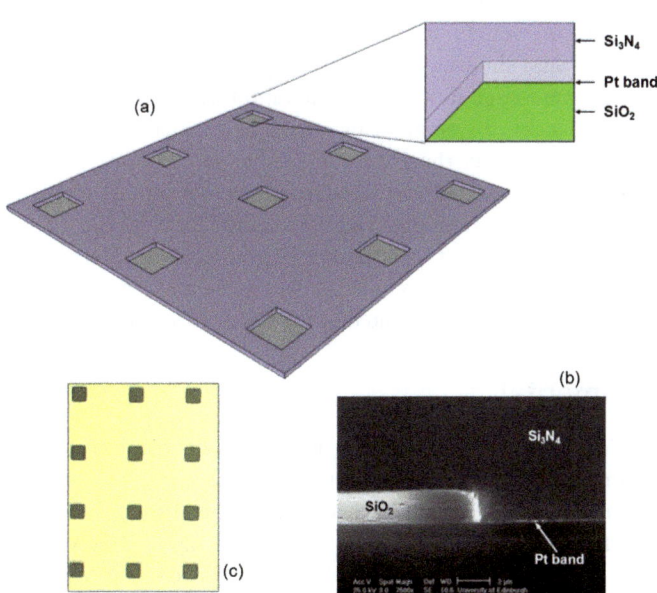

Fig. 1 (a) Schematic diagram of the microsquare nanoband edge electrode (MNEE) array architecture used in this work, with Si_3N_4 upper (purple) and SiO_2 lower (green) insulation sandwiching a Pt nanoband (grey) in each square aperture (inset). (b) an SEM image of a cleaved and polished cross section through a square aperture in the fabricated 10μ3D50n structure. The horizontal line of enhanced scattering indicating the 50 nm Pt band is highlighted, which as expected is seen to extend around the perimeter of the half aperture with SiO_2 at its base in the left of the picture. (c) an optical microscope image of the fabricated microsquare array.

of nanoscale metal catalyst centres and the nanotubes themselves are not uniform in length, diameter or orientation. In essence, the above methods all require extremely precise nanopatterning procedures and lack flexibility and control of both the electrode material and its size. A final fabrication approach that has been reported is nanoskiving,[48,49] which when producing arrays, requires addressing the challenges of first sectioning materials encapsulating nanowires, and then collecting, mounting and connecting them on a suitable support.

This paper utilizes our own nanoelectrode array design;[50] the microsquare nanoband edge electrode (MNEE, Fig. 1), which has been developed to address some of these issues and is fabricated using reliable, readily available and well characterized standard micron scale microfabrication techniques. The precise fabrication process developments required to produce these systems are detailed elsewhere,[51] as is the 1000 fold increase in detection sensitivity observed at an MNEE system compared to a single microelectrode of comparable electrode area.[52] In this work, the fundamental response of these systems is first considered to demonstrate that the resulting response is both quantifiable and reproducible as well as displaying the characteristics of a nanoscale electrode system of high effective area. These systems are then used to first measure and then consider the impact of nanoscale effects previously postulated and reported as occurring at other nanoelectrode systems.

2 The MNEE system

Fig. 1 shows the details of the electrode architecture, which was produced on 3-inch n-type silicon wafers using contact/proximity based photolithography, together with an SEM showing a section through one element of the array device. The square aperture edge dimension, separation between apertures and nanoband thickness has been systematically varied and is denoted by the nomenclature $L\mu HDMn$, where L is the edge dimension, in microns, H is the closest interaperture spacing ratioed to the edge dimension and M is the nanoband thickness. For example, exemplar characterisation has been carried out on a system with these dimensions set at 10 μm, 30 μm and 50 nm, denoted 10μ3D50n. The depth of the square aperture etched to expose the nanoband was controlled to ensure that the nanoband was located at the base of the etched hole.

3 Experimental

Electrochemical measurements were carried out in a Faraday cage using Autolab PGSTAT30 or PGSTAT128N potentiostats controlled *via* a PC and a frequency response analyser module (Windsor Scientific). General Purpose Electrochemical Software (GPES, version 4.9), Frequency Response Analysis (FRA, version 4.9) and NOVA (version 1.9) were used for data collection of cyclic voltammetry and electrochemical impedance spectroscopy (EIS) measurements. The fabricated nanoelectrode arrays were used as working electrodes, with a saturated calomel reference electrode and a platinum gauze counter electrode. All potentials, E, are quoted with respect to this reference electrode. Prior to experimentation, the glassware was rinsed in sulphuric acid (Chestech, 96% purity) to remove any organic contamination.

All solutions were prepared using deionized water (Millipore MilliQ). Aqueous 0.5 M potassium chloride (Fisher Scientific, 99.6% purity) solution was used for electrochemical cleaning. Redox measurements used 100 μM ferrocene carboxylic acid (FCA, Alfa Aesar, 98% purity) in either 0.1 M potassium chloride background electrolyte in a solution (pH = 7.0) prepared with 0.1 M citric acid (Fisher Scientific, 100.3% purity) and 0.3 M sodium hydroxide (Fisher Scientific, 97% purity), effectively 0.1 M trisodium citrate solution (FCA(I) solution) or the same concentration of FCA in a ten times diluted electrolyte solution (pH = 7.0) of 10 mM potassium chloride, 10 mM citric acid and 30 mM sodium hydroxide (FCA(I/10) solution). Hexaammineruthenium(III) (HAR) chloride (Aldrich, 98% purity) measurements were recorded at 100 μM concentration in 0.5 M (HAR(I) solution) and 10 mM (HAR(I/50) solution) potassium chloride electrolyte solutions that were deoxygenated using argon.

4 MNEE characterisation

4.1 COMSOL FE validation and modelling

Various methods for the modelling of diffusion limited chronoamperometric currents have been reported based on commercial[53] and in-house software.[54,55,56] They involve solving Fick's second law for the considered geometry using different explicit, implicit and semi-implicit techniques to solve the resulting matrices.

The advantages of using commercial finite element modelling (FEM) software packages such as COMSOL include their ready availability and their relatively

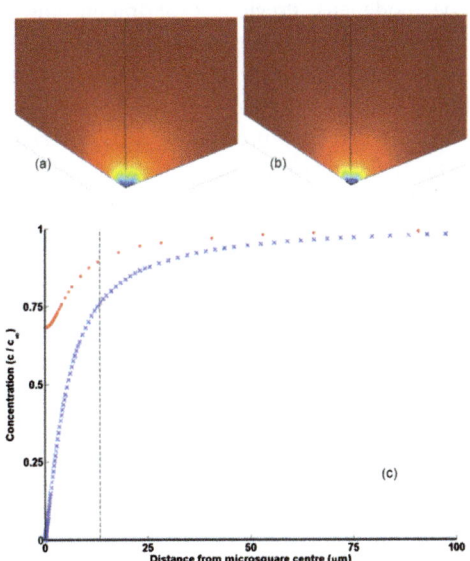

Fig. 2 Concentration profiles at the onset of a steady state current for (a) a quarter of a microsquare (edge length, $L = 10$ μm) and (b) a quarter of an MNEE ($L = 10$ μm, thickness 50 nm) placed in the near corner of the bottom face (the $x, y, 0$ plane) of the simulation cube and simulated using FE modelling. In this simulation the diffusion coefficient was set to 10^{-6} cm^2 s^{-1}, and the colours span a concentration range of $c = 0$ (blue) to $c = c_\infty = 1$ mM (red). (c) the concentration along the z axis normal to the microsquare centre for the microsquare (blue crosses) and the MNEE (red dots); values are normalized to the bulk concentration c_0. The concentration at the microsquare centre is not zero for the MNEE electrode as this is an insulator rather than an electroactive surface. The vertical line is the hemispherical radius within which complete redox conversion of FCA occurs, calculated from the volume, V, determined from C_N^{max} and eqn (7).

large user base, which has led to the validation of the resultant modelling with experimental measurement for a wide variety of applications. However, concerns have been raised about the validity of this approach for transient microdisc, microhemisphere and microband 3D simulations,[57,58] which has led to the development and assessment of the validity of other finite difference, finite element, finite volume and other modelling methods, many of which are not routinely available in COMSOL. These models have been reviewed in the context of their correspondence to microelectrode voltammetric theory.[59] This highlights the need for prior benchmarking and validation of any FEM modelling employed on a comparable system with known analytical response. In our previous steady-state COMSOL FEM simulations of single microsquare response,[53] we therefore first successfully benchmarked our modelling to the analytical response of a microdisc of similar dimension. In this work we extend this approach, first benchmarking the time-dependent response at steady-state to that obtained from a time-dependent COMSOL microdisc simulation.

Time-dependent three-dimensional FE simulations of the simulation cube were carried out using COMSOL 3.5a for Linux (COMSOL, Inc., Burlington, Massachusetts, USA). Although it is clear that microdisc simulation could have been reduced to a two dimensional problem by using rotational symmetry, this option was not chosen to keep the microsquare, microdisc and MNEE simulations as similar as possible for benchmarking. (However, taking advantage of the symmetry about the plane $(x = y)$ in the simulation cube, Fig. 2, only half the cube was required to be modelled). Previous experiments and 2D simulations on microdisc array electrodes have established negligible diffusion layer overlap between neighbouring array elements at steady-state when electrodes are separated by more than 24 times the disc radius,[60,61] which indicates steady-state diffusion is established when the diffusion layer thickness is of the order of 12 times the radius. Previous microsquare simulation also showed no boundary effects with the cube dimension set at 20 times the edge length.[53] The simulation cube edge length in this work was therefore set at 100 times the microdisc radius or 50 times the edge length, as appropriate, to ensure no boundary edge effects. The concentration at the upper face of the simulation cube and throughout at time zero was then fixed at the bulk concentration of the redox species.

The user-defined mesh was first optimised in a parametric study to find the combination of number of nodes and node distribution (linear, one- or two-sided logarithmic) for each boundary and volume that showed a converged current response. Subdomains were used to facilitate this process. The final mesh density reflected the reagent concentration gradient within each mesh box, which as expected decreased dramatically as a function of distance from the electrode surface. A parametric study was then used to analyse the various settings COMSOL offers for the solution of the matrix. This revealed very little effect on the modelling results but large variations in efficiency. After pretreatment with the *Algebraic Multigrid* (hierarchy quality 3) preconditioner, the system was solved using the *Conjugate Gradients* solver. The required tolerance levels for a converged response were determined as 0.01 (relative tolerance) and 0.001 (absolute tolerance).

Microdiscs of radius, $r = 5, 10, 15$ and 25 μm were simulated by stepping the redox ion concentration at the electrode boundary to zero at time zero, using 235 000, 348 000, 577 000 and 572 000 degrees of freedom, respectively. Values were obtained at times between 10^{-7} s and 10^5 s and from these the normal diffusive flux at the electrode surface was calculated. Integration of this flux over the electrode area, then multiplication by Faraday's constant, gave the electrode redox current. After about twenty seconds the current was observed to be constant to within 0.1% with time for all microdiscs, indicative of a steady-state current; these relatively small variations were considered to arise from rounding errors. The mean value of the current recorded after this time was reported as the steady-state current.

This steady-state current was then validated against the limiting current given by:

$$i_L = BNnFDc_\infty L \qquad (1)$$

where c_∞ is the bulk concentration and D is the diffusion coefficient of the redox species, F is Faraday's constant (96 485 C mol^{-1}), N is the number of electrodes in the array, n is the number of electrons transferred in the redox reaction and B is a constant depending on microelectrode geometry. For a single disc this means $B = 4$, $N = 1$ and $L = r$.[1,62,63] The difference between modelling and simulation values at steady-state was found to be 0.2% for the smallest r, rising to 3% for the largest. This indicates that we can determine steady-state currents for microsquare and MNEE electrodes of similar dimension using COMSOL transient FE modelling to an accuracy of the order of 1% using this approach.

Previous COMSOL FE steady-state simulation and experimental measurement[53] also showed that for an edge length, $L = 50$ μm single microsquare electrode system, the steady-state diffusion limited current, i_L, is larger than the microdisc of equivalent area, and given by eqn (1) with a value of $B = 2.34$.[53]

Time-dependent modelling of both the $L = 10$ μm microsquare and the $L = 10$ μm 50 nm thick single MNEE was then carried out using the approach described above. For the microsquare, the resulting mesh consisted of 195 000 degrees of freedom; whilst the MNEE mesh was made slightly denser particularly close to the electrode edges and at the top edge of the aperture, resulting in 352 000 degrees of freedom. This gave steady-state mass transport limited currents of 226 pA for a microsquare electrode and 92 pA for the MNEE (for a bulk concentration of redox species of 10^{-3} mol dm^{-3} and a diffusion coefficient of 10^{-6} cm^2 s^{-1}).

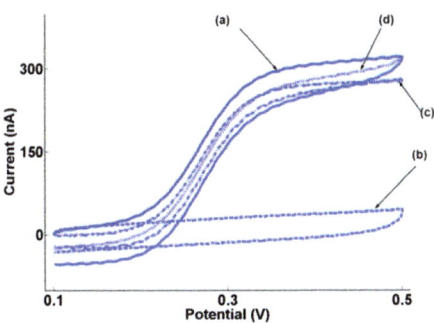

Fig. 3 (a) Typical MNEE array first sweep CV at $v = 100$ mV s^{-1} for the oxidation of 0.1 mM FCA in aqueous pH = 7.0 solution (0.1 M citric acid/0.3 M sodium hydroxide with 0.1 M potassium chloride; background electrolyte); and for (b) background electrolyte alone. (c) The difference between (a) and (b). (d) The mean of forward and backward scans of (a). The onset potential was 100 mV with the initial direction of scan to positive potentials.

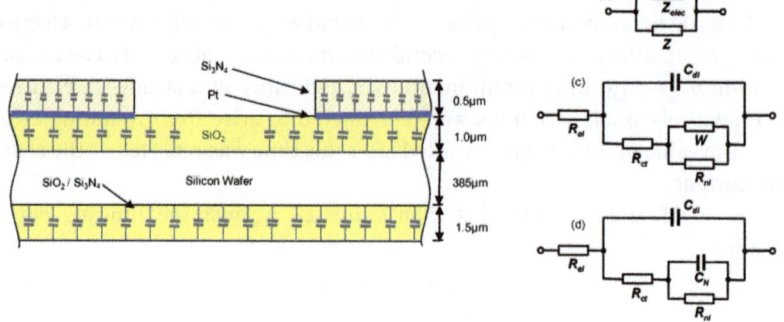

Fig. 4 (a) The origin of the parasitic capacitance associated with the MNEE array. (b) Equivalent circuit which describes the parasitic capacitance in parallel with the MNEE array response. (c) Established modified Randles equivalent circuit appropriate for a microsquare electrode array. (d) Equivalent circuit used for the MNEE array.

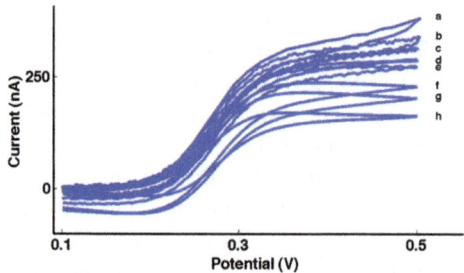

Fig. 5 Typical subtracted first sweep CVs (as in Fig. 3c) of the 10μ3D50n MNEE array for the oxidation of 0.10 mM ferrocene carboxylate (FCA solution) at v = (a) 5 V s^{-1}, (b) 1 V s^{-1}, (c) 500 mV s^{-1}, (d) 100 mV s^{-1}, (e) 50 mV s^{-1}, (f) 10 mV s^{-1}, (g) 5 mV s^{-1}, (h) 1 mV s^{-1}. The onset sweep potential was 100 mV with the initial direction of scan to positive potentials.

It is reassuring that the simulated single microsquare steady-state current is consistent with eqn (1), with $B = 2.34$, which provides further validation of this approach. It is also interesting that both the microsquare and MNEE current values are comparable (with $B = 0.956$ for MNEE, 41% of that for the microsquare). This indicates that the predominance of the contribution of edge diffusion to the limiting current in microsquares is such that the MNEE produces a comparable mass transport limiting current, with only 2% of the effective electrode area. This modelling also showed that, consistent with previous simulation,[53] placing the microsquare or MNEE at the bottom of the shallow aperture had little effect on the limiting current obtained. The calculated concentration profiles (Fig. 2) also indicate that the closed geometry of the MNEE produces hemispherical diffusion layers which is consistent with time-independent mass transport limited diffusion. It is advantageous and not unexpected that the MNEE response is more similar to a microsquare electrode response than to a semi-infinite linear band, for which pseudo steady-state conditions only are observed,[38,39,64–66] as similar behaviour has been previously observed and modelled for other closed systems e.g. microsquare, microdisc[53] and microring[67] electrodes.

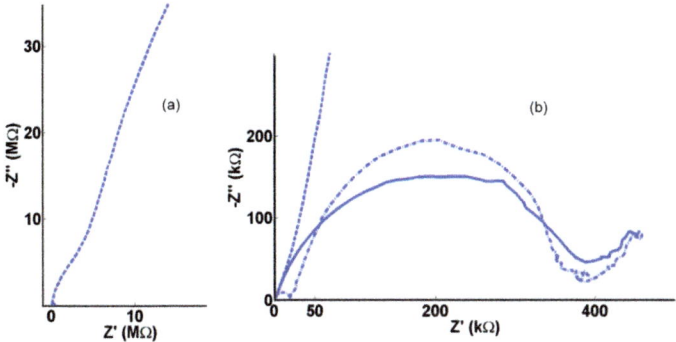

Fig. 6 (a) Nyquist Electrochemical Impedance Spectroscopy (EIS) data for oxidation of FCA in FCA(*l*) solution at the 10μ3D50n electrode at E_{dc} = +0.275 V, frequencies logarithmically spaced between 100 kHz and 0.01 Hz and an ac perturbation of 10 mV. (b) Same data on an expanded scale for background electrolyte (dashed line), redox solution (solid line) and the subtracted impedance (dash dotted line).

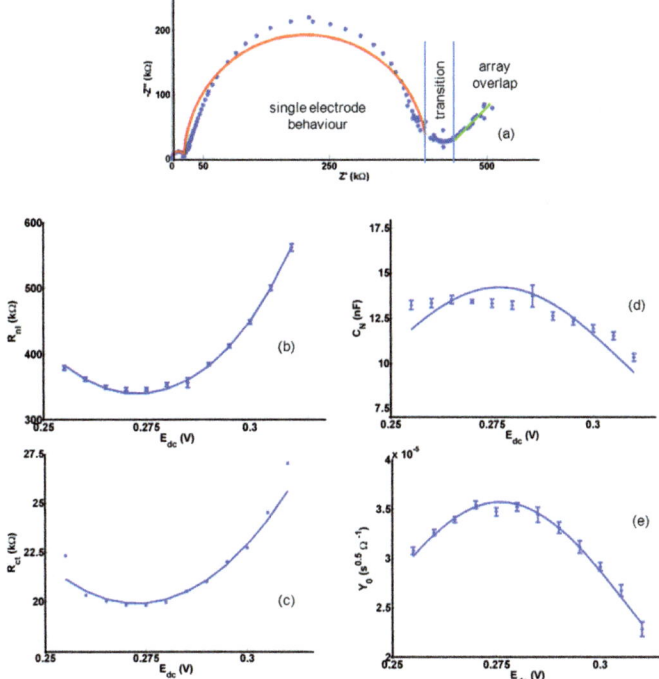

Fig. 7 (a) Typical subtracted Nyquist Electrochemical Impedance Spectroscopy (EIS) data at the 10μ3D50n electrode for oxidation of FCA(*l*) solution at E_{dc} = +0.290 V. Also shown are theoretical fits to the circuit in Fig. 4c and to the circuit (resistance in series with Warburg element) appropriate at low frequency when there is complete diffusional array overlap. Fits (solid lines) to the (b) R_{nl} (c) R_{ct} (d) C_N (e) Y_0 data extracted from iterative fitting to the EIS measurements at each E_{dc} performed using either Matlab (The Mathworks, Natick, Massachusetts, USA) or Nova (Version 1.9, Metrohm Autolab B.V., Utrecht, The Netherlands). The fits of these data to eqn (3), (8), (10) and (13), (lines shown in (b), (c), (d) and (e)) gave the following global values: E' = 0.273 ± 0.002 V; R_{ct}^{min} = 19.87 ± 0.06 kΩ; R_{nl}^{min} = 339 ±1 kΩ; C_N^{max} = 14.3 ± 0.3 nF, Y_0^{max} = 35.7 ± 0.2 μs$^{0.5}$ Ω^{-1}.

4.2 Experimental MNEE characterisation

Experimental MNEE voltammetry (Fig. 3a) shows a near sigmoidal response at $v = 100$ mV s^{-1} characteristic of the steady-state response expected from a nanoarray with $N = 15\,625$ independent electrodes. However, even at this sweep rate, a significant current can be observed (Fig. 3b) from parasitic capacitance, which arises from non-Faradaic charging (of the capacitance between the platinum plane and solution through the nitride and oxide insulator and silicon substrate, see Fig. 4a). The insensitivity of this additional CV current to potential is consistent with this non-Faradaic capacitance, which as expected linearly increases with v, which explains its growing importance at increasing sweep rates.

As this charging current is in parallel with the redox current, subtraction of the response without redox agent at any sweep rate leads to the MNEE array redox response (Fig. 5). In fact, the near constant non-Faradaic capacitance means that at fast sweep rates, when there is time-independent sigmoidal redox response, simple averaging at each potential of the data from forward and backward voltammetric scans essentially leads to this redox response (Fig. 3d). It is reassuring that comparing experiment and modelling results in a sensible value of $D = 2.2 \times 10^{-6}$ cm^2 s^{-1} being obtained from eqn (1).

However, at slower sweep rates, there is an increasing overlap of the growing hemispherical diffusion layers from neighbouring microsquares.[66] This causes a decrease in the redox current with time (and decreasing sweep rate) in the subtracted voltammograms (Fig. 5), where the time-independence of the redox response is lost and simple averaging cannot be employed.

The MNEE impedance response, as with the voltammetric data, also shows the backplane non-Faradaic capacitance to be in parallel with the redox response (Fig. 4b, eqn (2)). The subtraction of the observed impedance response in background electrolyte, Z_{elec}, from that observed with added redox agent, Z_{tot}, readily leads to the microsquare nanoband edge electrode array impedance, Z (Fig. 6). Such subtraction is obviously essential when non-Faradaic currents are significant compared to Faradaic currents ($Z \geq Z_{elec}$).

$$(Z_{tot})^{-1} = (Z_{elec})^{-1} + (Z)^{-1} \qquad (2)$$

Considering Z (Fig. 7a), two zones are apparent as shown; in the higher frequency zone the diffusion layer is thin compared to the microsquare spacing and each microsquare responds independently. In the lower frequency zone, the 45° Warburg line is diagnostic of complete overlap of neighbouring diffusion layers and linear diffusion to the entire electrode array. Previous work for a single microsquare electrode[53] has shown that, through the use of transmission line theory, the fundamental redox response reduces with a reasonable level of accuracy to the modified Randles circuit (Fig. 4c). The circuit elements then represent physical parameters: the electrolyte resistance (R_{el}), non-Faradaic microelectrode double-layer capacitance (C_{dl}), charge transfer resistance (R_{ct}) determined by the electron transfer rate of the redox reaction, the Warburg (W) element (whose parameter Y_0 is determined by the redox concentration and diffusion coefficient) and the non-linear resistance (R_{nl}) corresponding to the steady-state hemispherical diffusion limited current. It might be expected that in the higher frequency (single electrode) zone, when the electrodes behave independently ($Z_{single} = NZ$), the impedance data, Z, for the redox response of the N

Paper   Faraday Discussions

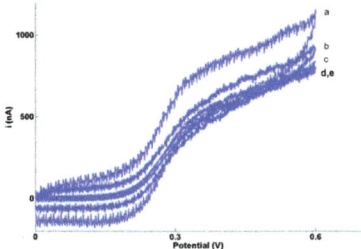

Fig. 8 Cyclic voltammograms recorded in highly stirred solution on 10µ3D50n for 0.1 mM FCA in FCA(/) without background subtraction. The solution was stirred chaotically by using a small electric whisk at high rotation speed. The scan rates were $v =$ (a) 200 mV s^{-1}, (b) 100 mV s^{-1}, (c) 20 mV s^{-1}, (d) 10 mV s^{-1} and (e) 2 mV s^{-1}.

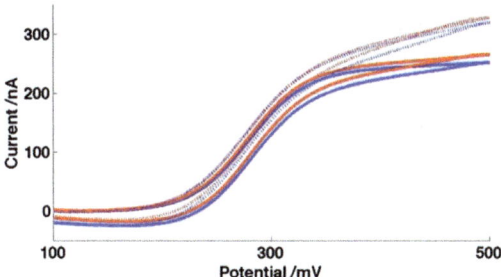

Fig. 9 CVs for oxidation on 20µ3D50n of 0.10 mM FCA in FCA(/) (solid) and FCA(//10) solution (dashed). These voltammograms are shown at sweep rates of 50 mV s^{-1} (blue) and 100 mV s^{-1} (red).

MNEE apertures could be modelled using this circuit. However, the data fit better to a variant of this circuit (Fig. 4d), where the Warburg element is replaced by a redox capacitance (C_N). The physical origins of this difference are interesting. We postulate this to be the enhanced non-linear diffusional transport to the nanoband edges, which gives rise to the rapid redox conversion of the volume of solution above each microsquare,[9–12,68] and which is best modelled in a single parameter by the capacitance, C_N. Steady-state hemispherical diffusional transport (determined by R_{nl}) then occurs into this volume as for the microsquare. Fig. 7b–d show the close fit of the impedance data to this relatively simple circuit, and to the predicted variation of the extracted parameters (see EIS theory section

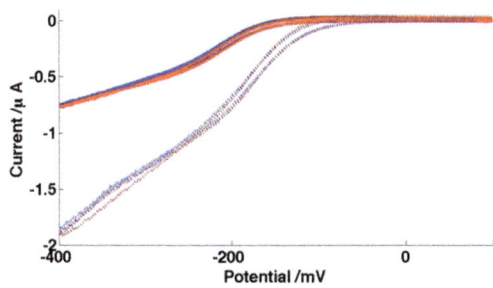

Fig. 10 CVs for reduction on 20µ3D50n of 0.10 mM HAR in HAR(/) (solid) and HAR(//50) solution (dashed). These voltammograms are again shown at sweep rates of 50 mV s^{-1} (blue) and 100 mV s^{-1} (red).

below) with the applied dc potential, E_{dc}. The main fit variance is in the C_N, R_{nl} near-semicircular region and can be attributed to the small deviation seen due to enhanced corner diffusion in microsquares[53] and the neglect of any diffusional contribution to C_N; these are approximations that have been made to restrict the number of fit parameters.

When considering fit parameter values, the insensitivity of C_{dl} to E_{dc} is expected, but it is interesting that C_{dl}, per unit geometric area, at 0.94 µF cm^{-2} is an order of magnitude smaller than that expected for a platinum macroelectrode double layer capacitance, which suggests a thicker nanoelectrode double layer. This could be a nano effect, as electrodes of smaller than 100 nm are thought to show screening over greater distances than larger electrodes, due to enhanced ion transport in the double layer; the change from planar to hemispherical geometry fields,[69] and/or the suggestion that ion size is no longer negligible.[70] The value of $E' = 273$ mV is also consistent with measured half-wave potential values at a rotating disc macroelectrode and previously reported measurements at similar pH.[66] Similarly, the value of D obtained from eqn (14) and EIS studies ($D = (2.13 \pm 0.02) \times 10^{-6}$ cm^2 s^{-1}) agrees well with that obtained from voltammetry. The C_N^{max} value is calculated to correspond to complete redox conversion of a hemispherical volume of solution of radius 13.5 µm centred on each microsquare, which is consistent with the diffusion layer thickness produced by simulation (Fig. 2c).

Previously, widely varying values for k^0 of FCA have been reported, ranging from 2.48 × 10^{-3} cm s^{-1} in 0.1 M potassium chloride and 0.2 M phosphate buffer (pH 7.4) on glassy carbon[71] to 1.4–3.6 cm s^{-1} in acetonitrile electrolyte on Pt microdisks.[72] A value of 0.850 ± 0.003 cm s^{-1} was extracted from the impedance data and eqn (9), which confirms that the enhanced MNEE diffusion, like other nanoelectrode systems, makes this system capable of the precise and quantitative measurement of rapid electrochemical kinetics. The enhanced diffusional transport to the MNEE system also makes it relatively insensitive to the effects of chaotic convection (Fig. 8), opening up the prospect of quantitative analytical measurement in such systems.

5 Nanoelectrode effects

Having established the fundamental response of the MNEE system, the production and characterisation of systems with systematic control of dimensional

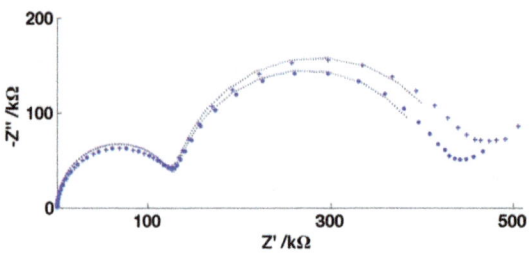

Fig. 11 EIS data for the 20µ3D50n electrode showing the oxidation of 0.1 mM FCA in FCA(I) solution (+) and FCA(I/10) solution (•) and the corresponding fits for $E_{dc} = E'$, $V_{ac} = 10$ mV, for the frequency range 100 kHz to 10 mHz (logarithmic distribution). Fits: FCA(I) solution: $R_{ct} = 135 \pm 1$ kΩ, $C_{dl} = 31.1 \pm 0.7$ nF, $R_{nl}^{min} = 309 \pm 3$ kΩ, $C_N = 1.34 \pm 0.02$ µF, FCA(I/10) solution: $R_{ct} = 135 \pm 1$ kΩ, $C_{dl} = 27.7 \pm 0.6$ nF, $R_{nl}^{min} = 283 \pm 2$ kΩ, $C_N = 1.33 \pm 0.02$ µF.

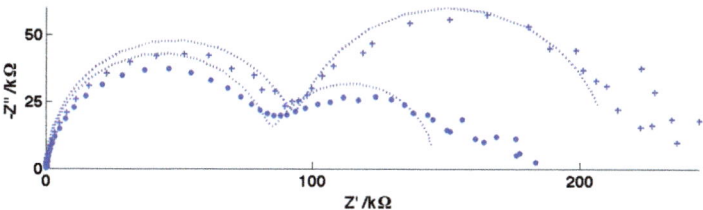

Fig. 12 Equivalent EIS data to Fig. 11 for the 20μ3D50n electrode for the reduction of 0.10 mM HAR in HAR(/) solution (+) and HAR(//50) solution (•) and the corresponding fits for $E_{dc} = E'$. Fit values: HAR(/) solution: $R_{ct} = 96 \pm 1$ kΩ, $C_{dl} = 31 \pm 1$ nF, $R_{nl}^{min} = 116 \pm 2$ kΩ, $C_N = 2.35 \pm 0.08$ μF, HAR(//50) solution: $R_{ct} = 86 \pm 1$ kΩ, $C_{dl} = 27 \pm 1$ nF, $R_{nl}^{min} = 60 \pm 2$ kΩ, $C_N = 3.0 \pm 0.2$ μF.

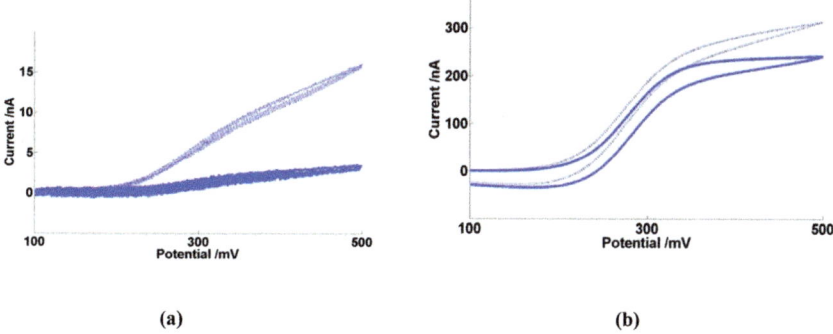

Fig. 13 Cyclic voltammetry in FCA(/) solution (solid line) and FCA(//10) solution (dashed line) at 20 mV s^{-1} of (a) 10μ4D5n (b) 10μ4D50n.

parameters has been employed to further probe their electroanalytical response, in particular determining whether and under what conditions nanoscale electrode effects are significant. Fig. 9 shows the typical response of a 20μ3D50n device when oxidizing FCA in FCA solution and in FCA(I/10) solution.

It is clear from these voltammograms that there is a small but significant difference, with an increased current seen in both voltammograms at lower ionic strength. All voltammograms also show a linear increase with potential in the region of mass transfer limited current at both low and high ionic strength. Equivalent voltammograms are also shown for this device for HAR reduction (Fig. 10); the effect is even more marked for this system, with the increase in current on reducing ionic strength and decreasing potential for this reduction being proportionally greater than seen for FCA oxidation.

This is consistent with potential dependent enhancement of FCA/HAR ion reagent transport to the electrode through migration (as a combination of the larger change in ionic strength and larger charge on the redox agent, 3+ for HAR reduction and 1− for FCA oxidation, for HAR than FCA, would explain the enhanced response). Migration effects have been considered previously[73] for nanoscale hemispherical electrodes. In the case when the Debye length, κ^{-1}, of the electrolyte approaches the size of the electrode, there is significant field in the diffusion layer, and migration enhances mass transfer, leading to a potential dependent increase in the current. This previous work considered such effects to be significant for 5 nm radius hemispherical nanoelectrodes when I was of the order of 200 μM and below, for which $\kappa^{-1} \sim 20$ nm. Krapf et al. suggested such effects occur at a higher value of $I = 0.5$ M.[74] In this work, for FCA(I) and FCA(I/10), $I = 0.70$ mol dm^{-3} and 0.070 mol dm^{-3} respectively, whilst for HAR(I) and HAR(I/50) solutions, $I = 0.50$ mol dm^{-3} and 0.01 mol dm^{-3} respectively, which result in values of κ^{-1} ranging between ~0.3 and 2.4 nm respectively, compared to the much greater nanoelectrode width of 50 nm. At first sight it seems puzzling that such relatively small values of Debye length compared to electrode dimension can give rise to the significant migrational contribution observed, but it should be remembered that unlike hemispherical electrodes, for which the current density is constant across the electrode surface, these nanoband electrode systems display markedly asymmetric current distribution, with the very highest current densities at the band edges. It is here that the depletion layer is at its thinnest, the field lines will be most crowded and the effects of migration will be most keenly felt. This work indicates that this results in significant migrational contribution to the transport and hence the observed current at these edges even for a 50 nm nanoband, and even at these relatively large values of I.

In order to confirm that the effect is predominantly an additional migrational contribution to the mass transfer and not a field effect on the charge transfer rate constant, Fig. 11 and 12 show the corresponding EIS spectra for FCA and HAR at $E_{dc} = E'$, along with their fits to the equivalent circuit, Fig. 4(d). It is clear for FCA these data at the two different values of I produce essentially the same fit parameters for all except R_{nl}^{min}, whose value has decreased by around 10% when changing from higher to lower I. For HAR, which has similarly rapid charge transfer kinetics, the difference in R_{nl}^{min} is greater, with a decrease of around 50% when changing from higher to lower I. It is reassuring that these values, when substituted into eqn (22) predict currents which correspond well to the currents observed in Fig. 9 and 10.

By contrast, Fig. 13a shows equivalent voltammetric data for oxidation of FCA at a 5 nm thickness nanoband, 10μ4D5n. By comparison with Fig. 13b, the equivalent 50 nm nanoband system, the currents observed at high ionic strength are ~98% lower. Similar dramatic differences have been observed previously[75,76] for linear nanoband electrodes of similar width. This has been explained by the fact that the dimension of the redox species is now comparable to the thickness of the nanoband, which results in a breakdown of applicability of the analytical equations which describe mass transport of redox species to the electrode. In this work, a model was proposed which accounted for this behaviour in terms of two effects; the first is a marked decrease in the diffusion coefficient near the electrode due to changes in the ion–solvent interaction, the solvent composition and solvent viscosity due to the fact that the electrode dimension, the depletion zone thickness and κ^{-1} are all comparable. The second is the fact that as the electrode dimension approaches the molecular dimension, the concentration gradient at and across the electrode surface can no longer be represented by a smooth analytical function, which has the practical effect of reducing this gradient and the resulting diffusive flux still further from that predicted analytically. In addition, however, the lower ionic strength data in Fig. 13a demonstrate a dramatic increase in the current for 10μ4D5n by a factor of around 5. It is interesting that this additional current is comparable in magnitude to that observed on decreasing I for 10μ4D50n (Fig. 13b), which is consistent with this enhanced migrational contribution being again due to migration at the band edges. The decrease in band width therefore appears to result in promotion of the importance of the band edges compared to diffusion to and reaction across the band centre (which has become a vanishingly small contribution to the current). It is also possible that there is at least some contribution to this increase in current with decreasing I which is due to the enhancement of the local electric field at the electrode with increasing potential. This could result in direct electron transfer between redox species and electrode becoming possible at greater distances from the electrode through an increase in the electron tunnelling rate,[77] which could relax the requirement for short range electron transfer at the electrode surface.

6 Conclusions

This work demonstrates the MNEE array to be a sensitive electroanalytical device with enhanced characteristics. The design enables the high diffusional rate and high signal-to-noise ratio available with nanoelectrode measurements (facilitating the measurement of fast electrode kinetics) to be combined with both the steady-state response found with microelectrodes and the high total current available with arrays. Another attraction is that the sensitive quantitative response coupled with the relative insensitivity to environmental convection makes these electrodes highly suitable for quantitative electroanalysis and monitoring in systems for which well-defined stirring is either undesirable or impractical. The use of CMOS compatible microfabrication techniques provides the possibility of integrating these devices with control and measurement circuitry, in the so-called "More than Moore" approach.[78,79] The photolithographic method of their fabrication also enables reproducible production and systematic variation of such factors such as band thickness (width) and placement, hole dimension and separation. This work also indicates that they provide a suitable platform for measuring and understanding nanoelectrode effects. Nanoband systems of 50 nm thickness are demonstrated to be sufficiently thin to provide enhanced nanoelectrode characteristics, albeit that their response shows some migrational transport contribution at all but the highest ionic strengths. Nanoband systems of 5 nm thickness show lower currents and increased relative migrational effects, consistent with previously postulated effects of finite redox ion size on the nanoband dimension.

7 EIS theory

Previous work has modelled the EIS behaviour of microelectrodes of varying sizes and geometries. A method to extract the diffusional impedance of a microdisc from the chronoamperometric current[80] was extended to microbands of various ideal and non-ideal geometries.[81,82] Gabrielli *et al.* used FEMLAB (a predecessor of COMSOL) to simulate the EIS response of a microdisc.[83]

This work develops an analytical description for the MNEE. Previous work[9–12,68] has shown the variation of C_N with the concentrations of the two forms of a one electron redox couple (*e.g.* reduced and oxidised FCA), a and b in the volume of solution above the microsquares should be given by

$$C_N = \frac{NAlF^2}{RT} \frac{ab}{c_\infty} \qquad (3)$$

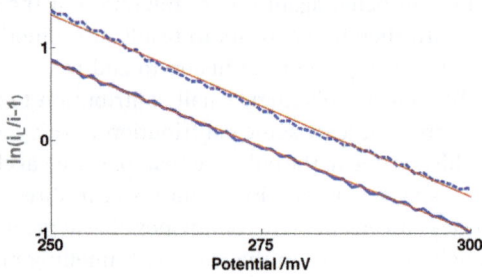

Fig. 14 Modified Tafel plot (eqn (15)) from CV, Fig. 4, for 10μ3D50n for 0.10 mM FCA oxidation in FCA(l) for oxidative (solid) and reductive (dashed) sweeps at v = 100 mV s⁻¹. The Tafel slopes are 37 and 39 V⁻¹ respectively by linear regression; the theoretical value is 39 V⁻¹ from eqn (15) at T = 298 K.

where l and A, the height and area combine to give the volume, V. As for a reversible one electron couple, near the electrode surface

$$E_{dc} = E' + \frac{RT}{F}\ln\left(\frac{b}{a}\right) \tag{4}$$

where E' is the formal redox potential for FCA; combining these equations results in

$$C_N = 4C_N^{max}(2 + e^{-P} + e^P)^{-1} = 4C_N^{max}\left(\cosh^2\left[\frac{P}{2}\right]\right)^{-1} \tag{5}$$

where

$$P = \frac{F(E_{dc} - E')}{RT} \tag{6}$$

and the maximum capacitance at $E_{dc} = E'$ is given by

$$C_N^{max} = \frac{NVF^2 c_\infty}{4RT} \tag{7}$$

The equations for the potential dependence of R_{nl}, and R_{ct} for single microsquares have been derived previously;[53] when combined with the fact that for an array of N parallel independent circuit elements, $R_{array} = R_{single}/N$; $C_{array} = NC_{single}$, this gives

$$R_{ct} = \frac{R_{ct}^{min}}{2}\left(e^{-\alpha_{ox}P} + e^{(1-\alpha_{ox})P}\right) \tag{8}$$

where α_{ox} is the transfer coefficient for the oxidation redox reaction and at $E_{dc} = E'$, the minimum charge transfer resistance is given by

$$R_{ct}^{min} = \frac{2RT}{NAF^2 c_\infty k^\theta} \tag{9}$$

from which k^θ, the standard rate constant for the redox reaction, can be obtained.

As shown previously[53]

$$R_{nl} = R_{nl}^{min}\cosh^2\left(\frac{P}{2}\right) \tag{10}$$

which again gives the minimum value, R_{nl}^{min}, at $E_{dc} = E'$. The value of Y_0 (the magnitude of the admittance at an EIS frequency, $\omega = 1$ rad s^{-1}) for the Warburg element seen at low frequency is

$$Y_0 = \frac{F^2 A_{tot}}{RT}\left(\frac{1}{\sqrt{D_O}b} + \frac{1}{\sqrt{D_R}a}\right)^{-1} \qquad (11)$$

Assuming that the diffusion coefficients of the oxidized, D_O, and reduced, D_R, redox species are equal and given by D

$$Y_0 = \frac{\sqrt{D}F^2 A_{tot}}{RT}\left(\frac{ab}{c_\infty}\right) \qquad (12)$$

where, due to complete diffusion layer overlap from the array electrodes, A_{tot} is the total array area. By analogy with eqn (3)–(5), this gives

$$Y_0 = Y_0^{max}\left(\cosh^2\left(\frac{P}{2}\right)\right)^{-1} \qquad (13)$$

with at $E_{dc} = E'$

$$Y_0^{max} = \frac{\sqrt{D}F^2 A_{tot} c_\infty}{4RT} \qquad (14)$$

from which D can be obtained.

In order to relate the mass transfer limited current, i_L, to R_{nl}, the modified Tafel equation, shown to be applicable to a reversible one electron transfer at these electrodes (Fig. 14)

$$\ln\left(\frac{i_L}{i} - 1\right) = \frac{F(E - E_{1/2})}{RT} \qquad (15)$$

is rearranged to give

$$i = \frac{i_L}{1 + \exp\left[\frac{F(E - E_{1/2})}{RT}\right]} \qquad (16)$$

where the half-wave potential, $E_{1/2} \approx E'$. A small change in potential under these conditions, $\Delta E = E_1 - E_2$, as in an EIS experiment, then results in a change in current $\Delta i = i_1 - i_2$, where

$$\frac{i_L}{i_1} - \frac{i_L}{i_2} = 1 + \exp\left[\frac{F(E_1 - E_{\frac{1}{2}})}{RT}\right] - \left(1 + \exp\left[\frac{F(E_2 - E_{\frac{1}{2}})}{RT}\right]\right)$$

$$= e^{P_2}\left(\exp\left[\frac{\Delta EF}{RT}\right] - 1\right) \qquad (17)$$

where $P_2 = (E_2 - E')F/RT$.

For this small potential change, eqn (17) becomes

$$\frac{i_L}{i_1} - \frac{i_L}{i_2} = \frac{i_L \Delta i}{i_1 i_2} = \left(\frac{\Delta EF}{RT}\right)e^{P_2} \qquad (18)$$

The resulting impedance is then the non-linear resistance under these conditions

$$\frac{\Delta E}{\Delta i} = R_{nl} = \frac{i_L RT}{i_1 i_2 F} e^{-P_2} \qquad (19)$$

When ΔE is small

$$i_1 i_2 \approx \left(\frac{i_L}{1+e^P}\right)^2 \qquad (20)$$

Combining eqn (19) and (20) then gives

$$R_{nl} = \frac{RT}{i_L F}(2 + e^{-P} + e^P) = \frac{4RT}{i_L F} \cosh^2\left(\frac{P}{2}\right) \qquad (21)$$

Thus when $\cosh^2(P/2) = 1$ ($P = 0$, or $E_{dc} = E_{1/2} = E'$), the value of R_{nl} is equal to R_{nl}^{min}, the minimum non-linear resistance, which is related to i_L through

$$i_L = \frac{4RT}{FR_{nl}^{min}} \qquad (22)$$

Acknowledgements

The School of Chemistry at Edinburgh is celebrating 300 years of Chemistry at Edinburgh and is part of the EaStCHEM joint chemistry research school with St Andrews. The Institute for Integrated Micro and Nano Systems is part of the Institute for Integrated Systems (a Joint Research Institute with Heriot-Watt University as part of the Edinburgh Research Partnership in Engineering and Mathematics (ERPem)). Support from the Scottish Funding Council is acknowledged for the ERPem and EaStCHEM initiatives, along with knowledge transfer funding. Support from EPSRC/IeMRC is also acknowledged and IS and HLW thank EPSRC and the University of Edinburgh for financial support.

References

1. J. Heinze, *Angew. Chem., Int. Ed. Engl.*, 1993, **32**, 1268–1288.
2. R. J. Forster, *Chem. Soc. Rev.*, 1994, **23**, 289–297.
3. K. Stulik, C. Amatore, K. Holub, V. Marecek and W. Kutner, *Pure Appl. Chem.*, 2000, **72**, 1483–1492.
4. A. J. Bard and L. R. Faulkner, *Electrochemical methods*, 2nd ed., Wiley, New York, 2001, pp. 168–176.
5. C. Batchelor-McAuley, E. J. F. Dickinson, N. V. Rees, K. E. Toghill and R. G. Compton, *Anal. Chem.*, 2012, **84**, 669–684.
6. C. Amatore, C. Pebay, L. Thouin, A. Wang and J.-S. Warkocz, *Anal. Chem.*, 2010, **82**, 6933–6939.
7. J. T. Cox and B. Zhang, *Annu. Rev. Anal. Chem.*, 2012, **5**, 253–272.
8. A. M. Bond, *Analyst*, 1994, **119**, 1R–21R.
9. Y. Tu, Y. Lin and Z. F. Ren, *Nano Lett.*, 2003, **3**, 107–109.
10. I. Heller, J. Kong, H. A. Heering, K. A. Williams, S. G. Lemay and C. Dekker, *Nano Lett.*, 2005, **5**, 137–142.
11. D. Wei, Y. Liu, L. Cao, Y. Wang, H. Zhang and G. Yu, *Nano Lett.*, 2008, **8**, 1625–1630.
12. D. W. M. Arrigan, *Analyst*, 2004, **129**, 1157–1165.
13. R. G. Compton, G. G. Wildgoose, N. V. Rees, I. Streeter and R. Baron, *Chem. Phys. Lett.*, 2008, **459**, 1–17.
14. R. W. Murray, *Chem. Rev.*, 2008, **108**, 2688–2720.
15. J. M. Pingarron and P. Yanez-Sedeno, *Curr. Top. Electrochem.*, 2003, **9**, 165–176.
16. R. Feeney and S. P. Kounaves, *Electroanalysis*, 2000, **12**, 677–684.
17. C. A. Amatore, *Spectra 2000*, 1990, **151**, 43–46.
18. M. Fleischmann and S. Pons, *Anal. Chem.*, 1987, **59**, 1391A–1399A.

19 J. O. Howell, *Curr. Sep.*, 1987, **8**, 2–16.
20 K. Gerasopoulos, X. Chen, J. Culver, C. Wang and R. Ghodssi, *Chem. Commun.*, 2010, **46**, 7349–7351.
21 M. A. Alonso-Lomillo, O. Rüdiger, A. Maroto-Valiente, M. Velez, I. Rodríguez-Ramos, F. J. Muñoz, V. M. Fernández and A. L. De Lacey, *Nano Lett.*, 2007, **7**, 1603–1608.
22 C. Koenigsmann, W.-P. Zhou, R. R. Adzic, E. Sutter and S. S. Wong, *Nano Lett.*, 2010, **10**, 2806–2811.
23 D. Zhan, J. Velmurugan and M. V. Mirkin, *J. Am. Chem. Soc.*, 2009, **131**, 14756–14760.
24 R. Hu, B. A. Cola, N. Haram, J. N. Barisci, S. Lee, S. Stoughton, G. Wallace, C. Too, M. Thomas, A. Gestos, *et al.*, *Nano Lett.*, 2010, **10**, 838–846.
25 Q. Li and R. M. Penner, *Nano Lett.*, 2005, **5**, 1720–1725.
26 D. Gu, H. Baumgart and G. Namkoong, *Phys. Status Solidi RRL*, 2011, **5**, 104–106.
27 I. Heller, J. Kong, H. A. Heering, K. A. Williams, S. G. Lemay and C. Decker, *Nano Lett.*, 2005, **5**, 137–142.
28 A. C. Barton, S. D. Collyer, F. Davis, D. D. Gornall, K. A. Lawa, E. C. D. Lawrence, D. W. Mills, S. Myler, J. A. Pritchard, M. Thompson and S. P. J. Higson, *Biosens. Bioelectron.*, 2004, **20**, 328–337.
29 M. S. Doescher, U. Evans, P. E. Colavita, P. G. Miney and M. L. Myricks, *Electrochem. Solid-State Lett.*, 2003, **6**, C112–C115.
30 R. M. Penner and C. R. Martin, *Anal. Chem.*, 1987, **59**, 2625–2630.
31 U. Evans, P. E. Colavita, M. S. Doescher, M. Schiza and M. L. Myricks, *Nano Lett.*, 2002, **2**, 641–645.
32 S. Viswanathan, C. Rani, S. Ribeiro and C. Delerue-Matos, *Biosens. Bioelectron.*, 2012, **33**, 179–183.
33 P. Ugo, L. M. Moretto, M. De Leo, A. P. Doherty, C. Vallese and S. Pentlavalli, *Electrochim. Acta*, 2010, **55**, 2865–2872.
34 M. Faustini, B. Marmiroli, L. Malfatti, B. Louis, N. Krins, P. Falcaro, G. Grenci, C. Laberty-Robert, H. Amenitsch, P. Innocenzic and D. Grosso, *J. Mater. Chem.*, 2011, **21**, 3597–3603.
35 S. P. Branangan, N. M. Contento and P. W. Bohn, *J. Am. Chem. Soc.*, 2012, **134**, 8617–8624.
36 M. Manheller, S. Trellenkamp, R. Waser and S. Karthäuser, *Nanotechnology*, 2012, **23**, 125302–125305.
37 M. E. Sandison and J. M. Cooper, *Lab Chip*, 2006, **6**, 1020–1025.
38 M. P. Nagale and I. Fritsch, *Anal. Chem.*, 1998, **70**, 2902–2907.
39 M. P. Nagale and I. Fritsch, *Anal. Chem.*, 1998, **70**, 2908–2913.
40 E. J. Menke, M. A. Thompson, C. Xiang, C. Yang and R. M. Penner, *Nat. Mater.*, 2006, **5**, 914–919.
41 C. Xiang, S.-C. Kung, D. K. Taggart, F. Yang, M. A. Thompson, A. G. Guell, Y. Yang and R. M. Penner, *ACS Nano*, 2008, **2**, 1939–1949.
42 F. Yang, D. Jung and R. M. Penner, *Anal. Chem.*, 2011, **83**, 9472–9477.
43 N. A. M. Said, K. Twomey, V. I. Ogurtsov, D. W. M. Arrigan and G. Herzog, *J. Phys.: Conf. Ser.*, 2011, **307**, 012052.
44 E. Jeoung, T. H. Galow, J. Schotter, M. Bal, A. Ursache, M. T. Tuominen, C. M. Stafford, T. P. Russell and V. M. Rotello, *Langmuir*, 2001, **17**, 6396–6398.
45 J. Koehne, J. Li, A. M. Cassell, H. Chen, Q. Ye, H. T. Ng, J. Han and M. Meyyappan, *J. Mater. Chem.*, 2004, **14**, 676–684.
46 K. Wang, H. A. Fishman, H. Dai and J. S. Harris, *Nano Lett.*, 2006, **6**, 2043–2048.
47 N. Yang, H. Uetsuka, E. Osawa and C. E. Nebel, *Nano Lett.*, 2008, **8**, 3572–3576.
48 M. D. Dickey, D. J. Lipomi, P. J. Bracher and G. M. Whitesides, *Nano Lett.*, 2008, **8**, 4568–4573.
49 Q. Xu, R. M. Rioux, M. D. Dickey and G. M. Whitesides, *Acc. Chem. Res.*, 2008, **41**, 1566–1577.
50 N. J. Freeman and A. R. Mount, *Microelectrode System*, PCT WO99/60392A1.
51 J. G. Terry, I. Underwood, I. Schmueser, A. S. Bunting, A. R. Mount, N. J. Freeman and A. J. Walton, *IET Nanobiotechnology*, to be submitted, 2013.
52 N. J. Freeman, R. Sultana, N. Reza, H. Woodvine, J. G. Terry, A. J. Walton, C. L. Brady, I. Schmueser and A. R. Mount, *Phys. Chem. Chem. Phys.*, 2013, **15**, 8112–8118.
53 H. L. Woodvine, J. G. Terry, A. J. Walton and A. R. Mount, *Analyst*, 2011, **135**, 1058–1065.
54 J. A. Alden and R. G. Compton, *J. Phys. Chem. B*, 1997, **101**, 8941–8954.
55 J. B. Flanagan and L. Marcoux, *J. Phys. Chem.*, 1973, 77, 1051–105.
56 J. A. Alden, J. Booth, R. G. Compton, R. A. W. Dryfe and G. H. W. Sanders, *J. Electroanal. Chem.*, 1995, **389**, 45–54.
57 I. J. Cutress, E. J. F. Dickinson and R. G. Compton, *J. Electroanal. Chem.*, 2010, **638**, 76–83.
58 J. A. Alden and R. G. Compton, *J. Electroanal. Chem.*, 1996, **402**, 1–10.
59 C. A. Basha and L. Rajendran, *Int. J. Electrochem. Sci.*, 2006, **1**, 268–282.
60 J. Guo and E. Lindner, *Anal. Chem.*, 2009, **81**, 130–138.

61 T. J. Davies and R. G. Compton, *J. Electroanal. Chem.*, 2005, **502**, 177–187.
62 D. Britz, K. Poulsen and J. Strutwolf, *Electrochim. Acta*, 2004, **50**, 107–113.
63 K. B. Oldham, *Electrochem. Commun.*, 2004, **6**, 210–214.
64 C. Amatore, *Physical Electrochemistry: Principles, Methods and Applications*, I. Rubinstein, Ed., Marcel Dekker, New York, 1995, pp. 131–208.
65 K. Aoki, *Electroanalysis*, 1993, **5**, 627–639.
66 A. Berduque, Y. H. Lanyon, V. Beni, G. Herzog, Y. E. Watson, K. Rodgers, F. Stam, J. Alderman and D. W. M. Arrigan, *Talanta*, 2007, **71**, 1022–1030.
67 M. Fleishmann and S. Bandyopadhyay, *J. Phys. Chem.*, 1985, **89**, 5537–5541.
68 A. R. Mount and M. T. Robertson, *Phys. Chem. Chem. Phys.*, 1999, **1**, 5169–5177.
69 R. He, S. Chen, F. Yang and B. Wu, *J. Phys. Chem. B*, 2006, **110**, 3262–3270.
70 C. P. Smith and H. S. White, *Anal. Chem.*, 1993, **65**, 3343–3353.
71 M. D. Osborne, B. J. Seddon, R. A. W. Dryfe, G. Lagger, U. Loyall, H. Schäfer and H. H. Girault, *J. Electroanal. Chem.*, 1996, **417**, 5–15.
72 A. M. Bond, T. L. E. Henderson, D. R. Mann, T. F. Mann, W. Thormann and C. G. Zoski, *Anal. Chem.*, 1988, **60**, 1878–1882.
73 J. J. Watkins, B. Zhang and H. S. White, *J. Chem. Educ.*, 2005, **82**, 712–719.
74 D. Krapf, B. M. Quinn, M.-Y. Wu, H. W. Zandbergen, C. Dekker and S. G. Lemay, *Nano Lett.*, 2006, **6**, 2531–2535.
75 R. B. Morris, D. J. Franta and H. S. White, *J. Phys. Chem.*, 1987, **91**, 3559–3564.
76 J. D. Seibold, E. R. Scott and H. S. White, *J. Electroanal. Chem.*, 1989, **264**, 281–289.
77 R. J. White and H. S. White, *Langmuir*, 2008, **24**, 2850–2855.
78 M. Lapisa, G. Stemme and F. Niklaus, *IEEE J. Sel. Top. Quantum Electron.*, 2011, **17**, 629–644.
79 A. J. Walton, J. T. M. Stevenson, I. Underwood, J. G. Terry, S. Smith, W. Parkes, C. Dunare, H. Lin, Y. Li, R. Henderson, D. Renshaw, B. Rae, K. Muir, M. Desmulliez, D. Flynn, M. J. MacIntosh, W. S. Holland, A. F. Murray, T. B. Tang and A. S. Bunting, *SAIEE Res. J.*, 2010, **10**, 3–10.
80 A. Bezegh and J. Janata, *J. Electroanal. Chem.*, 1986, **215**, 139.
81 R. G. Compton and J. Winkler, *J. Phys. Chem.*, 1995, **99**, 5029.
82 J. A. Alden and R. G. Compton, *Electroanalysis*, 1996, **8**, 30–33.
83 C. Gabrielli, M. Keddam, P. Rousseau and V. Vivier, Numerical Simulation of the Electrochemical Impedance of a Microelectrode using FEMLAB, *Excerpt from the Proceedings of the COMSOL Multiphysics User's Conference*, 2005, Paris.

Faraday Discussions

DISCUSSIONS

General discussion

DOI: 10.1039/C3FD90030D

Professor Compton opened the discussion of the paper by Professor Pumera: Please can you say a little more about the conditions used for insonation? Did you use a sonic horn or a bath? Did you measure the ultrasound power and was this above or below the cavitation threshold?

Professor Pumera answered: We used ultrasonication bath with well defined ultrasonication power. However, we have not varied the ultrasonication power. This is very good idea for follow-up project, many thanks.

Professor Compton asked: Can I ask about the possible mechanisms for exfoliation? Do you believe that cavitation is essential? Or will the mechanical forces caused by application of ultrasound be sufficient? Could cavitation cause the formation of defects in the graphene?

Professor Pumera replied: I do not recall any systematic study on the intensity of the ultrasonication and on the cavitation. I am of the opinion that cavitation should cause formation of the defects.

Professor Barbero remarked: It is now quite common to use ultrasound to accelerate organic chemistry reactions. Do you think graphene oxide suffers a chemical reaction promoted by ultrasound which could create redox groups active in the electrocatalysis of analytes?

Professor Pumera answered: This is an excellent question. I usually do not comment on our unpublished work but, yes, indeed, we observed changed inherent electrochemistry of graphene oxides upon different sonication times. This is a subject of our coming paper, hopefully published soon.

Dr Batchelor-McAuley said: Recently, there has been some fantastic work coming out of the Unwin group in Warwick (S. C. S. Lai *et al.*, *Angew. Chem., Int. Ed.*, 2012, **51**, 5405–5408). In contrast to much of the literature, they demonstrated that at the microscopic scale the rate of electron transfer to basal plane graphite sites is surprisingly high. I would like to take a step back on your work and ask; given that you have demonstrated that impurities in graphite may have a significant influence on the electrochemical response of the formed graphene, do you think it is possible that occluded metal particles in HOPG may also have a significant role in its electroactivity?

Professor Pumera answered: This is an interesting question! I can provide only my opinion, with a chance to be correct or wrong of 50/50, like all of my opinions! It is without question that the metallic impurities are present in graphite. However, we have not observed any electrocatalysis on graphite or BPPG HOPG electrodes when using macroelectrodes. This is likely to be due to the fact that most of the metallic impurities are hidden between the graphitic layers and the amount present on the electrode surface is insufficient to dominate the electrochemistry of the macroelectrode. However, it is possible (and here I am likely wrong) that on the microscale, when scanning with electrochemical microscope at very local area, the resulting electrocatalytic effect can dominate the observed electrochemistry. I think you should ask Pat Unwin; although not present at this meeting he is the best person to answer this question.

Dr Kataky remarked: What solvent do you use to extract impurities from graphene oxide? How do you dry and weigh GO?

Professor Pumera replied: This depends on the project, we use DMF or water. In this case, we used water. We dry GO in the oven before further processing. It is well documented, even from the pre-graphene era, that swelling of graphite oxide depends on the oxidation method and on humidity.

Professor Compton commented: You form your graphene modified electrodes by drop-casting material onto the surface of the electrode to be modified. To what extent is this reproducible? Is there the possibility for the graphene sheets to re-assemble? Certainly we find the drop casting of nanoparticles to be highly unsatisfactory for quantitative work (C. C. M. Neumann et al., *ChemElectroChem*, in press).

Professor Pumera answered: The reproducibility of the deposition of graphene related materials on the electrode depends in part on the quality of the dispersion of graphene in the solvent. This is certainly a challenging issue, RSD of measurements performed on graphene electrodes are certainly higher that these done on GC bulk electrodes. However, the RSD are not usually so bad to prohibit quantification. To address the second part of the question, indeed, the individual graphene (oxide) particles stack on the surface, randomly, blocking the materials below. Craig Banks has very nice paper on this topic in *RSC Advances* (2011); we have contributed to this as well in *Chem. Asian. J.*, *Electrochem. Commun.* and *Anal. Chem.* (2010), showing that for many analytes and for capacitance measurements, it does not matter if you use single, few or multilayer graphene. I do not think many people followed this.

Professor Compton asked: Can you compare the relative electroanalytical merits of single sheet graphene with an electrode modified with a layer of carbon nanotubes?

Professor Pumera answered: The proper answer is more complicated than it seems. Provided that the SWCNTs are absolutely pure and graphene is perfect and pure, the difference should be only in the amount of electroactive edges, which is higher in the case of graphene, and can be considered as an open

SWCNT. However, I have not seen gram quantities of pure SWCNT or graphene.

Professor Schuhmann requested: Please give us your personal definition of 'graphene'.

Professor Pumera replied: Well, we have already a proper definition of graphene by IUPAC, and I do not wish to add to it. Indeed, the word 'graphene' is often misused and materials which have a structure quite different from graphene are still called graphene. I wrote article called "Will the real Graphene please stand up?" criticizing this practice which was published in *Materials Views* http://www.materialsviews.com/will-real-graphene-please-stand-up/ but I am not sure how many people listened.

Professor Schuhmann continued: How can we define 'graphene' solidly? Is reduced graphene oxide graphene? How many layers of graphene sheets can still be called graphene? I always assumed that graphene is distinguished from HOPG only if the physical properties unique for a single graphene sheet are still measurable (probably up to 4 to 10 graphene layers). I would suggest that one should name the materials not based on wishful thinking of 'graphene' but correctly, based on physical and chemical properties, including defects; *e.g.* reduced graphene oxide *etc.*

Professor Pumera responded: I completely agree. The way in which one prepares a final graphene-related materials (we can call them graphenes for simplicity of this discussion; here goes my definition of the word including graphene and graphene oxides of various forms. Language is tool and not dogma so please allow me this definition of a 'new' word for this class of materials) defines its properties. There should be stated not only the reduction method, *i.e.* chemical or electrochemical reduction (see, *e.g.*, Ambrosi *et al.*, *Chem.–Eur. J.*, 2011, **17**, 10763.) but also an oxidation method because it strongly influences the C/O ratio, density of defects and the resulting electrochemistry of reduced graphene oxides (see, *e.g.*, Poh *et al.*, *Nanoscale*, 2012, **4**, 3515). Many of these materials are closer to amorphous carbon than to the IUPAC definition of graphene (see Wong *et al.*, *Nanoscale*, 2012, **4**, 4972). In addition, graphene oxides prepared by three different classical methods, such as Staudenmaier, Hummers and Hofmann exhibit significantly different inherent electrochemistry, that is electrochemical reduction of oxygen-containing groups on the surface of graphene oxide. Clearly, the oxidation methods are not equivalent. You can find more details in our Chua *et al.*, *Chem.–Eur. J.*, 2012, **18**, 13453, and upcoming *Chem.–Eur. J.*, 2013. I completely agree with your comment, as the produced 'graphenes' are complex materials defined by their history, there should always be given a proper description of how they were prepared and proper characterization. These graphenes should be indeed called according the method of preparation, *e.g.* 'Staudenmaier thermally reduced graphene oxide'. We do this in our papers and I ask as referee that authors to do so as well; however slowly I am giving up, as the resistance is huge.

Professor Bond said: In the increasingly common electrode kinetic studies being undertaken at impure graphene electrodes of the kind described in this paper, is it the uncertainties in electrode density of states, point of zero charge, or alteration of the double layer or other factors, that give rise to the uncertainties in significance of published heterogeneous charge transfer rate constants? Is there a major problem because we have no reference value available as to what is the true value of the point of zero charge at a pure graphene electrode?

Professor Pumera replied: The problem of reference material is major problem of graphene research, as it had been in case of carbon nanotubes. In addition, quite often the graphene-related materials are poorly characterized and any of the points you listed can contribute to the uncertainties.

Professor Schuhmann opened the discussion of the paper by Professor Kanoufi: What evidence do you have that diazonium chemistry does definitively lead to an Au–C bond?

Professor Kanoufi replied: The discussion on the bond strength of the aryl layer and the surface has been reviewed by Pinson and Bélanger.[1] Briefly, direct spectroscopic evidences of the surface–aryl bond was obtained for iron, copper and glassy carbon substrates.[2] For other surfaces the strength is rather related to the resistance of the grafting to various stimuli. In the case of an Au surface, the Au–aryl bond strength was compared to the Au–thiol bond from both DFT calculations[3] and experiments.[4] DFT calculation showed that the Au–aryl bond is comparable or stronger than the Au–thiol one and the Au–C bond has a higher ionic character than the Au–S one: the Au^+–C^- bond formation is accompanied by partial charge transfer of 0.45 e^- from Au. On the other hand, experiments demonstrated that contrary to Au–thiol layers the Au–aryl layers are poorly sensitive to potential activation from −2 V to 1.8 V (no potential-activated desorption) or to desorption by thiol displacement.

1 J. Pinson and D. Bélanger, *Chem. Soc. Rev.*, 2011, **40**, 3995.
2 (a) K. Boukerma *et al.*, *Langmuir*, 2003, **19**, 6333; (b) M.-C. Bernard *et al.*, *Chem. Mater.*, 2003, **15**, 3450; (c) T. Itoh and R. L. McCreery, *J. Am. Chem. Soc.*, 2002, **124**, 10894.
3 (a) D. E. Jiang *et al.*, *J. Am. Chem. Soc.*, 2006, **128**, 6030; (b) E. de la Llave *et al.*, *J. Phys. Chem. C*, 2008, **112**, 17611.
4 (a) G. Liu *et al.*, *J. Electroanal. Chem.*, 2007, **600**, 335; (b) D. M. Shewchuk and M. T. McDermott, *Langmuir*, 2009, **25**, 4556.

Dr Bohn asked: Your images show significant roughness at the electrode surface. Can you expand on the modeling efforts you used to develop a theoretical description of the optical reflectance data?

Professor Kanoufi responded: The movie I presented in the oral presentation was used to illustrate qualitatively the optical effects that can be observed. Indeed the electrode used presented significant scratches, as is the one depicted in Fig. 4c. The behaviours observed have been actually generalized to <2 nm roughness gold-coated microelectrodes. For a roughness much smaller than the light wavelength, λ, typically <10 nm, the model proposed here is valid. It is possible to account theoretically for the effects of higher roughness on the optical response and this was developed for optical methods such as reflectometry and ellipsometry.[1,2] However these models might be difficult to validate experimentally.[3] Therefore, in a first approximation we have not taken into account the electrode roughness in the proposed optical model and preferentially considered the optical responses of the regions of the electrodes presenting no apparent defects (scratches).

1 R. A. Azzam and N. M. Bashara, *Ellipsometry and Polarized Light*, North Holland, Amsterdam, 1977.
2 I. Ohlidal et al., *J. Phys. Coll.*, 1977, **38**, 77.
3 D. A. Ramsey and K. C. Ludema, *Rev. Sci. Instrum.*, 1994, **65**, 2874.

Professor Compton commented: Concerning the mechanism you propose, are you certain that the aryl radicals are free in solution or could they be adsorbed on the electrode surface? What fraction of the aryl radicals formed become attached to the electrode? What is the fate of the others – by what species are they scavenged? Could they dimerise?

Professor Kanoufi replied: From an optical point of view we detect any species which is adsorbed on the electrode surface. Actually, during the CCD camera acquisition timescale we are looking more at adsorption or deposition of material on already grafted surfaces. The grafting of a pristine Au surface would require faster acquisition CCD cameras. We then actually look at all the processes occurring within the grafted layer, adsorption, grafting or incorporation of intermediates within the layer. We then compare the amount of deposited material detected optically to the amount of reduced diazonium species obtained from coulometry. This is given in eqn (8) by n_e, the apparent number of electrons required for the reduction of one diazonium ion. For $n_e > 1$ clearly some generated radicals are not trapped in the layer growth. If we consider that all deposited material is aryl radical the fraction of attached radicals is $1/n_e$. The latter assumption may be erroneous, indeed recent[1] combined QCM–electrochemical–ellipsometric monitoring of the growth of aryl layers from diazonium electrografting suggests significant trapping of the supporting electrolyte and solvent in the layer, as is usually observed in electropolymerization processes. This would consequently decrease the fraction of grafted radicals.

The fates of the other radicals may be diverse. We considered here an apparent 1st order chemical transformation. This might be H-atom abstraction. Dimerization, a second-order process, is also likely and more favoured at high diazonium concentration. Such issues have been discussed in the understanding of SRN1 reactions obtained from aryl halide reduction.[2] Actually, owing to its high oxidizing potential, the aryl radical could undergo direct reduction by the electrode. This reduction step, which may occur at the electrode or in solution through mediated ET, consumes electrons. It will then increase the value of n_e and also consequently decrease the fraction of grafted radicals. Its role is increased as the electrode potential is polarized to more negative regions when reaching higher grafting flux regimes.

1 J. Vinther et al., *Langmuir*, 2013, **29**, 5181.
2 (a) C. Amatore et al., *J. Am. Chem. Soc.*, 1981, **103**, 6930; (b) C. Amatore et al., *J. Am. Chem. Soc.*, 1985, **107**, 3451.

Professor Barbero asked: The optical absorbance of your absorbing film is in the region of 400–500 nm? Could it be due to azo dye formation? While it is thought that diazonium coupling requires a highly activated aromatic (*e.g.* phenoxide) in surface reactions, where low yields still produce attached groups, even poorly activated aromatic rings can react with diazonium ion to form azo dyes.

Professor Kanoufi replied: The absorbance of grafted film in the 400–500 nm region may have different origins. The formation of highly conjugated oligomeric structure such as a poly(nitrophenylene) is a possibility. Indeed the formation of azo bonds during the grafting of multilayers from diazonium reduction has been reported in the literature.[1–4] Even if the proportion of aryl groups within the layer engaged in azo coupling is <20%, the high absorption coefficient ($\alpha \sim 5 \times 10^4$ cm^{-1}) of azo dyes[5] makes them another good possibility for the observed strong absorbance of the grafted layers in this optical region. The coupling of diazonium with surface activated phenoxides was postulated when the grafting takes place at carbon electrode surfaces.[1] At metallic electrodes, such surface phenoxides are less likely and then the coupling between a diazonium and a radical within the grafted layer has instead been invoked.[4]

1 C. Saby *et al.*, *Langmuir*, 1997, **13**, 6805.
2 B. L. Hurley and R. L. McCreery, *J. Electrochem. Soc.*, 2004, **151**, B252.
3 M.-C. Bernard, *et al.*, *Chem. Mater.*, 2003, **15**, 3450.
4 P. Dopplet *et al.*, *Chem. Mater.*, 2007, **19**, 4570.
5 *E.g.* H. M. El-Nasser, *Phys. B: Condens. Matter*, 2011, **406**, 1940.

Professor Xu opened the discussion of the paper by Dr Walcarius: The pH of the electrode surface can be changed by oxidation or reduction. Do you think oxidation or reduction is a better way to prepare self-assembled mesoporous silica films?

Dr Walcarius answered: Yes, both cathodic and anodic tuning of pH at an electrode surface can be basically used to deposit sol–gel films. In the present case, however, the applied cathodic potential not only plays the role of generating the hydroxyl ions required to accelerate polycondensation of silica precursors, but also to pre-assemble in a transient way the cationic CTAB surfactant hemi-micelles onto the electrode surface (around which the mesoporous silica film will grow according to a self-assembly condensation mechanism owing to favorable electrostatic interactions between the positively charged surfactant and the negatively charged silica precursors). So, if the goal is to get highly ordered and oriented mesoporous silica films generated by surfactant self-assembly under potential control, reduction is mandatory.

Professor Xu commented: Please tell me the diameter range of pore in mesoporous silica film?

Dr Walcarius replied: The pore diameter in the final material is controlled by the size of the surfactant template. Here, the films prepared using CTAB are characterized by a pore size of about 2 nm. By varying the length of the alkyl chain in the alkyl-trimethyl-ammonium template (*e.g.*, between C14 and C18), it is possible to slightly tune the pore diameter in the 2–3 nm size range (*i.e.*, lattice parameters in the range 3.8–4.6 nm).

Professor Xu returned the discussion of the paper by Professor Pumera: Although metal impurities in carbon nanomaterials may result in good electrocatalytic properties, Professor Compton has reported that pure carbon nanotubes show good electrocatalytic properties for some compounds (C. P. Jones *et al.*, *Langmuir*, 2007, **23**, 9501). Graphene is another kind of carbon nanomaterial, it has also been reported to show good electrocatalytic properties for some compounds. Some papers reported that electrocatalysis results from the edge, some papers reported that the basal plane also shows electrocatalytic properties. What are your ideas about the reasons of the good electrocatalytic properties of graphene?

Professor Pumera responded: According to our studies, more defects and higher C/O ratio lead to higher HET. We had a large comparative study some time ago in Ambrosi *et al.*, *Chem.–Eur. J.*, 2011, **17**, 10763. The fastest HET exhibits thermally reduced graphene oxide which has a structure similar to amorphous carbon (Wong *et al.*, *Nanoscale*, 2012, **4**, 4972). As for edge plane or basal plane, this depends on the analyte.

Professor Compton returned to the discussion of the paper by Dr Walcarius: At the end of your paper you describe the use of functionalised films for metal (*e.g.* silver) ion accumulation leading to enhanced stripping voltammetry. Can you explain the mechanism for this process and suggest where the metal particles are formed — is it in the film or on the electrode surface?

Dr Walcarius responded: The preconcentration electroanalysis process operates in two successive steps: a first open-circuit accumulation of the analyte, which is followed by medium exchange to an analyte-free solution where the previously accumulated metal ions are desorbed, immediately reduced onto the electrode surface (the cathodic potential is applied directly upon immersing the electrode in the electrolyte solution to avoid any leaching of metal ions in solution), and then detected by anodic stripping voltammetry. As far as metal ions are effectively reduced and then the metal particles re-oxidized in the stripping step, and considering the isolating character of the silica matrix, the metal deposits should be formed onto the electrode surface, at the bottom of the mesopore channels. Note that this corresponds to a very small amount of metal (one can evaluate, from integration of the stripping signal, a thickness of metal deposits no more than *ca.* 1 nm in the assumption that all mesopores are concerned with zero-valent metal deposition). On the other hand, to date, it remains difficult to fill totally the mesopores with metal nanowires, one of the best examples of electrodeposited mesoporous metals using mesoporous silica as a hard template being that reported by the Bartlett group (using supercritical fluids, see J. Ke *et al.*, *Proc. Natl. Acad. Sci. U. S. A.*, 2009, **35**, 14768).

Professor Compton asked: Considering the issue of metal pre-concentration (for stripping analysis) in your films is there a sensitive dependence on the strength of the binding of the metal? If the association is weak then there will be little benefit, whereas on the other hand if it is too strong the metal will not be released for electro-reduction to the metal. Is it possible to predict the optimal values?

Dr Walcarius answered: Metal pre-concentration is indeed connected to the binding strength of the metal ions to the active ligands in the material (even if the preconcentration step is performed under kinetic control (*i.e.*, diffusion-limited) and the thermodynamic equilibrium is almost never reached). You are completely right: the association should be strong enough to ensure good recognition properties (*i.e.*, for selectivity issues in the accumulation step) but not too strong to enable desorption in the detection medium (*i.e.*, effective reduction and anodic stripping). Rather than predicting optimal values (not always easy to determine stability constants for complexes formed in such a confined medium), we chose to work by selecting a compromise between a good recognition selectivity (high binding strength between the immobilized ligand and the metal ion) and a possible desorption in the detection medium (*i.e.*, by adding a competitive ligand in the medium, if necessary; for example thiourea in case of mercury analysis). Note that desorption is always made under cathodic potential application, acting as an additional driving force for recovery/reduction of the previously accumulated metal ions.

Dr Kataky returned to the discussion of the paper by Professor Pumera: Pristine graphene is hydrophobic. The hydrophobicity will be compromised by impurities and defects. Have you made any contact angle measurements on graphene composites?

Professor Pumera replied: No, we have not. It is an interesting idea for future works.

Professor Kornyshev asked: What is known about graphene–ion interactions? For instance what is the image force acting on a discrete charge on graphene?

Professor Pumera answered: This is not my area of expertise. I know that there are experimental and theoretical studies on metal NP–graphene interactions but I am not aware of any ion–graphene studies. However, this does not mean that they do not exist.

Professor Compton addressed Professor Pumera and Professor Kornyshev: I would expect that for a single sheet of graphene supported on an electrode the nature of the underlying support would become important.

Professor Pumera added: We have found that inherent electroactivity of graphene oxide on different surfaces, such as Pt, Au and GC, significantly differs (A. Bonanni and M. Pumera, *RSC Adv.*, 2012, **2**, 10575).

Professor Kornyshev responded: I expect that there will be an effect. If it is a dielectric material, like a glass, it will affect electric fields near this interface modifying the boundary condition.

Dr Freeman returned to the discussion of the paper by Professor Kanoufi: Concentric growth of rings was observed at the 25 micron diameter electrode using the reflectance technique you have developed. What did you see when you increased the length scale to the 500 micron diameter electrode – was the growth of concentric rings evident across all length scales?

Professor Kanoufi answered: The concentric growth presented in the oral communication was observed with a 250 µm diameter electrode. Similar concentric growth of optical rings is described in Fig. 6 in our paper for a 500 µm diameter electrode. The electrode edge is still detected as regions of higher reaction rate. As you suggested and, is shown in Fig. 6b, such an edge effect is constrained to regions closer to the electrode edge.

Professor Wang added: Just a (historical) comment, related to sonication and careful characterisation. Over 25 years we (along with Ted Kuwana) observed 'catalytic reactions' on GC electrodes that were attributed to alumina slurries used in the polishing step. Different sonication times, used to remove these alumina particles, and also influenced the electrochemical reactivity.

The observed 'catalytic' current of glutathione at CNT or graphene electrodes can be used for rapid and crude CV evaluations of the level and nature of metal impurities.

Professor Pumera enquired: To what other materials is this method applicable?

Professor Kanoufi responded: It actually depends on the optical properties of the substrate and the deposited material. The substrate must reflect the illuminating light. For optimal sensitivity it is better if the material reflectance R is between 0.2 and 0.7. The reflectance of the substrate can be evaluated from eqn (1) in our paper, given the substrate refractive indexes. Such values are tabulated and can be found here http://refractiveindex.info/.

The optimal sensitivity of material deposition on the substrate requires that R varies significantly with the deposited material optical thickness (n_L th, the product of the material refractive index and its thickness, respectively n_L and th in the manuscript). The value of $dR/d(n_L\ th) > 2 \times 10^{-3}$ nm^{-1} is sufficient.

In practice, the yellow reflecting metals such as Cu, Au will be particularly sensitive under blue illumination. Si, W, Pt, stainless steel and most of the materials that can be sensitively characterized by ellipsometry can be imaged through this method.

Professor Amatore remarked: Thank you dear Frédéric for this interesting and well appreciated contribution. But, I always have one concern about the mechanism of polycondensation of aryl radicals onto the electrode. Indeed, aryl radical attack on an aromatic moiety should lead to the release of a hydrogen atom to rearomatize the product. It is hard to believe that H-atoms may be released freely as such. So what is the species they react with? Another aryl radical? If so the decorating Faradaic yield should vary as a function of the presence of H-atom abstractors or of the current density (*viz.* of the aryl radical steady state concentration). Has that ever been observed or tried?

Professor Kanoufi responded: The mechanism of the layer growth implies radical coupling. This was confirmed independently from the use of radical traps either to propagate controlled radical polymerization[1] or to constrain the layer growth.[2] Indeed, the attack (2) of the electrogenerated (1) aryl radical on the surface immobilized aromatic yields the intermediate coupling radical.

$$N_2^+\text{—}\bigcirc\text{—}NO_2 + e^- \longrightarrow {}^\bullet\bigcirc\text{—}NO_2 + N_2 \tag{1}$$

$$\boxed{}\text{—}\bigcirc\text{—}NO_2 + {}^\bullet\bigcirc\text{—}NO_2 \longrightarrow \boxed{}\text{—}\underset{-H}{\overset{\bullet}{\bigcirc}}\text{—}NO_2 \tag{2}$$

with pendant $\bigcirc\text{—}NO_2$

The relatively fast growth of the polyaryl-like structure suggests electronic conductivity and then that the aromaticity is maintained in the layer. As you propose, the intermediate coupling radical may lose an H atom for rearomatization. The postulated mechanism for such rearomatization relies on the Pschorr reaction of intramolecular[3] substitution of one arene by an aryl radical during diazonium reduction. The rearomatization involves the reoxidation of the intermediate radical followed by its deprotonation.

$$\boxed{}\text{—}\underset{-H}{\overset{\bullet}{\bigcirc}}\text{—}NO_2 \longrightarrow \boxed{}\text{—}\bigcirc\text{—}NO_2 + H^+ + e^- \tag{3}$$

with pendant $\bigcirc\text{—}NO_2$

The same mechanistic path (3) is then expected to be operative in the layer growth by polycondensation. In the Pschorr mechanism, Cu(ɪ) and/or Cu(0) is used as a catalyst. Cu(0) is used to reduce the diazonium salt to generate the radical, the generated Cu(ɪ) then oxidizes the intermediate coupling radical. In this respect, an aryl-diazonium is likely to be a strong enough oxidant to do the job and oxidize/rearomatize the coupling intermediate, generating a new radical and a proton (4). It might be viewed as a redox propagation process of a radical polymerization, like those involved in Atom Transfer Radical Polymerization (ATRP).

$$\boxed{}\text{—}\underset{-H}{\overset{\bullet}{\bigcirc}}\text{—}NO_2 + N_2^+\text{—}\bigcirc\text{—}NO_2 \longrightarrow \boxed{}\text{—}\bigcirc\text{—}NO_2 + {}^\bullet\bigcirc\text{—}NO_2 + H^+ \tag{4}$$

From a Faradaic point of view, like in the SRN1 process, the biaryl coupling may be viewed as a catalytic scheme in which electrons are just needed to initiate the process, and also sustain the radical trapping in solution or in the layer. The Faradaic yield of the layer growth is then null in principle and should not affect the proposed reaction/diffusion model. Actually this assumption suggests also generation of a strongly acidic medium in the vicinity of the electrode, this consequence has not been evidenced so far.

1 H. Hazimeh et al., Chem. Mater., 2013, 25, 605.
2 T. Menenteau et al., Chem. Mater., 2013, DOI: 10.1021/cm401512c.
3 P. Hanson et al., J. Chem. Soc., Perkin Trans 2, 1999, 49.

Professor Amatore asked: Could it be that the propagating waves that you evidence on your grafting outcome are due to ohmic drop/capacitive effects? Indeed, we recently showed theoretically (C. Amatore et al., Anal. Chem., 80, 2008, 7957–7963) that such effects may generate potential gradients that move as waves from the edge of disk electrodes towards the center while a voltammetric wave is scanned. This effect was shown to be completely elusive in voltammetric traces unless unrealistic scan rates could be used. However, your method may install permanent markers of these waves on the electrode surface.

Professor Kanoufi answered: The propagating waves evidenced in Fig. 6 in our manuscript have a purely optical origin. They actually illustrate the change of reflectivity with the thickness of the grafted layer. Basically this is related to optical phenomenon as observed in the development of constructive or destructive interference fringes. As for ohmic drop/capacitive effects, the grafting proceeds here in concentrated solutions of supporting electrolyte.

Professor Williams addressed Professor Pumera: If I have interpreted your paper correctly, the electrochemistry of graphene is pretty well dominated by the impurities. Has anyone systematically varied the impurity content to determine whether there are specific electronic effects that can be understood as due to the special properties of graphene as distinct from graphene being a 'trendy' high-surface-area support for impurity nanoparticles which actually do all the work (Professor Compton commented that his work on carbon nanotubes had demonstrated that this was simply a trendy high surface area support, without any other special property).

Professor Compton added: To clarify, I commented that *in many cases* the role of carbon nanotubes (CNTs) was devoid of 'unique' or special properties, especially in the electrochemistry of relatively simple molecules. It was important to note the following, all of which potentially can give illusory 'electrocatalytic effects': (a) the presence of metallic impurities in the CNTs play a significant role in many cases); (b) nanographite impurites can behave similarly; (c) oxygen-containing groups such as are found at the ends of the nanotubes are influential; (d) the porous nature of the CNT films leads to a change of the mass transport regime from semi-infinite diffusion (on the unmodified electrode) to thin-layer type diffusion (within the porous) layer leading to increased currents and decreased overpotentials. The field has been admirably reviewed by Martin Pumera (M. Pumera, Chem. Rec., 2012, 12, 201–213).

Professor Pumera answered: We have looked at this issue for the case of carbon nanotubes and we have found such a threshold (M. Pumera and

Y. Miyahara, *Nanoscale*, 2009, **1**, 260). However, for graphene we do not have a systematic study, we have only few point studies on purified graphenes by Cl_2 treatment in the *Proc. Natl. Acad. Sci. U. S. A.*, 2012. Even at low (ppm) concentration the impurities dominate the electrochemistry of graphene. Professor Compton published a nice study on the careful doping of CNTs with metal oxide nanoparticles; he may wish to comment on this.

Professor Compton continued: Professor Williams asked about possible synergy between nanoparticles and graphene. In the case of metal impurities in multiwalled carbon nanotubes where we showed that iron oxide particles could play a significant role, for example in the detection of hydrogen peroxide (R. Compton *et al.*, *Angew. Chem., Int. Ed.*, 2006, **45**, 2533–2537); then the electrochemical behaviour was exactly that expected for the metal particles simply electrically 'wired' using the carbon nanotubes. No 'synergy' was seen.

Professor Compton then asked: Can you predict where the graphene electrochemistry field is going? Where will it be in one year's time? In five years' time? What are the real advantages of using graphene, if any, as compared to other electrode materials? In particular what are the important factors other than much-increased surface area?

Professor Pumera answered: It is extremely difficult to do so. I cannot even predict what my group will do next year! We do what we find interesting to do, and this evolves on-the-go. As for the direction, we find it very interesting to study the structure of graphene oxide, to study its inherent electrochemistry. We use this inherent electrochemistry as labels for biomolecules, in similar manner to which Au NPs are used. We find it interesting to change the electronic structure of graphene by doping it with electron donating/withdrawing groups, *i.e.* halogens, boron, nitrogen, and to study the resulting electrochemistry. This may be a dead-end in five-years' time but for now we really enjoy working on this. For the direction of the field in general, I can see a significant move towards the doping of the graphene structure, towards transparent electrodes based CVD graphenes. It will be indeed very interesting to look back at the field in five years' time.

Professor Williams said: I'd like to generalise the discussion of graphene electrochemistry and ask whether it is feasible (and reasonable to propose) to support single graphene sheets on a suitable single crystal support, systematically dope by vacuum deposition of sub-monolayer amounts of metals, and then explore their electrochemistry to see if there are any clear (and theoretically predictable) effects of the unusual electronic properties of graphene.

Professor Barbero replied: I agree that the experiment suggested would be quite interesting. Of a more fundamental interest, the measurement of the AC impedance of an undoped graphene layer would be very interesting to understand the double layer structure of graphene electrodes. I a more practical tone, would be interesting to adsorb small metallic (Fe, Ni, Co) nanoparticles onto few-layers graphene to produce the best catalytic electrodes based on graphene.

Professor Pumera responded: This will be indeed very interesting. We are now working on a similar topic.

Professor Xu addressed Professor Kanoufi: Radicals are generally less stable in aqueous solutions than in organic solvent, is it possible to map the fluxes of radicals in aqueous solutions from the combination of electrochemical activation and optical microscopy?

Professor Kanoufi answered: Aryl radicals can be generated in aqueous solutions from the reduction of diazonium salts, as was proposed here in acetonitrile. For the same nitrophenyl radical, its efficiency of grafting is indeed lower in aqueous than in acetonitrile solution which could attest indeed to its lower stability in aqueous solution. This combination of electrochemical activation and optical microscopy can be operated in the same manner to map radical fluxes. The same sensitivity of deposited material is obtained in both solvents, mainly because they have a very close refractive index.

Dr Tschulik spoke: Professor Walcarius, today the growth of ordered porous alumina by anodic oxidation can be adjusted very precisely in terms of *e.g.* degree of ordering, inter-pore distance, pore diameter and length. Would you predict that the ability to control the growth of silica will ever reach the current state of controllability in production of ordered alumina?

Dr Walcarius replied: The electro-assisted generation of hexagonally ordered and vertically oriented mesoporous silica films exploits the particular self-assembly of silica precursors and surfactants under potential control and it is therefore restricted to the particular case in which the hemi-micelle surfactant phase is likely to form onto the electrode surface and the sol-gel precursors are likely to interact with this surfactant phase to enable self-assembly condensation. This is the case of the CTAB–silica system (as pointed out here) or the sodium dodecyl sulfate–zinc oxide system (*i.e.*, to form lamellar mesoporous ZnO). On this basis, it is clear that electro-assisted deposition of mesoporous materials by self-assembly condensation would not reach the same level of control/tuning of film features as the anodization of alumina. Nevertheless, this latter approach is restricted to mesopore sizes typically down to 6 nm in the best cases (see, *e.g.*, C. X. Xu *et al.*, *J. Metastable Nanocryst. Mater.*, 2005, **23**, 75), while our electro-assisted self-assembly method enables to reach a pore size of *ca.* 2 nm, corresponding to an increase in the mesopore density (*i.e.*, the number of mesopore channels per unit surface) by almost one order of magnitude.

Professor Williams addressed Dr Walcarius and Professors Amatore and Compton: The lovely micrographs of your structures show 2-D 'grains', 'grain boundaries' and defects that interrupt the long-range order. One technique that has been used to preserve long-range order over greater distances in systems that used block-copolymers deposited on a surface to develop similar structures has been to pre-deposit a regular array of columns to act as 'pinning' sites. I wonder whether a similar strategy could be applied in your case. Thus, the electrode could be patterned with a regular array of micro-cones. The current flow would tend to focus on the cone tips, provided the ionic strength was low enough, thus imposing a spatial structure on the diffusion field of protons above the electrode and hence perhaps controlling the spatial pattern of the initial nucleation of the film.

First, **Professor Compton** answered: The behaviour of arrays of microcones have been simulated – see E. J. F. Dickinson *et al.*, *J. Phys. Chem. B*, 2008, **112**, 4059–4066.

Second, **Dr Walcarius** replied: Thank you for this suggestion. Indeed, the long-range order is disturbed by grain boundaries and small defects. The method that you described for block copolymers could be adapted to control better the long-range order of the mesoporous silica films, provided that a regular array of micro-cones could be patterned regularly on the electrode surface. To be efficient, these micro-cones will need to be smaller than the 2-D grains (typically a few tens of nm).

Third, **Professor Amatore** asked: Thank you, dear Alain, for this important contribution. I would like to just comment on your apparent questioning that the surface crystals order cannot be maintained over long radial distances. In fact this is due to entropy. We are used to crystals involving large enthalpies so that the ordering entropy cannot invert the Gibbs free enthalpy unless billions-upon-billions of crystal cells (possibly even much more, as in giant salt rock crystals) are packed. However when the crystalline enthalpy per unit cell is not very high *vs.* the thermal quantum (*viz.* RT), the accumulated entropic term, *i.e.*, $RTn \log(n)$, where n is the number of crystallographic cells, may quickly already compensate the cumulative enthalpy, $n\Delta H$, for rather small values of n. At this point, *i.e.* when $n° = \exp(-\Delta H/RT)$, the crystal must relax its order, which it does by creating a edge (grain joint) and starting a new crystal (see for example: C. Amatore, *Chem.–Eur. J.*, 2008, **14**, 8615–8623). This is akin to the "Blob" theory from de Gennes for polyelectrolytes, which explains why polyelectrolytes condense as pearl necklaces whose "pearl"-size is larger the larger the charge of the countercation or counteranion (*i.e.*, the larger the value of ΔH).

Dr Walcarius added: Thank you, dear Christian, for this valuable comment. As you know, I was aware of your nice paper discussing the question of long range order/disorder of surface crystals on a thermodynamic basis. Actually, I did not realize that this theory might also help in interpreting the observation of grain boundaries in our systems, so that I am deeply grateful to you for drawing our attention to that point.

Professor Amatore opened the discussion of the paper by Professor Mount: Upon decreasing the electrode size, the diffusion layer thickness also decreases almost proportionally for near-spherical or disk electrodes. On the other hand, the diffuse components of double layers have to remain approximately the same size as at larger electrodes because these do not obey diffusion, but thermal equilibrium between thermal quanta (disorder) and electrostatic ordering of ions in the electrical field. Hence, one has to reach a point where diffuse layers (*ca.* 1 nm in the presence of a good electrolyte) and diffusion layers become comparable. This has been investigated by Henry S. White for hemispherical nano-electrodes and by ourselves for ultrafast cyclic voltammetry (the diffusion layer at a microelectrode depends on $1/\sqrt{(Dt)}$ when Dt is small enough to provide planar diffusion; D is the diffusion coefficient and t the experiment duration). Furthermore, if one increases the density of nanoelectrodes in an array one may reach a point where diffuse double layers start to overlap (*i.e.*, the array starts to behave almost as a plane of surface area equal to that of the array) while the diffusion layers may not overlap. In such case all the benefit with regards to signal-to-noise (S/N) due to the use of micro-arrays will be lost (overlapping diffusion layers and diffuse layers) or even deteriorate (non-overlapping diffusion layers but overlapping diffuse layers). So the gain in S/N upon decreasing the electrode sizes and increasing the array density may not increase continually but have to pass through an optimum irrespective of the nanofabrication possibilities. However, I doubt that, today and in the future, one can reach such dimensions while having perfectly characterized devices (shape, size, conductivity, electroactivity, *etc.*); upon reaching nanometers one may already experience contamination of the signals by such perturbations.

Professor Mount replied: We agree this would be an interesting regime to explore, but also agree that nanofabrication of such devices would be extremely challenging.

Professor Compton asked: If you make measurements of the capacitance per unit area of your microsquare nanoband edge electrodes can you rationalise the values in terms of the expected values from macroelectrodes?

Professor Mount answered: As discussed in the paper, the capacitances per unit area are typically an order of magnitude lower than those observed for macroelectrodes; this has been observed for other nanosystems (references 69 and 70 in our paper) and has been attributed to a thicker double layer due to ion size/transport effects and/or the change to hemispherical fields. A thicker layer would also be consistent with the postulated enhanced migrational effects near the electrode.

Professor Bond asked: In Fig. 3 of your paper, the background current magnitude displayed when scanning in positive and negative potential directions seem to differ considerably. Is this simply a dc current instrumental artifact offset or is there is a fundamental issue involved in this outcome?

Professor Mount replied: We agree that *e.g.* in Fig. 3b there is a small dc current superimposed on the observed capacitive (non-Faradaic) charging. That this is a backplane capacitive effect, and not representative of the redox response, can be seen by the zero dc current observed at the foot of the subtracted voltammogram (Fig 3c).

Professor Bund enquired: The curve charge transfer resistance *vs.* potential in Fig. 7 goes through a minimum. Why is that?

Professor Mount answered: This is discussed more fully in our previous paper, reference 53 (*Analyst*, 2010, **135**, 1058) and its associated references. This is because the overall measured charge transfer resistance is determined by the larger of the two component charge transfer resistances (the resistance due to the oxidation reaction in the oxidation half cycle and the resistance due to the reduction reaction in the reduction half cycle of the ac perturbation). Which this is depends on the dc potential, E_{dc}. In the thermodynamic case where the electrode response is rapid, this fixes the concentration of reduced (a) and oxidised (b) species at the electrode surface; at the formal potential, E', $a = b = c/2$, where c is the total bulk concentration of redox species and there is the minimum charge transfer resistance R_{ct}^{min} as both oxidation and reduction reactions occur at the same rate. When E_{dc} is made more negative than E', b becomes progressively smaller, the reduction resistance rises and therefore the overall resistance rises; when E_{dc} is made more positive than E', a becomes progressively smaller, the oxidation resistance rises and the overall resistance again rises. In the kinetic case, the same behaviour is observed, primarily due to effect of E_{dc} on the oxidation and reduction rate constants k_{ox} and k_{red}, the smallest of which dominates. At E', k_{ox} and k_{red} are equal, given by the standard rate constant, and the overall charge transfer resistance is at a minimum (eqn (9) in our paper); when E_{dc} is more positive than E', k_{red} decreases and the overall charge transfer resistance increases, whereas when E_{dc} is more negative than E', k_{ox} decreases and the overall charge transfer resistance again increases.

Professor Williams said: The 5 nm thick electrodes are clearly electrically connected, but it is by no means obvious that such very thin films should everywhere be continuous; they might, for example, be a set of connected threads in a fine mesh, depending on how the evaporated metal wets the silicon oxide substrate. What effect might this have? The actual critical electrode dimension could be smaller than imagined, for example, and effects such as the double layer overlap from the surroundings described by Dr Bund could be more important than suspected. What evidence do you have that these very thin band electrodes are in fact continuous, since direct microscopic observation would be very problematic?

Professor Mount replied: One advantage of using established microfabrication methods (*e.g.* seed layers, layer deposition and etching protocols, silicon substrates) is that this maximises the reproducibility of the production of high-fidelity film structures, even when thin. Typically, we monitor accurately film mass (which is directly related to thickness for a coherent, flat film) during deposition with a vibrating quartz crystal. To confirm film structure and integrity, we have deposited and profiled films in multiple spatial locations to a length of 50 microns using a surface profilometer (which has a stated resolution of the order of 0.1 nm and which can certainly measure variation on the order of 1 nm; enough to assess film integrity). SEM and four-point-probe resistance measurements are also consistent with a coherent film. Once etched to form the bands, as shown in Fig. 1, we always see contiguous bright bands (of the expected dimension at 50 nm) using SEM, although for the thinnest band (5 nm) it is hard to assess accurately the dimensions by this method due to interference from the charging of the neighbouring insulators. We are therefore looking at using scanning probe methods to confirm the band widths. As to the effect of charge on the neighbouring insulator, we agree this could be one factor which could contribute to local fields and some of the migrational effects we discuss in the paper; the relative importance of this effect remains to be established.

Professor Williams enquired: The reliable fabrication of electrodes on the 5-nm size scale raises interesting questions about the timescale a molecule may spend within the vicinity of the electrode and thus be able to interact. Just using (half-width)^2/diffusion coefficient as a timescale estimate gives results typically on the scale of 10 ns. This timescale is, interestingly, similar to that for molecular rotations and raises the question whether fluctuations in electrochemistry associated with different molecular orientations in a collision with the electrode could be detected. The square edge length would perhaps have to be reduced and single cavities studied, so we are talking very small currents, but there are means for dealing with this – for example using a light-emitting diode as a current-to-photon converter, in association with a photon counter, as proposed by Fleischmann many years ago; single electron measurements are in principle possible.

Professor Mount noted: We agree that this raises an interesting possibility for further investigation. It will require further fabrication and measurement system development by us.

Professor Amatore remarked: Another problem which I foresee is related to the distance from the surface at which electron transfer (ET) may occur. For most usual molecules ET is considered to occur at the OHP, hence a few Angstroms. There, electronic coupling (see Hush, Hush–Marcus) between the substrate and the metal is sufficient to allow electron tunneling through the compact part of the double layer at a sufficient rate (note that this is the basis of Frumkin's correction of the electrochemical standard rate constants). However this view applies only when the diffusion layer size (*viz.*, the electrode size) may be regarded as infinite *vs.* the thickness of the compact layer. If this is not the case because the nanoelectrodes are too small, one may observe that electrons 'shoot' statistically over distances comparable or wider than the diffusion layer. Though this is a very important question, this will totally affect the meaning of electrochemical signatures.

Professor Mount responded: In the paper we have discussed the possibility of the increasing contribution of electron transfer at greater distances from the electrode (also greater than the diffusion layer thickness) as the electrode size decreases; we agree this is an effect on this length-scale which may contribute to the observed variation in response for our electrodes.

Dr Bohn queried: Can you exploit the high efficiency of your nanoband electrodes even further by coupling them to convective fluid delivery methods? For example, is it possible to stack multiple bands in the z-direction and control their potential independently?

Professor Mount replied: These results and our other recent work (N. J. Freeman *et al.*, *Phys. Chem. Chem. Phys.*, 2013, **15**, 8112–8118) show that the MNEE system has highly efficient mass transport. Any convective delivery methods would have to be very efficient to augment this (our results show relative insensitivity to even rapid convection, which of course is good for quantitative measurement, independent of flow, in a wide variety of systems). Nonetheless, we are investigating the subtle effects of rapid convection.

Yes, it is possible to stack bands in the z-direction and control their potential independently. We are actively investigating such devices and will report the results in the near future.

Professor Compton opened the discussion of the paper by Dr O'Riordan: In your paper you suggest that the reduction of the nitro groups on your target analyte proceeds via a four electron process. That has been shown often for the case where the reduction occurs at macro electrodes (see for example G. Wildgoose *et al.*, *ChemPhysChem*, 2005, **6**, 352–362) but with your nanowire-electrodes is there not a high probability that the finite kinetics of the coupled proton transfers will be outrun and thus rather less than four eelctrons will be transferred? Are there any analytical implications of this?

Dr O'Riordan replied: This is correct and the analytical implications are that a lower electrochemical signal will be recorded for each molecule and also that the full electro-reduction of the nitro groups to amine groups may not be possible. However, the overall kinetics at macro electrodes are diffusion limited. So, it's very likely that a trade-off exists between full (four) electron transfer but diffusionally limited regime at the macro electrode, and the enhanced mass transport but potentially incomplete (<four) electron transfer regime at the nanowire. Which regime dominates is unknown but it is certainly worth exploring further.

Professor Compton enquired: What is the limit of detection of the nitroaromatics and how does this relate to the practical determination of explosives, for example in effluent? What levels are needed to be applicable for security monitoring requirements?

Dr O'Riordan said: The limits of detection achieved for each of the nitroaromatic species tested were well below 50 ng ml^{-1}. We are prohibited from providing the exact amounts by the funding body. Our practicable operating levels are 1 μg ml^{-1} which is 150 times lower than the saturated concentration of TNT in dissolved aqueous solutions.

Professor Amatore noted: This is very interesting but what happens when the trace concentration drops too low? Then the electron flux drops accordingly and, as discussed with Professor Mount, the limit will be that of the experimental electron number that can be detected. However, before such a limit is achieved there is a area where detection will be technically possible (*i.e.*, sufficient current signal intensity) but too close to the noise. Then, you may think of feeding your device with pulsed rather than constant flow as can be easily observed when, for example, dogs try to detect faint smells and the direction of their source.

Professor Amatore then later communicated: Since my remark about the 'dog sniff' analogy seems to have been obscure I will try to make it more precise. For example, we (but this is even more visible on dogs) take pulsed breaths when we wish to identify a faint smell (you may try and observe how you pulse your sniffing periods and frequency). Here our brain is monitoring the change of signal intensity with the sniffing and retains only that with the same frequency. Furthermore the sensitivity is increased by modulating up and down the sniffing frequency (as, for example, frequency modulated radio waves (FM) are detected with less noise contamination than radio waves emitted at a single frequency). Evidently this is a simple natural reflex either for us or for dogs, but this may be worth trying to attempt with your devices.

Dr O'Riordan replied: We welcome these comments from Professor Amatore. This suggestion is a very interesting approach and one that we will explore further to enhance the limit of detection. As mentioned previously we are also employing differential analysis by subtracting signals from sensors not exposed to the test solution to enable more accurate measurement of small changes in signal

Professor Hutchings addressed the meeting: I have two questions: the first I asked in open session; and the second I asked Alan after the session. Both are for Alan O'Riordan. 1. Following on from the earlier discussion when the device is used in the field there are likely to be many molecules present which could interfere with the analysis. Some may give false positives, which may not be worrying, but some could mask the real effect of a positive result. Is it possible to comment on this and if it is the case how can you mitigate against this?

2. Is there something special about gold for the nanowire? In my field which is surface reactions in heterogeneous catalysis we find that alloying gold with other metals, *e.g.* palladium or platinum, greatly enhances the reactivity. In addition different morphologies can be effective. Is this something that can be explored in your system?

Dr O'Riordan responded: 1. Interfering species, either naturally occurring or deliberately introduced to thwart detection, do represent a challenge when developing a detection system. To mitigate against this three approaches are being adopted: Firstly, the sensors will be integrated with μHPLC to perform crude separation prior to analysis, thereby eliminating a number of interfering species. Secondly, the final electrochemical sensor devices will form part of an integrated system comprising other sensors, specific for the same target molecules, but with different sensing modalities, *e.g.*, fluorescence, mass *etc*. Therefore to mask detection of a positive result the interfering species must block all sensing modalities equally. Thirdly, chemometric genetic algorithms are being developed based on the output of the different sensor types to develop molecular patterns or 'fingerprints' to allow facile identification. The Principal Component Analysis (PCA) data provided in our paper is an initial demonstration of this approach.

2. Absolutely; nanowires fabricated from metal alloys would be of great interest. As Dr Hutchings rightly points out these alloys would act as heterogeneous catalysts for many different reactions, for instance nickel oxide nanostructures are being used for high sensitivity pH measurement while MnO structures are being applied to enzyme-free detection of glucose. We would welcome working with Dr Hutchings in this regard.

Professor Mount enquired: One would expect that an on-chip Pt pseudo-reference electrode would show significant changes in reference potential when placed in a solution containing a significant amount of different redox couples, as they should then determine the observed reference potential. Do you see any such effects and would affect this or limit the practical application of this system?

Dr O'Riordan answered: A change in potential of a redox couple is typically observed with respect to the pseudo reference electrode from the formal potential. For these electrodes we have used Pt, Ag and Ag/AgCl as reference electrodes and the working potential shifts were no more than ±50–100 mV. Currently all analytical explorations employing this nano electrochemical cell commence with electroanalysis in a model redox couple in order to estimate the difference between the formal potential and the experimental potential of the target analyte, so we do not envisage this being a problem for future applications.

Dr Falciola asked: Which is the advantage of using Principal Component Analysis in your work, considering that you are using only one variable? Why not using that single variable to describe the data set? In this context, the consideration in your paper "…only one PC can be computed and PC1 explains 100% of the data variance." (page 8, lines 23–25) is quite obvious.

Dr O'Riordan said: We agree with this comment, as there is only one variable in the presented data set. However, as we discussed at the meeting this work is on-going and we are now focused on resolving mixtures of analytes where we are employing other variables such as concentration and also including the outputs from other non-electrochemical sensors that will be integrated in the final system. This initial PCA work presented herein is to demonstrate that the electrochemical data can be directly imported and analysed using PCA software, challenged using an 'unknown compound', and the molecule correctly identified as in the case of the TNT where a military grade sample (including excipients and other impurities) was measured and the results clustered with the analytical standard obtained from Aldrich.

Dr Batchelor-McAuley enquired: I notice that in Fig. 6 of your paper the calibration plot seems to have a non-zero intercept. Do you have an explanation for this and may it indicate the presence of an adsorption process?

Dr O'Riordan responded: It is possible that the non-zero intercept may arise from an adsorption process occurring at the electrodes during electro-reduction of nitroaromatics. In this process, a nitroso intermediate followed by a 5-hydroxylamine intermediate are first formed *via* a $4e^-/4H^+$ electron and proton transfer system. The hydroxylamine group is then further reduced to an amine *via* a $2e^-/2H^+$ electron and proton transfer system. Also, as this is a military grade sample there are a number of unknown non-electroactive excipients present in solution which may also interact/adsorb on the electrodes.

Professor Compton asked: The square wave voltammograms presented in Fig. 4 are 'background subtracted' – what procedure do you use for this and what do the voltammograms look like without the subtraction?

Dr O'Riordan replied: A blank measurement of the phosphate buffer only was taken for use as a blank using the same experimental conditions and parameters. The previously recorded blank was then subtracted from the standard solution voltammograms using the potentiostat (CH Instruments 660a) software. For high concentrations the background subtraction had very little influence over the voltammograms, only appearing to shift the base line back up to 0 nA. For the lower concentrations the subtraction process significantly emphasised the peaks. While background subtraction isn't normally performed for SWV analysis, we applied this approach to explore the possibility of using differential sensing, *i.e.*, where the output of a second working electrode (which is in a controlled environment and not exposed to the test solution) is subtracted from the first electrode in order to permit measurement of very small changes in signal intensity.

Professor Williams addressed Dr O'Riordan and Professor Mount: I'd like to ask about costs and yields, since the processes used are fairly demanding. It would be interesting to compare with methods in the literature using printing, laminating, cutting and laser ablation (as for example shown by Hubert Girault and myself to make microband or embedded microring and microdisc electrodes). These printing methods have very low entry-costs and were developed because in the past the methods derived from semiconductor fabrication had a significant entry-cost and access problems. So now what has happened to change the situation? Do the methods described in the papers by Mount and O'Riordan offer an advantage over such printing techniques that justify the (apparently) increased costs?

First, **Dr O'Riordan** answered: To address the question concerning yield first, we have developed a robust and reproducible hybrid electron beam/optical lithography process with a process yield across a number of fabrication runs of ~97%. At present, we deposit twelve nanowire-based sensors per chip (patterning different designs and geometries as required) along with on-chip counter and reference electrodes resulting in fully integrated sensor devices. The devices are extremely stable and have been employed in aqueous media (acidic–basic organic), organic solvents and ionic liquids without any measured degradation in device quality or function. Furthermore, the reproducibility (including fabrication variation and experimental errors) calculated from electrochemical characterisation of devices ($n = 73$) from four separate fabrication runs over a period of approximately one year is ~8%. While the work as mentioned above is very elegant, we chose to develop nanowires at silicon chips as this will permit future on-chip integration driver circuitry. In addition, as each electrode is individually addressed this also enables multiplexed detection of a variety of analytes simultaneously. Concerning cost; at present we typically batch-fabricate three four-inch silicon wafers with twenty-one chips on each wafer. As mentioned above each chip contains twelve separate sensors. The fabrication cost per chip, including lithography metal deposition and chemical processing is about €10 so each sensor costs less than €1 and depending on the application these sensors may be re-usable. Electron beam lithography is a serial process and the cost does not reduce/scale with increased numbers of chips on a wafer. However, other parallel nanofabrication approaches, including nanoimprint lithography phase-shift or immersion lithography, are scalable and costs reduce. Therefore, the benefits of favourable economies of scale are available to reduce costs significantly if higher volumes of devices were to be fabricated for use as disposable devices. To put the cost of fabrication in context we recently purchased three commercial micro-disk electrodes for a total of €650. Access to semiconductor fabrication is now more widely available as a number of research institutes will undertake batch fabrication requests.

Professor Mount then also replied: Today, the majority of electronics companies don't fabricate their own product chips but rather follow the fabless model of paying foundries for fabrication of their designs, thereby negating the expense of cleanrooms and extensive toolsets. This model is also followed by microsystems technology and there are now many fabrication facilities with the capability to manufacture electrodes such as those of the MNEE architecture, at costs easily affordable to research institutes and start-up companies.

Regarding yields, mature IC process yields are typically >90% and it should be remembered this is for a 30+ mask process with deep sub-micron technologies. The MNEE architecture in this paper has 3 masks and critical dimensions of ~10 microns. This, together with the inherent redundancy in the design and the straightforward process flow explains the observed yields of essentially 100%.

For high-volume commercial applications the technology that meets an application's requirements at the lowest cost will almost certainly be selected by the market. Silicon-based technology is clearly necessary for those applications where complete integration of electrodes and electronics is required, and at present is required when tight dimensional control of electrodes is important. However, we acknowledge that in the fabrication of non-integrated electrodes, and those where relaxed dimensions and tolerances are acceptable, then other approaches (*e.g.* reel-to-reel, screen and inkjet printing, *etc.*) may well prosper.

Dr Freeman commented: Standard semi-conductor fabrication techniques are well suited to mass-manufacture given that somewhere in the region of five integrated circuits are manufactured for every human on the planet each year. In terms of manufacturing methodologies, the state of the art in terms of screen printing (*e.g.* the method used to produce the majority of blood glucose test strips) results in variability in the finished product of around 20%. The US Food and Drug Administration is likely to push for tighter manufacturing tolerances in the future. Therefore even in the field of millimolar glucose measurements there are pressures to improve reproducibility of manufacture. Whilst the industry will no doubt rise to the challenge, this may lead to opportunities in the IVD sector for the introduction of alternative manufacturing technologies. Given the integrated circuit market is mature, works to very high levels of reproducibility and has the capability to meet typical IVD volumes, it may well be that future IVD products utilise the skills and competencies of the semi-conductor industry more widely.

Professor Hillman returned to the discussion of the paper by Dr O'Riordan: Quite generally, the success of any analytical approach is dependent on sampling and the matrix in which the measurement is made. A number of (different) strategies for detection of explosives rely on sampling vapour-phase molecular species. On the one hand, these suffer from the low vapour pressures typical of explosive materials. On the other hand, the matrix (simplistically, air) provides relatively few challenges. Turning to your approach, the solubility of a number of explosive materials means that you do not have the same concentration-based challenges, but the liquid phase matrix may involve the presence of interferants and may present other challenges. For a practical application, could you give a brief overview of the relative positions of these two competing approaches?

Dr O'Riordan responded: As mentioned, detection of explosives in the vapour phase is extremely challenging due their low vapour pressures. A further complication is that these values assume an equilibrium between the solid and vapour phases has been achieved, which is practically impossible in real-world environments. Consequently, the concentration of explosives in vapour is usually significantly lower than the vapour pressure would suggest. To address this challenge analytically, pre-concentration approaches are generally employed where a sensor is allowed to react (rather than interact) with the vapour for long periods of time prior to analysis. An emerging approach is to filter the air actively and trap dust particles of explosive, arising from manual handling, which will have high local concentration within the detection range. Concerning aqueous based detection, in real work-environments the presence of any interfering compounds could seriously affect the detection. To address this challenge, chromatography (typically coupled with a solid microphase extraction pre-concentration step) is used to separate potential interferants prior to analysis. In our work we show: (i) highly sensitive detection of nitroaromatics without the need for pre-concentration; and (ii) the ability to identify different nitroaromatic molecules including structural isomers based on their respective electro-reduction voltage. Work is now in progress to integrate our sensors with μHPLC, employed to perform a crude pre-separation of potential interferants.

Professor Kanoufi addressed Dr O'Riordan and Professor Mount: In SECM one is often confronted by partial microelectrode surface poisoning from adsorption of non electoactive material. Such a problem is more often encountered when decreasing the SECM tip size to the nanometer scale. I wonder how such adsorption problems are encountered with your both hybrid nano/micro electrodes. What is the solution you propose to clean the electrode surfaces, or how well do they resist stringent (organic, acidic, piranha…) conditions?

First, **Dr O'Riordan** answered: Before dicing, wafers were coated with a protective resist layer. This protective layer and any residual were then removed by immersing the chips twice in semiconductor commercial cleaning solution R1165 (Shipley) for 40 min at 65 °C, in DI water for 10 min at room temperature and in isopropyl alcohol for 10 min at room temperature, followed by drying with a flow of N_2. Before each electrochemical experiment, we clean our devices by immersing them sequentially for 10 min each in acetone (bath 1), trichloroethylene, acetone (bath 2), isopropyl alcohol and DI water, followed by drying with a flow of N_2. Furthermore, only a very few (<10) cyclic voltammogramms in 0.1 M H_2SO_4 were required until a stable gold oxide reduction peak was obtained. All of these cleaning processes have not been observed to degrade the devices. However, immersion in piranha solution for longer than 5 min delaminates the electrodes from the chip surface.

Professor Mount then also replied: We discuss in the paper the enhanced catalytic activity of nanoelectrodes. The converse of this is that they are also, of course, more sensitive to adsorption or poisoning. It is therefore important to avoid contamination of solutions, glassware *etc.* to preclude these effects.

We have shown CV-cycling in clean aqueous sulphuric acid and KCl electrolytes to be effective in restoring the full electrode activity. As with macroelectrodes, as well as the effective mechanical and chemical cleaning caused by this cycling (which we have shown can even remove strongly bound species such as sulfide), the positions and charges passed in the oxide and adsorbed hydrogen formation peaks are good indicators of electrode cleanliness and active area. The MNEE system was also found to be resistant to stringent conditions (piranha, sulphuric, nitric, hydrochloric and citric acids, acetone, methanol, ethanol, DMSO, DCM, and hydroxide solutions).

Professor Amatore commented: These are exciting results, but I wonder if they represent the best strategy for achieving high performance in detection of 'odors' emitted at trace levels. Indeed, nowadays biologists can express any ion channel whose action is gate-modulated by a specific ligand recognizing molecularly a single specific molecule with an extremely high sensitivity and selectivity (as *e.g.*, occurs naturally in our synaptic communication). This in my view is the ultimate technology since the patch-clamp technique allows monitoring fA currents related to a single ion channel opening, and post-genomic knowledge allows virtually any specific receptor to be designed and expressed by cells.

I believe that the most important issue when we consider ultratraces is about the flux of target molecules that may 'visit' the sensor. A near-constant flux will produce a near-constant detection current whether the system is artificial as yours, natural, or results from modified genetic expression, so that small currents may be hidden by any spurious currents. Conversely, if one modulates the flux of air containing the target traces which arrives over the sensor, one may expect that phase detection will be possible and increase greatly the performance. This is for example what we spontaneously do when trying to smell a faint odor by pulsating the air flow by a series of brief saccades. Look at a dog and you will see that this is not specific to humans. Hence, I suggest that you examine such a possibility to improve the detection limit.

Professor Kornyshev commented: Currently, using self-assembled arrays of gold nanoparticle at liquid–liquid interfaces it was possible to reach very high nanoplasmonic enhancement of Raman signals of a large number of prototypes of 'bad' molecules with clear Raman fingerprints. The level of detection for some of them reached femtomolar concentrations (M.P. Cecchini *et al.*, Self-assembled nanoparticle arrays for multiphase trace analyte detection, *Nat. Mater.*, 2013, **12**, 165–171). This is only one of the recent examples. Various solid-state SERS-based nanoplasmonic platforms have been reported earlier, with excellent lowest levels of detection; for a review, see: *e.g.* S. Mahajan *et al.*, Understanding the surface-enhanced Raman spectroscopy 'background', *J. Phys. Chem. C*, 2009, **114**, 7242–7250. How do your detection platforms compete with Raman techniques?

Dr O'Riordan responded: The SERS results presented to date are very elegant and are pushing the envelope in the area of nanophotonics. Like every other technique, both SERS and electrochemistry have their strengths and weaknesses. The platform developed herein competes with the Raman techniques in the following ways: (i) the platform has very low maintenance requirements; (ii) very low power requirements – no laser; (iii) very small form factor; (iv) it is compatible with microfluidics-based continuous flow-through analysis; (v) it allows multiplexed detection of up to twelve different analytes simultaneously. These advantages allow the system to be installed and allow remote water sampling for months without the need for operator intervention.

Double layer effects at nanosized electrodes[†]

Andreas Bund and Clemens Kubeil*

Received 3rd April 2013, Accepted 18th April 2013
DOI: 10.1039/c3fd00050h

This paper discusses numerical simulations of double layer effects at shrouded electrodes with dimensions below 100 nm. Special focus is given to the surface charge on the shrouding material. The Poisson–Nernst–Planck equations are solved to study the effects on the limiting current arising from the electrical double layer of the shrouding.

1 Introduction

Nanosized electrodes combine various features which make them very interesting for fundamental electrochemical investigations.[1] This is due to the small size-disturbing effects from uncompensated Ohmic potential drops, and that capacitive currents are usually less severe than with macroscopic electrodes. On the other hand, the limiting current densities are high and convective effects usually can be neglected. Prominent applications are the determination of fast electron-transfer kinetics or electroanalytical measurements on very small length scales, *e.g.* biological cells or even single molecules. An overview on fabrication and applications of nanoelectrodes is given in a recent review by Oja *et al.*[2]

In general the flux of an electroactive species can be described with the Nernst–Planck eqn (1).

$$J_i = -D_i \nabla c_i - \frac{z_i F}{RT} D_i c_i \nabla \phi + c_i u \tag{1}$$

In eqn (1) D_i is the diffusion coefficient, c_i the concentration, z_i the charge of species i, ϕ the electrical potential, and u the velocity of the fluid. F, R and T are the Faraday constant, the gas constant, and the absolute temperature, respectively. It should be mentioned that eqn (1) is strictly valid only for diluted solutions and the correct choice of the reference system for the fluid velocity can be challenging (see Newman's textbook[3] for a detailed discussion).

The electric current associated with the reaction of an electro-active species Ox at an electrode surface ($x = 0$) is given by the value of the flux at the electrode surface, $J_{Ox}(x = 0)$. Generally it is assumed that in the presence of supporting electrolyte the gradients of the electric potential $\nabla \phi$ are negligible and the transport of redox species to and from the electrode is mainly driven by diffusion.

Technische Universität Ilmenau, Elektrochemie und Galvanotechnik, Ilmenau, Germany. E-mail: clemens.kubeil@tu-ilmenau.de; Fax: +49 3677 3151; Tel: +49 3677 3109

Note that for most cases $u = 0$ applies at the electrode surface (no slip condition). In the absence of supporting electrolyte migration becomes important and electrostatic effects can promote or inhibit the flux of charged redox species. Amatore treated this case and developed useful equations to estimate the changes of the limiting current at small electrodes.[4,5] His theoretical treatment yields good agreements with experimental data for electrode dimensions down to several hundred nm. However, for electrodes in the nanometer range (typically below 10 nm) significant deviations were obtained which could not be explained by migration effects.[6-10] If the characteristic length scale of the dimension of the electrode is in the order of the Debye length λ – which describes the characteristic length scale of the electrical double layer (EDL) – we can expect effects from the EDL on the charge transfer process. Generally speaking the observed phenomena at small electrodes may be caused by an overlapping of the electric field of the double layer with the diffusive flux of the ions and the electric field driving the charge transfer reaction at the electrode. It has been shown that the enhancement or inhibition of the limiting current at small electrodes in the absence of supporting electrolyte varies with the nature of the electro-active species. Several possible contributions are discussed like the dynamic diffuse double layer,[11] the Frumkin-effect,[12] changes in the dielectric properties of the double layer and electron transfer kinetics,[13] and ion-pairing.[14]

Refined numerical models have shown a significant penetration of the EDL into the concentration boundary layer (dynamic diffuse double layer effect).[11,13,15] The thus calculated steady-state voltammograms have the expected sigmoidal shapes. However, the limiting currents are off by 10–20% from the purely diffusion controlled value and the half-wave potentials are shifted by several hundred mV depending on the electrode size and charges of the electro-active species. Variations of the dielectric constant across the EDL only seem to play a minor role.

In many cases nanoelectrodes are fabricated by embedding the electrode material in an insulating shrouding, *e.g.* by melting a Pt wire in a glass capillary. Several insulating materials have been used to fabricate electrodes and they are supposed to be inert. But even if they are chemically inert, these materials will have a surface charge due to the adsorption, dissociation reactions of surface groups[16] or polarization effects of the shrouding material due to the electrode potential.[17] This surface charge may significantly affect the voltammetric behavior of the electrode provided that the electrode dimensions are in the order of the Debye length.

One prominent example for such an EDL effect is the rectification of ionic currents at conically shaped pores.[18] Such pores can be fabricated in polymers or glass with pore mouth radii in the order of nanometers. For a given absolute value of the electric potential of the pore the ionic flux depends on the polarity of this potential, *e.g.* for a transmembrane potential of $\Delta\phi$ the current is higher than for $-\Delta\phi$. This effect is called ion current (ICR) rectification and manifests itself in an asymmetric current voltage curve. Similar effects are observed at biological pores and therefore artificial conical pores are interesting model systems for their biological counterparts. Other applications are nanofluidic devices[19] and the detection of nanoparticles and single molecules.[20] The underlying mechanism for ICR seems to be a depletion or enrichment of ionic species in the lumen of the pore depending on the polarity of the transmembrane potential. In the case of enrichment the pore goes into a highly conducting state whereas for depletion it enters a low conducting state. The build-up of these concentration gradients is driven by asymmetric ion fluxes which are governed by the EDL at the pore mouth. Analytical[21,22] and numerical models[23–25] have been published to describe ICR. One goal of the researchers is to bring the size of the nanopores down to dimensions which would allow the electrochemical detection of single molecules. A promising application would be the sequencing of DNA by reading the current transient of a nanopore during the translocation of the biomolecule.[26]

It has also been shown that for recessed nanoelectrodes the EDL of the insulating material can govern the current responses at electrodes. As a result such systems can be gated by an external stimulus,[27,28] comparable to a solid state field effect transistor (but so far with much smaller gains). In the case of glass the surface charge can be varied over a wide range *via* the pH of the electrolyte. In a recent paper we discussed the voltammetric response of such systems.[29] In the present paper we extend the discussion to disk and hemispherical electrodes with a shrouding sheath.

2 Numerical procedure

For the numerical treatment we consider a one-electron reduction reaction in which an oxidized species Ox is reduced to R according to eqn (2).

$$\text{Ox}^z + \text{e}^- \rightarrow \text{R}^{z-1} \tag{2}$$

We focus the discussion on hemispherical and disk electrodes. Due to the radial symmetry the problem (the computation domain is shown in Fig. 1) can be treated in two dimensions with cylindrical coordinates and appropriate boundary conditions (Table 1).

The fluxes of all species J_i are governed by the Nernst–Planck eqn (1). As we are dealing with nanoelectrodes we neglect transport by convection, *i.e.* $u = 0$ in the whole domain.

The EDL effects are caused by space charge regions near the electrode. Therefore eqn (1) must be coupled to another differential equation relating space charge density and electrical potential, the Poisson eqn (3).

$$\nabla^2 \phi = -\frac{F}{\varepsilon_0 \varepsilon_R} \sum_i z_i c_i \tag{3}$$

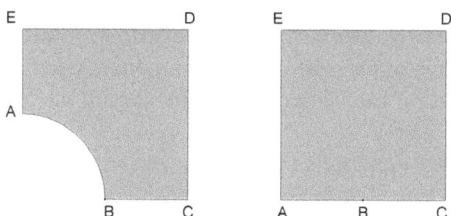

Fig. 1 Sketch (not to scale) of the computational domain for hemispherical and disk electrodes. Due to the axial symmetry of the systems the simulations can be treated in two dimensions. See Table 1 for the corresponding boundary conditions.

Table 1 Boundary conditions for the numerical solution of eqn (1) and (3) in the computation domain sketched in Fig. (1)

Boundary	Eqn (1)	Eqn (3)
AB	$c_{Ox} = 0, J_O = -J_R$	$n\nabla\phi = 0$
BD	$nJ_i = 0$	$n\nabla\phi = -\sigma/(\varepsilon_0\varepsilon_R)$
CD	$c_i = c_i^*$	$n\nabla\phi = 0$
DE	$c_i = c_i^*$	$\phi = 0$
EA	Axial symmetry	Axial symmetry

In eqn (3) ε_0 and ε_R are the dielectric constant of vacuum and the relative dielectric constant of the medium, respectively.

For a system consisting of N species we have N equations of the type (1) and one of type (3). This system of equations constitutes the Poisson–Nernst–Planck (PNP) equations. Together with the appropriate boundary and initial conditions they determine the electrochemical behavior of our system. Solving the PNP equation system yields the concentration fields of the N species and the electric potential, from which e.g. the electrical current can be calculated. We assume an electrolyte consisting initially only of oxidized species Ox, $c_{Ox}(x, t = 0) = c_{Ox}^*$ where c_i^* is the bulk concentration of the species i. Furthermore there will be a stoichiometric amount of counter-ions, A^+ or B^- depending on the charge of Ox. For the cases where supporting electrolyte is present, additional amounts of A^+B^- are added.

Far away from the electrode (boundary CDE in Fig. 1) bulk conditions prevail. In a preliminary series of simulations we identified a distance of 3 μm to be sufficient to model semi-infinite conditions. For this distance the concentration fields at the electrode are not directly affected by the condition at the boundary CDE. For limiting current conditions the concentration of the reactant Ox at the electrode surface was set to zero, $c_{Ox}(x = 0, t) = 0$. The flux of R is coupled to the flux of Ox via the relation $J_{Ox}(x = 0) = -J_R(x = 0)$ to ensure conservation of mass. A surface charge, σ, was applied to the surface of the shrouding material surrounding the electrode while the electrode surface charge was zero. This means that effects from electrical double layer (EDL) of the electrode itself are neglected (zero field approximation). In other words, the double layer effects discussed here are solely due to the surface charge on the dielectric embedding material. The zero-field approximation decouples the EDL from the diffusion layer. Dickinson et al. studied the applicability of this model and found that it is a good approximation for larger electrodes while deviations may occur for nanoelectrodes.[30] These deviations depend strongly on the size and charge of the electrode. Unfortunately the exact potential distribution within the EDL and the location of the charge transfer is difficult to predict. Therefore we refrain from a coupling of the EDL of the electrode and the shrouding surface and focus the discussion on the latter.

We solved the PNP equations for the steady state using the commercial finite elements package Comsol Multiphysics 3.5a. The computation domain was discretized into triangular elements and quadratic Lagrange polynomials were chosen as the element functions. Pitfalls regarding the accuracy of finite element methods especially due to poor meshings have been discussed by Cutress et al.[31]

Therefore we performed a careful mesh study. The preset "fine meshing" was adapted with regards to the geometry and physics of our systems. The mesh size was 0.1 nm at charged walls to resolve the EDL properly and between 0.01 nm and 0.1 nm at the electrode. The maximum element growth rate was set to 1.1. The model was validated by comparing analytical solutions of the PNP system (*i.e.*, Gouy–Chapman theory) with our numerical results as has been described in our previous papers.[24,29] To improve further the accuracy of the calculated fluxes Lagrange multipliers were used. The simulated diffusion limited current in a 5 mM solution at a 5 nm radius disk electrode was 6.473 pA. This is in excellent agreement with the calculated value (6.464 pA) according to eqn (4). From this we conclude that our precautions regarding the accuracy of the finite element method are sufficient. Unless otherwise stated, the following parameters were used: $D_{Ox} = D_R = 0.67 \times 10^{-9}$ m^2 s^{-1}, $D_{A^+} = 1.975 \times 10^{-9}$ m^2 s^{-1}, $D_{B^-} = 2.032 \times 10^{-9}$ m^2 s^{-1}, $T = 298$ K, $c^*_{Ox} = 5$ mM, $c^*_R = 0$ mM and $\varepsilon_R = 78$. These values would for examples describe an aqueous solution of Ru(NH$_3$)Cl$_3$ with KCl as supporting electrolyte.

3 Results and discussion

For the following discussion we normalized the obtained limiting currents with respect to the diffusion-controlled limiting currents, i_{dl} or the limiting current in the absence of surface charge, $i_{l,0}$. For reaction (2) these are given by eqn (4) and (5) for disk and hemispherical electrodes (radius r_0), respectively.

$$i_{dl,disk} = 4Fc^*_{Ox}D_{Ox}r_0 \quad (4)$$

$$i_{dl,hemisp} = 2\pi Fc^*_{Ox}D_{Ox}r_0 \quad (5)$$

As has been discussed above, deviations from the values predicted by eqn (4) and (5) will occur in the absence of supporting electrolyte.[4,32] In the following we will show that in the case of shrouded nanoelectrodes EDL effects will additionally affect the voltammetric behavior. The effect of a positive surface charge next to a disk and hemispherical electrode on the limiting current for a positively charged species Ox is shown in Fig. 2. A significant decrease of the limiting

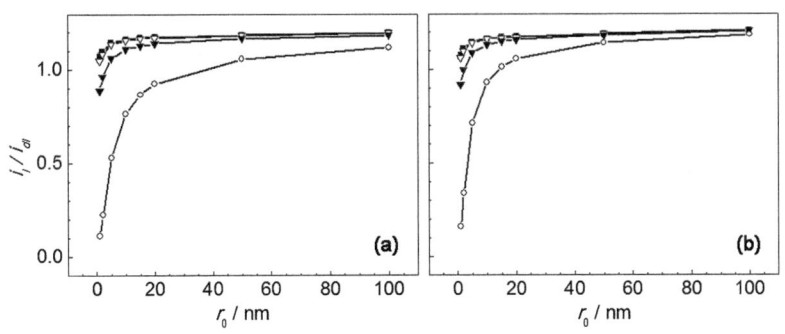

Fig. 2 Calculated limiting currents at nanometer-sized disk electrodes (a) and hemispherical electrodes (b) for different surface charges at the shrouding boundary. (■ 0 C m^{-2}, ▽ 1 × 10^{-4} C m^{-2}, ▼ 1 × 10^{-3} C m^{-2}, ○ 1 × 10^{-2} C m^{-2}).

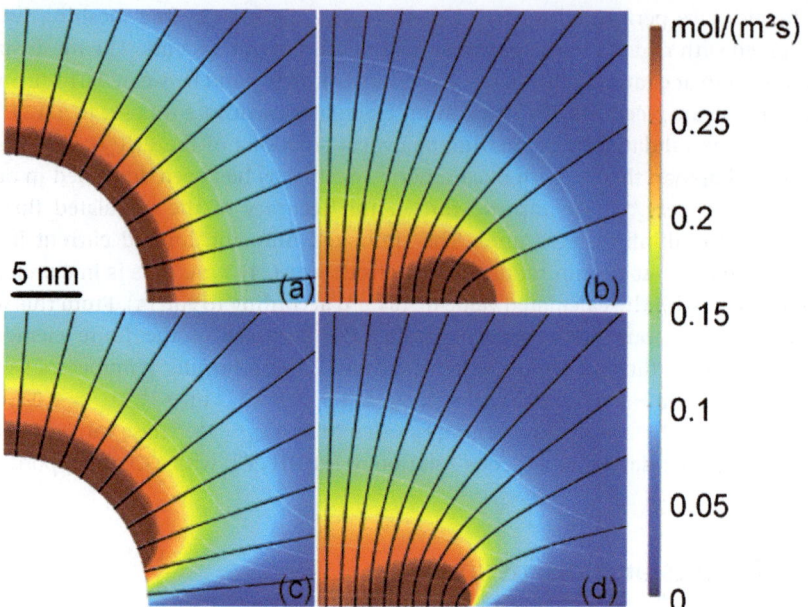

Fig. 3 Surface plot of the fluxes of the electro-active species ($z_{Ox} = +3$) under limiting current conditions for 10 nm-radius hemispherical (a,c) and disk (b,d) electrodes. The solid black lines show the streamlines of flux and the thin gray lines show the iso-concentration lines of Ox. In panels (a) and (b) there is no surface charge, and in panels (c) and (d) the surface charge on the shrouding material is 5×10^{-3} C m^{-2}.

current is observed for surface charges $\sigma \geq 1 \times 10^{-3}$ C m^{-2}. For example the limiting current at a 10 nm disk electrode is decreased by 5.1% (from $1.17 i_{dl}$ to $1.11 i_{dl}$) and by 3.4% ($1.17 i_{dl}$ to $1.13 i_{dl}$ for a 10 nm hemispherical electrode, respectively (quoted values are for a surface charge $\sigma = 1 \times 10^{-3}$ C m^{-2}). In general, the deviation from i_{dl} increases with decreasing electrode size (especially below 20 nm) and with increasing surface charge. At relatively large electrodes a

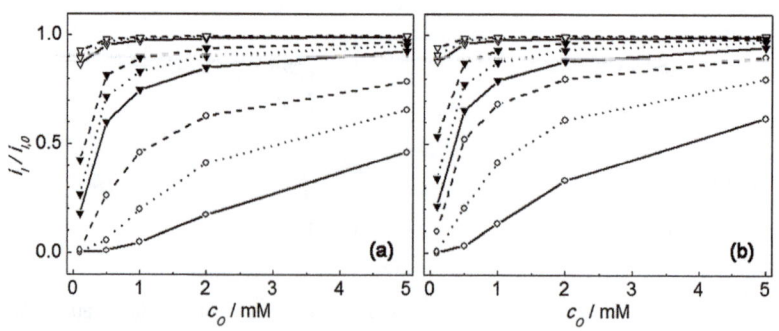

Fig. 4 Dependence of the normalized limiting current on the concentration of the electro-active species (z_{Ox}) for different sizes r_0 of disk (a) and hemispherical (b) electrodes (— 5 nm, --- 10 nm, ··· 20 nm) and different surface charges (▽ 1×10^{-4} C m^{-2}, ▼ 1×10^{-3} C m^{-2}, ○ 1×10^{-2} C m^{-2}). The limiting current i_l has been normalized to the limiting current in the absence of surface charge, $i_{l,0}$.

constant value of the limiting current $1.21i_{dl}$ is found. This is slightly higher than the value predicted by Amatore's theory[5] which would be $1.17i_{dl}$ for a one-electron transfer and $z_{Ox} = +3$. This could be due to the slightly different electrode geometry (Amatore's treatment is for cylindrical electrodes) and approximations in Amatore's theory (assumption of equal diffusion coefficients).

The size dependence of the limiting current seen in Fig. 2 is related to the thickness of the EDL at the shrouding surface BC (see Fig. 1). The thickness of the EDL can be estimated as 3λ where λ is the Debye length. A typical value for the Debye length for the systems discussed here ($c^*_{Ox} = 5$ mM and $z_{Ox} = +3$) is 1.75 nm. From a surface plot of the fluxes of the electroactive species it is obvious that for small electrodes the EDL will screen a part of the electrode surface (Fig. 3). The limiting currents at small hemispherical electrodes show a slightly weaker dependence on the surface charge. A hemispherical electrode rises above the surrounding EDL and therefore the influence of the EDL is weaker compared to a disk electrode flush with the shrouding material. This is clearly seen when comparing Fig. 2(a) and (b) for the size range 5 nm to 20 nm. For very small electrodes the shielding effect becomes comparable for both electrode shapes.

Such a strong size dependence of the limiting current has been found also in experiments.[7,8,14] Our model neither includes electron transfer kinetics nor EDL effects from the electrode itself (note that the surface charge at the boundary AB in Fig. 1 is zero). Nevertheless the effects that we see are comparable to the effects seen in the experiments (a detailed discussion follows below). This indicates the importance of taking into consideration the effect of the EDL of the shrouding medium.

For $\sigma = 0$ the limiting currents approach the values predicted by eqn (4) and (5) as the electrode size decreases. This is due to the fact that the electric potential in our treatment is only coupled to the diffusion-migration field in front of the electrode. For smaller electrodes the absolute value of this potential decreases and the migrational contribution to the current decreases.

In the absence of supporting electrolyte the thickness of the EDL is governed by the concentration of the electro-active species. Thus, the effects of the EDL on the limiting current will depend on this concentration. In Fig. 4 one can see that with increasing concentration the suppression of the limiting current decreases.

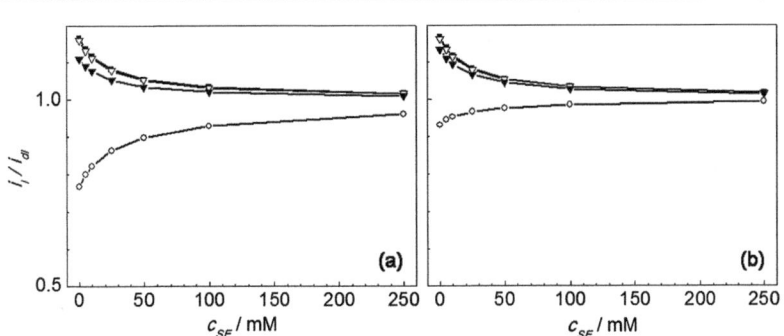

Fig. 5 Dependence of the normalized limiting current on the concentration of the supporting electrolyte for a disk (a) and a hemispherical (b) electrode with radius $r_0 = 10$ nm for $c^*_{Ox} = 5$ mM, $z_{Ox} = 3$ and different surface charges (■ 0 C m^{-2}, ▽ 1 × 10^{-4} C m^{-2}, ▼ 1 × 10^{-3} C m^{-2}, ○ 1 × 10^{-2} C m^{-2}).

Finally, the limiting current reaches the value for a neutral shrouding surface $i_{l,0}$ (not shown). Note, that decreasing the concentration results in a non-linear decrease of the limiting current. For small electrodes (e.g. 5 nm, solid line) this effect is much more pronounced than for larger electrodes (20 nm, dashed line). Of course, the strongest effect is obtained for the highest surface charges, but also for moderately charged shrouding surfaces ($\sigma = 1 \times 10^{-4}$ C m^{-2}), low concentrations, and small electrodes a clear deviation of the normalized current from unity can be observed. Note that such conditions can be readily encountered in experiments. The surface charge of glass (a prominent shrouding material) can be in the order of several mC m^{-2} depending on the pH and ionic strength of the electrolyte.[33]

In the presence of a supporting electrolyte the charges on the shrouding will be screened and the EDL effects should become weaker. In the case of a complete shielding of the surface charge one should find classical voltammetric behavior, corresponding to a purely diffusion controlled reaction, which is in agreement with our numerical results (Fig. 5). There are however reports in literature where in the presence of high concentrations of supporting electrolyte deviations from purely diffusion-controlled behavior have still been found. To explain this finding additional effects need to be considered. Recent theoretical treatments by Liu et al. include effects due to the dynamics of the diffuse double layer and electron-transfer kinetics.[13,15] Such extensions are beyond the scope of the present paper. The point that we want to make is that the limiting current in diluted electrolytes depends strongly on the surface charge on the shrouding surface of a nanoelectrode. It can be increased due to migration but also suppressed when shielding effects due to the surface charge are effective (Fig. 5).

Intuitively one would expect that for opposite charges of the electro-active species and the shrouding surface attractive forces will lead to enhancements of the limiting current. Our simulations show that this is indeed the case (Fig. 6). Increasing positive surface charges shield the electrode for a positively charged species and the limiting current decreases. Negative surface charges lead to increased limiting currents. Again the effects are somewhat stronger for disk electrodes than for hemispherical electrodes. Due to the asymmetric shape of the EDL near the rim of the electrode the shape of the i vs. σ curve is not symmetric.

Finally we would like to discuss the effect of the charge of the electro-active species itself. For positively charged species the limiting current will be suppressed in the presence of positive surface charges while it is increased in the case of negatively charged species (Fig. 7). The change in the limiting current can be rather high, e.g. from $1.3i_{dl}$ to $0.7i_{dl}$ for $z_i = 1$ at a disk electrode when the surface charge changes from 0 to 1×10^{-2} C m^{-2}. Chen and Kucernak[8] reported strong deviations from Amatore's predictions for various redox species with different charges, such as Fe(CN)$_6^{3-}$, Ru(NH$_3$)$_6^{3+}$ and Ir(Cl)$_6^{2-}$. They show that for the anionic species a much stronger suppression occurs compared to cationic species. Similar results have been published by Conyers and White.[7] Our numerical results are in agreement with these experimental results. Obviously the enhancement/suppression due to surface charges on a shrouding surface are not equal in magnitude.

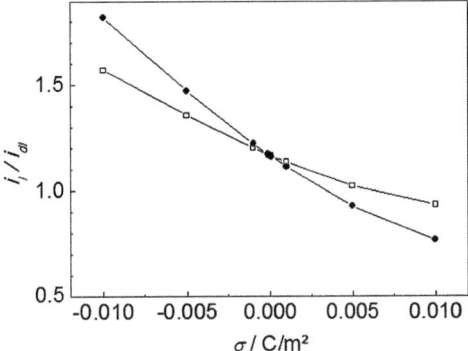

Fig. 6 Dependence of the normalized limiting current on the charge of the shrouding surface for a disk (●) and a hemispherical (○) electrode with radius $r_0 = 10$ nm for $c^*_{Ox} = 5$ mM, $z_{Ox} = +3$.

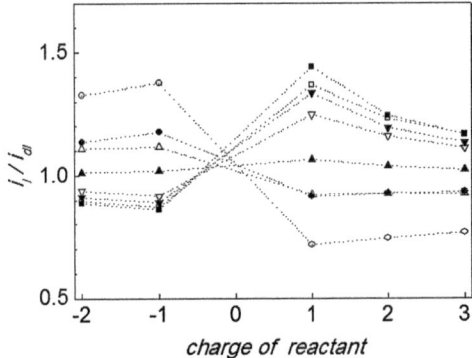

Fig. 7 Dependence of the normalized limiting current on the charge of the electroactive species for different values of the surface charge (□ 0 C m^{-2}, ▽ 1 × 10^{-3} C m^{-2}, △ 5 × 10^{-3} C m^{-2}, ○ 1 × 10^{-2} C m^{-2}) at disk (open symbols) and hemispherical electrodes (solid symbols) for $r_0 = 10$ nm and $c^*_{Ox} = 5$ mM.

4 Conclusion

A numerical model based on the Poisson–Nernst–Planck equations has been used to assess the effects of electrical double layers (EDLs) at the shrouding surfaces of disk and hemispherical nanoelectrodes. In a first approach the surface charge creating these EDLs has been assumed as constant. A refined model could take into consideration the variation due to chemical equilibria or polarization effects of the shrouding material as well as the EDL of the electrode itself. Even the simple model can reproduce some unexpected experimental findings such as the dependence of the limiting current on the size of the electrode for various charges of the electro-active species. It is shown that relatively small surface charges (1–10 mC m^{-2}) can affect the limiting current at small electrodes in the absence of supporting electrolyte. Therefore, theoretical models should include this feature to describe accurately charge transfer at sub-nanometer electrodes. A knowledge of the presence of a surface charge and its influences is needed for the interpretation of experimental data. This is crucial in order to decide whether kinetic or electrostatic effects are controlling the charge transfer.

Acknowledgements

Financial support from the Deutsche Forschungsgemeinschaft (grant BU 1200/13-1) is gratefully acknowledged.

References

1. C. Batchelor-McAuley, E. J. F. Dickinson, N. V. Rees, K. E. Toghill and R. G. Compton, *Anal. Chem.*, 2012, **84**, 669–684.
2. S. M. Oja, M. Wood and B. Zhang, *Anal. Chem.*, 2013, **85**, 473–486.
3. J. S. Newman and K. E. Thomas-Alyea, *Electrochemical Systems*, John Wiley & Sons, 3rd edn, 2004.
4. C. Amatore, M. R. Deakin and R. M. Wightman, *J. Electroanal. Chem.*, 1987, **225**, 49–63.
5. C. Amatore, B. Fosset, J. Bartelt, M. R. Deakin and R. M. Wightman, *J. Electroanal. Chem.*, 1988, **256**, 255–268.
6. C. P. Smith and H. S. White, *Anal. Chem.*, 1993, **65**, 3343–3353.
7. J. Conyers, J. L. and H. S. White, *Anal. Chem.*, 2000, **72**, 4441–6.
8. S. L. Chen and A. Kucernak, *J. Phys. Chem. B*, 2002, **106**, 9396–9404.
9. D. Krapf, B. M. Quinn, M. Y. Wu, H. W. Zandbergen, C. Dekker and S. G. Lemay, *Nano Lett.*, 2006, **6**, 2531–2535.
10. Y. Sun, Y. Liu, Z. Liang, L. Xiong, A. Wang and S. Chen, *J. Phys. Chem. C*, 2009, **113**, 9878–9883.
11. R. He, S. Chen, F. Yang and B. Wu, *J. Phys. Chem. B*, 2006, **110**, 3262–3270.
12. A. J. Bard and L. Faulkner, *Electrochemical Methods: Fundamentals and Applications*, John Wiley & Sons, 2nd edn, 2001.
13. Y. Liu and S. Chen, *J. Phys. Chem. C*, 2012, **116**, 13594–13602.
14. J. J. Watkins and H. S. White, *Langmuir*, 2004, **20**, 5474–5483.
15. Y. Liu, R. He, Q. Zhang and S. Chen, *J. Phys. Chem. C*, 2010, **114**, 10812–10822.
16. R. J. Hunter, *Foundations of Colloid Science*, Oxford University Press, 2nd edn, 2001.
17. N. Calander, *Anal. Chem.*, 2009, **81**, 8347–8353.
18. Z. S. Siwy, *Adv. Funct. Mater.*, 2006, **16**, 735–746.
19. F. H. J. vanderHeyden, D. J. Bonthuis, D. Stein, C. Meyer and C. Dekker, *Nano Lett.*, 2006, **6**, 2232–2237.
20. C. Dekker, *Nat. Nanotechnol.*, 2007, **2**, 209–215.
21. C. Wei, A. J. Bard and S. W. Feldberg, *Anal. Chem.*, 1997, **69**, 4627–4633.
22. D. Woermann, *Phys. Chem. Chem. Phys.*, 2004, **6**, 3130–3132.
23. J. Cervera, B. Schiedt and P. Ramrez, *Europhys. Lett.*, 2005, **71**, 35–41.
24. H. S. White and A. Bund, *Langmuir*, 2008, **24**, 2212–2218.
25. E. R. Cruz-Chu, A. Aksimentiev and K. Schulten, *J. Phys. Chem. C*, 2009, **113**, 1850–1862.
26. R. W. Murray, *Chem. Rev.*, 2008, **108**, 2688–2720.
27. G. Wang, A. K. Bohaty, I. Zharov and H. S. White, *J. Am. Chem. Soc.*, 2006, **128**, 13553–13558.
28. G. Wang, B. Zhang, J. R. Wayment, J. M. Harris and H. S. White, *J. Am. Chem. Soc.*, 2006, **128**, 7679–7686.
29. H. S. White and A. Bund, *Langmuir*, 2008, **24**, 12062–12067.
30. E. J. F. Dickinson and R. G. Compton, *Chem. Phys. Lett.*, 2010, **497**, 178–183.
31. I. J. Cutress, E. J. F. Dickinson and R. G. Compton, *J. Electroanal. Chem.*, 2010, **638**, 76–83.
32. K. B. Oldham, *J. Electroanal. Chem.*, 1988, **250**, 1–21.
33. S. H. Behrens and D. G. Grier, *J. Chem. Phys.*, 2001, **115**, 6716–6721.

Pulse electroanalysis at gold–gold micro-trench electrodes: Chemical signal filtering

Sara E. C. Dale and Frank Marken*

Received 21st February 2013, Accepted 11th March 2013
DOI: 10.1039/c3fd00022b

Bipotentiostatic control of micro- and nano-trench sensor systems provides new opportunities for enhancing signals (employing feedback currents) and for improved selectivity (by "chemical filtering"). In this study both phenomena are exploited with a gold–gold micro-trench electrode with ca. 70 μm width and ca. 800 μm trench depth. In "generator–collector mode", feedback current enhancement is demonstrated for the hydroquinone/ benzoquinone redox system. Next, a "modulator-sensor mode" experiment is developed in which one electrode potential is stepped into the negative potential region (employing the normal pulse voltammetry method) to induce an oscillating pH change locally in the micro-trench. The resulting shift in the hydroquinone/ benzoquinone reversible potential causes a Faradaic sensor signal (employing chronoamperometry). This method provides a "chemical filter" by selecting pH-sensitive redox processes only, and by showing enhanced sensitivity in the region of low buffer capacity. The results for the chemically reversible hydroquinone/ benzoquinone system are contrasted to the detection of the chemically irreversible ammonia oxidation.

1. Introduction

Generator–collector electrode systems[1] can be utilised as sensors to detect low concentrations of analyte[2] and may have many more uses in environmental or medicinal analysis.[3] There is considerable interest in particular in micro-gap or nano-gap electrode systems, as pioneered by Lemay and coworkers.[4,5] These consist of two closely spaced working electrodes, where one is the generator and one the collector electrode (Fig. 1). The purpose of the generator electrode is to perform either an oxidation or reduction upon the redox species. The function of the collector electrode is providing feedback to enhance the current reading and/or to give a much cleaner current response free of capacitive background.[6] Significant enhancements are possible in particular for sub-micron generator–collector gap or trench systems.[7]

A second mode of operation is possible with the first electrode ("modulator") locally changing the pH in the micro-trench, which then leads to an oxidation at the "sensor" electrode (Fig. 1C). The geometry and the gap-width between the two

Department of Chemistry, University of Bath, Bath BA2 7AY, UK. E-mail: F.Marken@bath.ac.uk

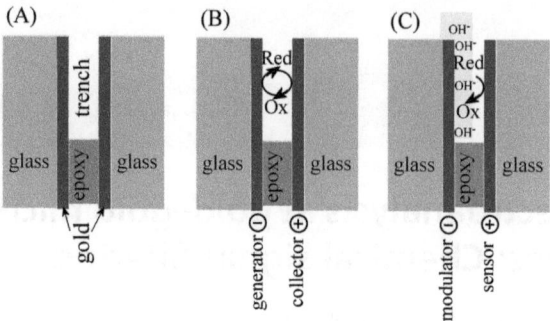

Fig. 1 Schematic drawing of (A) the gold–gold micro-trench electrode system, (B) the operation in "generator–collector" mode, and (C) the operation in "modulator–sensor" mode.

electrodes is of key importance with smaller gap sizes being beneficial to faster feedback. Gold–gold double-hemisphere electrodes have been suggested[8] but due to the small active area these electrodes are of limited use in electroanalysis. Recently, it has been demonstrated with tin-doped indium oxide (ITO) plate electrodes that ITO–epoxy–ITO micro-trench designs allow excellent collection efficiencies due to the electro-generated species remaining within the confined space between the two electrodes.[9] Here a gold–gold micro-trench electrode is investigated for operation in pulse-mode.

Chronoamperometry and pulse electroanalysis has been studied before with generator–collector electrode systems. Work on "time of flight" studies in chronoamperometry mode has been performed for example at interdigitated electrodes[10] and at paired hemispherical electrodes under bipotentiostatic control.[11] The time it takes for the redox species to diffuse across the gap is determined and the diffusion coefficient of the redox species or charge carrier calculated.[12] Differential pulse voltammetry at hemispherical gold–gold electrode junctions grown by electro-deposition has been reported.[13] In previous work by Rassaei et al. a "modulator-sensor" pulse technique for gradually changing the local pH was reported for glucose detection.[3] The "modulator electrode" was pulsed to negative potentials to reduce water and produce hydroxide ions to increase the pH within the diffusion field. The oxidation of glucose on gold was then detected as a Faradaic response at sufficiently high pH with fixed applied potential. Here, this methodology is developed further to demonstrate the "chemical filtering" effect.

In this report a novel gold-epoxy-gold micro-trench electrode with coplanar gold electrodes is formed (by Piranha etching of the epoxy spacer layer; see Fig. 1). The co-planar gold–gold micro-trench electrode system with *ca.* 70 μm gap (by SEM) is investigated for pulse voltammetry electroanalysis. Examples for feedback enhancement (hydroquinone oxidation) and chemical control *via* localised pH shift (hydroquinone oxidation and ammonia detection) are reported. The feasibility of nano-trench electrode experiments is discussed.

2. Experimental

2.1. Reagents

Hydroquinone, phosphoric acid, sodium hydroxide, nitric acid, hydrochloric acid, sulphuric acid, hydrogen peroxide and ammonium nitrate were all purchased from Sigma-Aldrich (UK) and were used as received without further purification. All solutions were made with ultra-pure water (Barnstead) with a resistivity of not less than 18 MΩ.

2.2. Instrumentation

Electrochemical measurements were carried out on a Biologic SP-300 bipotentiostat (Biologic, France). A four electrode arrangement was employed where the reference electrode was a KCl-saturated calomel electrode (SCE, Radiometer REF 401), the counter electrode a platinum disk and two working electrodes were the generator and collector of the gold–gold micro-trench electrode. Normal pulse voltammetry was carried out on the generator (or modulator) electrode with a step height of 5 mV, a pulse width of 1 s and a step time of 2 s. The collector electrode was held at a constant potential and the current recorded every 0.1 s. SEM images were obtained with a JEOL SEM6480LV system and samples were coated in 5 nm of chromium prior to imaging.

2.3. Procedure: Fabrication of gold–gold piranha junction electrodes

Microscope slides with a 100 nm thick layer of gold were purchased from Sigma-Aldrich (UK) and were sliced into 1.0 cm × 2.5 cm sections using a diamond cutter (Buehler Isomet 1000). Kapton (Farnell) tape was then used to mask off a strip of gold in the centre of the electrode (0.5 cm × 2.5 cm). Gold was then etched from the exposed areas using aqua regia (1 : 3 v/v nitric acid : hydrochloric acid; *WARNING: this solution is highly aggressive and appropriate precautions are needed*) and the slide dipped into the solution for 3 min. Once the slide was removed from the aqua regia solution, rinsed with water and the Kapton tape removed, the slides were put into a furnace for 30 min at 500 °C to oxidise the remaining conducting titanium under-layer and make it fully insulating. Two etched gold slides were then stuck together with epoxy (SP106 multi-purpose epoxy system, SP Gurit; a 1 h delay after mixing the epoxy was necessary to avoid short-circuiting of the two gold surfaces) and left in a home-made press overnight to cure. The end of the micro-trench was sliced off to reveal a surface which was then polished flat (SiC paper P320, Buehler). The epoxy in between the electrode was etched out using Piranha solution (1 : 5 v/v hydrogen peroxide : sulphuric acid; (*WARNING: this solution is highly aggressive and appropriate precautions are needed*) to give the trench (see Fig. 2).

3. Results and discussion

3.1. Gold–gold micro-trench electroanalysis I: Generator–collector voltammetry for the oxidation of hydroquinone

Cyclic voltammetry for hydroquinone oxidation was used as a test system to calculate/confirm the trench depth/width between the two gold plate electrodes. The oxidation of hydroquinone to benzoquinone takes place *via* a 2-electron 2-proton process (eqn (1)[14]) with diffusional transport between the two plate electrodes.

$$\text{hydroquinone(aq)} \rightleftarrows \text{benzoquinone(aq)} + 2\ e^- + 2\ H^+(aq) \quad (1)$$

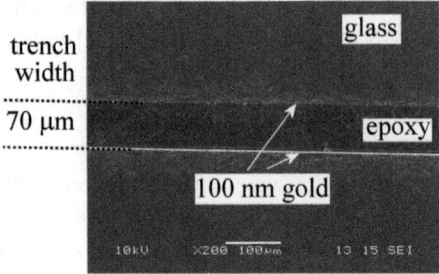

Fig. 2 SEM image of the gold–gold micro-trench electrode with a trench width of 70 μm, electrode length ca. 5 mm, and trench depth ca. 1 mm (estimated from SEM focal planes) used in pulse voltammetry experiments.

This process is pH dependent[15] with a dependence given approximately by E_{mid} vs. SCE/V = 0.46–0.059 pH under the conditions used in this study. With a buffer pH of 7 the reversible potential should therefore be at E_{mid} = 0.05 V vs. SCE. This is confirmed by data in Fig. 3.

Fig. 3A shows the cyclic voltammogram for the generator electrode (with $E_{collector}$ = −0.2 V vs. SCE fixed) and the peak at 0.1 V vs. SCE indicating the formation of benzoquinone. A slight peak at ca. 0.5 V vs. SCE is probably due to the surface oxidation of gold. Simultaneous chronoamperometry on the second electrode was carried out with the collector electrode set at a fixed potential of −0.2 V vs. SCE (Fig. 3B) and at different scan rates. Hysteresis effects (diffusional lag, see Fig. 3C) were observed for the collector signal with the effect becoming more apparent with higher scan rates. The trench depth can be estimated from the steady state limiting current and the approximate Nernst layer eqn (1).[16]

$$|I_{generator}| = |I_{collector}| = \frac{nFDdwc}{\delta} \quad (2)$$

Here I is the current, n is the number of electrons transferred per molecule, F is the Faraday constant, D is the diffusion coefficient of hydroquinone (ca. 0.74 × 10^{-9} m^{2}s^{-1}[17]), d is the trench depth, w is the trench width, c is the concentration of hydroquinone, and δ is the inter-electrode gap. From the data (I = 0.75 μA) the trench depth is estimated as 800 μm (in agreement with SEM data in Fig. 2). From the hysteresis effect in the collector current response (see Fig. 3C) the trench width is estimated as $\delta = \sqrt{\frac{\Delta E_{hysteresis} DRT}{0.0071 vF}} = 90$ μm,[18] (with $\Delta E_{hysteresis}$ determining a half height and v the scan rate; it is assumed that in first approximation the diffusion coefficients for reduced and oxidised species are equal) which is slightly higher when compared to the value obtained by SEM (Fig. 2).

Next, in order to explore the benefits of pulse electroanalysis in micro-trench systems, normal pulse voltammetry was utilised (applied to the generator electrode with the collector electrode in chronoamperometry mode with fixed potential) to study the same hydroquinone/ benzoquinone redox system. The generator was pulsed from −0.2 V to 0.6 V vs. SCE with increments of 5 mV for each pulse, pulse time was 1 s with the step time 2 s. Fig. 4A shows the generator electrode response, which is independent of the applied potential at the collector

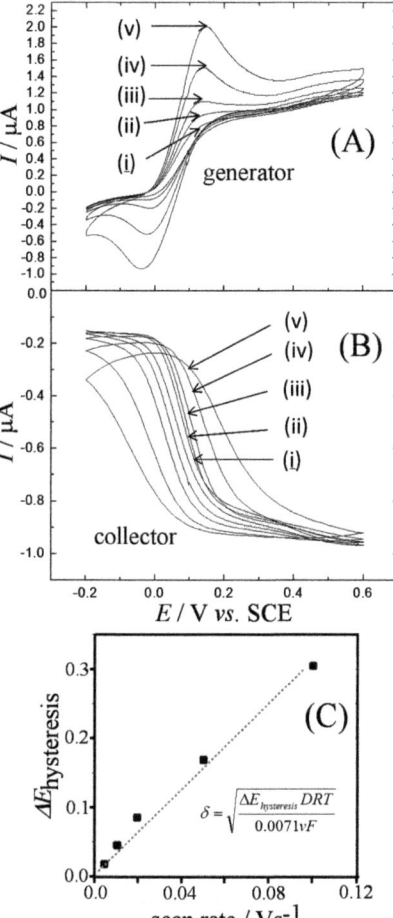

Fig. 3 (A) Cyclic voltammetry (generator signal, $E_{collector} = -0.2$ V, scan rate (i) 5 mV s^{-1}, (ii) 10 mV s^{-1}, (iii) 20 mV s^{-1} and (iv) 50 mV s^{-1} (v) 100 mV s^{-1}) for the oxidation of 1 mM hydroquinone in 10 mM PBS pH 7 using the gold–gold micro-trench electrode. (B) The corresponding collector current responses. (C) Plot of the collector current hysteresis versus potential scan rate with a dotted line indicating $\delta = 90$ μm (see text).

over a range of −0.2 V to 0.1 V vs. SCE. Fig. 4B shows the collector response (in chronoamperometry mode at fixed potential of −0.2 V vs. SCE), which is dominated by an oscillating current signal (see Fig. 4C).

A plot of the limiting reduction current during pulses (corresponding to close to mass transport limiting conditions) is shown in Fig. 4D. Here, the potential applied to the collector electrode clearly matters and the best response is observed at $E_{collector} = -0.2$ V vs. SCE where the response appears to plateau.

3.2. Gold–gold micro-trench electroanalysis II: Modulator-sensor voltammetry for the reversible oxidation of hydroquinone

In a different type of measurement mode one electrode (modulator, in normal pulse voltammetry method) can be employed to modulate the micro-trench pH

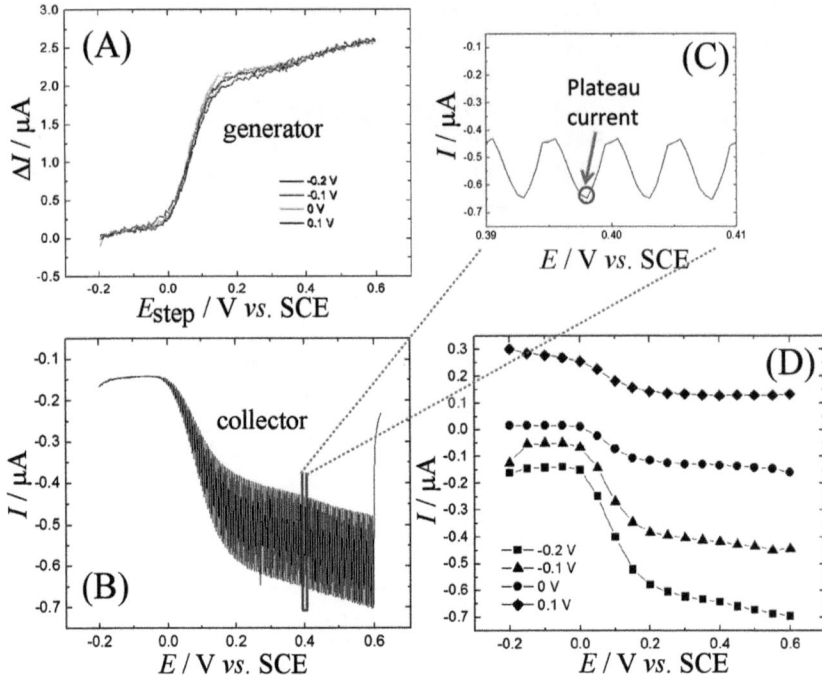

Fig. 4 (A) Normal pulse voltammetry generator response (from −0.2 V to 0.6 V, 5 mV steps, pulse time 1 s, step time 2 s, collector at −0.2 V/−0.1 V/ 0.0 V/ +0.1 V vs. SCE) for the oxidation of 1 mM hydroquinone in 10 mM PBS pH 7. (B) Collector current for −0.2 V vs. SCE. (C) Magnified plot of the collector current response. (D) Plot of collector plateau currents taken every 0.05 V for the collector electrode held at −0.2 V/ −0.1 V/ 0.0 V/ 0.1 V vs. SCE.

with the second electrode at fixed potential (sensor, in chronoamperometry method) responding to the shift in local pH. This kind of method was suggested recently for the determination of glucose[3] and is investigated here in more detail for the chemically reversible hydroquinone/ benzoquinone system. As the modulator potential is stepped more negative, a shift in pH to more alkaline conditions (depending on buffer type and capacity) is expected, associated with an oxidation at the sensor electrode in cases where the oxidation is pH dependent. A "chemical filter" effect is observed because redox systems that are not responsive to pH variation would remain undetected.

The reduction of water (in the presence of buffer) to produce hydroxide ions was achieved through normal pulse voltammetry at the modulator electrode whilst the sensor was held at a fixed potential. Sufficient local pH changes were observed once the modulator potential had reached −1.2 V vs. SCE (as shown in Fig. 5A) by the change in sensor plateau current at this potential (the corresponding generator current is not informative and therefore not shown). Changing the pH means a shift in the midpoint potential with higher pH conditions to more negative potentials. When the sensor potential is held at −0.1 V vs. SCE, there is a large step in the current compared to the other sensor potentials (see inset in Fig. 5A). This effect can be understood based on the buffer capacity in relation to the midpoint potential for the hydroquinone/ benzoquinone redox system. A midpoint potential of −0.1 V vs. SCE corresponds

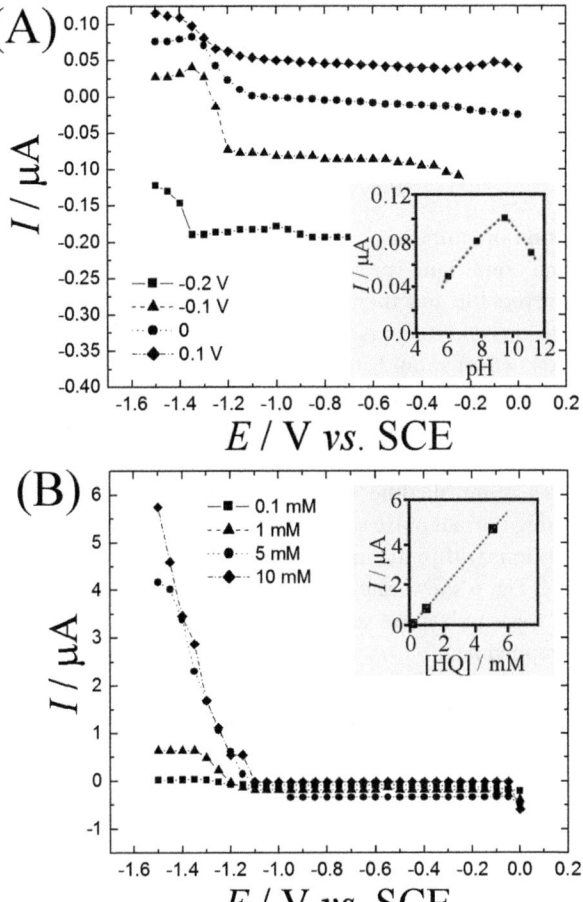

Fig. 5 (A) Chronoamperometry signal (plateau currents measured every 0.05 V) for the collector electrode *versus* generator step potential for the oxidation of 0.1 mM hydroquinone in 10 mM PBS pH 7 held at varying fixed potentials whilst normal pulse voltammetry was applied to the generator electrode to change the trench pH. The inset shows a plot of hydroquinone oxidation current *versus* midpoint pH (see text). (B) Chronoamperometry signals for the collector at different concentrations of hydroquinone with potential fixed at −0.1 V *vs.* SCE. The inset shows current signal *versus* concentration.

to a pH 9.5,[15] which is the point of minimum buffer capacity for phosphate buffer (with $pK_{A1} = 2.12$ at 18 °C, $pK_{A2} = 7.21$ at 18 °C, and $pK_{A3} = 12.67$ at 25 °C[19]). Setting the sensor potential to 0.0 V or −0.2 V *vs.* SCE moves in both cases the hydroquinone/benzoquinone midpoint potential into a more effective buffer regime, and therefore the pH change generated in the trench produces a smaller current response. In the maximum buffer range (and when increasing the buffer concentration) the signal is diminished.

Once the optimum sensor potential has been identified at −0.1 V *vs.* SCE, the effect of hydroquinone concentration is investigated. The concentration of hydroquinone in the gold–gold micro-trench electrode was varied from 0.1 mM to 10 mM (see Fig. 5B) and it was found that current increases linearly with concentration (see inset).

3.3. Gold–gold junction pulse-electroanalysis III.: Modulator-sensor voltammetry for the irreversible oxidation of ammonia

It is interesting to apply the "chemical filter" concept to an irreversible redox system, here the oxidation of aqueous ammonia (eqn (3)[20,21]).

$$2\ NH_3(aq) \rightarrow N_2(gas) + 6\ H^+(aq) + 6\ e^- \quad (3)$$

The oxidation of ammonia is known to be pH dependent[22] and effective only under alkaline conditions (pK_A = 9.23 at 20 °C [19]). However, this process is chemically irreversible and therefore not able to give a current-feedback effect. Therefore the sensor currents are expected to be lower. Nevertheless, the "chemical filter" effect should still be effective due to pH dependence.

Fig. 6 shows typical sensor data for an applied potential of 0.4 V *vs.* SCE and for the modulator electrode stepping in normal pulse voltammetry mode from 0.0 to −2.0 V *vs.* SCE. The phosphate buffer pH 7 solution stops the oxidation of NH_3, which requires more alkaline conditions. Ammonia oxidation was therefore carried out using normal pulse voltammetry on the modulator electrode to create high pH conditions within the micro-trench and the sensor electrode was set at a fixed potential. Fig. 6 shows the sensor response when the potential was held at 0.4 V *vs.* SCE. Clear pulses are seen even at −0.5 V *vs.* SCE, but as the modulator potential is stepped past −1.7 V *vs.* SCE, additional current spikes are seen at the start of each pulse segment. It is likely that a low level background process such as the gold surface oxidation (which, due to interfacial oxide formation, is also pH dependent) at the sensor electrode is observed in addition to the ammonia oxidation, which appears to be active at −1.7 V *vs.* SCE.

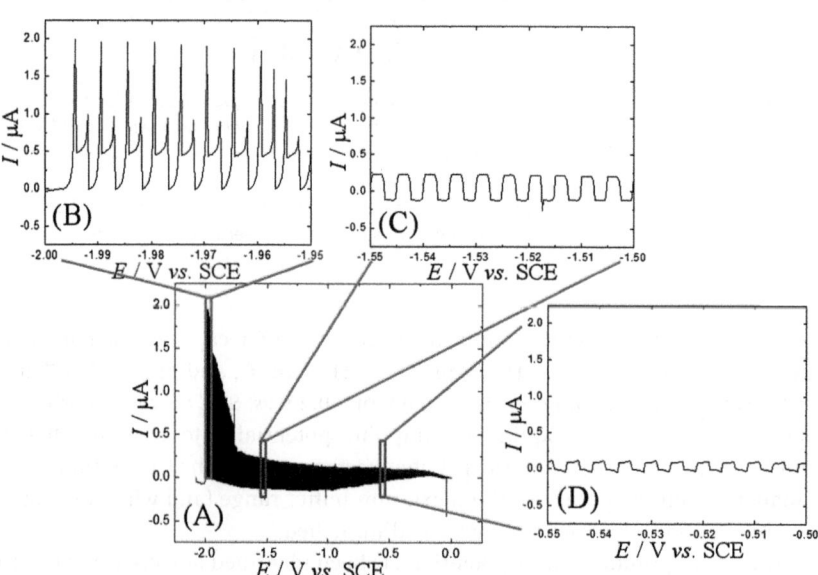

Fig. 6 Chronoamperometry data for the collector electrode (modulator electrode in normal pulse voltammetry mode stepping from 0.0 to −1.0 V *vs.* SCE in 5 mV steps, pulse time 1 s, step time 2 s) held at a potential of 0.4 V in aqueous 10 mM NH_4NO_3 in 10 mM phosphate buffer pH 7 with insets (B), (C) and (D) showing a zoomed in region.

The effect of ammonium nitrate concentration was investigated as shown in Fig. 7A. Although the current has not plateaued at −2 V vs. SCE, an increase in current can be seen with increasing ammonium nitrate concentration. Fig. 7B shows the plateau current taken for the last pulse at −2 V vs. SCE, where the pH is at its highest and shows the variation in current with collector potential and concentration. A second electrode potential of 0.4 V vs. SCE gives the highest collector current, indicating that this is the optimum potential to oxidise ammonia. The future potential of this modulator-sensor methodology for

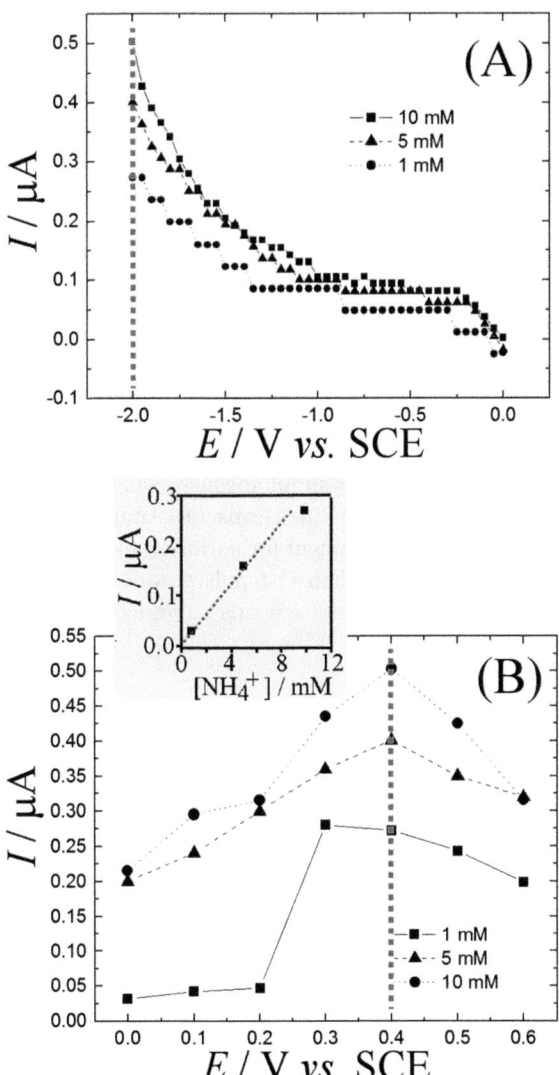

Fig. 7 (A) Plateau currents for the sensor (held at 0.4 V vs. SCE) with the modulator being swept negative from 0.0 to −2.0 V vs. SCE using normal pulse voltammetry (5 mV steps, pulse 1 s, step duration 2 s) for different concentrations of NH_4NO_3. (B) Plateau current for the sensor at generator potential of −2 V vs. SCE for varying sensor potentials and concentrations of NH_4NO_3. The inset shows a plot of sensor current versus NH_4^+ concentration (see text).

electroanalytical applications remains to be further investigated and optimised. In particular, work in much smaller trench electrodes with sub-micron gap will be desirable to improve sensitivity and speed. However it has been demonstrated in this study that, by employing a wider gap, control of the local pH in the micro-trench is possible and effective in modulating redox chemistry.

4. Conclusions

Bipotentiostatic control of processes in micro-trench electrodes allows: (i) feedback enhancement of analytical current responses; and (ii) "chemical filter" effects to be exploited. In particular chemically reversible processes for analytes such as hydroquinone are affected where both effects occur simultaneously. However, even for chemically irreversible processes such as ammonia oxidation, the methodology can be applied and a concentration dependent current signal can be obtained "*in situ*" with pH adjustment only locally within the micro-trench electrode system. Currently, the new methodology has many weaknesses in particular due to the gold film electrode being electrochemically active and sensitive to corrosion and degradation. Improved designs will be possible in the future, for example with boron-doped diamond[23] micro-trench electrodes or with flow-through micro-trench electrode systems associated with flow injection analysis.

The use of smaller nano-trench electrode systems will be beneficial in that feedback effects are much more pronounced. The control of pH in the trench will be possible with lower current pulses, which should further improve the methodology. A potential instrumentation problem/limit arises due to the need to control the two working electrodes simultaneously, which in particular for pulse conditions can lead to short instabilities (capacitive coupling of electrodes). This is expected to become more prominent for narrower and deeper trench systems where slower transients rather than fast pulses are more likely to give good electroanalytical performance. Work towards sub-micron trench systems is in progress.

Acknowledgements

S. E. C. D. thanks EPSRC for support (EP/I028706/1). We thank Dr John M. Mitchels and Ursula Potter for assistance with electron microscopy.

References

1 E. O. Barnes, G. E. M. Lewis, S. E. C. Dale, F. Marken and R. G. Compton, *Analyst*, 2012, **137**, 1068.
2 R. W. French, S. N. Gordeev, P. R. Raithby and F. Marken, *J. Electroanal. Chem.*, 2009, **632**, 206.
3 L. Rassaei and F. Marken, *Anal. Chem.*, 2010, **82**, 7063.
4 L. Rassaei, K. Mathwig, E. D. Goluch and S. G. Lemay, *J. Phys. Chem. C*, 2012, **116**, 10913.
5 M. A. G. Zevenbergen, B. L. Wolfrum, E. D. Goluch, P. S. Singh and S. G. Lemay, *J. Am. Chem. Soc.*, 2009, **131**, 11471.
6 R. W. French and F. Marken, *J. Solid State Electrochem.*, 2008, **13**, 609.
7 L. Rassaei, R. W. French, R. G. Compton and F. Marken, *Analyst*, 2009, **134**, 887.
8 R. W. French, A. M. Collins and F. Marken, *Electroanalysis*, 2008, **20**, 2403.
9 S. E. C. Dale, C. E. Hotchen and F. Marken, *Electrochim. Acta*, 2012, DOI: 10.1016/j.electacta.2012.08.121.
10 B. J. Feldman, S. W. Feldberg and R. W. Murray, *J. Phys. Chem.*, 1987, **91**, 6558.

11 L. Rassaei, M. Herrmann, S. N. Gordeev and F. Marken, *J. Electroanal. Chem.*, 2012, **686**, 32.
12 C. Y. Cummings, G. A. Attard, J. M. Mitchels and F. Marken, *Aust. J. Chem.*, 2012, **65**, 65.
13 G. E. M. Lewis, S. E. C. Dale, B. Kasprzyk-Hordern, E. O. Barnes, R. G. Compton and F. Marken, *Electroanalysis*, 2012, **24**, 1726.
14 X. B. Ji, C. E. Banks, D. S. Silvester, A. J. Wain and R. G. Compton, *J. Phys. Chem. C*, 2007, **111**, 1496.
15 R. A. Webster, F. J. Xia, M. Pan, S. C. Mu, S. E. C. Dale, S. C. Tsang, F. W. Hammett, C. R. Bowen and F. Marken, *Electrochim. Acta*, 2012, **62**, 97.
16 R. G. Compton, C. E. Banks, *Understanding Voltammetry*, Imperial College Press, London, 2011, p. 94.
17 R. N. Adams, *Electrochemistry at Solid Electrodes*, Marcel Dekker, New York, 1969, p. 220.
18 A. Vuorema, H. Meadows, N. Bin Ibrahim, J. Del Campo, M. Cortina-Puig, M. Y. Vagin, A. A. Karyakin, M. Sillanpää and F. Marken, *Electroanalysis*, 2010, **22**, 2889.
19 D. R. Lide, *Handbook of Chemistry and Physics*, CRC Press, London, 1993, pp. 8–47.
20 X. B. Ji, C. E. Banks and R. G. Compton, *Analyst*, 2005, **130**, 1345.
21 R. A. Webster, J. D. Watkins, R. J. Potter and F. Marken, *RSC Advances*, 2012, **2**, 4886.
22 N. J. Bunce and D. Bejan, *Electrochim. Acta*, 2011, **56**, 8085.
23 R. G. Compton, J. S. Foord and F. Marken, *Electroanalysis*, 2003, **15**, 1349.

Faraday Discussions

RSCPublishing

PAPER

Effects of adsorption and confinement on nanoporous electrochemistry†

Je Hyun Bae, Ji-Hyung Han, Donghyeop Han and Taek Dong Chung*

Received 13th February 2013, Accepted 5th March 2013
DOI: 10.1039/c3fd00014a

Characteristic molecular dynamics of reactant molecules confined in the space of the nanometer scale augments the frequency of collisions with the electrified surface so that a given faradaic reaction can be enhanced at nanoporous electrodes, the so-called nano-confinement effect. Since this effect is grounded on diffusion inside nanopores, it is predicted that adsorption onto the surface will seriously affect the enhancement by nano-confinement. We experimentally explored the correlation between adsorption and the confinement effect by examining the oxidation of butanol isomers at platinum and gold nanoporous electrodes. The results showed that electrooxidation of 2-butanol, which is a non-adsorption reaction, was enhanced more than that of 1-butanol, which is an adsorption reaction, at nanoporous platinum in acidic media. In contrast, the nanoporous gold electrode, on which 1-butanol is less adsorptive than it is on platinum, enhanced the electrooxidation of 1-butanol greatly. Furthermore, the electrocatalytic activity of nanoporous gold for oxygen reduction reaction was improved so much as to be comparable with that of flat Pt. These findings show that the nano-confinement effect can be appreciable for electrocatalytic oxygen reduction as well as alcohol oxidation unless the adsorption is extensive, and suggests a new strategy in terms of material design for innovative non-noble metal electrocatalysts.

1 Introduction

Developing better catalytic electrode materials has been one of the most crucial tasks in fuel cell research in recent decades. A myriad of reports have been published in this field; the consensus is that pure platinum (Pt) and a few Pt alloys are the best in terms of catalytic power at present. However, Pt is problematic because of its cost and limited deposits on Earth. Numerous researchers have been making tremendous efforts to find alternative electrode materials to Pt or seek strategies involving as small an amount of Pt as possible, e.g. atomic Pt monolayers[1,2] and alloys.[3-5] Pt alloys can show improved oxygen reduction

---

Department of Chemistry, Seoul National University, Seoul 151-747, Korea. E-mail: tdchung@snu.ac.kr; Fax: +82-2-887-4354; Tel: +82-2-880-4362

† Electronic supplementary information (ESI) available: Reaction pathway in the oxidation of 1-butanol and 2-butanol on a Pt electrode in acidic solution. See DOI: 10.1039/c3fd00014a

reaction activity compared with a pure Pt catalyst.[6] Recently, there have been a few reports[7,8] on highly electrocatalytic non-Pt, non-noble metal, more general electrode materials based on carbon, which are comparable with dispersed Pt/C electrodes. The possible answers that have been proposed were commonly based on adsorption of the reactants or interfering species onto the catalytic electrode surface. For example, the elements and composition of the surface can modify the d-band structure of the Pt atoms of the outermost layer, leading to enhanced interaction between the adsorbed reactants and the electrode surface, or reduced adsorption of spectators such as hydroxyls blocking the Pt surface sites.[3,6,9]

Nanoporous electrodes, which have been gaining increasing attention in recent years owing to their high catalytic activity, are strong candidates that are expected to provide a solution for the non-Pt issue. However, the studies on nanoporous materials have been focused largely on fabrication methods and applications of nanoporous electrodes rather than the mechanism underlying the nanoporous catalytic activity.[10-13] In terms of catalytic activity, the nanoporous electrode materials developed to date were evaluated mostly for oxygen reduction[14-16] and alcohol oxidation,[17-19] both of which involve adsorption on the electrodes, significantly affecting overall faradaic reaction rate as well as poisoning the surfaces. At the beginning of the research, the enhanced electrochemical reaction at the nanoporous electrodes was explained on the basis of the enlarged surface area.[20,21] But acknowledging that the enhancement by nanoporous structures was clearly greater than could be explained by the surface area enlargement, interpretation based on crystalline facets[22] and surface defects[23] has been attempted. In spite of a sizable number of reports on nanoporous electrochemistry and its applications, the exact origin of electrocatalysis at nanoporous electrodes has not yet been revealed.

It was very recently that the enhanced electrochemical reactions at nanoporous electrodes started to be considered in view of their morphological features, rather than surface characteristics such as surface area, crystalline facets, and defects.[24-26] The nanoporous structure has an extremely high surface-to-volume ratio, with every limited space surrounded by the inner wall. Reportedly, this confined space can affect the electrochemical reaction inside the nanopores *via* electronic coupling and molecular dynamics. The former, electronic coupling, was suggested by Balbuena and co-workers who carried out density functional theory (DFT) calculations to theoretically investigate the electronic effect on the molecules confined between transition-metal surfaces.[25] They found that the interaction of the atoms in the molecule with the top and bottom metallic surfaces of the confined system was not negligible, but capable of modifying the electronic structure of the molecule, possibly bringing about facilitation of the reaction. However, this effect was not proven experimentally and the two surfaces needed to be separated from each other by a very small distance of several Å,[27] which was too short to explain the electrochemical phenomena at the conventional nanoporous electrodes.

The confined space of the nanoporous structures provides a spatial environment that makes reactant molecules remain close to the electrode surface and prevents them from escaping to the bulk solution. We called this phenomenon 'confinement effect' in a previous paper.[24] The reactant molecules trapped in a porous structure must be close to the electrode surface and thereby have a greater chance to undergo the electrochemical reactions during a given period. As a result, the electron transfer rate is expected to increase compared to that on a flat surface. On the other hand, the reactant molecules are more likely to escape from the flat surface to the bulk solution. Thus, the time for residing near the electrode surface may not be sufficient for the electrochemical reactions. The confinement effect was theoretically substantiated by Monte Carlo (MC) simulation,[24] in which the effects of electrochemical reaction kinetics and adsorption on current amplification were investigated. The faster the electrochemical reaction kinetics at the electrode surface, the more probable the depletion of reactant molecules and the less significant the confinement effect. Conversely, the confinement effect enhanced sluggish reactions more effectively, because the increased frequency of encounter between the reactants and the electrode surface should play a bigger role in augmenting the rate of the electrochemical reaction. These theoretical results let us predict the effect of adsorption on the electrode. As for the electrochemical reaction involving strong adsorption of the reactants, the step determining the overall reaction rate is not the encounter of reactant molecules with the surface. Therefore, augmentation of collision frequency between diffusive reactant molecules and the electrode surface would hardly contribute to the faradaic enhancement. In good agreement with the prediction, the MC simulation showed that greater adsorption led to a reduced confinement effect, indicating that nanoporous electrodes may not be helpful for highly adsorptive reactants.

There have been a few studies that experimentally observed and interpreted the enhanced reactivity in the nano-confined space in terms of the confinement effect.[24,28–30] Assembly of cytochrome c (cyt c) within sulfonated graphene nanosheets showed high catalytic activity compared with free cyt c. It was explained by allowing more chances for the substrates to interact with cyt c.[29] Erlebacher and coworkers reported that oxygen reduction reaction activity on nanoporous Ni/Pt nanoparticles was considerably higher than that on solid Ni/Pt nanoparticles, and this was ascribed to the confinement effect within the nanoporous nanoparticles.[30] Han et al. showed that electrochemical activity on the nanoporous Pt was higher than that on flat Pt for oxygen reduction reaction, H_2O_2 reduction, and potentiometric response to pH.[24] However, these results are about apparent enhancement of electrochemical reactions at nanoporous electrodes, but failed to offer direct evidence for the presence of the confinement effect and its underlying mechanism. With regard to this issue, there was a meaningful report that chemical reduction reactions catalyzed inside the cavity of the hollow nanoparticles exhibited higher catalytic power than that on solid nanoparticles.[28] The factor of collision frequency with the inner wall due to confinement was calculated to explain why the catalysis arose. This was a more quantitative approach concerning this issue, and another case supporting the explanation of the nanoporous electrocatalytic behaviour based on molecular dynamics in the nanopores. However, the hollow nanoparticles in this study were not electrified nanoporous conducting materials.

In this work, we experimentally explored the effect of adsorption on the confinement effect at nanoporous electrodes. If a confinement effect based on molecular dynamics exists, a non-adsorption reaction should show larger enhancement than that of an adsorption reaction. In order to examine the dynamic confinement effect in the absence of other factors, the reactant molecules should have an identical functional group, which is electrochemically active. Moreover, they should be the same size, since variations may significantly influence the diffusion coefficient, leading to unwanted collision frequency changes.[26,31] We selected two butanol isomers, primary and secondary alcohols, for a model system.

On Pt surfaces, 1-butanol is an adsorptive molecule whereas 2-butanol is less adsorptive.[32,33] How adsorption affected the confinement effect was investigated as a function of the electrode materials and pH values. Following the sample study employing 1-butanol and 2-butanol isomers, the correlation between adsorption and confinement effect at nanoporous electrodes was presented by an oxygen reduction reaction that involves chemisorption as the rate determining step.

2 Experimental

Reagents

All chemicals, including hydrogen hexachloroplatinate hydrate (Kojima Chemicals Co., Ltd., Japan), *t*-octylphenoxypolyethoxyethanol (Triton® X-100, Sigma), sodium chloride (Daejung Chemical & Metals Co., Ltd., Korea), potassium dicyanoargentate(I) (Sigma), potassium dicyanoaurate(I) (98%, Sigma), sodium carbonate monohydrate (≥99.5%, A.C.S. reagent, Sigma), oxalic acid (98%, Sigma), 1-butanol (99.8%, anhydrous, Sigma), 2-butanol (99.5%, anhydrous, Sigma), perchloric acid (70%, A.C.S. reagent, Sigma), sulfuric acid (A.C.S. reagent, J.T.Baker), potassium phosphate monobasic (Daejung Chemical & Metals Co., Ltd., Korea), potassium phosphate dibasic (Daejung Chemical & Metals Co., Ltd., Korea) and potassium hydroxide (Sigma) were used without further purification. All the aqueous solutions in this experiment were prepared with ultrapure deionized water produced by a Barnstead Nanopure (Thermo Scientific).

Fabrication of nanoporous Pt and Au

The nanoporous Pt (np Pt), denoted L_2-ePt, was prepared by electroplating of Pt in reverse micelle solution as described in our previous report.[34] Hydrogen hexachloroplatinate hydrate (5 wt%), 0.3 M sodium chloride (45 wt%), and Triton X-100 (50 wt%) were mixed and heated to 60 °C. The mixture as made was transparent and homogeneous. The temperature of the mixture solution was maintained around 40 °C using a thermostat (WCB-11H, Daihan Scientific Co., Ltd.). L_2-ePt was electrochemically deposited on a Pt disk electrode (dia. 1 mm) at −0.16 V vs. Ag/AgCl (3 M NaCl, BAS, Inc.) to minimize the crack.[35] The resulting np Pt electrode was immersed in distilled water for 1 day to extract the Triton X-100 and this procedure was repeated 3–4 times. Following this, the electrode was electrochemically cleaned by cycling potential between −0.7 and +0.85 V vs. Hg/Hg_2SO_4 in 1 M H_2SO_4 until reproducibly identical cyclic voltammograms were obtained. The surface roughness of the np Pt was determined by measuring the area under the hydrogen adsorption peak of the cyclic voltammogram in 1 M H_2SO_4, using a conversion factor of 210 µC cm^{-2}.[36]

The nanoporous Au (np Au) electrode was prepared by electrodeposition of Ag–Au alloy layers followed by selective dissolution of Ag in concentrated nitric acid.[22,37] In detail, the Ag–Au alloy layer was electrochemically deposited on a Au disk electrode (dia. = 1.6 mm) at −0.95 V vs. Ag/AgCl from a solution containing 27 mM $KAg(CN)_2$, 15 mM $KAu(CN)_2$, and 0.25 M Na_2CO_3. The alloy layers were transferred to a concentrated nitric acid solution to dissolve Ag. The surface roughness of the np Au was determined by measuring the area under the reduction peak of electrochemically oxidized gold oxide during the potential cycling, using a conversion factor of 400 µC cm^{-2}.[36]

The nanoporous black Au (npb Au) electrode was prepared by anodization of Au electrodes in oxalic acid according to previously reported procedures.[38] The Au disk electrode (dia. = 1 mm) was anodized in 0.3 M oxalic acid with a magnetic stirrer using a three electrode system. The potential of the Au electrode was swept from 0 V to 1.8 V vs. Hg/Hg$_2$SO$_4$ at a rate of 1 mV s^{-1} and then retained at 1.8 V for 18 min. During anodization, the temperature was maintained at 0 °C using a thermostat (WCR-p6, Daihan Scientific Co., Ltd.). The surface roughness of the npb Au was determined in the same way as for np Au.

Electrochemical measurements

The voltammetric techniques employed in this study were conducted in a three electrode system, using Model CHI750 (CH Instruments, Inc.) as the electrochemical analyzer. Hg/Hg$_2$SO$_4$ (saturated K$_2$SO$_4$, RE-2C, BAS, Inc.) or Hg/HgO (1 M NaOH, CH Instruments, Inc.) was used as the reference electrode in acid and alkaline solution, respectively. The potential values in the figures were converted by normal hydrogen electrode (NHE). A Pt gauze was used counter electrode. The conventional Pt and Au disk electrodes (dia. 1.6 mm, BAS, Inc.) were used as flat working electrodes to compare with np Pt and np Au, respectively. The solution was purged with high purity nitrogen gas for about 20 min prior to use and a nitrogen environment was maintained over the solution throughout the experiments. Reference 600 equipped with EIS300 electrochemical impedance spectroscopy software (Gamry Instruments) was used for the electrochemical impedance spectroscopy (EIS) experiments. EIS was performed by applying 10 mV AC amplitude at frequencies between 0.05 Hz and 100 kHz. All experiments were carried out at room temperature.

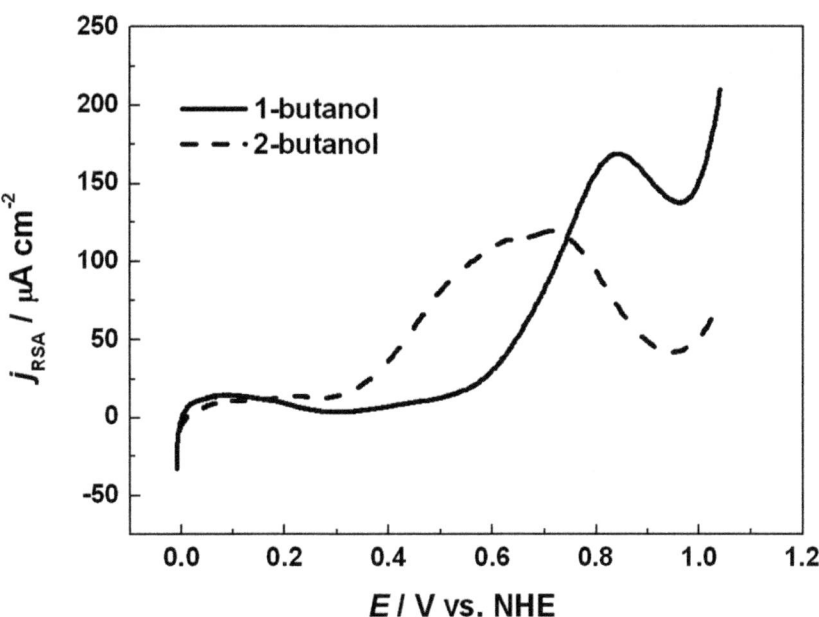

Fig. 1 Linear sweep voltammograms normalized by the real surface area (RSA) for the electrooxidation of 0.1 M 1-butanol and 0.1 M 2-butanol solutions in 1 M HClO$_4$ on a flat Pt at 10 mV s^{-1}.

3 Results and discussion

Voltammetric responses of 1-butanol and 2-butanol

In a previous study, we suggested the confinement effect as a reason for electrocatalytic enhancement at nanoporous electrodes.[24] If reactants are adsorbed on the electrode surface, the collision frequency in nanopores is expected to make less contribution to faradaic enhancement. Rather than instant collisions by diffusing molecules, the electron transfer preferentially occurs between the adsorbed molecules and the electrode surface in this case. Thus, more encounters with the electrode surface within a given period would not lead to significant increase of faradaic current. This was supported by the MC simulations.[24]

In this work, two butanol isomers were employed as a model system. Referring to the literature, the reaction mechanisms of 1-butanol and 2-butanol on Pt in acidic solution are summarized in Scheme S1, ESI.†[32,33] In brief, 1-butanol is very likely to be adsorbed on Pt accompanied by facile dehydration or dissociative chemisorption. As the applied potential is increased in the positive direction, the adsorbed 1-butanol undergoes electrooxidation to produce intermediates, including CO_{ad}, and the final product, CO_2. The active sites on the electrode are regenerated only after the adsorbed species are completely oxidized. Simultaneously, the non-adsorptive 2 electron transfer reaction occurs producing 1-butanal. Despite the parallel reactions, the adsorptive process is predominant for 1-butanol oxidation. On the other hand, 2-butanol is far less likely to be adsorbed on the Pt surface than 1-butanol. Fig. 1 shows that 2-butanol is electrochemically oxidized at less positive potentials than 1-butanol. Unlike 1-butanol, the non-adsorptive 2 electron transfer reaction of 2-butanol is overwhelming, generating butanone.

The electrochemical redox behaviour of 1-butanol at np Pt was examined by linear sweep voltammetry in 1 M $HClO_4$ solution. The apparent faradaic current at the np Pt was still larger than that at the flat Pt due to the enlarged surface area, as shown in Fig. 2(a). However, the current density normalized by real surface area in which contribution of the enlarged surface area was corrected at the np Pt was much smaller than that at the flat Pt (Fig. 2(b)). The current did not increase as much as the electrode surface area, indicating that the entire enlarged electrode surface inside the nanopores did not seem to participate in the faradaic reaction.

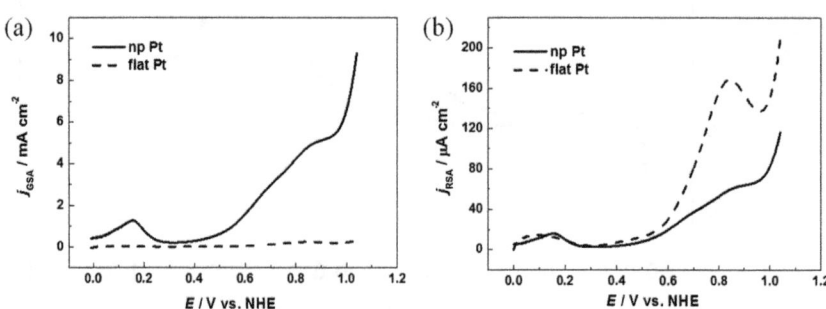

Fig. 2 Linear sweep voltammograms in which the current data were normalized by the (a) geometric surface area (GSA) and (b) real surface area for the electrooxidation of 0.1 M 1-butanol solution in 1 M $HClO_4$ on np Pt and flat Pt at 10 mV s^{-1}. The roughness factor of np Pt is 79.9.

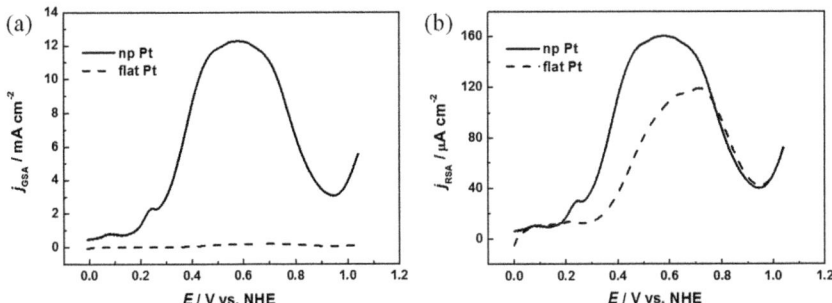

Fig. 3 Linear sweep voltammograms in which the current data were normalized by the (a) geometric surface area and (b) real surface area for the electrooxidation of 0.1 M 2-butanol solution in 1 M HClO$_4$ on np Pt and flat Pt at 10 mV s^{-1}. The roughness factor of np Pt is 76.5.

Taking into account that strong adsorption of 1-butanol on Pt occupied the active sites, more frequent encounters between the electrode surface and diffusive reactant molecules were hardly expected. As a consequence, the kinetic enhancement due to the confinement effect based on molecular dynamics in the nanopores was negligible. This result is consistent with what the simulation study suggested.[24] It is difficult for the faradaic reaction involving strong adsorption to benefit from nanoporous electrodes.

The absence or lack of the confinement effect due to strong adsorption can explain why extra enhancement of 1-butanol oxidation was not observed in the nanoporous electrode. However, it cannot justify the decrease of current density, which was normalized by real surface area. If the electrode surfaces were poisoned by strongly adsorbed species such as CO$_{ad}$ to the same extent regardless of porosity, we should have seen no drop in current density. It should be considered that the nanoporous electrodes have not only the enlarged surface area but also many narrow channels and interstices through which reactants and products have to move. Compared to the flat surface, the nanoporous structures can impede the mass transport to maintain the faradaic reaction, which is equivalent to the pore resistance possibly determined by ac impedance analysis.[26,39] Accordingly, it is less probable that the inner wall in the deeper region of the

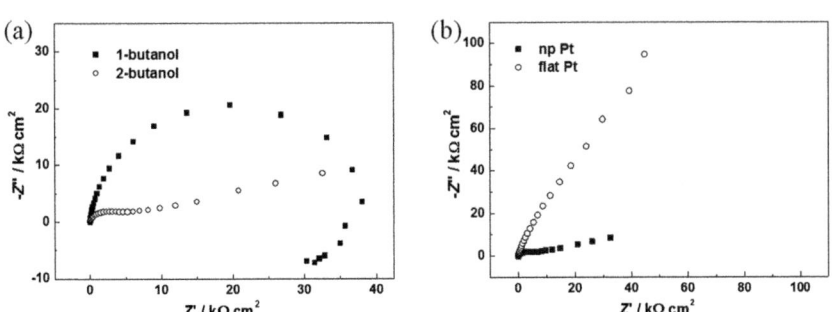

Fig. 4 Nyquist plots of the impedance data normalized by the real surface area of (a) np Pt in 0.1 M 1-butanol and 2-butanol/1 M HClO$_4$ solutions (b) np Pt and flat Pt in 0.1 M 2-butanol/1 M HClO$_4$ solution. The applied DC potentials for 1-butanol and 2-butanol were 0.64 V and 0.34 vs. NHE, respectively. The roughness factor of np Pt is 69.1.

electrode contributed to the faradaic current, and this effect would be more significant when higher faradaic current flows. Such a situation was previously observed in glucose oxidation at nanoporous Pt.[35] The current density of glucose oxidation gradually decreased as the roughness factor, which was the thickness of the nanoporous Pt film, increased.

The electrochemical behaviour of 2-butanol at np Pt is shown in Fig. 3. Interestingly, the voltammetric change of 2-butanol caused by the nanoporous structure is in contrast to that of 1-butanol. The anodic current normalized by the real surface area at the np Pt started to flow at less positive potential, indicating the nanoporous structure itself, not the enlarged surface area, acted as an origin of electrocatalysis. In addition to the onset potential, the peak current was higher than the flat Pt. Since 2-butanol is non- or much less adsorptive, it is believed that surface characteristics on np Pt, if any, would not have brought about such contrasting behaviour. Furthermore, there is no meaningful difference in the crystalline facets between flat and np Pts.[24] Considering the diffusion of the non-adsorptive reactants, 2-butanol, inside the nano-confined space on the np Pt, we can expect that the collision frequency with the electrified inner walls augmented the probability of heterogeneous electron transfer. Fig. 2 and 3 clearly reflect how significantly the adsorption process can affect the faradaic enhancement caused by the structures of nanoporous electrodes.

AC impedance analysis of 1-butanol and 2-butanol oxidation

AC impedance behaviour offers more insight into the enhanced electrochemical reaction in the absence of strong adsorption. Fig. 4(a) shows the Nyquist plot of 1-butanol, which is markedly different from that of 2-butanol at the np Pt. It exhibited a typical pseudoinductive behaviour, featuring a large arc in the high frequency regime and another small arc in the fourth quadrant at low frequencies. The pseudoinductive impedance data are characteristically observed when a desorption process is involved as the rate-determining step in a series of electrochemical reactions of adsorbed molecules on the electrode surface.[40,41] As for 1-butanol oxidation, it is predicted that a large proportion of the reaction sites on the Pt surface are covered with strongly adsorbed intermediates, including CO (Fig. S1, ESI†). As the frequency was lowered, the electrochemical potential that could drive the faradaic processes was effectively applied for a longer period. The pseudoinductive behaviour strongly implies that the previously adsorbed intermediates left the surface as a result of a faradaic reaction and such a desorption step was rate-determining for the overall processes.

On the other hand, impedance behaviour of 2-butanol corresponded to a typical Randles circuit including Warburg impedance, indicating a non-adsorption reaction.[42] The semicircle for 2-butanol was remarkably small compared with that for 1-butanol. This is in good accordance with what the voltammetric experiments suggested; that is, the fact that the charge transfer kinetics of 2-butanol oxidation is faster than that of 1-butanol.

Fig. 4(b) shows the Nyquist plots for 2-butanol oxidation at the np Pt and flat Pt, in which the current was normalized by the real surface area. Even after the correction of the electrode surface area, charge transfer resistance (R_{ct}) of np Pt was much smaller than that of flat Pt. The np Pt showed diffusion limited behaviour in the low frequency region, where the impedance from the flat Pt did

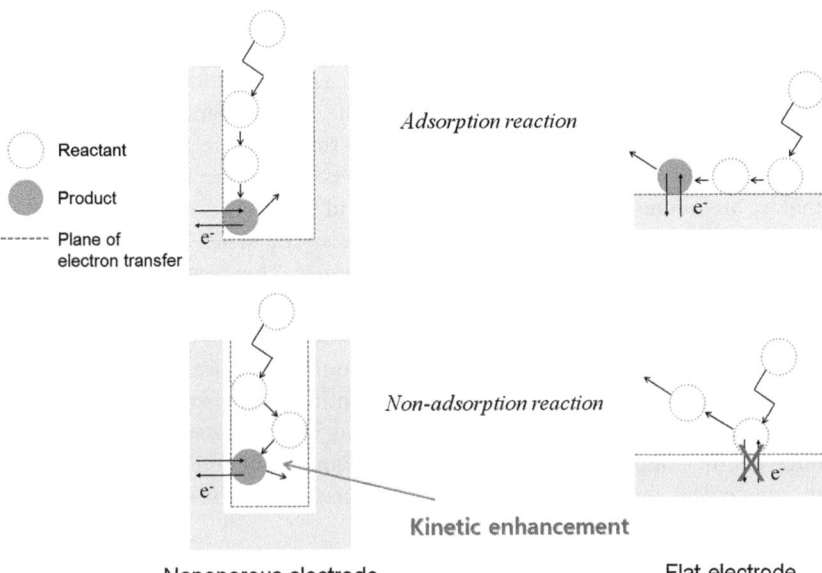

Fig. 5 Adsorption for the confinement effect.

not yet reach the Warburg regime. Considering that R_{ct}^{-1} is associated with the overall reaction rate at a given potential,[43] the electron transfer rate constant of the np Pt is thought to be much larger than that of the flat Pt. The impedance analysis confirms that the electrocatalytic power at np Pt arises neither from the surface area enlargement nor specific adsorption of reactants or intermediates on the electrode surface. Again, we can see that the observed non-adsorptive enhancement was caused by the morphological features of the nanoporous electrodes: the confinement effect.

Even for 1-butanol, a non-adsorptive process producing 1-butanal reportedly makes a significant contribution to the overall current at $E > 0.9$ V, where the adsorbed species are completely oxidized and the active sites on the electrode start to become available for reactant molecules dissolved in the solution (Scheme S1, ESI†). Therefore, it is predicted that the enhancement by the confinement effect at np Pt can be observed more or less even for 1-butanol oxidation, depending on the range of potential applied.

The presence and significance of the confinement effect

If electrochemical reactants are adsorptive at any rate on the electrode surface, many factors related to the surface, such as crystalline facets and defects, can be responsible for the kinetic enhancement. The theoretical and experimental results unequivocally suggest that stronger adsorption of the reactants is associated with reduced contribution of the confinement effect based on molecular dynamics of diffusive species, and *vice versa*. As for any non-adsorption reaction, an instant encounter between the diffusing reactant and the electrode surface is the only chance for heterogeneous electron transfer to take place. Among many potential factors for faradaic enhancement at nanoporous electrodes, anything other than the confinement effect can be imagined as the candidate for the enhanced current density from non-adsorptive reactants. Such a confinement effect is not negligible for weakly adsorption reactants as shown in Fig. 3 and 5. This finding proposes the possibility that the confinement effect provides a complementary electrocatalytic mechanism to the factors relying on the surface properties such as crystalline facets and defects. The electrochemical reactions that we are interested in may involve weakly and strongly adsorptive reactants or intermediates. Even a given reaction may consist of a series of adsorptive steps with a variety of characteristic adsorption energies. The conventional strategies try to make the reactants adsorb as strongly as possible for more orbital overlapping and facile bond cleavage by all means. The substrates developed based on such a concept are likely to suffer from poisoning or passivation by irreversible adsorption of intermediates or products, preventing a sustainable process. In this regard, the nanoporous electrodes that allow as weak an adsorption as possible offer another opportunity. Owing to the enhancement induced by their own morphological features, the confinement effect proposes electrocatalysis for weakly adsorptive reactants.

Apparently enhanced kinetics behaviour at the nanoporous electrodes is usually characterized by shifts of onset and peak potentials and narrow peak-to-peak separation. Two different approaches can be made to explain this behaviour. Firstly, this behaviour results from the change in mass transport and not at all from the electrode kinetics, which are unchanged. Compton et al. showed that the apparent enhanced kinetics at multiwalled carbon nanotubes was possibly explained on the basis of thin layer diffusion instead of semi-infinite linear diffusion.[44-46] They examined fast faradaic reactions such as ferrocyanide[44] and dopamine[46] at multiwalled carbon nanotubes. Considering the diffusion coefficients at room temperature, the thin layer diffusion model can be valid for macroporous electrodes (pore diameter > 50 nm).

Secondly, we can also see the real enhanced electrode kinetics in view of morphological features of nanoporous electrodes, including the confinement effect. As for the confinement effect, it should be noted that extremely limited space is given to the electroactive molecules in such nanopores before

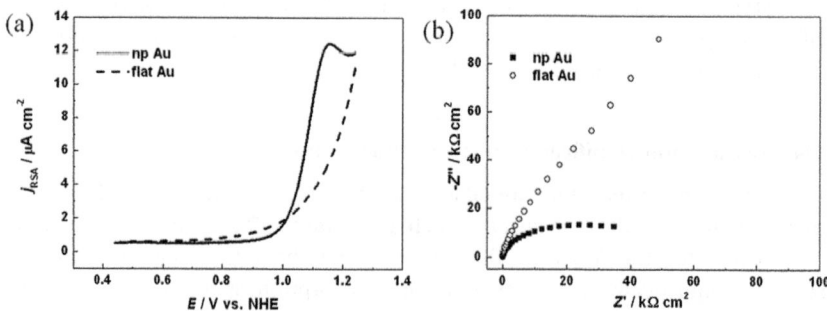

Fig. 6 (a) Cyclic voltammograms normalized by the real surface area for the electrooxdiation of 0.1 M 1-butanol solution in 1 M $HClO_4$ on np Au and flat Au at a scan rate of 10 mV s^{-1}. The roughness factor of np Au is 71.6. (b) Nyquist plots of the impedance data normalized by the real surface area of np Au and flat Au in 0.1 M 1-butanol solution and 1 M $HClO_4$. The applied DC offset potential is 1.19 V vs. NHE. The roughness factor of np Au is 42.4.

encountering the wall on the other side, because of wall-to-wall distances of less than several nanometers, especially in microporous electrodes. The channel space within a few nm scale seems to be sufficiently narrow to consider Knudsen diffusion, which is widely accepted as an appropriate model to describe molecular behaviour in the gas phase. It expresses well how molecules in the gas phase move in very small concave space like a narrow capillary whose width is narrower than the mean free path length.[47] Reportedly, catalytic effects coming from the geometrical features of nanoporous electrodes, such as irregularity and roughness, can be closely elucidated on the basis of molecular dynamics in the Knudsen regime. Although the electrocatalytic system in this work is not a solid–gas but a solid–liquid interface, and thus the mean free path of a molecule is predicted to be a few Å in solution, it is believed that the Knudsen diffusion model still works. Previously, enhanced current density of the proposed fuel cells exploiting the nanoporous catalysts was explained by Knudsen diffusion.[48] Considering its valuable benefits of simplicity and being a well established model in terms of theory as well as experiments, Knudsen diffusion provides a reasonable basis for understanding what is observed at nanoporous electrochemical interfaces.[26,48] Inspired by this effect, we can imagine potential utility of chemically inert surfaces for the same purpose.

Electrooxidation of 1-butanol on gold

As described earlier, 1-butanol, an adsorptive reactant on Pt, can hardly benefit from the confinement effect in terms of electrochemical oxidation. It is true that most primary alcohols are likely to be strongly adsorbed on Pt. However, the

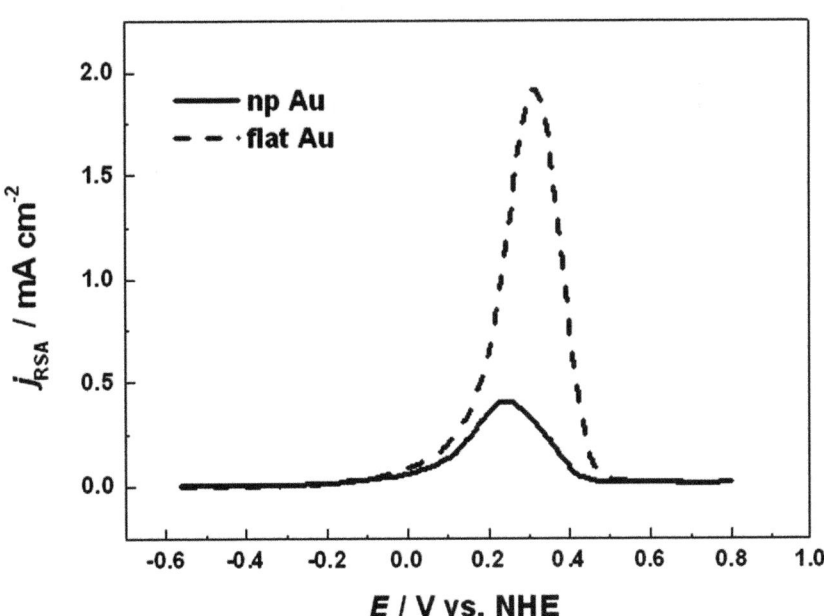

Fig. 7 Cyclic voltammograms normalized by the real surface area for the electrooxidation of 0.1 M 1-butanol solution in 1 M KOH on np Au and flat Au at a scan rate of 10 mV s^{-1}. The roughness factor of np Au is 33.5.

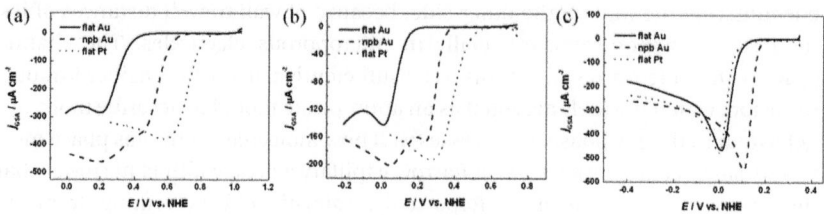

Fig. 8 Cyclic voltammograms normalized by the geometrical surface area for oxygen reduction obtained with bulk flat Au, npb Au, and flat Pt electrodes in an O_2-saturated (a) 1 M $HClO_4$, (b) 1 M phosphate buffer (pH 7), and (c) 1 M KOH solution at a scan rate of 10 mV s^{-1}. The curves were corrected for background measured in a N_2-saturated solution. The roughness factor of npb Au is 36.4.

degree of adsorption can be adjusted by replacing the electrode material or varying the pH. Unlike Pt, Au has a relatively inert surface to adsorption.[49] This is why catalytic activity of Au is poor in acidic solutions.

Fig. 6 presents the voltammetric behaviour of 1-butanol on np Au and flat Au. The background current level of flat Au was corrected to match that of np Au. Although it is well known that alcohol oxidation on Au in acidic solutions is mostly sluggish,[50] 1-butanol electro-oxidation on np Au is somewhat different from conventional behaviour. An anodic peak in the current density corrected by the real surface area appeared at 1.14 V. In agreement with the suggested reason for the kinetic enhancement on np Pt, weak adsorption of 1-butanol on Au led to a confinement effect at the np Au. It should be noted that these results were obtained under the same experimental conditions (compare with Fig. 2(b)). The current density of 1-butanol oxidation at the np Au was higher than that of flat Au, whereas np Pt gave lower current density than that of flat Pt.

According to the previous studies, smaller pores should be advantageous in terms of the confinement effect.[26] The np Au employed in this work had much larger pores, 16 (±2) nm,[22] than the np Pt, 1–3 nm.[34] However, the amplification of current density at the np Au over the flat Au was larger than that at the np Pt over the flat Pt. 1-butanol oxidation on the nanoporous Au shows that the confinement effect is a sensitive function of adsorption onto the electrode surface, leading to faradaic enhancement.

The ac impedance analysis supports the suggested explanation for the previously obtained results. Fig. 6(b) shows the Nyquist plots for the np Au and the flat Au. Although the impedance data were normalized by the real surface area, the semicircle of np Au was much smaller than that of flat Au. The semicircle of the flat Au was not well defined due to extremely slow kinetics. The ac impedance data also indicate faster apparent heterogeneous electrooxidation of 1-butanol on the np Au than on the flat Au. Adsorption, which was weakened by alternative electrode material, also resulted in kinetic enhancement by the confinement effect.

Electrooxidation of 1-butanol on np Au and flat Au in alkaline solutions provided additional evidence for the correlation between adsorption and faradaic enhancement. Hydroxyl ions are likely to be adsorbed on the Au surface in alkaline solutions[51] and promote adsorption of primary alcohols. As a result, Au electrodes generally become more effective catalysts for electrochemical alcohol oxidation at a high pH.[52] Current at a given area of the flat and nanoporous electrodes increased as overpotential was raised. Because the surface in the deeper region of the nanoporous electrode participated less in the faradaic reaction due to depletion of the reactants, its current density divided by the real surface area should be lower than that at the flat electrode.[35] The linear sweep voltammograms for electrooxidation of 1-butanol on np Au and flat Au in alkaline solutions in Fig. 7 shows that the current density normalized by the real surface area of flat Au was higher than that of np Au. Furthermore, the faradaic current for 1-butanol oxidation rapidly decreased at more positive potential than about 0.4 V, where the Au surface itself began to be oxidized. The oxide surface layer appeared to be responsible for the inhibition of the faradaic oxidation of 1-butanol.[53]

Oxygen reduction reaction (ORR)

Proton exchange membrane fuel cells (PEMFCs) have attracted great interest as potential power sources for next-generation vehicles due to their high energy conversion efficiency and low pollutant emission. One of the key issues in the technology is to develop cathode electrocatalysts at low cost and with high electrocatalytic activity for the ORR. It has been widely accepted that pure Pt or a few Pt alloys can currently provide the best electrocatalytic activity for the ORR in acidic media, thereby being the most commonly used cathode electrocatalyst in PEMFCs.

ORR activity on the metal in acidic media is governed by the first electron transfer to O_2(ad) to form HO_2(ad).[54] It implicitly assumes that adsorption of oxygen into an activated chemisorbed state is the rate limiting step.[48] Therefore, the conventional approach is to make O_2 adsorb as strongly as possible for electron transfer followed by bond cleavage. However, the presence of the confinement effect at nanoporous electrodes proposes another approach of utilizing chemically inert, especially with respect to chemisorption, electrode materials for ORR. As is well documented in the literature, the ORR activity at Au in acidic media is poor. Recently, it was reported that oxygen was reduced to H_2O *via* an effectively 4-electron route at np Au fabricated by dealloying of Ag.[55] This is an important result showing that 4 electron transfer of ORR is possible on the Au surface by introducing a nanoporous structure. However, the issue of residual material remains unclear, especially Ag, which may contribute to the electrocatalysis.[56,57] In this study, we used npb Au, which was prepared by direct electrochemical etching, to eliminate the possibility of such a residual effect. The npb Au had numerous pores smaller than 20 nm, which were roughly as large as those of np Au.[38]

The nanoporous structure of the npb Au electrode enhanced ORR activity as recognized by the positive shift of onset potential to approximately 0.25 V in Fig. 8(a). The waves of ORR at npb Au were split into two. Presumably the preceding peak at less negative potential came from the 2-electron reduction of O_2 to H_2O_2 and the peak following at more negative potential was due to the further 2-electron transfer for H_2O_2 to H_2O.[58] At the flat Au, the reduction of O_2 stopped at H_2O_2 after 2-electron transfer in the identical potential range. In contrast, ORR activity of npb Au was much enhanced in terms of onset potential as well as faradaic current. The shift of the onset potential was prominent so that significant faradaic current began to flow at a more positive potential than for the flat Pt by less than 0.2 V. Although the kinetic enhancement at npb Au failed to reach that at flat Pt, it was remarkably improved compared to the flat Au and as good as np Au.

At higher pH, ORR activity at npb Au was more greatly improved compared with that at flat Pt. In neutral solution at pH 7, the onset potential at npb Au was closer to that at flat Pt as shown in Fig. 8(b), while the overall voltammetric behaviors in acidic and neutral media were similar. In alkaline solution, oxygen at the npb Au started to be reduced at less negative potential than flat Pt. The shapes of the voltammograms from flat Au, npb Au, and flat Pt were similar, although their own onset potentials differed. However, the single reduction peaks in Fig. 8(c) are different from those observed in acidic and neutral media in Fig. 8(a) and (b). This is consistent with the fact that the ORR mechanism is different from that in acidic and neutral solutions.[59] Due to the lack of protons, ORR activity of flat Au was similar to that of flat Pt, and the number of electrons transferred at Pt was not more than that at Au. Fig. 8(c) shows the ORR activity at npb Au exceeded that of the flat Pt in basic media. In alkaline solutions, it is predicted that the strong adsorption of oxygen molecules is not favored any more, even on the Pt surface. Since smaller pores are expected to benefit from a larger confinement effect at a given temperature,[26] any nanoporous electrodes with less adsorptive surfaces and smaller pores than the npb Au employed in this work would lead to higher ORR activity. This implies the possibility that the optimal pore size of nanoporous electrodes made of cost-effective materials would enable sustainable ORR with comparable catalytic activity to Pt, even in acidic or neutral solutions, at relatively low temperature in the future.

4 Conclusions

The confinement effect based on molecular dynamics was experimentally confirmed. Adsorption of the reactants, 1-butanol in this study, on the inner wall of the nanoporous electrodes deteriorated the electrocatalytic activity. Our study demonstrated how significantly adsorption can influence the confinement effect at nanoporous electrodes. The experimental results showed that kinetic enhancement by the confinement effect was greater for weaker adsorption, which could be controlled by reactant molecules, electrode materials, and solution pH. ORR at the nanoporous black gold electrode added additional strong evidence supporting the presence and significance of the confinement effect at nanoporous structures. The electrocatalytic activity on the Au surface for ORR was greatly improved by introducing a nanoporous structure, even suggesting the possibility of comparable activity to that at flat Pt. The voltammetric study on oxidation of butanol isomers and reduction of oxygen gives a clear conclusion that weaker adsorption involved in the electrochemical reaction of interest leads us to expect a greater confinement effect in nanopores. This finding leads to a new strategy for better catalytic nanoporous electrodes. In terms of fuel cells, chemically inert non-Pt nanoporous materials would possibly evolve into innovative electrodes equipped with catalytic activity as high as Pt and capability of long term operation with controllable passivation at low temperature by optimizing the pore size and morphology. The confinement effect provides a fundamental clue for this perspective.

Acknowledgements

This work was supported by the Global Frontier R&D Program on Center for Multiscale Energy System (No. 2012M3A6A7055873), the SRC Program (No. 2007-0056334), and the Nano-Material Technology Development Program

(No. 2011-0030268) funded by the National Research Foundation under the Ministry of Science, ICT & Future Planning, Korea and by the IT R&D program of MKE/KEIT (10041596, Development of Core Technology for TFT Free Active Matrix Addressing Color Electronic Paper with Day and Night Usage).

References

1 K. P. Gong, D. Su and R. R. Adzic, *J. Am. Chem. Soc.*, 2010, **132**, 14364.
2 M. H. Shao, A. Peles, K. Shoemaker, M. Gummalla, P. N. Njoki, J. Luo and C. J. Zhong, *J. Phys. Chem. Lett.*, 2011, **2**, 67.
3 V. R. Stamenkovic, B. Fowler, B. S. Mun, G. F. Wang, P. N. Ross, C. A. Lucas and N. M. Markovic, *Science*, 2007, **315**, 493.
4 S. Koh and P. Strasser, *J. Am. Chem. Soc.*, 2007, **129**, 12624.
5 J. Greeley, I. E. L. Stephens, A. S. Bondarenko, T. P. Johansson, H. A. Hansen, T. F. Jaramillo, J. Rossmeisl, I. Chorkendorff and J. K. Norskov, *Nat. Chem.*, 2009, **1**, 552.
6 V. R. Stamenkovic, B. S. Mun, M. Arenz, K. J. J. Mayrhofer, C. A. Lucas, G. F. Wang, P. N. Ross and N. M. Markovic, *Nat. Mater.*, 2007, **6**, 241.
7 K. P. Gong, F. Du, Z. H. Xia, M. Durstock and L. M. Dai, *Science*, 2009, **323**, 760.
8 W. Xiong, F. Du, Y. Liu, A. Perez, M. Supp, T. S. Ramakrishnan, L. M. Dai and L. Jiang, *J. Am. Chem. Soc.*, 2010, **132**, 15839.
9 M. Mavrikakis, B. Hammer and J. K. Norskov, *Phys. Rev. Lett.*, 1998, **81**, 2819.
10 E. Seker, M. L. Reed and M. R. Begley, *Materials*, 2009, **2**, 2188.
11 A. C. Chen and P. Holt-Hindle, *Chem. Rev.*, 2010, **110**, 3767.
12 A. Kloke, F. von Stetten, R. Zengerle and S. Kerzenmacher, *Adv. Mater.*, 2011, **23**, 4976.
13 S. Park, H. C. Kim and T. D. Chung, *Analyst*, 2012, **137**, 3891.
14 P. R. Birkin, J. M. Elliott and Y. E. Watson, *Chem. Commun.*, 2000, 1693.
15 J. H. Jiang and X. Y. Wang, *Electrochem. Commun.*, 2012, **20**, 157.
16 J. H. Shim, Y. S. Kim, M. Kang, C. Lee and Y. Lee, *Phys. Chem. Chem. Phys.*, 2012, **14**, 3974.
17 J. T. Zhang, P. P. Liu, H. Y. Ma and Y. Ding, *J. Phys. Chem. C*, 2007, **111**, 10382.
18 L. X. Ding, A. L. Wang, G. R. Li, Z. Q. Liu, W. X. Zhao, C. Y. Su and Y. X. Tong, *J. Am. Chem. Soc.*, 2012, **134**, 5730.
19 Y. Yamauchi, A. Tonegawa, M. Komatsu, H. J. Wang, L. Wang, Y. Nemoto, N. Suzuki and K. Kuroda, *J. Am. Chem. Soc.*, 2012, **134**, 5100.
20 S. H. Joo, S. J. Choi, I. Oh, J. Kwak, Z. Liu, O. Terasaki and R. Ryoo, *Nature*, 2001, **412**, 169.
21 R. Szamocki, S. Reculusa, S. Ravaine, P. N. Bartlett, A. Kuhn and R. Hempelmann, *Angew. Chem., Int. Ed.*, 2006, **45**, 1317.
22 B. Seo and J. Kim, *Electroanalysis*, 2010, **22**, 939.
23 T. Fujita, P. Guan, K. McKenna, X. Lang, A. Hirata, L. Zhang, T. Tokunaga, S. Arai, Y. Yamamoto, N. Tanaka, Y. Ishikawa, N. Asao, Y. Yamamoto, J. Erlebacher and M. Chen, *Nat. Mater.*, 2012, **11**, 775.
24 J.-H. Han, E. Lee, S. Park, R. Chang and T. D. Chung, *J. Phys. Chem. C*, 2010, **114**, 9546.
25 J. M. M. de la Hoz and P. B. Balbuena, *J. Phys. Chem. C*, 2011, **115**, 21324.
26 J. H. Bae, J.-H. Han and T. D. Chung, *Phys. Chem. Chem. Phys.*, 2012, **14**, 448.
27 G. E. Ramirez-Caballero and P. B. Balbuena, *J. Phys. Chem. C*, 2009, **113**, 7851.
28 M. A. Mahmoud, F. Saira and M. A. El-Sayed, *Nano Lett.*, 2010, **10**, 3764.
29 B. Y. Hua, J. Wang, K. Wang, X. Li, X. J. Zhu and X. H. Xia, *Chem. Commun.*, 2012, **48**, 2316.
30 J. Snyder, I. McCue, K. Livi and J. Erlebacher, *J. Am. Chem. Soc.*, 2012, **134**, 8633.
31 K. Honda, M. Yoshimura, T. N. Rao, D. A. Tryk, A. Fujishima, K. Yasui, Y. Sakamoto, K. Nishio and H. Masuda, *J. Electroanal. Chem.*, 2001, **514**, 35.
32 J. L. Rodriguez, R. M. Souto, L. Fernandez-Merida and E. Pastor, *Chem.–Eur. J.*, 2002, **8**, 2134.
33 R. M. Souto, J. L. Rodriguez and E. Pastor, *Chem.–Eur. J.*, 2005, **11**, 3309.
34 S. Park, S. Y. Lee, H. Boo, H.-M. Kim, K.-B. Kim, H. C. Kim, Y. J. Song and T. D. Chung, *Chem. Mater.*, 2007, **19**, 3373.
35 S. Park, Y. J. Song, J.-H. Han, H. Boo and T. D. Chung, *Electrochim. Acta*, 2010, **55**, 2029.
36 S. Trasatti and O. A. Petrii, *J. Electroanal. Chem.*, 1992, **327**, 353.
37 C. X. Ji and P. C. Searson, *J. Phys. Chem. B*, 2003, **107**, 4494.
38 K. Nishio and H. Masuda, *Angew. Chem., Int. Ed.*, 2011, **50**, 1603.
39 J. M. Elliott and J. R. Owen, *Phys. Chem. Chem. Phys.*, 2000, **2**, 5653.
40 D. A. Harrington and B. E. Conway, *Electrochim. Acta*, 1987, **32**, 1703.
41 J. T. Muller, P. M. Urban and W. F. Holderich, *J. Power Sources*, 1999, **84**, 157.

42 A. J. Bard and L. R. Faulkner, *Electrochemical Methods Fundamentals and Applications*, John Wiley & Sons, Inc., 2001.
43 J. Otomo, X. Li, T. Kobayashi, C. J. Wen, H. Nagamoto and H. Takahashi, *J. Electroanal. Chem.*, 2004, **573**, 99.
44 I. Streeter, G. G. Wildgoose, L. D. Shao and R. G. Compton, *Sens. Actuators, B*, 2008, **133**, 462.
45 M. J. Sims, N. V. Rees, E. J. F. Dickinson and R. G. Compton, *Sens. Actuators, B*, 2010, **144**, 153.
46 M. C. Henstridge, E. J. F. Dickinson, M. Aslanoglu, C. Batchelor-McAuley and R. G. Compton, *Sens. Actuators, B*, 2010, **145**, 417.
47 P. Levitz, *J. Phys. Chem.*, 1993, **97**, 3813.
48 M. K. Debe, *J. Electrochem. Soc.*, 2012, **159**, B54.
49 M. Beltowskabrzezinska and J. Heitbaum, *J. Electroanal. Chem.*, 1985, **183**, 167.
50 E. Pastor, V. M. Schmidt, T. Iwasita, M. C. Arevalo, S. Gonzalez and A. J. Arvia, *Electrochim. Acta*, 1993, **38**, 1337.
51 L. D. Burke and W. A. Oleary, *J. Appl. Electrochem.*, 1989, **19**, 758.
52 M. BeltowskaBrzezinska, T. Luczak and R. Holze, *J. Appl. Electrochem.*, 1997, **27**, 999.
53 A. V. Tripkovic, K. D. Popovic and J. D. Lovic, *Electrochim. Acta*, 2001, **46**, 3163.
54 N. Ohta, K. Nomura and I. Yagi, *J. Phys. Chem. C*, 2012, **116**, 14390.
55 R. Zeis, T. Lei, K. Sieradzki, J. Snyder and J. Erlebacher, *J. Catal.*, 2008, **253**, 132.
56 A. Wittstock, V. Zielasek, J. Biener, C. M. Friend and M. Baumer, *Science*, 2010, **327**, 319.
57 L. V. Moskaleva, S. Rohe, A. Wittstock, V. Zielasek, T. Kluner, K. M. Neyman and M. Baumer, *Phys. Chem. Chem. Phys.*, 2011, **13**, 4529.
58 M. S. El-Deab and T. Ohsaka, *Electrochem. Commun.*, 2002, **4**, 288.
59 W. Jin, H. Du, S. L. Zheng, H. B. Xu and Y. Zhang, *J. Phys. Chem. B*, 2010, **114**, 6542.

Faraday Discussions

PAPER

Gold nanowire electrodes in array: simulation study and experiments†

Amélie Wahl, Karen Dawson, John MacHale, Seán Barry, Aidan J. Quinn and Alan O'Riordan*

Received 26th February 2013, Accepted 15th April 2013
DOI: 10.1039/c3fd00025g

Recent developments in nanofabrication have enabled fabrication of robust and reproducible nanoelectrodes with enhanced performance, when compared to microelectrodes. A hybrid electron beam/photolithography technique is shown that permits discrete gold nanowire electrode arrays to be routinely fabricated at reasonable cost. Fabricated devices include twelve gold nanowire working electrode arrays, an on-chip gold counter electrode and an on-chip platinum pseudo reference electrode. Using potential sweep techniques, when diffusionally independent, these nanowires exhibit measurable currents in the nanoAmpere regime and display steady-state voltammograms even at very high scan rates (5000 mV s^{-1}) indicative of fast analyte mass transport to the electrode. Nanowire electrode arrays offer the potential for enhancements in electroanalysis including increased signal to noise ratio and increased sensitivity while also allowing quantitative detection at much lower concentrations. However, to achieve this goal a full understanding of the diffusion profiles existing at nanowire arrays is required. To this end, we simulate the effects of altering inter-electrode separations on analyte diffusion for a range of scan rates at nanowire electrode arrays, and perform the corresponding experiments. We show that arrays with diffusionally independent concentration profiles demonstrate superior electrochemical performance compared to arrays with overlapping diffusion profiles when employing sweep voltammetric techniques. By contrast, we show that arrays with diffusionally overlapping profiles exhibit enhanced performance when employing step voltammetric techniques.

Introduction

Recent developments in fabrication of robust and reproducible nanoelectrodes have opened the door to a new and exciting area of electrochemistry.[1,2] Typically these electrodes demonstrate enhanced performance, when compared to micro-electrodes, due to improved mass transport occurring at the nanoelectrode

---

Tyndall National Institute – University College Cork, Lee Maltings, Dyke Parade, Cork, Ireland. E-mail: alan. oriordan@tyndall.ie; Tel: +353 21 2346403

† Electronic Supplementary Information (ESI) available. See DOI: 10.1039/b000000x/

thereby offering the potential for faster and more sensitive electroanalysis and improved kinetic measurement of electrochemical processes.[3,4] Smaller electrode dimensions permit fabrication at higher densities on silicon chips enabling much greater information-gathering capability per device. Finally, nanoelectrodes typically exhibit shorter response times, lower RC constants, low analyte depletion and significantly reduced sample volumes.[3]

To date, a significant portion of the studies employing nanoscale electrodes on silicon chips have been undertaken using nanodisks, nanopores or short inlaid nanobands.[5] However, large arrays of these structures are required to obtain reasonable measurable currents, i.e., in the nanoAmpere regime. By contrast, high aspect ratio nanostructures such as nanowires and nanobands, employed as electrochemical electrodes, exhibit relatively large currents (nA) due to the long length of the electrode (typically > 40 μm) but they also benefit from radial diffusion profiles to the electrode arising from their nanometre scale critical dimensions.[3] As such, these electrodes exhibit steady-state voltammograms indicative of fast analyte mass transport to the electrode even at very high scan rates (5000 mV s^{-1}). Furthermore, advances in nanofabrication techniques pioneered by the micro/nanoelectronics industry including: electron beam, nanoimprint and phase-shift lithographies, have made fabrication and integration of robust and reproducible one dimensional (1-D) nanostructure-based devices routinely achievable at reasonable economic cost.[6,7]

Consequently, 1-D nanostructures have been employed to explore heterogeneous electron transfer rate constants with rates up to two orders of magnitude higher than previously measured, which are being reported.[8-11] In addition, electrochemical devices employing discrete nanowire sensors demonstrating highly sensitive detection of key biomolecules including dopamine,[7] hydrogen peroxide[12] and glucose[13] amongst others have recently been reported. Electrodes based on nanowire arrays offer the potential for further enhancements in electroanalysis including: increased signal to noise ratio and increased sensitivity while also allowing quantitative detection at much lower concentrations.[14] However, to achieve this goal, a full understanding of the diffusion profiles existing at nanowire arrays is required. During electron transfer processes, electroactive species around the electrode are depleted creating depletion zones known as Nernst diffusion layers, δ, which thickness varies considerably with the electrode dimensions and geometry. In sweep voltammetry a diffusion-limited and time-dependent response (planar diffusion) generally occurs with larger electrodes (e.g. macroelectrodes), while an ideal steady-state and time-independent response (radial diffusion) is typical of much smaller electrodes (e.g. ultramicroelectrodes).[3,15,16]

At present, there is no analytical equation to describe diffusion to an electrode geometry where an electrode has sharp edges and protrudes from a planar substrate, i.e., a nanowire. For this reason, analytical solutions based on similar geometries such as hemicylinders or nanobands have been proposed and adopted to estimate voltammetric/mass transport behaviour at nanowire electrodes.[17] In this regard, the thickness of a Nernst diffusion layer, δ, at a hemicylindrical electrode approaching steady-state conditions can be estimated via the following equation:

$$\delta = r_0 \left(\ln \frac{2(D_o t)^{\frac{1}{2}}}{r_0} \right) \tag{1}$$

where D_o, is the diffusion coefficient, t is time and r_0 is the electrode radius. For a nanowire $r_0 = w/4$ where, w is the width of the electrode.[17–19] In this manner diffusion independence is theoretically maintained by ensuring that the separation between neighbouring electrodes, s, is greater than twice the diffusion thickness ($s > 2\delta$). However, we and others have recently shown that this is not the case at the nanoscale and significantly larger separations are required to maintain diffusional independence between neighbouring electrodes and this separation is significantly affected by the applied scan-rate.[14,20]

In this work, we undertake finite element analysis of diffusion profiles existing at arrays of nanowire electrodes (Comsol Multiphysics) and explore the effects of altering inter-electrode separations on diffusional independence for a range of scan rates. Although a range of scan rates are modelled, we are particularly interested in very high scan rates (5000 mV s^{-1}) since this allows rapid (sub 1 s) data capture required for, *e.g.*, biomedical, environmental and pharmaceutical diagnostic applications. To confirm our simulations experimentally, electron beam lithography, metal evaporation and lift-off techniques were employed to fabricate fully integrated nanowire array devices on silicon chips. These devices are fabricated in a manner as to include on-chip counter and pseudo reference electrodes. Finally, we show that arrays that are diffusionally independent demonstrate superior electrochemical performance when employing sweep voltammetric techniques compared to arrays with overlapping diffusion profiles. By contrast, arrays with diffusionally overlapping profiles exhibit enhanced performance when employing step voltammetric techniques.

Experimental

Finite-element simulations

Analyte concentration profiles for the oxidation of a redox molecule at electrode arrays containing three nanowires are simulated using the commercial finite element software package Comsol Multiphysics 4.1. (COMSOL, SE). These simulations are based on the single electron oxidation process of FcCOOH in solution at the nanowire electrodes surface, corresponding to:

$$O + e^- = \leftrightarrows R \qquad (2)$$

The objective of these simulations is to assess the distance required between adjacent nanowires in array to allow independent diffusional mass transport to each nanowire electrode, at high scan rates. As electrochemical experiments are undertaken in presence of excess supporting electrolyte, in static un-agitated environments at constant temperature and over short time periods, mass transport effects arising from migration and convection are assumed to be negligible. Thus emphasis is put on diffusional mass transport alone and only Fickian diffusion to a nanoband electrode in two dimensions is considered:[19,20]

$$\frac{\partial C_i}{\partial t} = D_i \left(\frac{\partial^2 C_i}{\partial x^2} + \frac{\partial^2 C_i}{\partial y^2} \right) \qquad (3)$$

where C_i and D_i represent the concentration and diffusion coefficient of the redox species in solution, respectively. In these models, the redox analyte concentration is 1 mM FcCOOH in 10 mM PBS, with an associated diffusion coefficient of 5.4×10^{-6} cm^2 s^{-1}.[21] Given the electron transfer process is reversible, both species O and R are soluble in solution and their diffusion coefficients are assumed to be equal ($D_O = D_R$), the concentration of FcCOOH (species R) at the nanowire electrodes surface at a given time interval for a potential sweep method may be expressed by the following Nernstian boundary conditions:[4,19]

$$\text{at } t = 0: C_R(t) = C_R^* \qquad (4)$$

$$\text{at } t > 0: C_R(t) = \frac{C_R^*}{1 + \exp\left[\dfrac{nF}{RT}(E^\circ - E(t))\right]} \qquad (5)$$

where t is time (s), C_R^* and $C_R(t)$ respectively are concentration (mol m^{-3}) of the reductant species R in the bulk, and at the electrode surface with respect to time, n is the number of electrons exchanged, F is Faraday's constant (96485 C mol^{-1}), R is the gas constant (8.314 J mol^{-1} K^{-1}), T is temperature (K), E° is the formal potential of the redox couple (V). $E(t)$ is the applied potential (V) defined as:

$$E(t) = E_{\text{int}} + vt \qquad (6)$$

where E_{int} is the initial voltage (V) of the potential sweep and v is the scan rate (V s^{-1}). From experimental results, a value of 0.155 V is used for E° and simulations are carried out in a potential window of −0.15 to 0.45 V for scan rates of 20, 100 and 5000 mV s^{-1}.

Nanowire fabrication

Gold nanowire array electrodes are fabricated using a hybrid electron beam/photo-lithography process on four inch wafer silicon substrates bearing a ~300 nm layer of thermally grown silicon dioxide (Si/SiO$_2$). In this approach, nanowires and alignment marks are patterned in resist (ZEP 520 Nippon Zeon) by direct electron beam writing (50 kV beam voltage, 100 pA beam current and 120 μC cm^{-2} beam dose). Following resist development, gold layers (Ti/Au 5/50 nm) are blanket deposited by metal evaporation (Temescal FC-2000 E-beam evaporator) and removed from un-patterned areas using standard lift-off techniques to yield nanowire stacked structures. Using electron beam written alignment marks, interconnection tracks are then overlaid onto the nanowires' termini by optical lithography, metal evaporation (Ti/Au 10/90 nm) and lift-off. Peripheral electrical contact pads, interconnection tracks and two central half-disk gold electrodes (all Ti/Au 10/90 nm) are deposited using the same procedure. A further metal deposition (Ti/Pt 10/90 nm) step is then performed onto one half-disk electrode. In this manner, the central half-disk electrodes may be employed as a gold counter and a platinum pseudo reference electrode, respectively.

To prevent unwanted electrochemical reactions occurring between metal interconnection tracks and electrochemically active species, a silicon nitride passivation layer (~500 nm) is then deposited by plasma-enhanced chemical vapour deposition (PECVD) onto the wafer surface. Photolithography and dry etching are then employed to selectively open windows (~45 × 100 μm) in the passivation layer directly above the gold nanowire working electrodes to allow exclusive contact between them and an electrolytic solution. Openings are also maintained above the counter, and reference electrodes along with peripheral contact pads. Following fabrication, wafers are diced into 16 × 16 mm chips. Each chip contains twelve individually contacted gold nanowire working electrode arrays, an integrated gold counter electrode and a platinum pseudo-reference electrode.

Structural and electrical characterisation

Optical micrographs of nanowire electrode arrays are acquired using a calibrated microscope (Axioskop II, Carl Zeiss Ltd.) equipped with a charge-coupled detector camera (CCD; DEI-750, Optronics). Structural characterisation is undertaken using scanning electron microscopy (SEM); images captured using a field emission SEM (JSM-6700F, JEOL UK Ltd.) operating at beam voltages between 5 and 10 kV. As a quality control check, to confirm electrical functionality, two-point electrical measurements are performed using a probe station (Model 6200, Micromanipulator Probe Station) in combination with a source meter (Keithly 2400) and a dedicated LabVIEW™ V8.0 program. In these current–voltage (I–V) measurements, the source electrode is grounded, a bias sweep up to ±10 mV is applied to the drain electrode, and the current through the nanowire is measured.

Electrochemical analysis

All electrochemical studies are performed using a CHI660a Electrochemical Analyzer and a Faraday Cage CHI200b (CH Instruments) connected to a PC. All experiments employ a standard three-electrode cell configuration using gold nanowire arrays as the working electrodes, *versus* the on-chip gold counter electrode and the on-chip platinum pseudo-reference electrode. Cyclic voltammetry (CV) is conducted in 10 mM phosphate buffer saline (PBS, pH 7.4) and in 1 mM ferrocenemonocarboxylic acid (FcCOOH) in 10 mM PBS in the voltage range of −0.15 V to 0.45, for a variety of scan rates (5, 10, 20, 50, 100, 200, 500, 1000, 2000 and 5000 mV s^{-1}). Square wave voltammograms are undertaken in 0.1 to 5 mM FcCOOH in 10 mM PBS solutions and in 10 mM PBS only in the voltage range of −0.15 to 0.45 V with an incremental potential of 0.001 V, a potential amplitude of 0.025 V and a frequency of 25 Hz. All chemicals are purchased from Sigma Aldrich and used as received. All solutions are freshly prepared using deionized water (18.2 MΩ.cm, ELGA Pure Lab Ultra). Prior to electrochemical experiments, all electrodes are cleaned by sequential immersion for 10 min in acetone (bath 1), trichloroethylene acetone (bath 2), iso-propyl alcohol, followed by a thorough rinse with deionized water and dried in a stream of nitrogen.

Results and discussion

Due to the high aspect ratio of the nanowires, a simplified two dimensional model known as the diffusion domain approach is adopted.[15,20] To perform simulations, a cross sectional plane through an array comprising of three nanowires is defined where each nanowire electrode within an array is assigned its own area; see Fig. 1(a). In this manner, each nanowire is outlined by a 50 nm × 100 nm rectangle (height × width) located centrally at the bottom of a much larger rectangle (space domain). To ensure accuracy of the simulations, the domain area is selected so as to be large enough to ensure bulk-like conditions at the boundaries

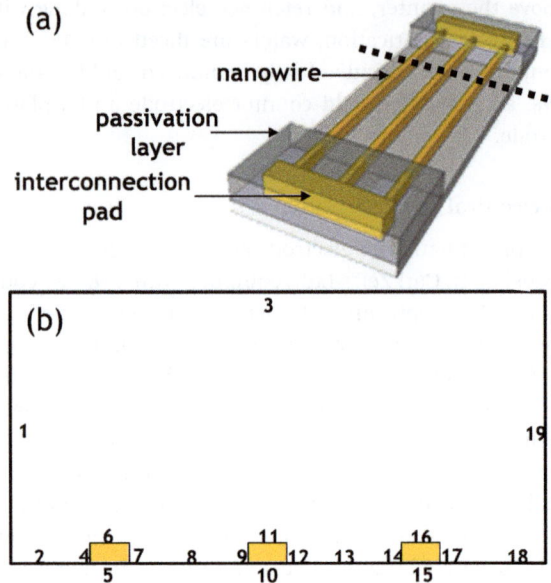

Fig. 1 (a) Schematic diagram of three nanowire electrodes in array. (b) 2D representation of three nanowire electrodes in array, used for simulation of diffusion at nanowire electrodes. Boundaries 1, 3 & 19 define the bulk concentration, C^*. Boundaries 2, 8, 13 & 18 correspond to flux = 0. No conditions are applicable at boundaries 5, 10 & 15. Boundaries 4, 6, 7, 9, 11, 12, 14, 16, & 17 are the concentration present at the electrode surface with respect to the time step of the electrolysis.

remain unaffected by the electrochemistry occurring at the electrodes. Fig. 1(b) depicts the two-dimensional geometry of the model employed with boundary conditions for flux = 0 and for the concentration of the reductant species R in the bulk and at the electrode surface with respect to time, respectively C_R^* and $C_R(t)$. During simulations the mesh is duly refined and simulations are allowed to iteratively resolve until a convergence error less than 2% is achieved.

We have previously demonstrated that radial diffusion of 1 mM FcCOOH in 10 mM PBS occurs at single gold nanowire electrodes (50 nm high, 100 nm wide and 45 μm long) using cyclic voltammetry for scan rates ranging from 20 mV s^{-1} to 5000 mV s^{-1}.[3] At electrode arrays, the inter-electrode separation is the critical factor that determines whether diffusional profiles overlap when other parameters including electrode width, solution composition (buffer type, concentration, *etc.*) and scan rate are kept constant. In this work, arrays containing three nanowires are chosen as they contain only one inner electrode competing for diffusional species on either side. Models are built for three nanowire arrays with increasing inter-electrode separations, *i.e.*, 5, 10, 15 and 20 μm for which diffusion profiles using 1 mM FcCOOH in 10 mM PBS under cyclic voltammetric conditions are calculated. The nanowire dimensions are maintained as 100 nm in width and 50 nm in height with $D_R = 5.4 \times 10^{-6}$ cm^2 s^{-1}, $E^0 = 0.155$ V, $T = 298.15$ K. As can be clearly seen in Fig. 2(a) and (b) overlap of adjacent diffusion layers occurs for nanowire arrays with inter-electrode separation of 5 and 10 μm. By comparison, in Fig. 2(c) and (d), for inter-electrode distances of 15 and 20 μm, the diffusion layers at each electrode in the array are completely independent from each other and radial in shape,

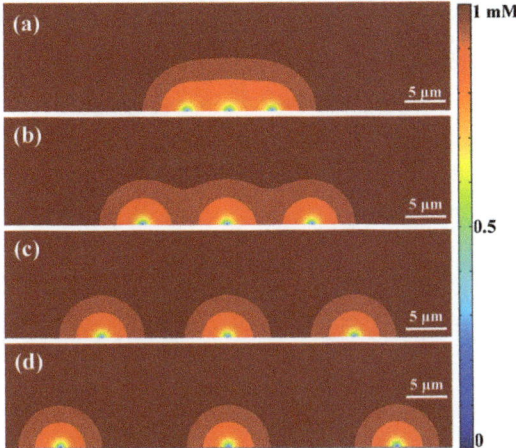

Fig. 2 2D simulations of FcCOOH concentration profiles at three nanowire arrays separated by: (a) 5 μm, (b) 10 μm, (c) 15 μm and (d) 20 μm at 5000 mV s^{-1}.

suggesting that each electrode in the array should behave as an individual electrode and that the current response of the whole array should be equivalent to three times that of a single nanowire.

To confirm simulation results, gold nanowire arrays are fabricated using a hybrid electron beam/photo-lithography process at Si/SiO$_2$ substrates as described in the Experimental section. Following fabrication, each electrode is structurally characterised using optical microscopy and SEM. Fig. 3(a) shows an optical micrograph of a gold counter electrode (left) and a pseudo-reference platinum electrode (right) integrated at a chip surface. An optical image of a fully fabricated, integrated and passivated single nanowire device is shown in Fig. 3(b). The gold squares at both nanowire termini are defined during the electron beam lithography step to minimise contact resistance when overlaid by the metallic interconnection tracks.

Fig. 3(c) to (f) are optical micrographs of fully fabricated, integrated and passivated three nanowire electrodes in array separated by 5, 10, 15 and 20 μm, respectively. The gold bars at the nanowire termini are again employed to minimise contact resistance. The width of the passivation window (central dark rectangle) defines the exposed nanowire length at 45 μm for both single nanowires and nanowire arrays. Visual inspection using SEM microscopy reveals excellent registration between nanowire electrodes and the overlaid micron-scale interconnection tracks, ensuring nanowires are well electrically contacted and uniform in width; see Fig. 3(g). A statistical analysis at multiple locations across seventeen individual nanowire structures ($n = 340$) from different chips yielded an average nanowire width of ~98 ± 5 nm (variation of ~5.1%), as presented in Fig. 3(h). This confirmed that proximity effects did not arise during the electron beam lithography process and consequently nanowire broadening did not occur.

Nanowire electrode devices were electrically characterised using standard two-point I–V, see the ESI.† All functioning nanowire array devices display linear Ohmic responses confirming good electrical contact to nanowires by the interconnections tracks, see Fig. S1.†

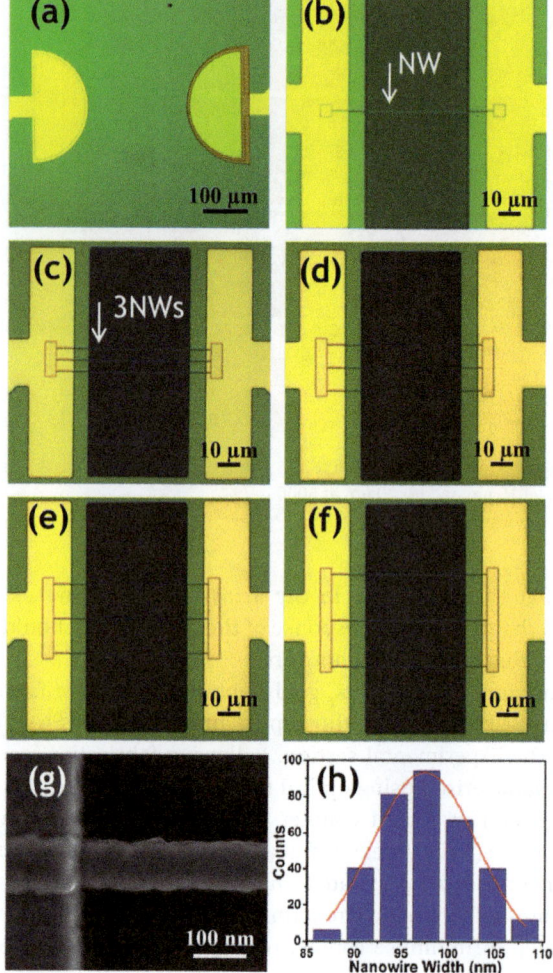

Fig. 3 Optical micrographs of fully integrated and passivated (a) gold counter (left) and platinum pseudo-reference electrodes (right), (b) single gold nanowire electrode. Nanowire electrode arrays containing three nanowires separated by (c) 5 µm, (d) 10 µm, (e) 15 µm and (f) 20 µm. The darker rectangle in the middle corresponds to the trench selectively opened in silicon nitride above the nanowires to allow contact with an electrolyte solution. (g) High magnification SEM micrograph of a region of passivated and fully exposed nanowire. (h) Histogram showing the distribution of widths of nanowires obtained from SEM analysis. Solid red line is a Gaussian fit to the data.

CV experiments corresponding to the simulations presented in Fig. 2 are performed by applying a potential range of −0.15 V to 0.45 V to nanowire arrays in 1 mM FcCOOH in 10 mM PBS, pH 7.4, at 5000 mV s^{-1}. CVs recorded at multiple nanowire arrays from different chips ($n = 15$) are highly reproducible across multiple nanowire arrays and typical data obtained at three nanowires separated by 5 µm and 15 µm are shown in Fig. 4. The forward sweep of the 5 µm spaced array exhibits steady-state behaviour, however, a diffusive peak is evident on the return reduction sweep indicative of diffusional overlap which is in agreement with the simulation results. By contrast both the forward and backward sweeps of

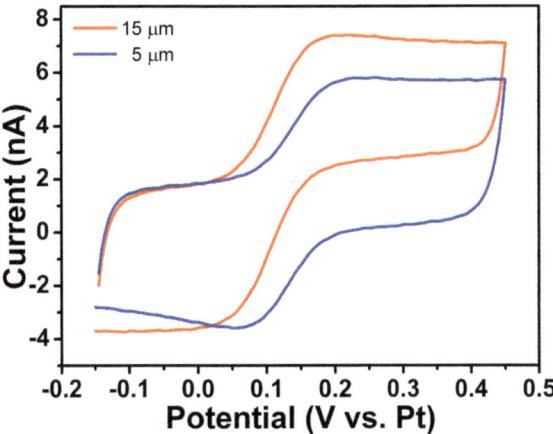

Fig. 4 Cyclic voltammograms obtained at 5000 mV s^{-1} scan rate in 1 mM FcCOOH in 10 mM PBS at arrays comprising three nanowires separated by 5 and 15 μm.

the 15 μm spaced array exhibit steady-state behaviour, also in agreement with the simulation results. These data strongly suggest that these electrodes are fully diffusionally independent. This is further supported by the observed increase in measured faradaic peak current I_p = 5.26 nA for 15 μm spaced arrays compared with I_p = 3.12 nA for the 5 μm spaced arrays, an increase of ∼60%. The magnitude of the steady-state currents for all nanoelectrode devices are found to be highly reproducible for measurements from separate nanowire arrays on different chips (n = 15 for each array). The magnitude of the average steady-state current measured at nanoelectrodes separated by 5 μm, 5.9 ± 0.2 nA, is the same (within experimental error) to the average measured at nanoelectrodes separated by 10 μm, 6.2 ± 0.3 nA (data not shown), while the magnitude of the average currents measured for nanoelectrode arrays separated by 15 μm and 20 μm is 7.0 ± 0.5 nA and 6.9 ± 0.3 nA (data not shown), respectively; again the same within experimental error. It is therefore likely that this increase in observed current measured for the latter two arrays arises from improved mass transport occurring at diffusion independent electrodes. At low scan rates (5 mV s^{-1}) all nanowire arrays exhibit steady-state behaviour with very low hysteresis as observed between the forward and reverse sweeps of the CVs (data not shown).

Steady-state CVs allow kinetic information for an electronic transfer process occurring at an electrode surface such as the heterogeneous rate of electron transfer k^0 to be determined. Historically, experiments undertaken to determine k^0 using macro and ultra-microelectrodes have been diffusion limited as such that low values have been reported.[8–11] The advent of nanoscale electrodes have eliminated diffusion effects and permitted higher (truer) values to be experimentally determined.[22]

Kinetic analysis is undertaken on steady-state CVs of a single electron oxidation of FcCOOH at arrays of three nanowires separated by 5 and 15 μm. In order to reduce artefacts caused by capacitive currents, we deduced the rate transfer from CVs captures at 5 mV s^{-1}. In this regard, the oxidative experimental voltammograms at 5 mV s^{-1} are plotted in Origin Pro 8.5 (OriginLab Corporation, USA). Assuming identical diffusion coefficients of the oxidized and reduced species and a uniformly accessible electrode surface, the steady-state voltammogram of an uncomplicated quasi-reversible one-electron oxidation reaction may be expressed by Butler–Volmer type kinetics:[12,22]

$$i_{bv} = \frac{i_{mt}}{1 + \exp\left[\frac{-F(E - E^{0\prime})}{RT}\right] + \frac{m}{k^0}\exp\left[\frac{-F(1-\alpha)(E - E^{0\prime})}{RT}\right]} \quad (7)$$

where i_{mt} is the mass transfer diffusion-limited current, F is the Faraday constant, R is the molar gas constant, T is temperature (K), E is the applied potential, $E^{0\prime}$ is the formal potential of the redox-couple, α is the transfer coefficient, k^0 is the standard heterogeneous rate constant (cm s^{-1}) and m is the mass transfer coefficient (cm s^{-1}). Eqn (7) was incorporated in Origin Pro 8.5 as a non-linear least square fit equation employing the Levenberg–Marquandt algorithm. The fitting parameters are the ratio k_m of the mass transfer coefficient m and the standard heterogeneous rate constant k^0 ($k_m = m/k^0$), α which was restricted between 0.3 and 0.7, and i_{mt}. The mass transfer coefficient may then be expressed as $m = i_{mt}(AFC^*)^{-1}$, where A is the electrode surface area and C^* is the bulk concentration of the electroactive species. Fixed parameters are $E^{0\prime} = 0.155$ V, $C^* = 1$ mM and $A = 2.7 \times 10^{-7}$ cm^2 (3 times the geometric area of a single nanowire). The fits are allowed to resolve until experimental CVs may be described excellently by eqn (7), i.e., until a correlation factor of $R^2 \geq 0.999$ is obtained, as shown in Fig. 5. This is repeated at least three times for every type of electrodes. Average values of $k_{m(5\mu m)} = 0.053 \pm 0.020$ and $k_{m(15\mu m)} = 0.054 \pm 0.015$, $\alpha_{5\mu m} = 0.58 \pm 0.16$ and $\alpha_{15\mu m} = 0.63 \pm 0.09$, and $i_{mt(5\mu m)} = 1.55 \pm 0.02$ nA and $i_{mt(15\mu m)} = 1.85 \pm 0.02$ nA are extracted from the best generated fits. These values permit the deduction of the standard heterogeneous rate constant: $k^0 = i_{mt}(AFC^*k_m)^{-1}$, yielding average values of $k^0_{(5\mu m)} = 1.14 \pm 0.03$ cm s^{-1} and $k^0_{(15\mu m)} = 1.32 \pm 0.03$ cm s^{-1}. This latter value is in excellent agreement to the average value determined using single (diffusionally independent) nanowires 1.29 ± 0.03 cm s^{-1} [3] confirming the diffusional independent nature of the 15 μm spaced arrays. In comparison, the value of 1.14 ± 0.03 cm s^{-1} is ~12% lower, which confirms the presence of diffusional overlap at 5 μm spaced arrays.

In square wave voltammetry, molecules are repeatedly oxidised and reduced as the voltage is stepped positively and negatively (in the appropriate voltage range, frequency, etc.).[23–26] During this process, molecules continuously diffuse to and from an electrode surface. Diffusion thereby limits the achievable signal as oxidised (or reduced) molecules diffuse from the electrode prior to being reduced (or oxidised). It is hypothesised that by fabricating nanowire arrays with overlapping diffusion profiles, a molecule oxidised (reduced) at one nanowire may diffuse to a neighbouring nanowire within an array (rather than to bulk solution) where it may consequently be reduced (oxidised) during a subsequent voltage step. This approach should result in an increased signal-to-noise and consequently sensitivity in direct contrast with sweep voltammetry.

To confirm this hypothesis, square wave voltammetric measurements were undertaken for FcCOOH concentrations ranging from 0.1 mM to 5 mM in 10 mM PBS. Fig. 6(a) shows typical square wave voltammograms recorded for 5 mM FcCOOH at: (i) a single nanowire electrode; (ii) nanowire electrode arrays separated by 5 μm; and (iii) nanowire electrode arrays separated by 15 μm. All

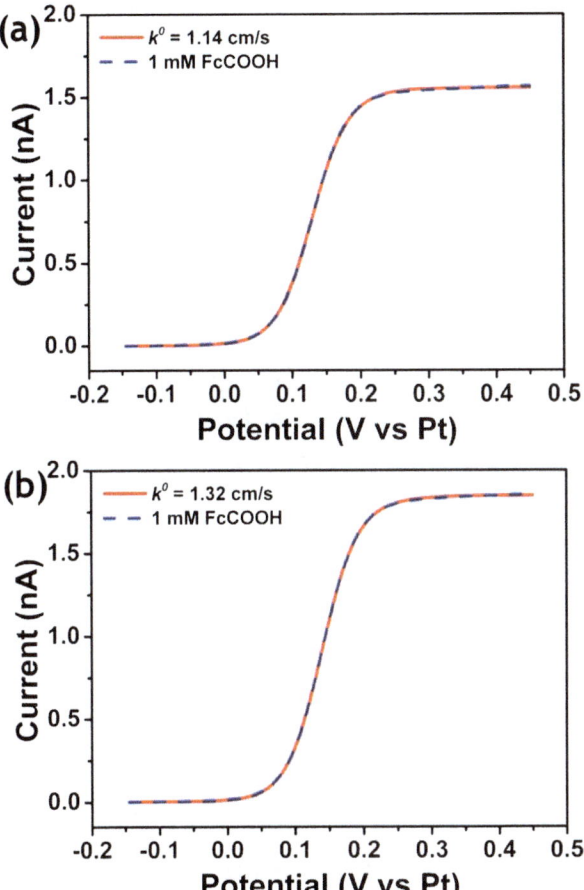

Fig. 5 Fits of the Butler–Volmer kinetic equation to an experimental cyclic voltammogram obtained for a single electron oxidation of FcCOOH at 5 mV s^{-1} at three nanowire electrodes separated by (a) 5 μm and (b) 15 μm. In both cases $R^2 = 0.999$.

recorded signals display current peaks in the voltage range of 0.15 to 0.20 V vs. the on-chip Pt pseudo-reference electrode. Clearly an increase in measured current signal is observed when increasing the number of nanowires from one to three within an array; (i) *versus* (ii) & (iii) in Fig. 6. Of note, is that a further increase in signal is observed when the inter-electrode separation is decreased from 15 μm (iii) to 5 μm (ii) confirming that diffusionally overlapped nanoelectrode arrays provide higher signal-to-noise than diffusionally independent variants. Fig. 6(b) shows the calibration plots obtained for FcCOOH concentrations ranging from 0.1 mM to 5 mM in 10 mM PBS for the single nanowire (i) and for the two different nanowire electrode arrays (ii) & (iii) as described above. These results are important to nanoscale electroanalysis as they suggest that a 'design for application' approach should be adopted prior to electrode fabrication in order to maximise sensitivity. Consequently this requires either *a priori* knowledge or at least a good appreciation of the diffusion regimes present at nanowire electrodes.

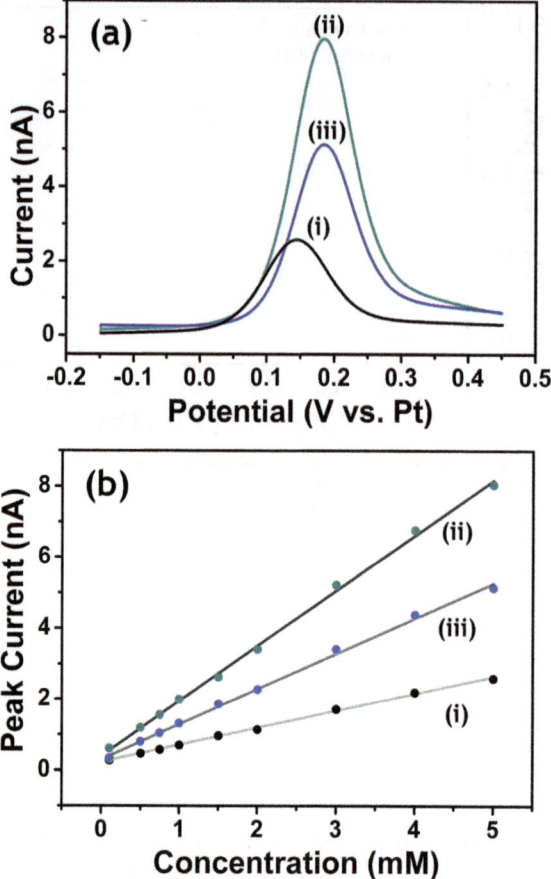

Fig. 6 (a) Typical square wave voltammograms of 5 mM FcCOOH in 10 mM PBS at: (i) a single nanowire electrode; (ii) three nanowire electrodes array separated by 5 µm; and (iii) three nanowire electrodes array separated by 15 µm. (b) Corresponding calibration plots for a range of FcCOOH concentrations 0.1, 0.5, 0.75, 1, 1.5, 2, 3, 4 & 5.mM in 10 mM PBS, $R_2 = 0.999$.

Conclusions

We show a hybrid electron beam/photolithography technique that permits gold nanowire array electrodes to be routinely fabricated at reasonable cost. Nanowire electrode arrays offer the potential for enhancements in electroanalysis including: increased signal to noise ratio and increased sensitivity while also allowing quantitative detection at much lower concentrations. However, to achieve this goal a full understanding of the diffusion profiles existing at nanowire arrays would be required. To this end, we simulated the effects of altering inter-electrode separations on analyte diffusion for a range of scan rates at three nanowire electrode arrays and we confirmed these results experimentally. Fabricated devices include twelve gold nanowire working electrode arrays, an on-chip gold counter electrode and an on-chip platinum pseudo reference electrode. CVs undertaken using these nanowire arrays exhibit measurable currents in the nanoAmpere regime and display steady-state voltammograms even at low scan rates (5 mV s-1) indicative of fast analyte mass transport to the electrode. We show that arrays that are sufficiently spaced so that they have diffusionally independent concentration profiles demonstrate superior electrochemical performance when employing sweep voltammetric techniques compared to arrays with overlapping diffusion profiles. By contrast, we also show that arrays with diffusionally overlapping profiles exhibit enhanced performance when employing step voltammetric techniques, e.g., square wave voltammetry. These results show that, to optimise sensitivity in nanoscale electroanalysis the final application needs to be taken into account when initially designing nanoscale electrode arrays.

Acknowledgements

The authors would like to thank the engineers of the Tyndall Central Fabrication Facility in particular, Mr D. O'Connell, Mr B. McCarthy, and Mr V. Djara for their assistance and advice. This work was supported by Science Foundation Ireland under the Research Frontiers Programme (SFI/09/RFP/CAP2455), by the European Commission under the FP7 Security Project CommonSense (261809) and FP7 ICT project "Nanofunction" (257375) and the Irish Higher Education Authority PRTLI programs (Cycle 3 "Nanoscience" and Cycle 4 "INSPIRE").

Notes and references

1 Y. H. Shao, et al., Nanometer-sized electrochemical sensors., Anal. Chem., 1997, **69**(8), 1627–1634.
2 J. L. Conyers and H. S. White, Electrochemical characterization of electrodes with submicrometer dimensions, Anal. Chem., 2000, **72**(18), 4441–4446.
3 K. Dawson, et al., Electroanalysis at single gold nanowire electrodes, J. Phys. Chem. C, 2012, **116**(27), 14665–14673.
4 J. P. Guerrette, S. J. Percival and B. Zhang, Voltammetric behavior of gold nanotrench electrodes, Langmuir, 2011, **27**(19), 12218–12225.
5 R. W. Murray, Nanoelectrochemistry: Metal nanoparticles, nanoelectrodes, and nanopores, Chem. Rev., 2008, **108**(7), 2688–2720.
6 E. J. Menke, et al., Lithographically patterned nanowire electrodeposition, Nat. Mater., 2006, **5**(11), 914–919.
7 P. Tyagi, et al., Patternable nanowire sensors for electrochemical recording of dopamine, Anal. Chem., 2009, **81**(24), 9979–9984.
8 R. Antiochia, et al., Single-wall carbon nanotube paste electrodes: a comparison with carbon paste, platinum and glassy carbon electrodes via cyclic voltammetric data, Electroanalysis, 2004, **16**(17), 1451–1458.
9 Y. Zhang, et al., Determination of electrochemical electron-transfer reaction standard rate constants at nanoelectrodes: Standard rate constants for ferrocenylmethyltrimethylammonium(III)/(II) and hexacyanoferrate(III)/(II), Electroanalysis, 2008, **20**(13), 1490–1494.
10 J. J. Watkins, et al., Zeptomole voltammetric detection and electron-transfer rate measurements using platinum electrodes of nanometer dimensions, Anal. Chem., 2003, **75**(16), 3962–3971.
11 J. J. Watkins and H. S. White, The role of the electrical double layer and ion pairing on the electrochemical oxidation of hexachloroiridate (III) Pt electrodes of nanometer dimensions, Langmuir, 2004, **20**(13), 5474–5483.
12 K. Dawson, et al., Single nanoskived nanowires for electrochemical applications, Anal. Chem., 2011, **83**(14), 5535–5540.
13 K. Dawson, M. Baudequin and A. O'Riordan, Single on-chip gold nanowires for electrochemical biosensing of glucose, Analyst, 2011, **136**(21), 4507–4513.
14 K. Dawson, et al., Electroanalysis at discrete arrays of gold nanowire electrodes, Electrochim. Acta, 2013, **101**(0), 169176.
15 T. J. Davies and R. G. Compton, The cyclic and linear sweep voltammetry of regular and random arrays of microdisc electrodes: Theory, J. Electroanal. Chem., 2005, **585**(1), 63–82.

16 C. Amatore, J. Saveant and D. Tessier, Charge transfer at partially blocked surfaces: A model for the case of microscopic active and inactive sites, *J. Electroanal. Chem. Interfacial Electrochem.*, 1983, **147**(1), 39–51.
17 R. M. Wightman and D. O. Wipf, Voltammetry at ultramicroelectrodes, *Electroanal. Chem.*, 1989, **15**, 267–353.
18 A. J. Bard and L. R. Faulkner, *Electrochemical Methods, Fundamentals & Applications*, 2011, 2nd edn, Wiley & Sons, New York.
19 A. Molina, *et al.*, Voltammetry of electrochemically reversible systems at electrodes of any geometry: A general, explicit analytical characterization, *J. Phys. Chem. C*, 2011, **115**(10), 4054–4062.
20 N. Godino, *et al.*, Mass transport to nanoelectrode arrays and limitations of the diffusion domain approach: Theory and experiment, *J. Phys. Chem. C*, 2009, **113**(25), 11119–11125.
21 C. G. Zoski, *Handbook of Electrochemistry*, 1st edn, Elsevier, Oxford, 2007.
22 I. Heller, *et al.*, Individual single-walled carbon nanotubes as nanoelectrodes for electrochemistry, *Nano Lett.*, 2005, **5**(1), 137–142.
23 D. P. Whelan, *et al.*, Square-wave voltammetry at small disk electrodes – Theory and experiment, *J. Electroanal. Chem.*, 1986, **202**(1–2), 23–36.
24 J. Odea, *et al.*, Square-wave voltammetry at electrodes having a small dimension, *Anal. Chem.*, 1985, **57**(4), 954–955.
25 J. G. Osteryoung and R. A. Osteryoung, Square-wave voltammetry, *Anal. Chem.*, 1985, **57**(1), A101.
26 L. Ramaley and M. S. Krause, Theory of square wave voltammetry, *Anal. Chem.*, 1969, **41**(11), 1362.

Faraday Discussions

PAPER

Nanoscale control of interfacial processes for latent fingerprint enhancement

Rachel M. Sapstead (nee Brown),[a] Karl S. Ryder,[a] Claire Fullarton,[a] Maximilian Skoda,[b] Robert M. Dalgliesh,[b] Erik B. Watkins,[c] Charlotte Beebee,[c] Robert Barker,[c] Andrew Glidle[d] and A. Robert Hillman[*a]

Received 15th April 2013, Accepted 17th May 2013
DOI: 10.1039/c3fd00053b

Latent fingerprints on metal surfaces may be visualized by exploiting the insulating characteristics of the fingerprint deposit as a "mask" to direct electrodeposition of an electroactive polymer to the bare metal between the fingerprint ridges. This approach is complementary to most latent fingerprint enhancement methods, which involve physical or chemical interaction with the fingerprint residue. It has the advantages of sensitivity (a nanoscale residue can block electron transfer) and, using a suitable polymer, optimization of visual contrast. This study extends the concept in two significant respects. First, it explores the feasibility of combining observation based on optical *absorption* with observation based on *fluorescence*. Second, it extends the methodology to materials (here, polypyrrole) that may undergo post-deposition substitution chemistry, here binding of a fluorophore whose size and geometry preclude direct polymerization of the functionalised monomer. The scenario involves a *lateral* spatial image (the whole fingerprint, *first level* detail) at the centimetre scale, with identification features (minutiae, *second level* detail) at the 100–200 μm scale and finer features (*third level* detail) at the 10–50 μm scale. However, the strategy used requires *vertical* spatial control of the (electro)chemistry at the 10–100 nm scale. We show that this can be accomplished by polymerization of pyrrole functionalised with a good leaving group, ester-bound FMOC, which can be hydrolysed and eluted from the deposited polymer to generate solvent "voids". Overall the "void" volume and the resulting effect on polymer dynamics facilitate entry and amide bonding of Dylight 649 NHS ester, a large fluorophore. FTIR spectra demonstrate the spatially integrated compositional changes. Both the hydrolysis and fluorophore functionalization were followed using neutron reflectivity to determine *vertical* spatial composition variations, which control image development in the *lateral* direction.

[a]*University of Leicester, Leicester LE1 7RH. E-mail: arh7@le.ac.uk*
[b]*Rutherford Appleton Laboratory, Didcot, Oxfordshire, OX11 0QX, UK*
[c]*Institut Laue Langevin, Grenoble, Cedex 9, France*
[d]*Glasgow University, Glasgow, G12 8LT, UK*

Introduction

Fingerprints provide an example of pattern formation in nature, carrying information that uniquely identifies an individual. Notwithstanding the rise of sophisticated genetic methods, they remain the cornerstone of many criminal investigations and have a number of non-criminal applications based upon identification of an individual.[1] The efficacy of this approach is in large measure associated with the complexity of a fingerprint and the consequent practical difficulty of forgery; powerful software tools for analysis and recognition facilitate exploitation of this potential. The power of fingerprint evidence for analytical purposes in general is anecdotally recognized by the use of the term in other contexts, such as the "fingerprint" region of IR spectra and DNA "fingerprinting". The challenge in realizing this analytical opportunity lies in visualization of the interfacial chemical transfer that constitutes a fingerprint, primarily in the case of *latent* fingerprints, for which the nature[2] and extent[3] of the deposit mean that they are not immediately visible to the eye. Here we explore how this can be accomplished by a novel electroanalytical approach based upon spatially selective deposition of electroactive polymers with variable optical properties. The intellectual novelty lies in the need to control the (electro)chemistry on different length scales and in both the lateral and vertical directions.

When a finger contacts a surface, exchange of material with the surface leaves behind a trace of this contact which resembles the pattern present on the finger. Dependent on the substrate and the nature of the contact, the fingerprint may be visible, latent or plastic.[4] Since they are not immediately visible to the eye, and thus less readily "wiped", latent fingerprints are the greatest source of forensic evidence. To give an indication of scale, in the UK on the order of 700 000 objects are fingerprinted per annum. In response to this demand, numerous methods and reagents have been developed to effect latent fingerprint visualization but, perhaps surprisingly, the operational success rate is only *ca.* 10%. While there may be local variations in demand and/or success rate for specific types of object, the global need for improvement is clear.

The existence of fingerprint patterns was recognized in ancient times:[5,6] they have been identified on hand-formed building materials in Jericho dating back to 7000 BC and on the reverse of Chinese clay seals from 300 BC. Much later they were used as accompaniments to signatures by citizens claiming damages following the siege of Londonderry in 1691[5] and by the engraver Bewick who used an engraved fingerprint as a signature on his work.[6] The first documented study of fingerprints was in the 17th century by Grew,[5] followed by attempts at pattern classification in 1823 by Purkinje,[5,6] and later by Henry.[5]

Development of these historical observations for forensic application was reliant upon three crucial deductions made during the 19th century. First, Herschel[7] made the critical observation – on his own hands – that fingerprints do not change during the life of an individual.[5] Indeed, the friction ridge skin pattern that constitutes a fingerprint persists after death, thereby enabling *post mortem* identification.[8] Second, by removing skin from the fingers and allowing it to re-grow, it was demonstrated that injury does not change fingerprint patterns.[5] Third, in 1892 Galton[9] estimated that the odds of two individuals having identical fingerprints were 64 billion to 1; thus, for all practical purposes, they are unique to an individual. Combination of these observations pointed to the value of fingerprints in criminal investigations.[10] The outcome of this, in the first decade of the 20th century, was establishment of the UK's first fingerprint bureau[5,7] and the use of fingerprint evidence to secure a murder conviction.[6]

Fingerprint patterns fall into three basic categories (so-called *first level* detail): loops, whorls and arches.[5,11] While differences at this gross level can clearly eliminate certain individuals, positive identification relies on the *minutiae* (or *second level* detail) within the pattern: these include such features as ridge endings, crossovers (bridges), short independent ridges, islands, bifurcations, spurs, dots and lakes.[5] The standard for matching a crime scene fingermark to one from a database varies with jurisdiction: in some cases there is a set minimum number of points of similarity (although the number is not universal) and in others (including the UK) it is decreed a matter for a recognized expert to decide. As a rule of thumb, identification of around 16 points of similarity can be expected to be considered conclusive.[7,12] There are also finer features (*third level* detail) present in a fingerprint image: these include the detailed shapes of the ridges and individual sweat pores. While not currently used in fingerprint identification, there is considerable research interest in third level detail, since it may in future permit analysis of smaller fragments of marks left by a finger, *i.e.* partial fingerprints.

Here we focus on latent fingerprints, since these are the primary source of forensic evidence. In general terms, the traditional approach has been to apply a reagent that interacted with the residues left by contact of the finger. We do not rehearse the diverse methods available, since these have been reviewed elsewhere recently,[13–15] but rather show how the apparently diverse methods used in fact have some similarities that limit their efficacy and motivate the novel approach developed here. The classical approach is to apply a powder (either dry or as a suspension) that adheres physically to the sweat residues; the powder may be fluorescent,[16–18] magnetic[19,20] or thermoplastic. In each case, this method is reliant upon there being sufficient residue present for adhesion of a powder with adequate optical contrast to the substrate.[21–26] More recent developments involve (by dipping, spraying or gas phase delivery, according to the chemistry) ninhydrin solution,[27] vacuum metal deposition,[21] small particle reagent,[28] physical developer,[28] cyanoacrylate ("superglue") polymerization in conjunction with a suitable dye,[29,30] S_2N_2 polymerization[31] and cadmium sulphide nanocomposites.[32] In some cases the interaction of the reagent may be relatively unspecific (physical adhesion of powders), in others it may involve moderately specific chemistry (CdS binding with fatty acids and amino acids generally found in fingerprint deposits) and in other cases it may be very specific (reaction with secreted drug metabolites of antibody-functionalised nanoparticles.[33] However, the common factor is that the reagent interacts with the deposited residue. This makes all these technologies vulnerable to loss or deterioration of fingerprint residue, *e.g.* as a consequence of ageing, environmental exposure or abuse (attempted washing).

The approach developed here is complementary to those described above, in that the "reagent" is applied to the bare substrate surface that lies between the deposited fingermark ridges. The means of accomplishing this is to use the fingerprint deposit as a "mask" or template, whose broadly insulating characteristics preclude electron transfer from a metal substrate to a solution precursor. Since electron tunnelling can only take place over very short distances, only a very thin layer (*ca.* 1–2 nm) of fingerprint residue is required – far less (by an order of magnitude) than required in the conventional strategies listed above. We have recently demonstrated proof-of-concept for this strategy[34,35] in the context of electropolymerization of aromatic solution precursors[36] to generate conducting polymer films. Other applications of electrochemically-based methods of latent fingerprint visualization include imaging using the SECM[38–41] and deposition of Au nanoparticles.[42]

In the case of electroactive polymer enhancement, after transfer to a background electrolyte, the electrochromic properties of these materials were exploited to adjust the visual contrast between the polymer and the metal substrate. Evaluation of the ability of this approach to visualize fingerprints subject to ageing in a range of environments suggests that there are practically relevant situations in which the methodology may be superior to currently employed methods, for example involving powders and cyanoacrylate.[37]

The aim of the present study is to extend the concept of the electrochromic polymer enhancement strategy in two significant respects. First, we wish to explore the possibility of combining observation based on optical *absorption* (as above) with observation based on *emission*. Since the advantages, notably sensitivity, of fluorescence detection are appreciated in fingerprint visualization, translation to practical application would be facilitated by existing instrumentation. In the context of "superglue" enhancement, subsequent use of fluorescent dyes is a necessity, since the cyanoacrylate polymer is not coloured. This leads naturally to our second generic goal, use of electropolymerized films permitting excellent spatial control but with sub-optimal optical properties. In particular, we wish to move from thiophene[35] and aniline[34] based materials to pyrrole-based systems, for which there is much greater opportunity for manipulation of properties by substituent chemistry on the >N–H functionality. Here we focus on the underlying fundamental (electro)chemistry of this approach.

The future viability of enhancement and analysis of latent fingerprints using this strategy relies on "writing" (electropolymerization and post-deposition reaction) and "reading" (absorption and emission observations) of spatial information at different length scales. The precise figures will vary from one individual to another, but typically this is at *ca.* 1 cm for first level detail (the whole print), at 100–200 μm for second level detail and at 10–50 μm (third level detail). These feature sizes set the chemical challenge: control of enhancement chemistries with commensurate *lateral* spatial resolution to the feature size. However, the requirement for *vertical* resolution is somewhat different. Typical fingerprint deposits may be a few microns thick as-deposited, but evaporative and other environmental losses will typically decrease this to 100s of nm before enhancement is undertaken. This defines the *vertical* resolution required for the (electro)chemistry: there is an optimum to be found between the lower limit of detection of deposited material and over-filling[35] of the trenches between fingerprint ridge deposits. To summarize, this singular analytical challenge of uniquely identifying an individual based on electroanalytical visualization of their fingerprint requires simultaneous control of (electro)chemical processes on length scales from the nanometre to the centimetre regime and in both lateral and vertical directions.

Experimental

Reagents, materials and electrodes

N-Cyanoethylpyrrole (PyCN), piperidine, tetrabutylammonium perchlorate (TBAP), perchloric acid, hydrogenous and deuterated acetonitrile were all used as supplied

(Sigma Aldrich). N-Aminopropylpyrrole (PyNH$_2$) was synthesised *via* reduction of PyCN (0.056 mol) with LiAlH$_4$ (0.26 mol) in anhydrous ether. The reaction was stirred for 2 h before excess LiAlH$_4$ was neutralised with water. The solution was dried over MgSO$_4$ before evaporating to dryness to yield a yellow oil (84% yield).

(9*H*-Fluoren-9-yl)methyl-3-(1*H*-pyrrol-1-yl)propylcarbamate (PyFMOC) was made using a Merrifield synthesis. FMOC-Cl (0.021 mol) in water–dioxane (60 ml, 1 : 1) was added to a solution of PyNH$_2$ (0.021 mol) and Na$_2$CO$_3$ (0.028 mol) in water–dioxane (60 ml, 1 : 1) in an ice bath. The solution was then stirred at room temperature for 2 h. PPyFMOC was extracted with ethyl acetate and dried over MgSO$_4$ before evaporating to dryness to yield a white solid (78% yield).

For the neutron reflectivity (NR) experiments, the working electrode was prepared by sputter coating gold onto a polished single-crystal quartz block (100 × 50 mm, Gooch and Housego) coated with a monolayer of 3-mercaptopropyltrimethoxysilane (MPTS) (Sigma Aldrich) to promote adhesion. The nominal Au film thickness was 20 nm. For other experiments, the electrodes were metal sheet, as indicated in the figure legends. The counter electrode was in each case a Pt gauze, of adequate size to ensure that the counter electrode reaction was not limiting. The reference electrode was a double junction Ag|AgCl|KCl (saturated) electrode. These were assembled into a standard three electrode cell configuration; for the NR measurements, the purpose built cell has been described elsewhere.[43,44]

Instrumentation

NR measurements were performed on FIGARO[45] and D17 at the Institut Laue-Langevin (Grenoble, France) and on INTER[46] at the ISIS Facility of the Rutherford Appleton Laboratory (Harwell Oxford, UK). Static neutron reflectivity measurements were performed *ex situ* in air ("dry") and *in situ* immersed in h$_3$- and d$_3$-acetonitrile before and after the film fabrication stage (see Fig. 3, below). Deuterated solvents were used to maximise contrast between the polymer and the electrolyte so that the solvation within the polymer could be probed. Kinetic measurements were recorded during the hydrolysis stage of the reaction (see Fig. 3), using time-of-flight instrumentation with λ ranges of 2–30 Å (ILL) and 1.5–16 Å (ISIS). Using different incident angles, this provides an accessible momentum transfer range of $0.004 < Q/\text{Å}^{-1} < 0.12$, where Q is defined as $(4\pi/\lambda)\sin\theta$; λ is the wavelength of the neutron and θ is the incident angle. (Strictly, this is Q_z but since we only consider specular reflection we use the simpler notation Q.) The collimation slits were set to give a beam footprint on the sample of 60 mm × 30 mm; they also define the $\Delta\theta/\theta$ resolution. The $\Delta\lambda/\lambda$ resolution is dependent on the source (spallation source at ISIS, reactor at ILL) and associated instrumentation (chopper settings at ILL). In both cases, the resultant resolution in momentum transfer was $\Delta Q/Q \sim$ 2–3%. Data acquisition times were *ca.* 1.5 h per static run and 10 min for kinetic runs.

Photographs were taken with a Canon A480 digital camera and were digitally enhanced using the GNU Image Manipulation Program 2.6.7. (G.I.M.P.). 3D profiles were recorded with a Zeta 200 Optical Profiler. Reflectance FTIR spectra were acquired with p-polarised radiation incident at a reflectance angle of 55° using a Spectra-Tech reflectance accessory mounted on a Bomem MB120 infra-red instrument.

Procedures

Poly(3,4-ethylenedioxythiophene) (PEDOT) films were deposited potentiostatically ($E = 0.90$ V) from an aqueous solution containing 0.01 M EDOT monomer in 0.01 M SDS–0.1 M H_2SO_4. Polypyrrole (PPy) films were deposited potentiostatically ($E = 0.90$ V) from an aqueous solution containing 0.01 M pyrrole monomer in 0.1 M $LiClO_4$. In both cases, the film thickness was controlled by deposition time. PPyFMOC and $PPyNH_2$ films were deposited, respectively, from 10 mM PyFMOC in 0.1 M TBAP–CH_3CN and 10 mM $PyNH_2$ in 0.1 M TBAP–$CH_3CN/HClO_4$ (pH 4.5). The films were grown potentiodynamically ($\nu = 20$ mV s^{-1}) in the potential range $0.3 < E/V < E_{max}$, where the anodic limit E_{max} (typically 1.15 V) was set to cap the anodic current to 4 mA. This procedure was designed to avoid uncontrollably rapid growth; the amount of deposited polymer was varied with adequately fine control by varying the number of deposition cycles. All measurements were made at room temperature, 20 ± 2 °C. PPyFMOC films were deprotected using a 30% v/v piperidine solution in CH_3CN to yield $PPyNH_2$. $PPyNH_2$ films were reacted with 0.01 M Dylight 649 NHS ester (hereafter referred to as "Dylight", for brevity) (Thermo Scientific) in DMSO–pH7 phosphate buffer in water (1 : 9 ml).

Data analysis

The principles of neutron reflectivity data analysis[47] and the issues arising for samples involving "wet" interfaces under electrochemical control[48] have been described elsewhere. The variation with momentum transfer of reflectivity from an interface, $R(Q)$, is determined by the depth profile of the scattering length density, Nb, where N represents the concentration of scattering atoms present and b is their scattering length; the value of b is isotopically unique and medium independent. The scattering length density of a composite medium, here a solvated polymer film, is a weighted sum of the Nb values of its components: film composition determines scattering length density and thence reflectivity. Experimentally, we invert the process and use reflectivity, R, to determine composition – in practice, the volume fractions of polymer and solvent – by model fitting the reflectivity profile, $R(Q)$. This was accomplished using the box-model approach, implemented in the Motofit software.[49]

Results

Electrochromic enhancement of latent fingerprints – basic observations in absorption mode

Before attempting to exploit the substitution chemistry opportunities presented by the >N–H function in pyrrole, it is first necessary to demonstrate that PPy films can in fact be deposited with spatial selectivity directed by a fingermark on a metallic substrate. Fig. 1 and Fig. 2 show representative images of two fingermarks on 304 stainless steel substrates, following enhancement by electrodeposition of PEDOT and PPy, respectively. In panel (a) of each figure, the paler (nominally white) regions correspond to the fingermark itself, *i.e.* the complex mixture of materials secreted from pores along the ridges on the fingertip. The darker regions (blue for PEDOT in Fig. 1 and black for polypyrrole in Fig. 2) correspond to polymer deposited on the bare metal *between* the fingerprint deposits. The PEDOT enhanced image acts as a control; this material has

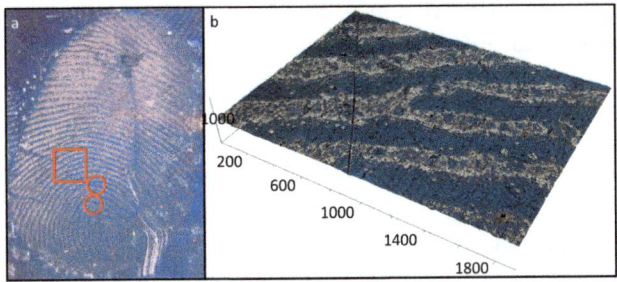

Fig. 1 PEDOT enhanced sebaceous fingerprint on 304 stainless steel. Deposition conditions as in main text; deposition time, t_{dep} = 3000 s. Panel (a): whole fingermark image. Dark (blue) regions correspond to PEDOT and lighter regions to fingerprint deposit. Circles highlight examples of second level detail, a bifurcation and a ridge ending. Panel (b): optical microscope image of selected area from panel (a) (as defined by the rectangle). Larger dark circular regions within light areas represent individual sweat pores. Vertical and horizontal distance scales (expressed in μm) are relative.

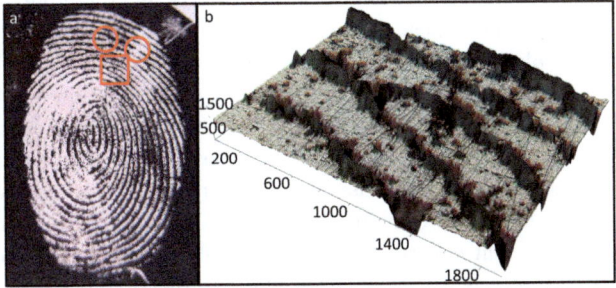

Fig. 2 PPy enhanced sebaceous fingerprint on 304 stainless steel. Deposition conditions as in main text; deposition time, t_{dep} = 3000 s. Panel (a): whole fingermark image. Dark regions correspond to PEDOT and lighter regions to fingerprint deposit. Circles highlight examples of second level detail, a crossover and a ridge ending. Panel (b): optical microscope image of selected area from panel (a) (as defined by the rectangle). Larger dark circular regions within light areas represent individual sweat pores. Vertical and horizontal distance scales (expressed in μm) are relative.

previously been demonstrated to provide good visualization of latent fingermarks, with high fidelity and controllable visual contrast.[35]

We have quite deliberately shown examples that might be typical of real evidence, rather than highly controlled ("groomed") model examples. Thus, one can see evidence of damage to the fingermark and of adjacent fingermarks in Fig. 1 and of smearing (for example, caused by motion of the finger on the surface) and variable amounts of residue in Fig. 2. These and other imperfections represent practical challenges to be addressed.

At the coarsest level of interpretation, *first level* detail, the fingerprints in Fig. 1 and Fig. 2 are, respectively, a loop and a whorl. While this is clearly not sufficient for identification purposes, it is clear that there is much *second level* detail present within these images. Panel (b) in each of Fig. 1 and Fig. 2 shows higher magnification optical images of the samples in panels (a). By any reasonable standard (see above), it would be possible to achieve an evidentially acceptable identification from images such as these using various combinations of *second level* detail.

To illustrate the principle, we simply identify two such features on each image. It is also possible to identify a large number of pores, seen as dots along the ridges where the absence of contact with the substrate permits polymer deposition; these are examples of *third level* detail.

Overview of extension to emission mode visual enhancement

The qualitative conclusion of the preliminary experiments shown above is that polypyrrole can be added to the set of electrochromic materials (to date, polyaniline[34] and PEDOT[35]) that permit visualization of latent fingermarks on metallic substrates by means of their optical absorption properties; note that the work of Bersellini[36] in this respect demonstrated the facility to electrodeposit polypyrrole, but did not go on to exploit its electrochromic properties. We therefore proceed to the more challenging goal of developing a polymeric system with suitable absorption *and* fluorescence characteristics. Conceptually, the simplest approach is to functionalise the nitrogen of the pyrrole ring with a fluorophore. In practice, the (necessarily) large size of most fluorophores creates such steric hindrance that the substituted pyrrole monomers cannot polymerize. We therefore arrive at the three step strategy schematically represented in Fig. 3.

The essential idea is to functionalize pyrrole units with the fluorophore post-deposition. However, in order to accomplish this, there is still a requirement to create sufficient free volume within the polymer film to accommodate the fluorophore units. The tactic employed is to polymerize not pyrrole itself, but an *N*-functionalized derivative that can readily be removed post-deposition. The balance to be struck is use of a substituent that is not so large as to preclude polymerization but that is large enough to create appreciable free volume. The functionality chosen was the widely used FMOC protecting group. Thus, we set out to polymerize the *N*-substituted FMOC derivative of pyrrole (PyFMOC), then hydrolyze and leach out the protecting group to leave an amine-functionalized polypyrrole film. The semi-fluid nature of the polymer film means that the

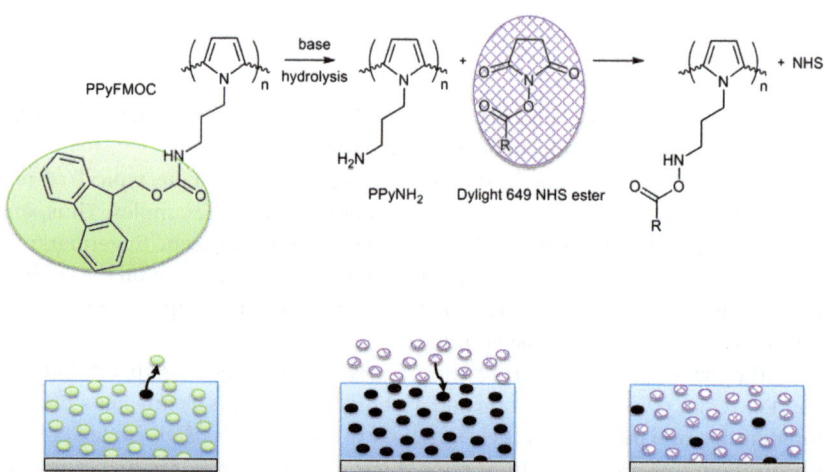

Fig. 3 Cartoon representation of (i) PPy-FMOC deposition, (ii) FMOC hydrolysis to create solvent "voids"; (iii) diffusion of fluorophore into voids and attachment to polymer. Stylised ellipsoids (simplistically, of similar size) represent FMOC (pale green), solvent (black) and fluorophore (diamond chequered).

randomly distributed free volume can aggregate to generate "voids" of sufficient size to accommodate larger fluorophore moieties. Choice of a fluorophore with an ester functionality provides the means of covalent bonding by amide formation (see Fig. 3).

In the experimental realisation of the strategy indicated in Fig. 3, critical issues are the completion of hydrolysis, the extent to which departing FMOC is replaced by solvent *cf.* film contraction, the vertical (perpendicular to the interface) spatial distribution of replaceable solvent and the penetration of fluorophore into the film. The outcomes, which will ultimately determine performance, are addressed using a combination of spectroscopic and neutron reflectivity measurements.

Optimization of film deposition

The limited aim of this study is establishment and assessment of the strategy of Fig. 3 on clean (*i.e.* non-fingermarked) metal surfaces. The deliberate absence of *lateral* spatial variation focuses attention on the required control of film composition in the *vertical* direction. Fig. 4 and Fig. 5, respectively, show the voltammetric responses of PPyFMOC and PPyNH$_2$ films on Au during electropolymerization (panels (a)) and after transfer to background electrolyte (panels (b)). Both sets of responses provide information relating to polymer deposition. In the former instance, the integrated current provides cycle-by-cycle monitoring to facilitate deposition of the chosen amount of polymer. In practice, use of the cathodic half cycle response is better, since this is not complicated by contributions from the (irreversible) anodic polymerization current contribution.

The coulometric assay described above is reliant upon complete film redox conversion on the experimental timescale; that this is accomplished is demonstrated by the data in Fig. 4b and Fig. 5b. In these measurements (following completion of film deposition) there is no monomer present in the solution, so the issue of distinguishing polymerization and film redox chemistry is irrelevant. For films of suitable coverage, *i.e.* appropriate both to the NR experiment (giving multiple well-defined interference fringes within the accessible Q range) and to future forensic exploitation (not so thick as to obscure all image detail when a fingermark is present), we find that the peak currents are linearly proportional to the potential scan rate (see Fig. 4c and Fig. 5c). This indicates complete redox

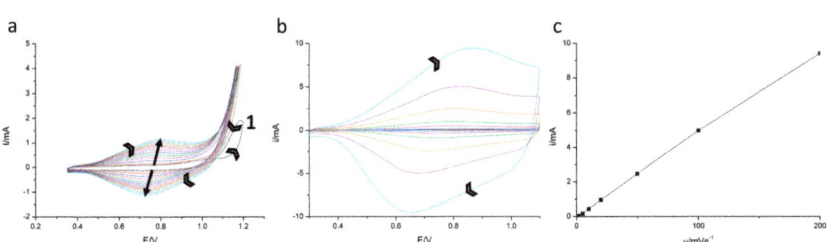

Fig. 4 *i–E* responses for a PPyFMOC film. Panel (a): during potentiodynamic electropolymerization ($v = 20$ mV s^{-1}); panel (b): after transfer to a monomer-free background electrolyte (subsequent to the final cycle of panel (a)), during cycling at $v = 1, 2, 5, 10, 20, 50, 100$ and 200 mV s^{-1} (increasing "outwards"); panel (c): variation of cathodic peak current (from curves in panel (b)) with scan rate. Solution compositions as described in the main text. In panels (a) and (b), chevron arrows indicate scan direction. In panel (a) large arrows indicate the time sequence and "1" indicates the first deposition cycle (note the nucleation loop).

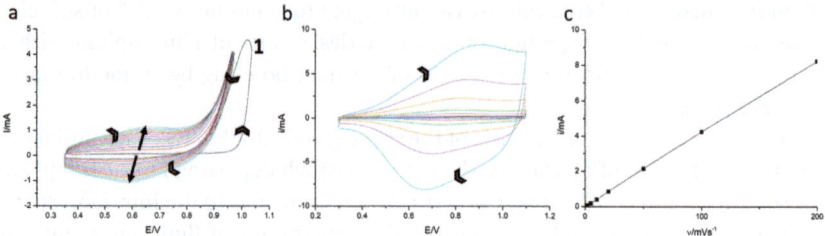

Fig. 5 i–E responses for a PPyNH$_2$ film. Panel (a): during potentiodynamic electropolymerization ($v = 20$ mV s^{-1}); panel (b): after transfer to a monomer-free background electrolyte (subsequent to the final cycle of panel (a)), during cycling at $v = 1, 2, 5, 10, 20, 50, 100$ and 200 mV s^{-1} (increasing "outwards"); panel (c): variation of cathodic peak current (from curves in panel (b)) with scan rate. Solution compositions as described in main text. In panels (a) and (b), chevron arrows indicate scan direction. In panel (a) large arrows indicate the time sequence and "1" indicates the first deposition cycle (note the nucleation loop).

conversion of the film on the experimental timescale, validating a coulometric assay of the spatially integrated surface population of polymer. On this basis, the surface coverage, Γ/mol cm^{-2}, is determined as q/nFA, where q/C is the charge, n is the number of electrons transferred ("doping level", $n = 0.33$)[50], A is the electrode area and F is the Faraday constant. For the data shown, the final polymer coverages are $\Gamma_{\text{PPyFMOC}} = 20$ nmol cm^{-2} and $\Gamma_{\text{PPyNH2}} = 19$ nmol cm^{-2}, where in both cases the surface population is expressed in terms of monomer units.

These surface coverages can be used to estimate a physical film thickness, as follows. The molar volume of monomer units, V_m/cm^3 mol^{-1} (*i.e.* the reciprocal of the volume concentration of monomer units), can be estimated as the quotient of monomer molar mass and density, RMM/ρ. The approximation here is that the monomer units in the polymer pack essentially the same as in pure monomer. For a compact, solvent-free ("dry") polymer, the film thickness $h^* = V_m \Gamma$. The RMM values of PyFMOC and PyNH$_2$ are 344 and 122 g mol^{-1}, respectively. Their respective densities are *ca.* 1.2 and 1.0 g cm^{-3}. Combining these with the values of Γ_{PPyFMOC} and Γ_{PPyNH2} from the previous paragraph, we estimate "dry" (*i.e.* collapsed, solvent-free) film thicknesses of *ca.* 600 Å and 240 Å for the PPyFMOC and PPyNH$_2$ films, respectively. Looking ahead to the NR part of the study, solvation might be expected to increase these values by 50–100%, since typical solvent volume fractions for electroactive polymer films are in the range $0.3 < \phi_S < 0.5$.[51-54] This overall scenario is appropriate both for a NR experiment (to explore structure) and filling of the inter-ridge "trenches" in a typical fingerprint deposit (to accomplish practical visualization). As a final observation, the different thicknesses of these two PPyFMOC and PPyNH$_2$ films containing essentially the same number of monomer units (irrespective of whether one compares two dry films or two solvated films) shows the potential for generation of free volume by FMOC elution from a film.

Observation of changes in surface composition using spectroscopic measurements

Before addressing the more sophisticated issue of spatial distribution of active components within the film, it is necessary to demonstrate that the strategy of Fig. 3 does indeed result in fluorophore immobilization. In principle, one might consider accomplishing this using either a fluorescence measurement or some other spectroscopic probe. Superficially, a fluorescence measurement has the attraction of also providing a more direct functional appraisal. However, while this could demonstrate the *presence* of the fluorophore, it might not do so if quenching were an issue; in either instance, it would not provide evidence of its *immobilization*. The concern here is that, unless covalent attachment to the polymer is achieved, facile entry of the fluorophore could just as easily be followed by its elution upon further exposure to electrolyte, thereby jeopardizing the entire surface synthetic strategy. Consequently, we sought a spectroscopic probe able to provide direct evidence of fluorophore–polymer binding. Qualitatively, this can be accomplished using vibrational spectroscopy, with particular focus on the presence (or absence) of carbonyl bands associated with the amide functionality, which will be present in an PPyFMOC film (at the start) and a successfully functionalized PPy-Dylight film (at the end), but should be absent in $PPyNH_2$ (following hydrolysis of the PPyFMOC film).

Representative data are shown in Fig. 6. Trace (a), representing a PPyFMOC film, has a strong absorption band at 1660 cm^{-1}. Trace (b), for the hydrolysed film, has no significant amide band, demonstrating removal of the FMOC functionality. Trace (c), for the hydrolysed film after exposure to Dylight solution, shows a strong band at 1630 cm^{-1}, showing the formation of an amide. Significantly, the last of these observations demonstrates not only the permeation of fluorophore into the film but also its chemical immobilization. Overall, these data show elution of FMOC and binding of fluorophore, but give no insight into the spatial distribution of the fluorophore or the factor(s) limiting its final population. We now address these issues using NR.

Determination of vertical spatial structure using neutron reflectivity

The second stage of the strategy in Fig. 3 is the creation of free volume in the PPyFMOC film by the hydrolysis and elution of the FMOC moieties. Importantly,

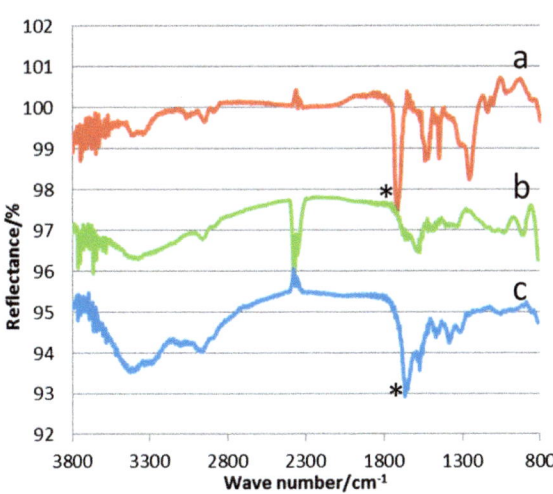

Fig. 6 Reflectance infra-red spectra for a PPyFMOC film subject to the reaction sequence of Fig. 3. Trace (a): PPyFMOC film, as deposited; trace (b): PPyFMOC film after hydrolysis, notionally a $PPyNH_2$ film; trace (c): film of panel (b) after reaction with Dylight solution. Traces are arbitrarily offset vertically for visual clarity. Reaction conditions as in main text. The asterisks on traces (a) and (c) indicate the amide peaks referred to in the main text.

we require this free volume to be created throughout the film. Fig. 7 shows neutron reflectivity data acquired at different times during the hydrolysis process, conducted in a deuterated solvent medium to optimise the contrast. The data shown in Fig. 7 in fact represent the second – and, in reaction terms, productive – phase of the experiment. Since there was no means *a priori* to predict the timescale of the hydrolysis–elution process, the film was initially exposed to a low (0.01 mM) concentration of piperidine, with the aim of slowing the reaction to a measurable rate. This approach turned out to be more than successful, in that the rate of hydrolysis was immeasurably slow. However, this had the advantage of providing a film solvation profile at a true "$t = 0$"; the outcome of this is cited below in the discussion of Fig. 7c. The hydrolysis solution was then exchanged to higher piperidine concentration (10 mM) and the data shown in Fig. 7a acquired.

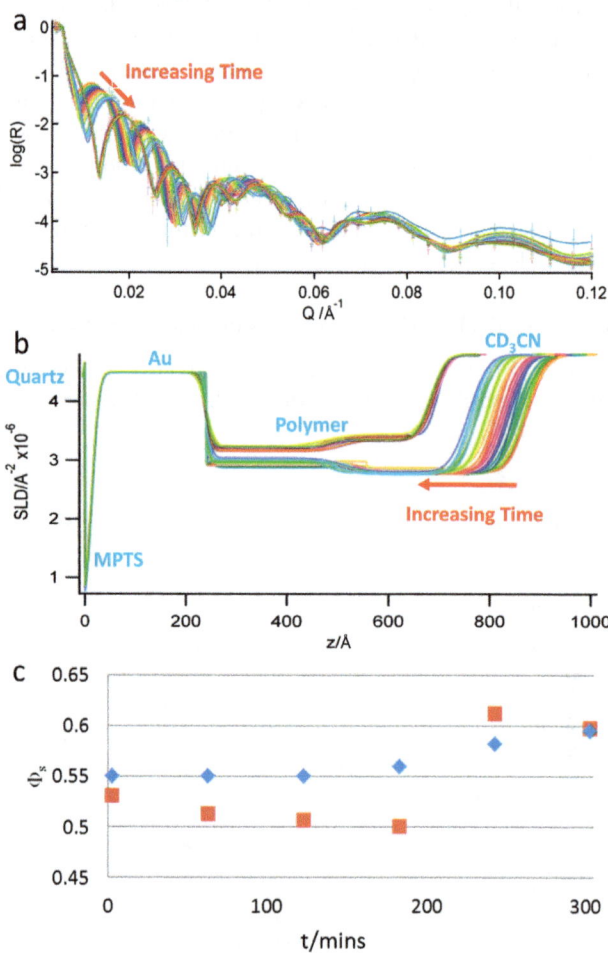

Fig. 7 Time-resolved NR experiment for PPyFMOC hydrolysis to give PPy-voids. Panel (a): $R(Q)$ profiles as a function of time (indicated by arrows); panel (b) model fitted scattering length density profiles as a function of time (from the data of panel (a)); panel (c): solvent volume fraction in the film at selected time intervals (see main text for comment on absolute values) during the hydrolysis. (■: inner polymer layer; ◆: outer polymer layer).

The time taken for cell mounting and alignment and for initiation of data acquisition was on the order of 30 min. Hence the time axis in Fig. 7c has an offset of this order; since there is no attempt to extract kinetic or diffusional parameters, this is not critical.

Broadly speaking, the $R(Q)$ profiles of Fig. 7a comprise three regions. At low Q (here, $Q < 0.0055$ Å^{-1}) there is total external reflection ($R = 1$); this is a consequence of the substrate material the neutrons are transmitted through (quartz) having a lower scattering length density than the material they are reflecting from (the deuterated solvent). Beyond this, there is a series of fringes whose periodicity is dictated by the thickness of the films present (Au electrode and polymer) and whose amplitude is dictated by the scattering length density contrast and sharpness of the interfaces between layers. At high Q (here, $Q > 0.06$ Å^{-1}), the fringes seen are attributable to the Au electrode. Although these are not central to the aims of the experiment *per se*, their quantitation does assist fitting of the full $R(Q)$ profile from which we extract the polymer film data. At intermediate Q, the Au-derived fringes are seen, but superimposed on these are fringes that result from the polymer film. The latter do not persist to high Q as a consequence of the diffuse polymer/solution interface.

Independent of any model, we can make four deductions from the data. Trivially, the presence of a critical edge shows that the film scattering length density is below that of the solution. The position of the critical edge, $Q^* = (16\pi \Delta Nb)^{1/2}$, where ΔNb corresponds to the difference in scattering length densities between the two bulk phases in the sample. In this instance, the relevant bulk phases are the quartz block supporting the electrode and the bathing solution to which the film is exposed; the scattering length densities are known for these materials, giving $\Delta Nb = 0.62 \times 10^{-6}$ Å^{-2}. Inserting this into the expression above, we estimate $Q^* \sim 0.0056$ Å^{-1}, consistent with the experimental data of Fig. 7. Secondly, the time invariance of the fringes at high Q allows these to be assigned to the Au electrode, the composition and thickness of which (necessarily) do not vary with time. Thirdly, an Au electrode thickness of 210 Å can be estimated from the periodicity of the fringes, $\Delta Q = 0.03$ Å^{-1}, *via* the equation $\Delta Q = 2\pi/d$ (where d = film thickness); this is entirely consistent with the nominal thickness of 200 Å from the sputtering process. Fourthly, the higher frequency fringes present only in the intermediate Q region can be seen progressively to stretch out with time. Since the momentum transfer, Q, is in reciprocal space, this stretching corresponds to some element of contraction of film thickness with hydrolysis of the FMOC groups. The quantitative question that follows is whether this contraction results in loss of some or all of the "void" volume generated by the FMOC departure.

To address this last question, detailed fitting of the data is required. Full details of this standard procedure are given in the Experimental section and elsewhere,[47–49] but the essential points are as follows. The scattering lengths of all the components present (quartz, Au, PPyFMOC, the departing FMOC, the remaining PPy-amine and CD$_3$CN solvent) are all known. The Au thickness is known (from a combination of fabrication protocol, bare electrode observations and the high Q data of Fig. 7). The unknowns are therefore the internal composition of the film (predominantly, solvation level), film thickness and the roughness of the polymer/solution interface.

The outcomes of the fitting process are shown in Fig. 7b and 7c. In the first of these, the scattering length density of the system, from the quartz block supporting the electrode through to the bulk solution, is shown as a function of distance, z (perpendicular to the interface). The values for quartz, Au and solvent represent bulk values for the pure components. The sharp dip between the quartz support and the Au electrode represents the MPTS bonding layer used to ensure good Au adhesion. For the purposes of this work, the region of interest is from the Au electrode outwards. The interface between the Au and the film is relatively sharp; a small amount of diffuseness (typically 10 Å) is required to account for the finite roughness of the Au surface. Interestingly, the polymer film is *not* compositionally homogeneous at any point in the process: there is a relatively diffuse outer region (a transition from "bulk" film to bulk solution over 40 Å) and the interior of the film comprises compositionally distinct inner and outer regions. A single polymer layer with high interfacial roughness was insufficient to model the film. The key outcome is a substantial shrinkage of the film as hydrolysis proceeds. For the example shown, the dry film thickness (measured in air) was 561 Å, and the solvated film thickness (measured upon exposure to solvent, prior to hydrolysis) was 639 Å, and the latter shrunk to 443 Å following completion of FMOC hydrolysis and elution.

The significance of this in solvation terms is shown in the solvent volume fraction data of Fig. 7c. While the fitting unquestionably demands that some film inhomogeneity be recognized (see Fig. 7b), the difference in scattering length density (and thence solvation) between the inner and outer regions of the film is relatively small. We therefore look at the average picture. The film solvent volume fraction at the outset, as solvated PPyFMOC exposed to a low piperidine concentration (as explained above) is 0.41. By the end of the hydrolysis process, this increases to 0.60, but it subsequently falls to 0.52. The latter decrease is the result of polymer relaxation, which occurs on a longer timescale than the FMOC hydrolysis and elution.

It is of course not possible to make the analogous measurements in hydrogenous solvent (CH_3CN) for the *same* film, since the hydrolysis reaction is a one-time process. However, such measurements were made on a nominally identical film and the outcome, in summary form for the analogue of Fig. 7c, was an initial solvent volume fraction of 0.35, rising to 0.58 immediately after hydrolysis and subsequently relaxing back to 0.44. The common conclusion from these experiments is that FMOC removal increases the film solvent volume fraction by $\Delta\phi_S \sim 0.2$ in the short term, but this increase is subsequently diminished to $\Delta\phi_S \sim 0.1$. While this may seem modest, we note that relatively small *changes* in solvent content can have profound effects on polymer chain mobility, for example as manifested in viscoelastic properties,[54,55] which would facilitate permeation of fluorophore reactant. In an absolute sense, a replaceable solvent volume fraction $\phi_S > 0.4$ is more than adequate to give a high (and thus visible) fluorophore population.

This leads to consideration of the final step in the scheme of Fig. 3, fluorophore functionalization of the $PPy-NH_2$ film. What reaction did occur – and the significant changes in $R(Q)$ profiles do unequivocally demonstrate change, quantified below – took place within the first 15 min of exposure to fluorophore solution. We attribute this to the more fluid-like environment of the film following hydrolysis (see above). From a mechanistic perspective, this removed the opportunity to follow the kinetics of the process (largely due to instrumental issues such as sample alignment prior to measurement), but from a practical perspective in future application it is obviously beneficial.

Consequently, on a separate (but nominally similar) film to that of Fig. 7, a separate set of measurements were made, as follows. First, $R(Q)$ data were acquired for a PPyFMOC film in the dry state (solvent-free) and exposed to h_3- and d_3-acetonitrile, prior to hydrolysis. Second, the film was hydrolysed (with no attempt to monitor the time-dependence of this process) and $R(Q)$ data acquired for the resultant PPyNH$_2$ film in the three environments (air, h_3- and d_3-acetonitrile). Finally, the film was Dylight functionalised and $R(Q)$ data acquired in the same three environments. This last part of the experiment is complementary to the hydrolysis step, in that the aim is entry of a large reactant to *consume* free volume, rather than elution of a large leaving group to *generate* free volume. As compensation for sacrifice of any kinetic information, this suite of measurements provided data in different solvent contrasts for the *same* film, which (through co-refinement) gives greater certainty in fitting. The resulting $R(Q)$ profiles are shown in Fig. 8a, grouped according to the film environment; the general form of the profiles is analogous to those of Fig. 7.

The key parameters of interest are film thickness and solvent content at each stage. To extract these, we need to consider the contributions of the polypyrrole spine, fluorophore and solvent components to the scattering length density. For the polypyrrole and solvent components, the scattering length and physical density are known. For the fluorophore, whose structure is commercially protected, this is not so straightforward, but acceptable approximations are possible,

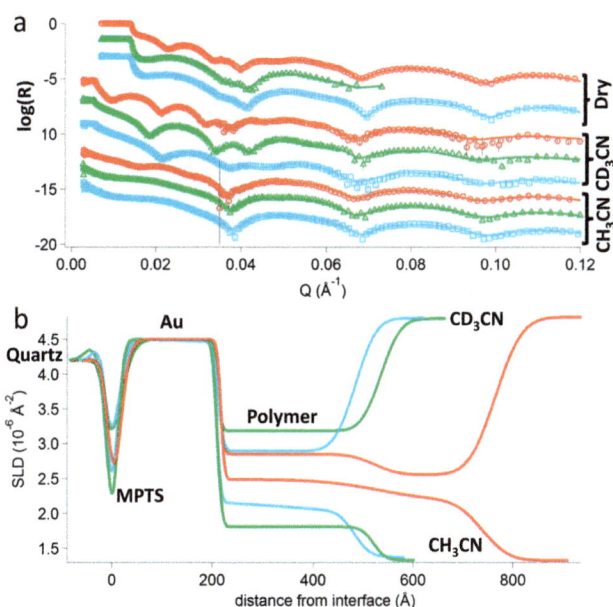

Fig. 8 Panel (a): $R(Q)$ data for a PPyFMOC film prior to hydrolysis (○), the PPyNH$_2$ film resulting from hydrolysis (△) and the PPy-Dylight film following exposure to the fluorophore (□). Points represent data; lines represent fits (see panel (b)). For visual comparison purposes, $R(Q)$ profiles are group according to the ambient medium (see annotations). Data are progressively offset downwards for presentational purposes; for the dry and d_3-acetonitrile exposed films, a critical edge is seen, below which $R = 1$. Panel (b): model fitted scattering length density profiles for the film exposed to h_3-acetonitrile and d_3-acetonitrile at each of the three stages of the process: nominally PPyFMOC (red traces), PPyNH$_2$ (green traces) and PPy-Dylight (blue traces).

as follows. The molar mass is on the order of 1000 Daltons and there are four sulfonate groups (to provide adequate solubility) on an essentially aromatic hydrocarbon skeleton. We thus have an entity that comprises ca. 320 Daltons of "SO_3^-" and ca. 680 Daltons of "CH"; we make the plausible approximation that the overall density is unity. With these physically reasonable approximations, combining the data sets for the h_3-acetonitrile and d_3-acetonitrile environments, we estimate that there is one fluorophore entity for every ca. five pyrrole monomer units. Physically, this is plausible, given the geometrical constraints of the cartoon representation of Fig. 3 and practically this is expected to be useful.

The model scattering length density profiles best fitting the data are shown in Fig. 8b and the $R(Q)$ fits (lines) are shown alongside the data (points) in Fig. 8a; the agreement is good across the accessible Q range. We particularly highlight four characteristics. First, the outer interfaces are diffuse. In the $R(Q)$ profiles, this accounts for the damping of the film-based fringes. In terms of reactivity, it undoubtedly contributes to the faster permeation of the fluorophore molecules. Second, once one progresses to the interior of the film, its scattering length density, and thus composition, shows at most only modest dependence on depth (actually, none for PPyNH$_2$). This indicates that diffusion of reactant into the film is not a limiting factor; if it were, then there would be a clear gradient of composition representing fluorophore penetration. Third, despite the *entry* of fluorophore into the polymer, the films *shrink* slightly during the process (compare the traces for PPyNH$_2$ and PPy-Dylight). This indicates that the volume of solvent expelled exceeds the volume of fluorophore entering, suggesting that transport processes are not so slow that mobile species (here, solvent) are trapped within the film. Finally, while the outer interface is slightly sharper for the dry film (profile not shown), the overall thickness is not much less than for the solvent-exposed film. Since (see Table 1) there is still appreciable solvent in the immersed films, this suggests that the film does not collapse upon emersion.

The characteristics of the model profiles are summarized in Table 1. Considering first the thickness data, the dramatic collapse of the PPyFMOC film upon hydrolysis (by ca. 40%) is accompanied by an *increase* in solvent content of the resulting PPyNH$_2$ film. This apparently counter-intuitive result is a consequence of the size of the FMOC group; recall the earlier estimations of film thickness accompanying the coulometry, when it was noted that the FMOC group constitutes ca. 60% of the film volume. Turning to the solvent volume fraction, the values for PPyFMOC ($\phi_S = 0.36$) and PPyNH$_2$ ($\phi_S = 0.47$) are satisfyingly consistent

Table 1 Summary of film thickness and solvent content values at each stage of assembly of the surface architecture represented schematically in Fig. 3. PPyNH$_2$ films could be modelled as a single layer (*i.e.* were internally homogeneous), so "inner" and "outer" regions are merged

Environment	Air ("dry")		Exposed to acetonitrile			
Parameter	Thickness/Å		Thickness/Å		Solvent volume fraction/φ_S	
Film region	Inner	Outer	Inner	Outer	Inner	Outer
PPyFMOC	173	303	312 (±18)	223 (±4)	0.34	0.38
PPyNH$_2$	257		318 (±8)		0.47	
PPy-Dylight	104	164	188 (±3)	84 (±3)	0.40	0.62

with those of $\phi_S = 0.38$ and $\phi_S = 0.48$ at the same points in the surface chemistry of Fig. 3 determined by averaging the outcomes of the two kinetic experiments of Fig. 7 and its h_3-acetonitrile counterpart.

Functional viability of the fluorophore-modified film

This report focuses on construction of the interfacial architecture and, *via* the NR measurements, on establishing spatial control of fluorophore immobilization within the polymer matrix; essentially, this represents the *compositional* and *structural* aspects of the strategy. In a subsequent phase of the work, the focus will shift to determination of the polymer chromophore and immobilized fluorophore properties; these represent the *functional* aspects of the strategy. The latter will involve a substantive programme of measurements, notably as functions of excitation and observational wavelengths and of polymer charge state (*i.e.* doping level, manipulated *via* applied potential). Nonetheless, in advance of such a future report, there is merit in a forward look to establish at a qualitative level fluorophore activity in the polymer film context. The technical issue here is whether (or not) proximity of the fluorophore sites to the underlying electrode results in fluorescence quenching.

Fig. 9 shows two images of PPy-based films representing two stages – simplistically, in the absence and presence of fluorophore, respectively – of the assembly process shown in Fig. 3. This preliminary observation does not attempt to address the spatial issue of imaging a full fingerprint, but focuses solely on the viability of fluorophore emission when in the film environment. The left hand image shows a section of a PPyNH$_2$ film deposited on an Au substrate (using the procedure of Fig. 5, terminating at the cathodic end of a potential cycle to establish the undoped redox state), removed from solution and viewed *ex situ* (dry) under illumination by light of wavelength 640 nm. This control observation shows a few brighter areas, but nothing systematic or substantive. The right hand image shows

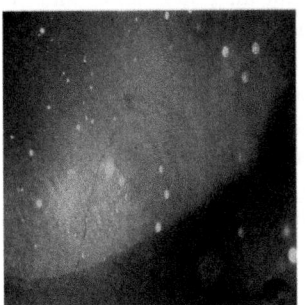 

Panel *a*   Panel *b*

Fig. 9 Panel (a): PPy film on Au (deposition procedure as in main text); this represents the control experiment, in the absence of fluorophore. Panel (b): PPyNH$_2$ film prepared by hydrolysis of a PPyFMOC film (deposition and subsequent treatment as in main text) after partial exposure to Dylight 649 NHS ester by contact with a droplet of fluorophore solution, and subsequent removal of excess fluorophore by rinsing. The top left part of the viewed region includes part of the droplet-exposed area. In both cases, the films were in the reduced (undoped) state and were viewed *ex situ* under illumination by light of wavelength 640 nm. In panel (b), the intensity was attenuated by a factor of 4.

a partially Dylight 649 ester functionalised film, prepared as follows. A PPyFMOC film was deposited on Au and hydrolysed (as discussed earlier with reference to Fig. 7), then a droplet of Dylight 649 NHS ester solution placed on one part of the surface, followed by rinsing with pure water (to remove unbound and surface/exterior fluorophore) and air drying (at room temperature for *ca.* 15 min). The resultant film was viewed *ex situ* under illumination with light of wavelength 640 nm. The top left part of the viewed region in panel (b) of Fig. 9 includes part of the droplet-exposed area. The sharply defined region of enhanced brightness is consistent with strong fluorophore emission; note that the intensity was attenuated (see legend) so the distinction between the images in panels (a) and (b) is significant. Although this is not a quantitative measure of the fluorescence efficiency and does not totally exclude quenching – perhaps from fluorophores sites closer to the electrode (see Fig. 8 for evidence of deep penetration of fluorophore) – it is clear that at least some fluorophore sites are sufficiently distant from the electrode that they are not vulnerable to quenching. Pragmatically, the practical viability of the interfacial (electro)chemical strategy is established.

Conclusions

A combination of spectroscopic, electrochemical and neutron-based techniques provides the capability to follow and quantify deposition and subsequent functionalization of electroactive polymer films relevant to latent fingerprint visualization. This approach has been used to explore the (electro)chemistry of pyrrole–FMOC electropolymerization and deposition, followed by hydrolysis of the FMOC leaving group, then permeation and bonding of the fluorophore Dylight 649 NHS ester. By revealing the presence, removal and reintroduction of amide functionalities, FTIR spectroscopy demonstrates qualitative success of this post-deposition functionalization strategy. Electrochemistry provides control over (and coulometric assay of) the surface population of polymer, the electrochromic matrix into which the fluorophore is introduced. Neutron reflectivity provides insight into the *vertical* spatial distribution of the permeating fluorophore and the changes in film population of the solvent that must leave to create space for it. Together, these techniques provide insights into film composition, structure and (in the cases of the electrochemical and neutron data) dynamics; simplistically, they address the tersely expressed questions "*what, how much and where?*"

We have demonstrated that the portfolio of materials suitable for electrochromic enhancement of latent fingerprints can be extended from the previously used aniline and thiophene (PEDOT) families to include pyrrole-based materials. Specifically, this was accomplished for the parent polypyrrole, *N*-propylamine functionalised pyrrole and FMOC functionalised pyrrole. In future, in addition to the substitution chemistry explored here, this will provide a wider colour palette with which to optimise latent fingerprint visual contrast against the substrate.

In the post-deposition functionalization of PPyFMOC, the hydrolysis process (leading to FMOC removal) is relatively slow (*ca.* 3 h under the conditions employed), which allowed the progress of the reaction to be monitored by neutron reflectivity. The subsequent entry of fluorophore (Dylight 649 NHS ester) is much more rapid – too rapid to follow readily using neutron reflectivity – which is attributed to the higher solvent volume fraction and thence greater fluidity of the film. During this sequence of events, the solvent volume fraction rises by *ca.*

0.2 immediately upon hydrolysis and FMOC elution, but then falls by *ca.* 0.1, presumably as a consequence of polymer relaxation in the now highly plasticised film. In absolute terms, the solvent volume fractions are 0.37 ($\pm$0.03) for PPyFMOC prior to hydrolysis, 0.59 ($\pm$0.01) immediately after hydrolysis and 0.47 ($\pm$0.04) after relaxation (where these data originate from both kinetic and non-kinetic experiments). After fluorophore entry, the film contracts slightly.

Future prospects for a combined absorption/fluorescence strategy in latent fingerprint enhancement appear promising. The next step is implementation of the strategy described here to fingerprinted surfaces. Having established control of reactivity and composition in the *vertical* direction at distance scales from 10–100 nm, this fine control can now be applied to the *lateral* direction. Since the fingerprint feature sizes are at the scale of >10 µm, the prospect of high resolution, high fidelity fingerprint images is excellent.

Acknowledgements

We thank the Institut Laue Langevin and ISIS Facility of the Rutherford Appleton Laboratory for provision of neutron beam time. R. M. B. thanks the University of Leicester for financial support. We thank Ines Carro-Muino (of Thermo Fisher Scientific) for helpful conversations relating to fluorophores and Zeta Instruments for helpful discussions regarding sample analysis using 3D optical profiling.

References

1 *Advances in Fingerprint Technology* (2nd edn), ed. H. C. Lee and R. E. Gaensslen, CRC Press, Boca Raton, 2001.
2 R. S. Ramotowski, in *Advances in Fingerprint Technology* (2nd edn), ed. H. C. Lee and R. E. Gaensslen, CRC Press, Boca Raton, 2001, p. 63.
3 R. D. Olsen and H. D. Lee, *Advances in Fingerprint Technology* (2nd edn), ed. H. C. Lee and R. E. Gaensslen, CRC Press, Boca Raton, 2001, p. 41.
4 P. Voss-De Haan, *Contemp. Phys.*, 2006, **47**, 209.
5 J. Berry and D. A. Stoney, in *Advances in Fingerprint Technology* (2nd edn), ed. H. C. Lee and R. E. Gaensslen, CRC Press, Boca Raton, 2001, p.1.
6 J. G. Barnes, in *The Fingerprint Source Book*, ed. A. McRoberts, National Institute of Justice, Washington, 2011, ch. 1. Available from http://www.nij.gov/pubs-sum/225320.htm.
7 W. J. Herschel, *Nature*, 1880, **23**, 76.
8 K. Barnett, in *Crime Scene to Court: the Essentials of Forensic Science* (1st edn), ed. P. White, Royal Society of Chemistry, Cambridge, UK, 1998, p. 98.
9 F. Galton, *Finger Prints*, Macmillan, London, 1892.
10 H. Faulds, *Nature*, 1880, **22**, 605.
11 M. Kucken, *Forensic Sci. Int.*, 2007, **171**, 85.
12 I. W. Evett and R. L. Williams, *J. Forensic Ident.*, 1996, **46**, 49.
13 *Advances in Fingerprint Technology* (3rd edn), ed. R. S. Ramotowski, CRC Press, Boca Raton, 2013.
14 S. Bell, *Annu. Rev. Anal. Chem.*, 2009, **2**, 297.
15 T. A. Brettell, J. M. Butler and J. R. Almirall, *Anal. Chem.*, 2011, **83**, 4539.
16 N. Akiba, N. Saitoh and K. Kuroki, *J. Forensic Sci.*, 2007, **55**, 180.
17 M. J. Choi, T. Smoother, A. A. Martin, A. M. McDonagh, P. J. Maynard, C. Lennard and C. Roux, *Forensic Sci. Int.*, 2007, **173**, 154.
18 J. Almog, G. Levinton-Shamuilov, Y. Cohen and M. Azoury, *J. Forensic Sci.*, 2007, **52**, 1057.
19 J. D. James, C. A. Pounds and B. Wilshire, *J. Forensic Sci.*, 1993, **38**, 391.
20 L. K. Seah, U. S. Dinish, W. F. Phang, Z. X. Chao and V. M. Murukeshan, *Forensic Sci. Int.*, 2005, **152**, 249.
21 V. Bowman, *Manual of Fingerprint Development Techniques* (2nd edn), Police Scientific Development Branch, Home Office, Sandridge, UK, 2004.
22 H. L. Bandey and T. Kent, *Superglue Treatment of Crime Scenes 30/03*, Police Scientific Development Branch, Home Office, Sandridge, UK, 2003.

23 A. A. Cantu, D. A. Leben, R. Ramotowski, J. Kopera and J. R. Simms, *J. Forensic Sci.*, 1998, **43**, 294.
24 C. Kauffman and K. Smith, *J. Forensic Ident*, 2001, **51**, 9.
25 Y. Migron and D. Mandler, *J. Forensic Sci*, 1997, **42**, 986.
26 G. S. Sodhi and J. Kaur, *Forensic Sci. Int.*, 2001, **120**, 172.
27 A. R. W. Jackson and J. M. Jackson, *Forensic Science*, Pearson Prentice House, Harlow, UK, 2004.
28 A. A. Cantu, in *Optics and Photonics for Counterterrorism and Crime Fighting III, SPIE-Int. Soc. Opt. Eng.*, ed. C. Lewis, 2007, vol. 6741, p. D7410.
29 P. Czekanski, M. Fasola and J. Allison, *J. Forensic Sci.*, 2006, **51**, 1323.
30 M. Colella, A. Parkinson, T. Evans, C. Lennard and C. Roux, *J. Forensic Sci.*, 2009, **54**, 583.
31 P. F. Kelly, R. S. P. King and R. J. Mortimer, *Chem. Commun.*, 2008, 6111.
32 K. K. Bouldin, E. R. Menzel, M. Takatsu and R. H. Murdock, *J. Forensic Sci.*, 2000, **45**, 1239.
33 R. Leggett, E. E. Lee-Smith, S. M. Jickells and D. A. Russell, *Angew. Chem., Int. Ed.*, 2007, **46**, 4100.
34 A. L. Beresford and A. R. Hillman, *Anal. Chem.*, 2010, **82**, 483.
35 R. M. Brown and A. R. Hillman, *Phys. Chem. Chem. Phys.*, 2012, **14**, 8653.
36 C. Bersellini, L. Garofano, M. Giannetto, F. Lusardi and G. Mori, *J. Forensic Sci*, 2001, **46**, 871.
37 A. L. Beresford, R. M. Brown, A. R. Hillman and J. W. Bond, *J. Forensic Sci.*, 2012, **57**, 93.
38 F. Cortes-Salazar, J. M. Busnel, F. Li and H. H. Girault, *J. Electroanal. Chem.*, 2009, **635**, 69.
39 M. Zhang and H. H. Girault, *Analyst*, 2009, **134**, 25.
40 F. Cortes-Salazar, M. Q. Zhang, A. Becue, J. M. Busnel, M. Prudent, C. Champod and H. H. Girault, *Chimia*, 2009, **63**, 580.
41 M. Q. Zhang and H. H. Girault, *Electrochem. Commun.*, 2007, **9**, 1778.
42 G. Qin, M. Zhang, Y. Zhang, Y. Shu, S. Liu, W. Wu and X. Zhang, *J. Electroanal. Chem*, 2013, **693**, 12.
43 A. Glidle, A. R. Hillman, K. S. Ryder, E. L. Smith, J. M. Cooper, N. Gadegaard, J. R. P. Webster, R. M. Dalgliesh and R. Cubitt, *Langmuir*, 2009, **25**, 4093.
44 A. Glidle, A. R. Hillman, K. S. Ryder, E. L. Smith, J. M. Cooper, R. M. Dalgliesh, R. Cubitt and T. Geue, *Electrochim. Acta*, 2009, **55**, 439.
45 R. A. Campbell, H. P. Wacklin, I. Sutton, R. Cubitt and G. Fragneto, *Eur. Phys. J. Plus*, 2011, **126**, 107, DOI: 10.1140/epjp/i2011-11107-8.
46 J. R. P. Webster, S. Langridge, R. M. Dalgliesh and T. R. Charlton, *Eur. Phys. J. Plus*, 2011, **126**, 112, DOI: 10.1140/epjp/i2011-11112-y.
47 J. Penfold, R. M. Richardson, A. Zarbakhsh, J. R. P. Webster, D. G. Bucknall, A. R. Rennie, R. A. L. Jones, T. Cosgrove, R. K. Thomas, J. S. Higgins, P. D. I. Fletcher, E. Dickinson, S. J. Roser, I. A. McLure, A. R. Hillman, R. W. Richards, E. J. Staples, A. N. Burgess, E. A. Simister and J. W. White, *J. Chem. Soc., Faraday Trans.*, 1997, **93**, 3899.
48 A. Glidle, J. Cooper, A. R. Hillman, L. Bailey, A. Jackson and J. R. P. Webster, *Langmuir*, 2003, **19**, 7746.
49 A. Nelson, *J. Appl. Crystallogr.*, 2006, **39**, 273.
50 S. H. Cho, K. T. Song and J. Y. Lee, in *Handbook of Conducting Polymers* (3rd edn), ed. T. A. Skotheim and J. R. Reynolds, CRC Press, Boca Raton, 2007, ch. 8, pp. 8-3.
51 I. Efimov and A. R. Hillman, *Anal. Chem.*, 2006, **78**, 3616.
52 A. Glidle, A. R. Hillman, K. S. Ryder, E. L. Smith, J. M. Cooper, N. Gadegaard, J. R. P. Webster, R. M. Dalgliesh and R. Cubitt, *Langmuir*, 2009, **25**, 4093.
53 A. Glidle, A. R. Hillman, K. S. Ryder, E. L. Smith, J. M. Cooper, R. M. Dalgliesh, R. Cubitt and T. Geue, *Electrochim. Acta*, 2009, **55**, 439.
54 A. R. Hillman, M. A. Mohamoud and I. Efimov, *Anal. Chem.*, 2011, **83**, 5696.
55 A. R. Hillman, I. Efimov and K. S. Ryder, *J. Am. Chem. Soc.*, 2005, **127**, 16611.

Faraday Discussions

DISCUSSIONS

General discussion

DOI: 10.1039/C3FD90031B

Dr Batchelor-McAuley opened the discussion of the paper by Dr Marken by commenting: Many of the experiments you have performed using your micro-trench electrodes can be undertaken at a single electrode by simply pulsing the potential. Importantly, in many ways the limiting process for your experimental system is the rate of mass-transport from one electrode to the other. However, in comparison, for a pulse technique – such as square-wave voltammetry – the sensitivity of the technique is limited by the cell's time constant. A direct comparison between the two techniques is not overly straightforward. However, it can be shown that in order for a micro-trench generator/collector system to exhibit a comparable enhancement in current to that found for square-wave, the electrodes would require a separation of 10–15 microns. This analysis is based on consideration of the expression you give in equation 2 in your paper and recognises that it is not unreasonable to expect a 5-fold increase in the measured current for a "diffusion only" square-wave voltammogram as compared to a simple linear sweep experiment. This calculated separation of 10–15 microns is much smaller than you have used experimentally. From an electroanalytical stand-point, what advantages do you believe the micro-trench electrodes have over a more experimentally facile one-electrode pulse system?

Addendum:

First, it must be recognised that one is attempting to compare a steady-state current (generator/collector) to that of a peak (square-wave). Second, the values reported in square-wave are the difference in current of the forward and reverse steps (*i.e.*, ΔI).

The peak current for the square-wave of a diffusion only redox species is given by[1] $\Delta I_p = FAD[i] \chi (f/D)^{0.5}$ where F is the Faraday constant (96 485 C mol^{-1}), A is the area of the electrode (m^2), D is the diffusion coefficient (m^2 s^{-1}), [i] is the concentration of species i (mol m^{-3}), χ is a constant which depends upon the amplitude and step size of the square wave and f is the frequency of the square wave. In order to make a fair comparison we need to consider the sum of the current at the generator and collector. From comparison of the square wave peak current with equation 2 we can write $\delta = (2/\chi)(f/D)^{-0.5}$. Taking relatively conservative values for the square wave parameters of a frequency of 100 Hz, an amplitude of 25 mV and a step size of 5 mV (here $\chi = 0.45$). We can show that for the diffusion coefficient given in the paper an inter-electrode distance of approximately 12 microns is needed in order to get a comparable faradaic signal.

1 A. Molina *et al.*, *J. Electroanal. Chem.*, 2011, **659**, 12.

Professor Marken responded: I fully agree with this assessment and with the comment. These types of trench electrodes become interesting at the sub-5 micron or more generally at the sub-micron level. One particular aspect of the reactivity at these electrodes that intrigued me is the separation of the "cavity" region and the "mouth" region with two important time constants associated with the transport into the trench and across the trench. This seems quite unique and for example measurements, when conducted in the presence of oxygen, can benefit from the chemically irreversible "*in situ* removal" of oxygen in the mouth region with no effect on the feedback chemistry in the cavity region. Two more general benefits are absence of capacitive charging on the collector or sensor electrode (depending on trench geometry and solution resistivity) and no need for high speed (assuming that the steady state responses in narrow trenches are the most interesting signals to study and employ). Square wave voltammetry is currently the main electroanalytical tool in academia and in industry and I don't think we are going to change this any time soon with micro-trench electrodes.

Professor Barbero said: I have two questions: 1) As far as I understood the width of the gap is defined by the viscosity of the epoxy that binds the two electrodes, flowing under pressure. Could it be possible to use some kind of spacer (*e.g.* polymer microparticles) mixed with the epoxy and higher pressure to define gaps?

2) As we are proposing in our own paper at this discussion, local pH changes could be produced by fast double layer charging of nanoporous electrodes, due to the slow mass transport of electrolyte ions inside the pores which seems to induce water dissociation. Do you think this effect could be used also for "chemical filtering" of analytes in single electrodes?

Professor Marken replied: These are very good points. In terms of the reproducible production of these trench electrodes, a lot of improvement is still required and we should in future avoid the crude epoxy method. Spacer particles or well-defined spin-coated polymer precursor films could be options for better protocols with pre-determined inter-electrode gap sizes.

The pH changes in small pores during double layer charging is an interesting phenomenon and application in electroanalysis may be possible. I could envisage a low concentration analyte "pre-concentration" effect into the pores and a pH drift in pores "moving" the response on the potential scale. These effects could be counter-intuitive and useful. Possible challenges will be posed by the capacitive background current and by distribution effects due to non-uniform pore geometries and pH gradients.

Professor Fielden queried: The RedOx cycling observed within your micro-trench electrodes is reminiscent of the dual electrode detection techniques used in electrochemical detection coupled to high performance liquid chromatography. Usually, a generator electrode is placed upstream of a detector electrode, whereby the products from the generator electrode are convectionally transported to the detector electrode. However, there was one variation where two electrodes were placed opposite each other, separated by a thin spacer, and polarised to promote RedOx cycling. This is sometimes named the "parallel opposed" configuration. If your micro-trench electrodes were to have the addition of an orthogonal flow, rather than the static mode you have reported, what do you think the effect would be on the current amplification? Could it offer any further experimental advantage?

Professor Marken answered: This is an interesting comment. Micro-trench electrodes could be of interest in post-column detection in particular due to the feedback currents being confined and counter electrode currents posing less uncompensated potential drop effects. Both flow modes, perpendicular or colinear with the trench, could be envisaged, with the latter probably causing additional hydrodynamic effects. The trench detector signal will depend on the ratio of detection time and liquid replacement time. Most interesting will be trench electrode systems with sub-micron trench width. In this case hydrodynamic effects will be minimal (diffusion-only), the detector volume will be very small, and therefore both detector time resolution and amplification effects will be very high. Given the electrochemical versatility of the generator-collector micro-trench detector, new applications seem feasible.

Dr Wain commented: One method of introducing hydrodynamic flow to the micro-trench electrode might be to attach it to a rotating disk. Modelling this would be quite difficult, but it ought to improve mass transport for analytical applications.

Regarding the pH modulation experiment, have you investigated the transient response or considered how the buffer capacity might affect the time resolution of the pH-sensitive measurement at the sensor electrode?

Professor Marken responded: One important point about these trench electrodes is that conditions in the "cavity" region are distinct from those in the "mouth" region. Feedback within the cavity and absence of capacitive charging on the collector electrode offer sensitivity and selectivity without the need for transport to the mouth. Introducing mass transport will affect the "cavity" to "mouth" spatial ratio up to the point that the trench becomes fully accessible (*e.g.* maybe with power ultrasound). For applications in electroanalysis this is not necessarily beneficial.

We recorded the collector transient responses in chronoamperometry mode and there is a considerable amount of information hidden in these transients. At this stage no attempt has been made to fully/further analyse the transients. In the future, micro-trench probes could be envisaged to measure buffer capacity.

Professor Compton opened the discussion of the paper by Professor Bund by asking: Please can you comment on the likely accuracy of Comsol Multiphysics for the simulations you report? We have reported[1] that two-dimensional problems are within an order of magnitude of accuracy of finite difference simulations and analytical solutions, as long as the problem is well defined in the software and care is taken with regards to appropriate meshing and boundary conditions. Three-dimensional simulations relating to microdiscs result in steady-state current values not quantitatively compatible with experimental observations or analytical solutions. What is your view with respect to the latter?

1 I. J. Cutress et al., *J. Electroanal. Chem.*, 2010, **639**, 76.

Professor Bund responded: I think that with a careful meshing procedure and some additional caution with the boundary conditions (*e.g.* by using Lagrange multipliers) good accuracy can be achieved also with 3D finite element (FE) problems. One can always run a problem for which an analytical solution is known (*e.g.* disk electrode) in 3D geometry. At first glance this might look like a waste of resources but it seems to be a suitable approach to assess the accuracy of FE methods. For a 3D problem with complicated boundary conditions for which no analytical solution is known (of course these are the systems of interest) the situation is more challenging. But that is an intrinsic problem of all numerical approaches. Physical intuition and comparison with similar known solutions must be our guide then.

Professor Compton opened the discussion of the paper by Professor Chung by asking: Can you please comment on how "nano-confinement" effects differ from those simply expected for thin layer diffusion type behaviour in porous layers? I would expect that an authentic "nano"-confinement effect would show effects related to altered properties (such as diffusion coefficients or rate constants) showing different values between bulk solution and the pore. Do you have experimental evidence for these?

Professor Chung answered: The fundamental difference between the nano-confinement effect and thin layer diffusion as models for analyzing voltammetric behavior is the scale of time and space. Thin layer diffusion describes macroporous electrochemical systems, for which the continuum model is still valid. The previous studies were mostly to explain the voltammograms of electroactive species undergoing relatively fast electron transfer, *e.g.* ferrocyanide, dopamine at CNT electrodes. For instance, the pore diameter among CNTs is around 60 nm, which belongs to the macroporous regime.[1] Suppose that we have a 10 mM reactant solution in a cylindrical pore 60 nm wide (diameter) and 200 nm long. The number of reactant molecules in the pore is expected to be *ca.* 3400, which can be treated on the basis of mass transport by diffusion along the concentration gradient, namely, classical Fick's law. If the pores are randomly interconnected and thus the average pore length is short, the inbound flux from the bulk solution into the pore should be high, meaning that the overall electrochemical behavior will be more likely to be explained by the concentration gradients inside the pore. That is why we can simulate and fit the voltammogram from CNT electrodes with the features such as the decrease of peak potential separation, we believe. On the other hand, the L_2-ePt we employed for this study has a diameter of 1–3 nm. There should be only *ca.* 8 molecules for 10 mM reactant solution in the pore assuming the same pore length, 200 nm, and the same shape. In this geometric environment, the molecule in the middle of the pore needs about 1 ns to reach the electrode surface by random walk at room temperature assuming the diffusion coefficient of 1×10^{-5} cm^2 s^{-1}. The average number of molecules and the time scale for encountering the inner wall of the pore strongly indicate that we should see this condition in view of stochastically individual molecular dynamics. As recent papers and researchers have started to find, the channel space within a few nm scale or smaller seems to be sufficiently narrow to understand on the basis of

Knudsen diffusion. Reportedly, catalytic effects coming from the geometrical features of nanoporous electrodes such as irregularity and roughness can be elucidated by molecular dynamics in the Knudsen regime.[2] Although the electrocatalytic system in this work is not a solid–gas but a solid–liquid interface and thus the mean free path of a molecule is predicted to be a few Å in solution, only a few reactant molecules with inert solvent molecules are diffusing and encountering the electrode surface, where faradaic reaction can occur, with exceptionally high frequency. Therefore, the overall behavior is pretty well understood by a Knudsen-like diffusion model. Previously, enhanced current density of the proposed fuel cells exploiting the nanoporous catalysts was explained by Knudsen diffusion.[3] As a simple and well established model in terms of theory as well as experiments, Knudsen diffusion provides reasonable basis for understanding what is observed at nanoporous electrochemical interfaces. The nanoconfinement effect was proposed in this sense. The main concern of this effect is electrocatalysis, making a sluggish faradaic process facile by employing nanoporous structures. We are particularly focusing on the fact that the reactant molecules on the electrode surface should be present within a limited distance by confining them in pores, resulting in much higher collision frequency and thus enhanced probability of faradaic reaction for a given period.

Under the experimental conditions in this study, it is difficult to determine the diffusion coefficient because no diffusion controlled reaction is expected inside the nanopores. Most faradaic reactions are facile enough for the reactant molecules to be completely depleted in the pores of a few nm diameter. Reactants which undergo sluggish processes exhibit electrochemical behavior that should not be interpreted by Fick's laws but by molecular dynamics of individual molecules. Therefore, further effort is required to estimate the diffusion coefficient based on a single molecule study.

As for the rate constant, we determined the apparent overall rate constants of 1,4-benzoquinone in 1 M $HClO_4$ solution at the equilibrium potential. The rate constants at flat Pt and L_2-ePt (nanoporous) were 2.1×10^{-4} and 2.6×10^{-3} cm s^{-1}, respectively, which were calculated from R_{ct} by electrochemical impedance analysis. This shows that the rate constant of 1,4-benzoquinone increases by about 10 times at nanoporous Pt compared with flat Pt. Not only quinone but also 2-butanol in 1 M $HClO_4$ solution gave the apparent overall rate constants of 9.8×10^{-8} and 7.5×10^{-7} cm s^{-1} for flat Pt and L_2-ePt, respectively.

As presented in the manuscript, the current density of 2-butanol oxidation at nanoporous Pt is significantly high compared with flat Pt. Plus, the onset potential of the current density in the voltammograms at L_2-ePt is unambiguously less positive than that at a flat electrode (Fig. 3(b) of our paper).

1 I. Streeter et al., Sens. Actuators, B, 2008, 133, 462.
2 J. H. Bae et al., Phys. Chem. Chem. Phys., 2012, 14, 448.
3 M. K. Debe, J. Electrochem. Soc., 2012, 159, B53.

Professor Compton commented: It is possible to predict voltammetric behaviour in well defined cylindrical pores on the basis of Fick's Laws and Butler–Volmer kinetics.[1] Thus in principle a comparison of the behaviour observed with that predicted using bulk solution parameters (e.g. for diffusion coefficients) is possible but in practice the quantitative application to the ill-defined porous layers such as those formed by electro-deposition is not immediately viable.

1 K. R. Ward et al., *J. Electroanal. Chem.*, **702**, 2013, 15.

Professor Chung replied: As discussed in the previous response, the thin layer diffusion model can successfully explain what is observed for diffusion controlled electrochemical reactions at macroporous electrodes. However, it is hard to apply it for kinetic controlled faradaic process in far narrower pores, several nm in diameter. Rather, molecular dynamics concepts should be introduced to elucidate the observation ,as seen in single molecule electrochemistry.

Professor Xu continued the discussion of the paper by Dr Marken by asking: As the distance between two gold micro-trench electrodes decreases, will molecules move in different ways?

Professor Marken responded: This is a fascinating question and probably going to the heart of the recently developing field of stochasticity[1] in nano-electrochemical systems. Not only transport phenomena, but also chemical reactivity of molecules will depend on concentration gradients and therefore on trench size.

1 P. S. Singh et al., *ACS Nano*, 2012, **6**, 9662.

Professor Xu enquired: What is the minimum distance between two gold micro-trench electrodes that you have ever obtained?

Professor Marken replied: For gold–gold dual-hemisphere electrodes we estimated an average of 300 nm,[1] but for the micro-trench electrodes[2] we still have more work to do to reach submicron dimensions.

1 R. W. French et al., *J. Electroanal. Chem.*, 2009, **632**, 206.
2 S. E. C. Dale et al., *Electrophoresis*, 2013, **34**, 1979.

Professor Kornyshev commented: How pore size affects the diffusion is an important problem in micro/nanofluidics. In large pores one commonly distinguishes friction due to the scattering of diffusing particles from generally rough pore walls, the so-called Knudsen diffusion, from the bulk-like random Brownian diffusion of the same particles. In narrow nanopores, clear separation between these two regimes becomes no longer possible. What can be said about this problem as a result of the present study?

Professor Chung responded: In small pores, a given chemical species seems to follow Knudsen diffusion. This notion came from the observation that faradaic current sensitively depended on the thickness of the nanoporous film, *i.e.* length of the pores. It is more indicative of the fact that the apparent catalytic activity augmented remarkably and proportionally as pore length increased. Compared with Brownian diffusion, Knudsen diffusion indicates that the number of collisions between the wall and a diffusive object is proportional to the surface area, which is directly controlled by pore length in nanoporous systems. Longer pores obviously mean larger surface area of the inner wall. If the mass transport was mostly done by Brownian diffusion, the current should have been only a function of real surface area. The apparent current increased as the surface area was enlarged, which is no wonder. But we got rid of that effect by normalizing the faradaic current with the real surface area in the present study. Therefore, the augmentation of normalized current did not come from enlargement of the surface area itself, but from more chances of electron transfer in that reactant molecules can get in narrow pores. This implies that mass transport in narrow pores (1–3 nm dia.) cannot be understood by a Brownian diffusion model only, and requires a Knudsen diffusion concept to explain the electrochemistry at nanoporous electrodes.

For this argument, we need an assumption that the reaction is free of adsorption. The experimental results in the present study are about comparison between adsorptive and non-adsorptive electrochemical reactions of small isomeric alcohol molecules which have same molecular weight, functional group, and almost identical size. It turned out that the adsorptive reaction step, which clearly involved the interaction of reactant molecules with the inner wall, played a pivotal role for electrocatalytic behavior of nanopores. Only non-adsorptive reactant exhibited electrocatalytic oxidization. Our results strongly indicate enhancement of collision frequency originating from the geometry, *i.e.* porous structure, and the importance of non-adsorptive Knudsen diffusion for electrocatalysis based on the nano-confinement effect.

Professor Barbero remarked: 1) I could not find details about the diameter and orientation, with respect to the substrate electrode surface, of the pores. Could you describe the size and orientation of the pores in your electrodes?

2) The alcohol oxidation reaction mechanism could be complex. Would it not be easier to study the effect using a simple reaction with slower heterogeneous kinetics (*e.g.* ferrous ion oxidation)? We have used that to uncover the "thin layer" effect in carbon nanotube films (D. F. Acevedo *et al.*, *Electrochim. Acta*, 2008, **53**, 4001).

Professor Chung responded: 1) We did not fabricate new nanoporous electrodes, but employed two different kinds that are expected to be appropriate for examining the nano-confinement effect.

The size of the nanoporous electrodes used for this study is described in page 12 and 13 of our paper. The nanoporous Au employed in this work had much larger pores, 16 (±2) nm, than the nanoporous Pt, 1–3 nm. The nanoporous black Au had numerous pores smaller than 20 nm, which were roughly as large as those of nanoporous Au.

The structure of the nanoporous electrode was sponge-like having numerous 3 dimensionally interconnected pores. For more detailed information, you can refer to a few papers for nanoporous Pt,[1] nanoporous Au,[2] and nanoporous black Au.[3]

1 S. Park *et al.*, *Chem. Mater.*, 2007, **19**, 3373.
2 B. Seo and J. Kim, *Electroanalysis*, 2010, **22**, 939.
3 K. Nishio and H. Masuda, *Angew. Chem., Int. Ed.*, 2011, **50**, 1603.

2) As described in the reply to a previous question, the nano-confinement effect is believed to be more pronounced for the sluggish reaction of non-adsorptive reactant molecules following a simple mechanism. The reason we

used alcohols was to see how critical adsorption onto the electrode surface was for the nano-confinement effect. We have to agree with the argument that alcohol oxidation is still complex for this purpose.

The redox couple of Fe^{2+}/Fe^{3+} was also considered as a candidate in this study. It is quite slow at carbon electrodes including glassy carbon: the redox peak separation (ΔE_p) is 0.5–0.6 V at 100 mV s^{-1}.[1] However, it is fast at Pt as shown by a separation of only 0.1 V at 100 mV s^{-1} at flat Pt. We did some experiments using H$_1$-ePt (pore diameter: 2.5 nm, roughness factor: 78) to obtain the voltammograms of 10 mM Fe^{3+}/1 M HClO$_4$. No kinetic current but only semi-infinite diffusion behavior was observed. That is not surprising, taking into account the fact that reduction of Fe^{3+} is too facile on the surface of the Pt electrode.

1,4-benzoquinone could be another potential choice. Its redox process was indeed affected by the nanoporous structure, indicating that it came into the pore and underwent faradaic reactions inside the pores.[2] However, its current quickly transitted to semi-infinite diffusion from the bulk solution as seen from the result that the peak current of 1,4-benzoquinone reduction varied linearly with scan rate$^{1/2}$. Therefore, the redox kinetics of 1,4-benzoquinone are too fast to investigate the nano-confinement effect as well. The thin layer model was not able to describe the voltammetric behavior of 1,4-benzoquinone. Noting the dimension of the nanopores employed in this study, a few nm in diameter (refer to the reply to a previous question), we can see it is not surprising.

1 L. Tang et al., Adv. Funct. Mater., 2009, **19**, 2782.
2 J. H. Bae et al., Phys. Chem. Chem. Phys., 2013, **15**, 10645.

Dr Kataky addressed Professor Chung and Professor Marken: Molecules confined in molecular nanopores such as cyclodextrins, cyclophanes, crown ethers show very different catalytic properties due to stereo restrictions, van der Waals forces, H-bonding, dipolar interactions. Is it likely that the nanoporous structures you use in your work are subject to similar effects? I include some of our earlier publications where we've seen electrocatalytic effects of cyclodextrins:

1 G. Grancharov et al., Analyst, 2005, **130**, 1351.
2 E. Morgan and R. Kataky, Biosens. Bioelectron., 2003, **18**, 1407.
3 R. Kataky et al., Analyst, 2001, **126**, 2015.

Professor Chung replied: We went through all of the papers, which were valuable and helpful. We do not think that the electrochemical anomaly that was observed at the nanoporous electrode surface is exactly the same as that found for the redox active molecule captured by cyclodextrin. Unlike the molecule in cyclodextrin, the reactant in a nanopore keeps colliding with the inner wall of the electrode, moving along the "nano-cave". That is why we should introduce molecular dynamics to understand the behavior at such a nanoporous electrode with pores of a few nm diameter. Nonetheless, there is an important common underlying principle, "grabbing the reactant within a limited distance of the active site".

Professor Mount asked: To what extent in the bulk linear sweep voltammogram and cyclic voltammetry experiments would you expect significant changes in the concentration of ions in the pores and the amount of water in the pores to cause significant activity effects? What effect(s) would this have on the observed responses?

Professor Chung answered: It is unclear yet how much the ion concentration or solvent property is modified in the nanopores. But there are some leads in the recent literature, which may allow us to expect its contribution. The electrolyte concentration was 1 M, which is high enough to neglect the ion concentration change in the pores compared to the bulk solution.[1] In spite of the concern about the concentration difference in the pore from that in the bulk solution, no further study or discussion was given in this report. With regards to the solvent density, the DFT calculation revealed that there is little difference between that in the pores of greater than 0.75 nm diameter and in the bulk solution.[2] Overall, some parts of previous reports imply the presence of the pore effect, especially in small pores of several nm diameter, but there is neither experimental evidence nor quantitative information. We think there is an increase in encounter rate between reactant molecules and the inner wall of the nanoporous electrode. For reactant molecules undergoing a sluggish faradaic process and thus requiring more opportunities to obtain or release electrons, we can expect electrocatalytic activity, which does not come from crystalline facets or defects on the electrode surfaces; the nano-confinement effect.

1 D. B. Robinson *et al.*, *J. Electrochem. Soc.*, 2010, **157**, A912.
2 D.-e. Jiang *et al.*, *J. Phys. Chem. Lett.*, 2012, **3**, 1727.

Dr Kanoufi commented: Recent studies[1,2] have shown that Pt micro- and nanoelectrodes can be significantly etched by HO˙ radicals generated during O_2 reduction. Such etching can be as effective as several tens of nm. In nanoconfined environments such as those you are observing such Pt dissolution would have a dramatic effect and would result in the loss of nanoconfinement. Have you observed such an effect? You likely have at hand a system allowing to evidence the confinement (or its loosening) from transient interrogation of the O_2 reduction process, or to rebute the Pt dissolution process.

1 J. M. Noel *et al.*, *J. Am. Chem. Soc.*, 2012, **134**, 2835.
2 J. M. Noel *et al.*, *Langmuir*, 2013, **29**, 1346.

Professor Chung responded: Important point... We carefully checked out the papers listed above. First of all, we agree with your argument that ˙OH radical generated during ORR may dissolve the Pt electrode. To see how much the dissolution effect can modify the electrochemical behavior and electrode geometry, we used a nanoporous Pt deposited on a Pt disk electrode (diameter = 1 mm) and performed an ORR experiment under the same conditions (100 1-second-long potential pulses from 0 to −1 V *vs.* Ag/AgCl in an aqueous solution of 0.1 M KNO_3) as the previous work.[1] If Pt dissolved, we should have seen a remarkable decrease of the real surface area of the nanoporous electrode as well as a confinement effect. The result is that we saw negligible change in the area; the measured roughness factors (real surface area/apparent geometric area) were 23.36 and 22.74 before and after ORR, respectively. A significant shift in neither onset potential nor current density was found, indicating that ORR electrocatalytic activity owing to a nano-confinement effect was not appreciable. At least, much fewer ˙OH radicals seemed to be created, or its activity etching the Pt surface was apparently suppressed. Although it is a preliminary test, it is interesting enough to investigate in more detail concerning this issue.

1 J.-M. Noël et al., Langmuir, 2013, **29**, 1346.

Dr Plowman addressed Professor Chung: In your paper you discussed the influence of nano-confinement on electrocatalysis, but could you please comment on the crystallographic orientation of the nanostructured materials compared with the flat surfaces? This may also influence the electrocatalytic activity of the nanostructured materials, particularly in relation to the oxygen reduction reaction.

Professor Chung replied: We checked the crystalline facets of flat and nano-porous Pt surfaces that were exposed to the solution and participated in the electrochemical process by obtaining cycling voltammograms in 1 M H_2SO_4. The adsorptive peaks are due to the redox process of the hydrogen atoms that are adsorbed onto the specific crystalline facets. This is easy but reproducible and reliable evidence to see the crystalline facet distribution on the Pt electrode surface of an electrochemical system.[1] The shape and height of the characteristic peaks revealed that there was no difference between the flat and nanoporous Pt electrode surfaces in terms of crystalline facets.

1 J. Solla-Gullón et al., Phys. Chem. Chem. Phys., 2008, **10**, 1359.

Professor Barbero continued the discussion of the paper by Professor Marken by commenting: You show that it is possible to specifically detect pH sensitive redox systems (such as quinone/hydroquinone) by modulation of the pH inside a gap between two electrodes using proton reduction. We have suggested in our contribution that local pH changes occur inside the pores during double layer charging in unbuffered solutions. Additionally, it has been shown in the contribution from Miss Wahl that single electrodes behave as a generator/collector arrangement during square wave voltammetry experiments. Do you think it is possible to chemically filter pH sensitive redox couples using single porous electrodes and square wave voltammetry?

Professor Marken responded: The idea of employing single electrodes is very interesting (see also ref. 1) and I am sure there are ways of developing a new filter strategy, for example based on the suggested porous structure (e.g. with an additional pH controlling potential step). As a general comment, it can be stated that methods at "single electrodes" will result in analytical currents which are transient in nature whereas methods at the "trench electrode" could also be used to give analytical currents steady state in nature (for electrode geometries with high aspect ratio). Both approaches could have beneficial consequences in particular sensing applications.

1 M. C. Henstridge et al., Langmuir, 2010, **26**, 1340.

Professor Xu continued the discussion of the paper by Professor Chung by asking: The electrochemistry of some compounds may change upon binding with cyclodextrin; did you observe any change in electrochemistry of some compounds upon confinement in nanoporous electrodes?

Professor Chung replied: We tried 1,4-benzoquinone, oxygen and many alcohols including sucrose to see if apparent kinetic enhancement appeared. As a consequence, we found that most of them benefitted from the nanoporous structure, not crystalline facets, in terms of the redox process. From the comparison studies hitherto, our current conclusion is that the nano-confinement effect is more pronounced for sluggish reactions of non-adsorptive reactant molecules following a simple mechanism (refer to my previous reply to the initial question from Professor Compton).

Professor Compton queried: In an answer to an earlier question you mentioned that you had observed up to 20 electrons per molecule being transferred in the voltammetry seen at your nanoporous layers. Under what conditions was this seen and how did you calculate the number of electrons transferred?

Professor Chung responded: We carried out bulk electrolysis with a multiple potential step method referring to Parpot's paper.[1] Multiple potential steps were required to prevent poisoning of electrode surface for steady electrolysis. The potential scheme was −0.1 V vs. NHE for 100 s, 0.75 V vs. NHE for 2 s (electrode reactivation) and −0.6 V vs. NHE around the hydrogen adsorption region for 10 s (refreshing the Pt surfaces). The three step scheme was repetitively applied for 3 hours. Subtracting background charges in 100 mM KOH without sucrose, we calculated the total charge that flew due to sucrose oxidation. The amount of consumed sucrose was quantified by liquid chromatography. At present, the number of electrons transferred from single sucrose (n) was calculated to be 20. But it was from only one measurement and there might have been errors or mistakes in the experiments. So we are going to undertake experimental confirmation and hope we will be able to report experimentally reproducible and thus reliable results in the near future.

1 P. Parpot *et al.*, *J. Appl. Electrochem.*, 1997, **27**, 25.

Professor Barbero continued the discussion of the paper by Professor Bund by commenting: Besides nanoelectrodes built by holding nanowires inside a dielectric shroud, arrays of nanoelectrodes can be easily built by covering a flat electrode with a very thin dielectric film (down to few nanometers) which have nanometric sized holes. This can be done by several methods including nanosphere lithography, liquid crystal templating and hole formation on a previously deposited film (*e.g.* by electron beam lithography). In deep pore arrays, the relationship between analyte charge and surface charge of pore walls has been previously used to "chemically filter" charged analytes by tuning the charge of the pore walls using pH (*e.g.* Bartlett *et al.*, *Anal. Chem.*, 2001, **73**, 2855). Using shallow pores, do you think the model you propose for shrouded electrodes with flat geometry could be used to simulate such a system?

Professor Bund responded: The systems that you describe are very interesting. Our model could be applied to understand their current–voltage behaviour in more detail.

Dr Tschulik asked: Commonly used mass transport and hydrodynamic equations are based on continuum approaches. In your paper you use the Nernst–Planck approximation and grid sizes as low as 0.01 nm. Down to which characteristic length scale can continuum approach based equations be usefully applied, especially in the light of the emerging field of micro- and nanofluidics requiring smaller and smaller length scales to be modelled?

Professor Bund answered: I think continuum models like ours work well down to feature sizes of 10 nm. For smaller systems one has to resort to molecular dynamics.

Mr Kubeil commented: Corry et al.[1,2] showed that continuum theory is in good agreement with Brownian dynamics when the pore radius is larger than 2 Debye lengths. Thus, PNP-models may be applied to features (pores, electrodes) down to 2–3 nm.

1 B. Corry et al., *Chem. Phys. Lett.*, 2000, **320**, 35.
2 B. Corry et al., *Biophys. J.*, 2000, **78**, 2364.

Professor Mount continued the discussion of the paper by Professor Chung by communicating: Small amplitude ac impedance produces much smaller changes in redox concentration and therefore should not suffer as much as LSV and CV from gross effects of large changes in ion and water concentration. Have you measured the ac impedance response as a function of dc potential? If you have, what variation have you seen and how do you rationalise this?

Professor Chung responded: The observed kinetic enhancement at nanoporous electrodes did not originate from concentration differences of solvent or ions. In this study, electrochemical impedance was recorded to measure the charge transfer resistance, R_{ct}, when a redox species was present in the solution. Varying the dc potential led to a similar propensity of electrocatalytic activity enhancement compared to that at the flat electrodes.

Dr Bohn continued the discussion of the paper by Professor Marken by asking: Have you done experiments in which you have coupled your nanoscale trenches to nanoscale delivery geometries that might permit you to deplete electroactive species in an efficient way and improve the sensitivity of redox cycling in the trench?

Professor Marken replied: The idea of coupling well-defined delivery systems to the trench electrode is very interesting and there could be important applications once fabrication challenges are overcome. Important and pioneering work of this type is currently progressed by the Lemay group at Twente.[1,2]

1 L. Rassaei et al., *J. Phys. Chem. C*, 2012, **116**, 10913.
2 S. G. Lemay et al., *Acc. Chem. Res.*, 2013, **46**, 369.

Mr Partington continued the discussion of the paper by Professor Chung by commenting: In your paper, you describe that there is inhibition of the faradaic oxidation of 1-butanol as a result of the Au surface itself beginning to oxidise at potentials more positive than +0.4 V. Have you carried out any characterisation studies to determine the extent of the surface self-oxidation and how this affects the electrocatalytic activity of the nanoporous system described?

Professor Chung responded: The linear sweep voltammogram in Fig. 7 in our manuscript shows a typical wave due to alcohol oxidation in alkaline solutions. At less positive potential, OH^- anions adsorb onto the Au surface and trigger heterogeneous alcohol oxidation. As more positive potential is applied, OH^- anions adsorbed onto the Au surface react with Au surface atoms, yielding a partially oxidized Au surface and thus increased anodic current from alcohol oxidation. However, at some degree of positive potential, Au oxide film is generated, which inhibits alcohol oxidation and diminishes anodic current. In our experiments, 1-butanol oxidation was suppressed at the potential where the oxide monolayer on Au started to form in 1 M KOH (*ca.* 0.3 V *vs.* NHE). This can be found in many previous publications. For further details about alcohol oxidation on Au, please refer to Z. Borkowska *et al.*, *Electrochim. Acta*, 2004, **49**, 1209.

Dr Wain continued the discussion of the paper by Professor Marken by enquiring: Is there a chromium adhesion layer, or similar, between the gold and the glass of the micro-trench? If so, this would be exposed at the top of the trench – do you observe any evidence of this in the electrochemistry?

Professor Marken answered: This is a good point. These electrodes have a 10 nm titanium adhesion layer and this will certainly be present as a thin layer below the gold. Effects from the titanium on the observed voltammetric responses should be very small.

Dr Wain asked: In the pH modulation experiments the electrode is stepped to quite extreme potentials in order to generate hydroxide ions. How stable are the electrodes to this treatment and do you see delamination due to hydrogen evolution?

Professor Marken responded: These gold electrodes have been stable when applying negative potentials, however, the films are sensitive to degradation in the positive potential range. The work reported in this study was performed with a single electrode and there was no sign of aging or degradation under these conditions. Once the experimental work was completed, the electrode was metallised and submitted to electron microscopy.

Professor Mount continued the discussion of the paper by Professor Bund by asking: How realistic is the assumption of constant surface charge over the dielectric (particularly near the electrode edges)? Might a reduction of this charge density due to electrostatic attraction/repulsion and/or local field effects affecting surface protonation/adsorption equilibria act to mitigate the predicted field effects?

Professor Bund replied: In a real system surface charges are discrete and localized at particular sites on the surface. To capture such effects – which should become important for systems with nanosized dimensions – a more sophisticated model is needed. But indeed the effects you address might contribute to the electric double layer effects – either mitigating or amplifying. We will try to refine our model and check how large such effects are. A good starting point might be the paper of Calander (*Anal. Chem.*, 2009, **81**, 8347).

Professor Amatore continued the discussion of the paper by Professor Chung by commenting: I am surprised that you chose such a very complex system for testing your device, and even more that you selected systems whose electroactivation mechanism is not even known with certainty. In my view to test any idea or device, one takes the most simple system already perfectly characterized so as to validate the concept. Hence, at best your model (which is by the way very interesting) has to remain extremely speculative. Could you comment?

Professor Chung responded: Absolutely, I agree with you concerning the need for a simple model to prove the concept first of all. Theoretically, the nano-confinement effect is expected to be pronounced for sluggish reactions of non-adsorptive reactant molecules following a simple mechanism (refer to my reply to the initial question from Professor Compton). We scrutinized the non-adsorptive and redox active species that undergo a sluggish reaction involving a simple mechanism, only single electron transfer if possible. Our efforts to find an appropriate redox species meeting these conditions have not yet been very successful. To date, we have tried 1,4-benzoquinone, oxygen and many alcohols, as well as sucrose, which are subject to complex mechanisms, but have sufficient significance to conduct comparison studies in terms of electrocatalytic activity because of immense literature to refer to. Seeking better species is ongoing, and any suggestions would be welcome.

Professor Amatore opened the discussion of the paper by Miss Wahl by commenting: I appreciate this work and the efforts it represents but the problem of interactions between micro–nanobands in an array has been solved many years ago. In fact this is published in one of my first papers.[1] Indeed, this paper solves the problem of communication between active electrodes in a regular array of disks and bands. The part about disks in this paper is extremely well used and highly cited but this may have overshadowed the part relative to arrays of micro–nanobands.

1 C. Amatore *et al.*, *J. Electroanal. Chem.*, 1983, **147**, 39.

Miss Wahl replied: Many thanks for this comment; we have now referenced your work in our manuscript (see ref. 16). Effectively, your work enabled us to have a better approach to solve and understand the results we obtained for nanowire electrode arrays of varying inter-electrode spacings.

Professor Amatore said: It is not crystal clear to me how you may deduce the rate of transfer from your data and modeling. Could you provide additional precisions?

Miss Wahl answered: We would like to thank you for this question as we agree that the description in the original manuscript was terse. In brief, the oxidative experimental voltammogram at 5 mV s^{-1} recorded at three nanowires in array separated by 5 and 15 μm were plotted in Origin Pro 8.5 (OriginLab Corporation, USA). The oxidative current may be described by the Butler–Volmer model of electrode kinetics (eqn 7 in our paper), and therefore we incorporated the corresponding equation in Origin Pro 8.5 for fitting purposes. From the best generated fits, we deduced average heterogeneous rate constant using the equation $k^0 = i_{mt}(AFCk_m)^{-1}$. We have now amended the manuscript accordingly.

Professor Amatore remarked: I may have not understood correctly but I do not understand why your currents intensities in Fig. 4 in your manuscript are in this order, unless the electrodes are of different sizes or the overall surface area of the array was changed. For an array of electrodes of identical sizes the diffusional overlap at any time should decrease with the width of the inter-electrode gap.[1] If the array has a constant overall surface area, at maximum overlap (infinite experimental time duration) the current for any gap size should be the same since the array will perform as if its whole surface was active. Conversely at shorter times, the overlap is minimized, so the currents tend to reflect the total surface area of electrodes which is larger for smaller gaps. Hence the value for a 5 μm gap should always be larger than for a 15 μm one. Conversely, if the number of electrodes is kept constant (*i.e.*, the overall surface area increases upon increasing the gap size), the above phenomenon is reversed. At short times the current intensities for the two arrays are equal (same number of electrodes of the same size), and tends to be larger for the larger gap at complete overlap between diffusion layers (*i.e.*, at sufficiently long times) because then the current depends on the overall array surface area which is larger for larger gaps.

Hence, unless I misunderstood, your results could not be discussed as you did. According to the situation (constant overall surface area of the array *vs.* constant electrode number in the array) the current magnitude order will be opposed. This is an important caveat in generalizing results such as those you presented.

1 C. Amatore *et al.*, *J. Electroanal. Chem.*, 1983, **147**, 39.

Miss Wahl responded: The cyclic voltammogramms (CVs) in Fig. 4 of our paper were recorded at arrays of three nanowire electrodes of the same dimensions (100 nm wide × 50 nm high × 45 μm long) but with different inter-electrode distance (5 μm *vs.* 15 μm), *i.e.*, the overall footprint of the arrays increases with increasing inter-electrode gap, while the overall nanowire surface area remains the same. The CVs were highly reproducible ($n = 15$) and typical data is shown in Fig. 4. We are thus very confident in the experimental results shown, and have now highlighted that point in our manuscript. However, in order to avoid any confusion or misunderstanding, we should have emphasised the fact that the number of nanowires is the same in each array but the inter-electrode distance varies, *i.e.*, the situation is variation of the overall surface area of the array *versus* constant electrode number in the array. This corresponds to the second case you are describing in your question.

Professor Amatore asked: What is the origin of the very large capacitive currents?

Miss Wahl replied: Analysis at single nanowires, nanowire arrays (all nanowires were 100 nm wide × 50 nm high × 45 μm long) and in the absence of nanowires was undertaken to quantify the presence of large charging capacitive currents. In this regard, we determined capacitive contributions experimentally using cyclic voltammetry in 10 mM PBS buffer only, at a range of scan rates (5–5000 mV s^{-1}). In all cases, it was observed that the charging current increased linearly with scan rate. Furthermore, it was found that no overall increase in charging current was observed with increasing number of nanowires, implying that the charging contributions may not be exclusively due to the nanowires. Effectively, we showed that the passivation layer (~500 nm SiN$_x$) permitted interfacial capacitance above the insulated interconnection tracks. Consequently, a double layer would form at the chip surface directly above the interconnection tracks which contribute to the experimentally measured charging current. Since the calculated area of the interconnection metallisation present in the sample reservoir is ~1.9×10^{-2} cm^2, the overall contribution would be significant and may explain why the measured capacitance at nanowire arrays is not scaling with increasing electrode area. Although overall the presence of this extra capacitance at the nanowire arrays may not be problematic, it may be overcome in the future by fabricating devices with thicker insulating dielectric layers.

Dr Batchelor-McAuley opened the discussion of the paper by Professor Hillman by commenting: I note that the samples you have used so far are relatively small and have flat surfaces. Many real life objects are large and irregularly shaped. Do you believe that controlling the electrodeposition of the polymer, in terms of thickness and quality, will be more challenging on these larger and unevenly shaped objects?

Professor Hillman responded: Our approach has been stepwise so, as you note, we begin with surfaces that are flat and relatively small. However, for the neutron reflectivity experiments it is worth noting that although the surface must be flat (a feature of a reflectivity measurement), the probed surface is not so small, typically *ca.* 30 cm^2. In general, the extension to large and irregularly shaped objects is a familiar one in the metal finishing industry. For the polymer deposition stage of the process, we have followed common practice, for example surrounding a knife (operating as the working electrode in the electrochemical cell) by a large cylindrical counter electrode (typically a gauze, so the operator can view the time course of polymer deposition). In the subsequent stages of hydrolysis and fluorophore binding, the processes are purely chemical, so the issues of potential and current distribution are not relevant. From this perspective, the fluorescent extension we propose here introduces no new challenges.

Dr Batchelor-McAuley enquired: Given that you have raised the point about knives, my guess is that the most important fingerprints on a knife are on the handle and that these handles are commonly plastic or rubber coated. Is your technique really fully applicable for use in cases of knife crime?

Professor Hillman answered: It would be bold to claim that any fingerprint visualization technique is "fully applicable"; the low success rate for all methods combined attests to this fact. Therefore, we make no such claim. However, the situation is not quite as bleak as you might imagine, for two reasons. First, particularly in the case of a struggle between perpetrator and victim, fingerprints from both individuals are commonly found in multiple locations on an object, *i.e.*, the perpetrator may touch both handle and blade. Second, many knives are all metal; this is particularly the case for those used in a domestic setting (*e.g.* kitchen knives), where a weapon of opportunity may be involved. We have developed fingerprints from such objects, readily purchased at many retail outlets and widely used.

Dr Batchelor-McAuley asked: As a final question, how sensitive is your technique to the cleanliness of the original surface?

Professor Hillman replied: There are two aspects to this question. In general terms, surface contaminants may function in much the same way as the fingerprint deposit, *e.g.* grease may also act as a mask though with no spatial selectivity. More specifically, our experience for surfaces that were not deliberately cleaned prior to fingermark deposition has been that this effect tends to be more pronounced in the region around the fingerprint. In such a situation there is little problem, but we recognize that other situations may arise in the uncontrolled circumstances of a crime scene.

Mr Wright enquired: Does the smearing of fingerprints affect the results you are getting?

Professor Hillman replied: At this stage of our work, we have focused on the *vertical* transport of species (FMOC leaving group and entering fluorophore), so the direct answer to your question is "no". However, as we move forward to apply these principles to surfaces decorated with fingermarks, then *lateral* distribution of material becomes important also. Under these circumstances, the answer to your question will become "yes". However, this is not a feature of the present methodology: it is quite general to any technique that is used, since they are designed to give a faithful reproduction of the surface distribution of material. Your question is very relevant to practical application, since the viscoelastic nature of skin means that movement of the finger during gripping of an object may generate some smearing; this is an area of current interest in the field.

Mr Wright asked: Would half a fingerprint still allow accurate results to be determined?

Professor Hillman answered: In reality, many fingerprints are "partial", *i.e.*, the complete image is not present, so the short answer to your question is "yes, in favourable circumstances". The more detailed answer requires recognition of how the identification is made; ref. 1 and 2 of our paper provide considerable detail, but the abridged version is as follows. Fingerprint analysis is based on the identification of second level detail features (*minutiae*) – such as ridge endings, crossovers, bifurcations, spurs, *etc.* – and the spatial relationship between them.

So long as sufficient of these can be found (see main text for a discussion of the jurisdiction-dependent criteria here), it is largely immaterial whether they are found within a complete fingermark or a partial one.

Dr Kanoufi addressed Professor Mount and Miss Wahl: Both of you used (pseudo)steady-state amperometric measurements for the estimate of the FcCOOH diffusion coefficient. It should be more convenient to analyse transient amperometric responses (long and short time). Could the 2-fold difference in both D values be due to the real electroactive surface area of your electrodes (electrode poisoning)?

Professor Mount answered: No, in our case, as steady-state amperometric responses are not the only method we have used to determine D. Steady-state amperometric measurements depend quantitatively on D (as do our independent R_{nl} measurements), whereas the Warburg impedances depend quantitatively on the square root of D. We are convinced from the combination of these different measurements and their different D dependencies that an accurate and consistent measurement of D has been determined, independent of area and concentration. We agree that similar information can be extracted from steady-state and transient amperometric measurements. The observed variation is more likely to reflect a difference in the temperature for the two sets of experiments, as our measurement was made at a relatively low temperature (16 °C).

Miss Wahl responded: In this work the diffusion coefficient for the 1 mM FcCOOH in 10 mM PBS, pH 7.4, N_2 sat, was determined using the classical approach of successive cycles of the redox couple at a pristine commercial disk electrode of diameter 3 mm, across a broad range of operational scan rates, (5–5000 mV s^{-1}), in triplicate. The magnitude of the resultant characteristic oxidation and reduction peak currents were found to have a linear correlation with the square root of the scan rate in accordance with the Randles–Sevcik equation:[1] $I_p = 0.4463 \ (n^{3/2} \ F^{3/2})/(RT)^{1/2} \ A D_O^{1/2} \ C^* \ v^{1/2}$. From this an average diffusion coefficient, D_O, was obtained as 5.4×10^{-6} cm^2 s^{-1} for $T = 293.15$ K, $C^* = 1.03 \times 10^{-5}$ M and $A = 3.14$ mm^2. This value agrees closely with the D_O value reported by Arrigan et al.[2] which was determined using almost identical experimental procedures and conditions, with reagents from the same suppliers. Unfortunately, we cannot explain why results obtained by ourselves and Mount et al. differ, having not seen their work. However we can suggest some experimental conditions that may vary such as the following:

1. Difference in temperature. Our experiments were not thermostatted; however the temperature of the lab environment is routinely monitored. At the time we undertook the measurements, the ambient temperature was around 20 °C ± 2 °C.
2. Difference in the background electrolyte; our 10 mM PBS solutions are prepared from PBS tablets from Sigma Aldrich (as were those of Arrigan et al.[2]) rather than prepared from scratch.
3. Difference in the pH, i.e., our experiments were conducted at pH 7.4.
4. Presence of oxygen; all of our solutions were deaerated by bubbling with N_2 for up to 40 minutes prior to usage.

Concerning surface area contamination or poisoning, we can rule this out as sulphuric acid cycles performed between the triplicate runs have indicated no reduction in the electroactive surface area of the electrodes. The employment of amperometric transient responses to better estimate the diffusion coefficient was also suggested by Professor Compton. In this regard, we recorded double potential step chronoamperograms at a 12.5 μm diameter gold micro disk electrode immersed in a 1 mM FcCOOH, 10 mM PBS solution using the following parameters: $E_1 = 0$ V, $E_2 = 0.6$ V and $E_3 = 0$ V (all vs. sat Ag/AgCl, $t_1 = t_2 = 10$ s). These potentials were chosen so that the analyte is oxidised at a diffusion controlled rate between E_1 and E_2, and the analyte redox product is reduced back to the starting analyte material at a diffusion controlled rate between E_2 and E_3. The experimental current $I(t)$ was normalised with respect to the experimental steady-state current I_{ss}; the ratio of $I(t)/I_{ss}$ plotted against $1/\sqrt{t}$ was linear. From the slope of the linear relationship, we estimated the diffusion coefficient to be around 4.7×10^{-6} cm^2, which is in reasonable agreement with the value obtained previously. Note that the experimental I_{ss} obtained using chronoamperometry (1.31 ± 0.09 nA) was the same within experimental error as that obtained using cyclic voltammetry (1.36 ± 0.12 nA) at 5 mV s^{-1}.

1 A. J. Bard and L. R. Faulkner, *Electrochemical Methods, Fundamentals and Applications*, Wiley, 2001.
2 Y. H. Lanyon et al., *Anal. Chem.*, 2007, **79**, 3048.

Professor Schuhmann continued the discussion of the paper by Professor Hillman by commenting: Fingerprints are usually found on the surface of a large variety of materials; *e.g.* steel, stainless steel, alloys, copper *etc*. Are you not afraid that the polymer formation may fail if the material is passivated at the necessary high electrode potentials for the electrochemically induced polymer formation process?

Professor Hillman responded: Naturally, this is always a concern for non-noble metal surfaces. In practice, the deposition conditions are separately optimized for each metal. Simplistically, ignoring for the moment the well-known nucleation issues and the mechanistic complexities, the rate of growth can be considered as the product of a rate constant and a monomer concentration. Consequently, the selected rate can be achieved by a range of combinations of rate constants and concentrations. For those surfaces where high anodic potential is a problem, this means that the rate constant will necessarily be lower, so a higher monomer concentration should be used.

Professor Schuhmann enquired: How do you deposit your polymer film? Constant potential, constant current or potentiostatic pulse methods? We have previously used potentiostatic pulse methods for the very uniform deposition of polypyrrole and especially copolymerization of pyrrole and N- or 3-substituted pyrrole derivatives.[1] Did you consider the use of potentiostatic pulse based polymer deposition?

1 K. Habermüller, A. Ramanavicius, V. Laurinavicius and W. Schuhmann, *Electroanalysis*, 2000, **12**, 1383.

Professor Hillman responded: The direct answers to your questions are given in the Experimental section of our paper: PEDOT and PPy films were deposited potentiostatically and PPyFMOC films potentiodynamically. The more general answer to your question is that the optimum electrochemical control function will depend on the metal surface in question. In practical terms, we envisage two requirements. First, the chosen control function must generate films whose thickness is uniform across the surface on the scale of the fingerprint deposits. Second, whichever method is selected, the control parameters must be such that deposition is not too rapid, since it is likely that the deposition time (whether time directly or number of potential cycles) will be determined empirically, *i.e.*, the operator will view the deposition process on the exhibit until the required contrast is developed.

Professor Compton asked: Concerning the possible practical use of your method for real world imaging of fingerprints, it would be necessary to be sure of securing a usable image at the first attempt otherwise the "evidence" would be irreversibly altered. To what extent is a general protocol capable of application to all metal surfaces and all fingerprints likely to emerge from your work?

Professor Hillman replied: The rate and nature of the polymer deposition process varies significantly with the chemical nature of the surface; this is widely documented in the literature. Consequently, there is no prospect of a universal deposition protocol for all forensic exhibits. Instead, one would have a range of individually optimised protocols for different metal surfaces; we have explored this to some extent in other work. At first sight, there is the requirement for "first time" success. However, for a wide range of surfaces, there are well established sequences of treatments in which the first option is selected such that it does not preclude success of secondary options further along the sequential treatment chain. Our approach of using the bare metal as the target deposition area is beneficial in that the polymer is not interacting with the same region of the surface as conventional reagents. Consequently, there is the opportunity to use, for example, polymer deposition on the bare metal parts of the surface and cyanoacrylate ("superglue") treatment on the fingerprint deposits in a sequential manner. As part of the translation to operational procedures, we have explored this issue for the electrochromic technology with some success. The end result is that one is not restricted to a "single shot" situation.

Dr O'Rorke continued the discussion of the paper by Miss Wahl by stating: I disagree with Professor Amatore's point that the voltammograms in Fig. 4 of your manuscript are labelled the wrong way around, as a larger current would be expected for a nanowire separation of 5 microns than of 15 microns, for the following reason. If the nanowire separation and scan rate are sufficient for overlapping diffusion layers of neighbouring electrodes, a maximum signal will be achieved, owing to an increase in the effective surface area of the electrode. If, however, the nanowire separation is reduced further, there will be a corresponding reduction in measured current owing to a reduction in the effective surface area of the electrode.

The FE results in Fig. 2 indicate overlapping diffusion layers at a separation of 10 microns. Therefore, the current will be expected to decrease as the nanowire separation is reduced below 10 microns. As such, it is reasonable that the measured current for a nanowire separation of 15 microns would be larger than that for a nanowire separation of 5 microns, as in Fig. 4, contrary to Professor Amatore's earlier comment.

Professor Amatore remarked: I believe that the process used by Dr O'Rorke is not correct. I did not make any remark about the paper, in which the arguments were correctly presented, but only about those and the answer he gave during the oral presentation. His comment about mine is perfectly correct but this was not what he said orally, which prompted my comment.

A system having, for example, two parallel microbands of width w and separated by d will behave approximately as a microband of width $(2w+d)$, hence will produce a larger current when d increases at constant w. However, to obtain that one needs to have a full overlap of the diffusion layers, *i.e.*, on the one hand, d should be much smaller than the diffusion layer thickness generated by one microband alone, *viz.*, $\sqrt{(Dt)} \gg d$, where D is the diffusion coefficient and t the experiment duration. On the other hand, this overlap must occur over a range such as the concentration gradients remain comparable to the average gradient at one microband surface when performing alone. The two conditions are necessary (see J. E. Bartelt *et al.*, *Anal. Chem.*, 1988, **60**, 2167, for a first theoretical and experimental demonstration).

Professor Compton addressed Professor Mount and Miss Wahl: Concerning the issue of the diffusional independence of arrays of nano- or micro-electrodes can I please first comment that the easiest way of predicting this is to use the Einstein equation relating the distance, d, diffused in a time t to the diffusion coefficient D: d approximately equals square root of the product of D and t. In the case of data shown in Fig. 4 in Miss Wahl's manuscript this gives a value of the order of 10 microns suggesting that the conditions are close to the transition between diffusional independence and diffusion overlap (cases 2 and 4 – see T. J. Davies and R. G. Compton, *J. Electroanal. Chem.*, 2005, **585**, 138). Note that the condition is in terms of an absolute diffusion distance NOT a ratio of electrode separation to electrode radius.

Turning to the issue of the significantly different diffusion coefficients for the same molecule between your experiments and those of Professor Mount, can I please ask both groups if the experiments were thermostatted? Note that diffusion coefficients are strongly temperature sensitive, often showing Arrhenius behaviour to a good approximation (see S. R. Jacob *et al.*, *J. Phys. Chem. B*, 1999, **103**, 2963).

Professor Mount answered: We agree with the comments on the relationship of D, t and the degree of array overlap. This is entirely consistent with our analysis and our observed MNEE array response. Our experiments were not thermostatted, as we did not have this capability within our Faraday cage measurement system. The measured temperature for these experiments was 16 °C; we agree this relatively low temperature is likely to explain the low value of D we obtained.

Miss Wahl replied: Many thanks for this comment; we have already referenced this work in our manuscript, indeed this work enabled us to better understand the results we obtained for nanoelectrode arrays of varying interelectrode spacings. Furthermore thank you for highlighting the Einstein equation, we had elected to use the equation for diffusion to a hemicylinder as we thought it best described the shape of the electrode, in future we will use the Einstein relationship. Based on the data achieved throughout this work, it is our belief that nanowire electrodes separated by 5 μm would correspond to case 3 where diffusion layers are overlapped, the so-called transition regime and nanowire electrode separated by 15 μm would correspond to case 2, where independent diffusion profiles would be achieved at each nanowire. Although our experiments were not thermostatted, the temperature of the lab environment is routinely monitored. At the time we undertook the measurements, the ambient temperature was around 20 °C ± 2 °C, with the solutions at room temperature. While we appreciate that a difference in temperature could be one reason to explain the significant difference between the two values, we don't however think that the temperature alone would explain the huge difference observed. Please see our response to Dr Kanoufi's question on this matter for further details.

Professor Compton addressed Professor Mount and Miss Wahl: The diffusion coefficients of both species (if chemically stable on the voltammetric timescale) in a redox couple can be easily, accurately and independently measured using double potential step chronoamperometry at a microdisc electrode (see O. V. Klymenko et al., *J. Electroanal. Chem.*, 2004, **571**, 211; C. A. Paddon et al., *Electroanalysis*, 2007, **19**, 11) so that there is no need to use these "fitting parameters"!

Professor Mount responded: We agree this is a convenient method for obtaining D values; we point out the equivalent information can also readily be obtained from ac impedance, from a measurement of the variation of the Warburg impedances and non-linear resistances as a function of E_{dc}.

Miss Wahl replied: Many thanks for this advice. We have now recorded double potential step chronoamperograms at a 12.5 μm diameter gold micro disk electrode immersed in a 1 mM FcCOOH, 10 mM PBS solution using the following parameters: $E_1 = 0$ V, $E_2 = 0.6$ V and $E_3 = 0$ V (all vs. sat Ag/AgCl , $t_1 = t_2 = 10$ s). These potentials were chosen so that the analyte is oxidised at a diffusion controlled rate between E_1 and E_2, and the analyte redox product is reduced back to the starting analyte material at a diffusion controlled rate between E_2 and E_3. The experimental current $I(t)$ was normalised with respect to the experimental steady-state current I_{ss}; the ratio of $I(t)/I_{ss}$ plotted against $1/vt$ was linear. From the slope of the linear relationship, we estimated the diffusion coefficient to be around 4.7×10^{-6} cm^2, which is in reasonable agreement with the value obtained previously. Note that the experimental I_{ss} obtained using chronoamperometry (1.31 ± 0.09 nA) was the same within experimental error as that obtained using cyclic voltammetry (1.36 ± 0.12 nA) at 5 mV s^{-1}.

Professor Barbero continued the discussion of the paper by Professor Hillman by asking: In the paper you discuss the use of polypyrrole to be able to functionalize it with fluorophores. Why not use polyaniline? We have shown that polyaniline thin (< 200 nm) films can be easily functionalized by diazonium ion coupling,[1] or nucleophile addition.[2] The latter method can even be controlled electrochemically. Additionally it is possible to use quantum dots to react with the added groups revealing the microtopography.[3]

1 D. F. Acevedo *et al.*, *Electrochim. Acta*, 2011, **56**, 3468, and references therein.
2 C. Barbero *et al.*, *Electrochim. Acta*, 2004, **49**, 3671.
3 D. F. Acevedo *et al.*, *J. Phys. Chem. B.*, 2009, **113**, 14661.

Professor Hillman replied: At this point, we have only had the opportunity to demonstrate the concept of the combined electrochromic/fluorescence approach to one combination of polymer (polypyrrole) and fluorophore. It is our intention to extend this to other polymers and fluorophores (provided suitable immobilization chemistry is available); polyaniline, a material with which we have considerable experience, presents an obvious opportunity. We had not previously considered the use of quantum dots, but this is an interesting idea and we thank Professor Barbero for his insightful contribution.

Professor Barbero commented: I would like to point out that it is possible to produce conductive polymers (polyaniline, polypyrrole, PEDOT, *etc.*) by electropolymerization on any common metal (even aluminum) choosing the conditions and specifically the counterions. Therefore, the method proposed in the paper can be very general, as far as the fingerprints are present on metals.

Professor Hillman responded: This is a very relevant comment. For the purposes of this fundamental study, we have deliberately confined our attention to a noble metal surface, thereby minimizing issues of surface (electro)chemistry associated with the substrate. However, for practical applications, a much wider range of metal surfaces is of interest. With this in mind, we have successfully applied the electrochromic technology (represented by the data of Fig. 1 and 2 of our paper) to platinum, gold, stainless steel, lead, brass and nickel surfaces; these are representative of objects that might be intrinsically valuable, implicated in crimes of violence or relevant to metal theft. We share Professor Barbero's optimism that the fluorophore-based extension of the methodology will be applicable to all these (and other) metal surfaces.

Professor Barbero continued the discussion of the paper by Miss Wahl by remarking: The results about electrodes with overlapping diffusional profiles with SWV are very interesting because several fabrication techniques based on interference (*e.g.* direct laser interference patterning) could produce periodic arrays of microelectrodes but spaced at a distance comparable with its size. However, this was usually considered a disadvantage because direct techniques (*e.g.* cyclic voltammetry) usually, as in your case, do not show enhanced diffusion except at fast scan rates. Questions are: i) do you think this is an effect of nanoelectrodes or does it also apply to micrometric sized electrodes? ii) It seems to me that the difference should not be restricted to CV *vs.* SWV but to any direct method (*e.g.* chronoamperometry) compared to reverse techniques (*e.g.* double pulse voltammetry). Is that the case? iii) reverse techniques have been extremely useful to detect and measure chemical reactions coupled to the electrochemical step. Do you think the effect you find could be used to evaluate coupled chemical reactions?

Miss Wahl and **Dr O'Riordan** answered: (i) Using SWV techniques, the effect should also apply to ultra-microelectrodes which are separated by gaps smaller than the diffusion thickness, such as those achieved by direct laser interference patterning. However, it would be expected that the effect would decrease with increasing electrode size as planar diffusion to the top surface of the electrode dominates over edge effects. Indeed work has been recently published by Laborda et al.[1] investigating the linear sweep *vs.* pulsed techniques at electrodes modified with porous layers. (ii) We agree that this effect would not be purely limited to CV *vs.* SWV but more broadly to any fixed potential/potential sweep methods *vs.* pulsed potential techniques. As developing novel nanoelectrode systems for the detection of key, target analytes is the main goal of our research, cyclic voltammetry and square wave voltammetry are the key electroanalytical techniques that we have employed to date. (iii) As most of our work involves the characterisation of novel nanoelectrodes our electroanalytical studies have been limited to simple, single step reactions, such as ferrocenemonocarboxylic acid and other model redox couples. In this regard we can only hazard a guess but we feel it would be interesting to investigate this possibility in the future.

1 E. Laborda *et al.*, *Electrochim. Acta*, 2012, **73**, 3.

Professor Compton addressed Professor Barbero and Miss Wahl: The predictions of Professor Barbero concerning the relative sensitivity of nano-, micro- and macro-electrodes towards pulse or square wave voltammetry have been confirmed by simulation both for flat electrodes and for electrode modified with porous layers (E. Laborda *et al.*, *Electrochim. Acta*, 2012, **73**, 3).

Mr Wright continued the discussion of the paper by Professor Hillman by asking: Does the thickness of your polymer film affect the results you are getting?

Professor Hillman answered: There are two aspects to this issue, one fundamental and one practical. We have been guided by the practicalities of the end application. In particular, if one considers the basic strategy of depositing the coloured "reagent" (polymer, here polypyrrole) in between the fingerprint deposits, then there is a trade-off between analytical sensitivity and spatial selectivity. If the film is too thin, then the optical (visual, in a practical application) signal may be too small. If the film is too thick, then it will not only fill the "valleys" between the fingerprint deposits but will also spread over the tops of the "hills", resulting in an apparently uniform coverage across the surface, such that the image is lost. Consequently, the optimum polymer film thickness will be in the order of 50–75% of the height of the fingerprint deposit. There will be variations in the thickness of fingerprint deposits – according to the individual, the circumstances, the age of the fingerprint and the environment to which is has been exposed – but this is most likely to be of the order of a few hundred nanometres. We have therefore chosen to operate with films whose thickness is of this order. It is also convenient (and necessary) that most conducting polymer films of this thickness have a significant absorbance in the visible region.

From a fundamental perspective, there are three other film thickness considerations relevant to our study. The first is that the redox switching times for films of polypyrrole (and other conducting polymers) in this thickness regime are not too long, typically a few seconds, dependent on the electrolyte. Second – and for related reasons – transport times for leaving group exit and fluorophore entry are not too long. Finally, films in this thickness regime are amenable to study by neutron reflectivity with current instrumentation: the resultant fringe width is large enough to give good resolution and small enough to give more than one fringe within the accessible Q-range. We have not yet explored film thickness as a variable, but increasing it would give greater analytical signal (more optical absorbance and greater capacity for fluorophore uptake), but at the cost of longer reaction times (for hydrolysis and fluorophore binding) and possible over-filling of the "valleys" between the fingerprint deposits. We speculate that the latter would be the limiting factor, since it leads to decreasing lateral contrast of ridge detail, *i.e.*, jeopardizes image veracity.

Professor Fielden asked: Apart from the intended application to enhance barely visible fingerprints, could this technique of growing conducting polymer in the fingerprint gaps be used to enhance the performance of the techniques that seek to exploit the properties of the fingerprint itself? For example, could the enhanced edge definition afforded by this technique offer improvement in the spatial resolution of fluorescence based imaging techniques, through the reduction of localised scattering?

Professor Hillman responded: This is an interesting idea and one that we have not considered to date. Speculatively, one way this might work could involve the deposition of polymer subsequent to application of another fluorescence based method, such that the polymer covered fluorophores located outside their intended (fingermark deposit) region. Thus, emission by these incorrectly located fluorophores might be attenuated by the overlaying polymer, thereby effecting some element of damage limitation.

Faraday Discussions

PAPER

Closing remarks: looking back and ahead at 'nano' electroanalytical chemistry

David E. Williams*

Received 5th August 2013, Accepted 5th August 2013
DOI: 10.1039/c3fd00106g

One issue with summarising this meeting is that the term 'nanoscale' seems to be infinitely flexible. Certainly from this Discussion one can conclude that, at least in some dimension, pretty well all electrochemistry is 'nanoscale'. It has always been the aim of a Faraday Discussion to bring out the principal themes in contemporary thinking in an area, and to critique these rigorously. One role of the concluding remarks is to try to pull together the themes and try to fulfil the aim of bringing out the big picture. This is the third Faraday Discussion at which I have contributed the Concluding Remarks. I closed the 2000 conference on Bio-electrochemistry with the question "Bioelectrochemistry: beyond cartoons?"[1] and the 2002 conference on "The Dynamic Solid–Liquid Interface" with the thought "Towards 4th Generation Electrochemistry".[2] Perhaps it is worth reflecting a little on these comments before I embark on reviewing the present meeting.

In 2000, I commented that "it now seems relatively straightforward to use electrochemistry to examine rather complex systems of biological molecules and to probe biological mechanisms"; that "one could look forward to advances in controlled fabrication and manipulation of well-ordered and precisely characterised biolayers, including controlled and ordered multilayer structures"; that "methods which probe structure at the electrode interface will necessarily need to be applied" – these would include spectroelectrochemistry and photo-electrochemistry, and this would lead the subject "beyond cartoons". I also noted that the use of electrochemistry coupled to biological molecules to make useful sensors then seemed rather mature.

In 2002, I remarked that electrochemistry had passed through three distinct generations, each of which had been marked by a new set of experimental methods and hence a new general direction dictated by new capabilities. I foresaw a "4th generation electrochemistry" marked by widespread use of advanced structure-determination tools, perhaps more concentrated around large collaborations and central facilities, revisiting practical problems whose description by the earlier generation methods was incomplete. The present Faraday Discussion

MacDiarmid Institute for Advanced Materials and Nanotechnology, School of Chemical Sciences, University of Auckland, Private Bag 92019, Auckland 1123, New Zealand. E-mail: david.williams@auckland.ac.nz

has been the first explicitly devoted to electroanalytical chemistry. Indeed some of the advances previewed can be discerned.

Electrochemistry developed differently in the USA than in Europe. With of course notable exceptions, in Europe the emphasis was at first predominantly on electrode kinetics, structure and mechanism[3] whilst in the USA the emphasis was at first predominantly on the practical application of electrochemistry to problems of chemical analysis: electroanalysis.[4] Haber had been the first to note the pH sensitivity of the potential difference across a glass membrane[5] and some of the earliest practical applications of glass electrodes for pH measurement was at University College London,[6] but it was Beckman who developed the pH meter and made the glass electrode practical.[7] Similarly, Heyrovsky developed polarography but it was Laitinen and Kolthoff[8] who understood the use of a well-defined diffusion geometry at a solid electrode as a practical means for solution of electroanalytical problems, though they used the term 'microelectrode' to mean an electrode which did not perturb bulk solution concentration, rather than meaning a steady-state electrode.[9] The people who pushed and developed this field were in the USA: Delahay, Osteryoung and Reilley, followed by Royce Murray, Allen Bard, Mark Wrighton and Henry White. They recognised the importance of the work of Geoffrey Barker and built the panoply of methods that we use today.[9] Their influence can be seen strongly in the discussion that we have just had. A recent perspective by Bard and Murray highlights the interlinked importance of the development of instrumental techniques and theory.[10] The opening speaker, Joe Wang, is of course thoroughly imbued in this tradition. We saw wonderfully ingenious and intriguing ideas brought to fruition (DOI: 10.1039/C3FD00105A). Truly 'electroanalysis at the nanoscale'.

In the UK, electroanalysis was revitalised by W. John Albery from the 1980s and his influence remains very strong. The key principle is to set up the experiment so that the convection–diffusion boundary conditions are defined, so that a model can be rigorously calculated, the results compared with theory and discrepancies explored. That approach is exemplified in a number of the contributions to this Discussion.

In the present Discussion, two general themes can be discerned.

1. Push the limits of measurement, but be sure that the measurement is well-understood. One must go beyond the 'gee-whiz' stage. That, as we saw particularly in the papers from Amatore (DOI: 10.1039/C3FD00028A) and from Schuhmann (DOI: 10.1039/C3FD00011G), is hard. One part of the subject that is particularly notable is the maturing and development of the field of scanning electrochemical microscopies (SECM). As Schuhmann (DOI: 10.1039/C3FD00011G) and others[11] have shown, it is now possible to control the tip–substrate separation across a complex topography whilst simultaneously measuring redox electrochemistry at the tip. This is leading to new results on metabolism, membrane transport and storage in single living cells. Other new developments concern the use of scanning ion-conductance microscopy in conjunction with SECM, and electrochemical measurement and processing within a tiny droplet confined in the contact zone between a pipette tip and a substrate.[12,13] This has enabled exploration of electrochemistry in localised zones on small size scales, and to scan a sub-micrometer-scale electrochemical cell over tiny structures such as single nanotubes.[14,15]

2. Push the limits of fabrication, so if small is good then smaller is better: 'micro' leads inevitably to 'nano'.[16] The aim is to be able to make things reliably and at scale, to address the central problem of analytical chemistry, which is to secure repeatability. We saw particular examples in the fabrication of nm-scale electrodes.

In the 1980s and -90s, printing and laminating methods were developed for fabrication of electrodes and electrochemical cells. This has been very successful indeed. To make microelectrodes, laser ablation was applied to blow out micro-discs[17] and rings, and edge-cutting was used for microbands.[18,19] What we have seen in the papers from Mount (DOI: 10.1039/C3FD00038A) and O'Riordan (DOI: 10.1039/C3FD00025G, 10.1039/C3FD00027C) is the maturing of the use of vacuum deposition and fabrication techniques from the semiconductor industry, that originally appeared in use in electrochemistry in the late 1980s.[20] These methods give very good control over sealing and interfaces so nm-scale electrodes now seem readily available. As a consequence, we have seen the emergence of single-molecule electrochemistry, as exemplified in the paper by Bohn (DOI: 10.1039/C3FD00013C). The thought that one can measure the electrochemical kinetics of individual molecules, and directly observe the variation of electrochemistry between one molecule and another dependent on as-yet-unknown factors, is truly exciting. We saw that the electrochemistry of a single molecule can be very different from that of a large ensemble, and we heard speculations that the size of a system could alter the observed ensemble average. We also heard how the surround of an electrode could critically influence the electrochemistry when the electrode is very small, as a consequence of the interaction of electrical double layer and diffusion boundary layer. We also learned to beware of the impact of impurities in electrode materials when electrode size scales are reduced and 'defect' properties become dominant. Reducing the size-scale of electrochemical cells, using scanned pipette techniques to localise electrochemistry at particular features seems one way to resolve some of the questions that arise.[15] Another aspect of fabrication is structures specifically designed to enhance analytical performance – that is, structures that are designed to confine a system or shape an electrode in a specified way and therefore enable better measurement. We saw several examples. Here is where considerable ingenuity has come into play: templating, surface manipulation and 'designer' particle fabrication are all examples, as is the use of single nanopores[21,22] to measure the passage of and characterise nanoscale objects and discriminate between differently functionalised such objects for example to detect DNA damage.[23]

Where might all this be leading? I will make some speculation based on an example. The Ion-Selective Field-Effect Transistor (ISFET) was invented in the early 1980s and there was an explosion of work in this area. It seemed to go nowhere, largely because of problems of insulation and sealing around the active area, and work largely stopped. ISFET pH-chips are now available but have not in any way displaced glass electrodes. A recent breakthrough has transformed prospects for this device. When a DNA polymerase attaches a nucleotide to an extending chain of DNA, protons are released. The resultant small change in pH in a tiny reaction volume can easily be detected by an integrated ISFET.[24] The development of this idea is the "Ion Torrent" instrument, which is based on massively parallel measurement using disposable microfabricated chips implementing the principle of detecting pH change when a base is attached to an extending DNA chain.[25] Such an instrument will sequence a genome on a timescale of hours at very low cost – a real breakthrough technology that capitalises on advances in technology to bring together disparate ideas to solve an important problem.

What lesson might perhaps be drawn as a conclusion to this Discussion? We have seen how to make beautiful, tiny containers, how to manipulate surfaces, to manipulate particle fabrication and combine ideas with new electrode types.

Particles can be built with very sophisticated chemistry attached – a trend which can be termed 'particles as laboratories'. By combining the containers, the particle laboratories and the electrodes, one can envisage an era of ever more ingenious, effective and diverse analytical applications. The 'transcriptome' and the 'metabolome' await?

References

1 D. E. Williams, *Faraday Discuss.*, 2000, **116**, 353–353.
2 D. E. Williams, *Faraday Discuss.*, 2002, **121**, 463–465.
3 K. J. Vetter, *Electrochemical kinetics: theoretical and experimental aspects*, Academic Press, New York, 1967.
4 P. Delahay, *New instrumental methods in electrochemistry*, Interscience, New York, 1954.
5 F. Haber and Z. Klemensiewicz, *Zeitschrift Fur Physikalische Chemie-Stochiometrie Und Verwandtschaftslehre*, 1909, **67**, 385–431.
6 P. T. Kerridge, *Biochemical Journal*, 1925, **19**, 611–617.
7 http://en.wikipedia.org/wiki/Arnold_O._Beckman.
8 H. A. Laitinen and I. M. Kolthoff, *J. Phys. Chem.*, 1941, **45**, 1061–1079.
9 A. J. Bard and L. R. Faulkner, *Electrochemical Methods: Fundamentals and Applications*, 2nd edn, Wiley, New York, 2001.
10 A. J. Bard and R. W. Murray, *Proc. Natl. Acad. Sci. U. S. A.*, 2012, **109**, 11484–11486.
11 R. A. Lazenby, K. McKevey and P. R. Unwin, *Anal. Chem.*, 2013, **85**, 2937–2944.
12 K. McKelvey, M. A. O'Connell and P. R. Unwin, *Chem. Commun.*, 2013, **49**, 2986–2988.
13 C. Laslau, D. E. Williams and J. Travas-Sejdic, *Prog. Polym. Sci.*, 2012, **37**, 1177–1191.
14 A. G. Guell, N. Ebejer, M. E. Snowden, K. McKelvey, J. V. Macpherson and P. R. Unwin, *Proc. Natl. Acad. Sci. U. S. A.*, 2012, **109**, 11487–11492.
15 M. E. Snowden, M. A. Edwards, N. C. Rudd, J. V. Macpherson and P. R. Unwin, *Phys. Chem. Chem. Phys.*, 2013, **15**, 5030–5038.
16 R. W. Murray, *Chem. Rev.*, 2008, **108**, 2688–2720.
17 B. J. Seddon, Y. Shao, J. Fost and H. H. Girault, *Electrochim. Acta*, 1994, **39**, 783–791.
18 D. H. Craston, C. P. Jones, D. E. Williams and N. El Murr, *Talanta*, 1991, **38**, 17–26.
19 J. S. Rossier, M. A. Roberts, R. Ferrigno and H. H. Girault, *Anal. Chem.*, 1999, **71**, 4294–4299.
20 R. L. McCarley, M. G. Sullivan, S. Ching, Y. N. Zhang and R. W. Murray, in *Microelectrodes: Theory and Applications*, ed. M. I. Montenegro, M. A. Queiros and J. L. Daschbach, NATO ASI Series, Series E: Applied Sciences, Kluwer, Dordrecht, 1991, vol. 197, pp. 205–226.
21 D. Kozak, W. Anderson, R. Vogel, S. Chen, F. Antaw and M. Trau, *ACS Nano*, 2012, **6**, 6990–6997.
22 S. R. German, L. Luo, H. S. White and T. L. Mega, *J. Phys. Chem. C*, 2013, **117**, 703–711.
23 N. An, A. M. Fleming, H. S. White and C. J. Burrows, *Proc. Natl. Acad. Sci. U. S. A.*, 2012, **109**, 11504–11509.
24 C. Toumazou, L. M. Shepherd, S. C. Reed, G. I. Chen, A. Patel, D. M. Garner, C.-J. A. Wang, C.-P. Ou, K. Amin-Desai, P. Athanasiou, H. Bai, I. M. Q. Brizido, B. Caldwell, D. Coomber-Alford, P. Georgiou, K. S. Jordan, J. C. Joyce, M. La Mura, D. Morley, S. Sathyavruthan, S. Temelso, R. E. Thomas and L. Zhang, *Nat. Methods*, 2013, **10**, 641–646.
25 J. M. Rothberg, W. Hinz, T. M. Rearick, J. Schultz, W. Mileski, M. Davey, J. H. Leamon, K. Johnson, M. J. Milgrew, M. Edwards, J. Hoon, J. F. Simons, D. Marran, J. W. Myers, J. F. Davidson, A. Branting, J. R. Nobile, B. P. Puc, D. Light, T. A. Clark, M. Huber, J. T. Branciforte, I. B. Stoner, S. E. Cawley, M. Lyons, Y. Fu, N. Homer, M. Sedova, X. Miao, B. Reed, J. Sabina, E. Feierstein, M. Schorn, M. Alanjary, E. Dimalanta, D. Dressman, R. Kasinskas, T. Sokolsky, J. A. Fidanza, E. Namsaraev, K. J. McKernan, A. Williams, G. T. Roth and J. Bustillo, *Nature*, 2011, **475**, 348–352.

Poster titles

Finite element modelling: a new approach to understanding electrochemical impedance spectra from self-assembled monolayers, **Richard D. O'Rorke, Irene Zaccari, R. Sharma, Dominka Nowak, A. Giles Davies and Christoph Wälti**, *University of Leeds, UK*

Reduced graphene oxide reinforced with polypyrrole nanoparticles for biosensing application, **H. N. Lim and Y. S. Lim**, *Universiti Putra Malaysia, Malaysia*

Direct electrochemistry and catalysis of Hemoglobin immobilized on nanostructured polyaniline films coated/self assembled monolayer modified gold electrode, **Gülçin Bolat, Filiz Kuralay and Serdar Abaci**, *Hacettepe University, Turkey*

One-step electrodeposition synthesis of silver-nanoparticle-decorated graphene on indium-tin-oxide for enzymeless hydrogen peroxide detection, **Amir M. Golsheikh and Nay-Ming Huang**, *University of Malaya, Malaysia*

Electropolymerization of thiophene on gold nanoparticle modified electrode in aqueous media, **Ozge Surucu, Gulcin Bolat and Serdar Abaci**, *Hacettepe University, Turkey*

A chronopotentiometric approach for measuring chloride ion concentration using transition time measurement, **Yawar Abbas, D. B. de Graaf, Wouter Olthuis and Albert van den Berg**, *University of Twente, The Netherlands*

Voltammetric oxidation of ascorbic acid mediated by PEDOT/MgB_2-MWCNT modified glassy carbon electrode, **Yusran Sulaiman, Wee T. Tan and Darlene Banan**, *Universiti Putra Malaysia, Malaysia*

Nanofluidic electrocatalysis: Coupling electroosmosis and electrocatalysis under nano-confinement, **Nicholas M. Contento, Sean P. Branagan and Paul W. Bohn**, *University of Notre Dame, USA*

Electrocatalytic activity of gold and carbon nanoparticulate films toward direct electrooxidation of glucose and ascorbic acid, **Aleksandra Karczmarczyk and Marcin Opałło**, *Polish Academy of Sciences, Poland*

Electrocatalytic properties of suspended carbon nanoparticles in flow, **Dominika Ogończyk, Aleksandra Karczmarczyk and Marcin Opałło**, *Polish Academy of Sciences, Poland*

Preparation of SECM nanotips using heating coil puller, **Justyna Jedraszko, Wojciech Nogala, Martin Jönsson-Niedziolka and Marcin Opałło**, *Polish Academy of Sciences, Poland*

The effect of temperature and surfactants on soiled fabrics, **Mehrin Chowdhury, Sharon J. Cooper and Ritu Kataky, Richard Thompson, Paul Hodgkinson and John Girkin,** *Durham University, UK*

Salt-free electrolysis in thin layer cell, **Jingyuan Chen, Koichi J Aoki, Chunyan Li and Chaofu Zhang,** *University of Fukui, Japan*

Random arrays of vitamin K1 modified electrodes for pH sensing, **Monika Schoenleber, Jay Wadhawan, John MacFie, Robert Singh, Linda Shields, John Greenman and Barbara Elliott,** *University of Hull, UK*

Deposition of electrochemically addressable metals in Ta oxide nanotubes, **Corie A. Horwood and Viola I. Birss,** *University of Calgary, Canada*

Incorporation of a nanoelectrode in a microfluidic flow cell, **Nicola J. Kay, Reshma Sultana, Ilka Schmueser, Anthony J. Walton, Jonathan G. Terry, Neville J. Freeman and Andrew R. Mount,** *Nanoflex Limited, UK*

Characterization and electroanalytical applications of GC electrodes modified with Nafion® stabilized silver nanoparticles, **Valentina Pifferi, Valeria Marona, Mariangela Longhi and Luigi Falciola,** *Università degli Studi di Milano, Italy*

Study of poly(Brilliant Green) on the performance of different electrode architectures based on poly(3,4-ethylenedioxythiophene) and carbon nanotubes, **Valentina Pifferi, Madalina M. Barsan, M. Emilia Ghica, Luigi Falciola and Christopher M. A. Brett,** *Università degli Studi di Milano, Italy*

Novel bio-inspired sensory material for implantable tissue applications, **Nyasha Ntola, Rui Campos, Fiona Shenton, Susan Pyner and Ritu Kataky,** *Durham University, UK*

Spectroelectrochemical studies of lignosulphonate-modified graphene, **Mohammed N. Haque, Katherine B. Holt, Jingping Hu and John S. Foord,** *University College London, UK*

Size and pH dependent redox activity of undoped diamond nanoparticles, **Meetal Hirani and Katherine B. Holt,** *University College London, UK*

Development of a microfluidic device for assessment of tumour associated procoagulant activity, **Yuehua Dou, Kathryn Date, Etienne Joly, Ian Bell, Leigh Madden, Stephen Haswell and John Greenman,** *University of Hull, UK*

Inside–outside nanoparticle decoration of carbon nanofibers: A strategy to improve utility of Pt and nanofibers for fuel cell applications, **Baljit Singh and Eithne Dempsey,** *Institute of Technology Tallaght (ITT Dublin), Republic of Ireland*

Penetration of gold nanoparticles into bilayer and monolayer modified electrodes, **Anne Krol and Ritu Kataky,** *Durham University, UK*

List of participants

Mr Yawar Abbas, University of Twente, *The Netherlands*
Professor Christian Amatore, *UMR CNRS 8640, France*
Mr Je Hyun Bae, *Seoul National University, Korea*
Dr Craig Banks, *Manchester Metropolitan University, United Kingdom*
Professor Cesar Barbero, *Universidad Nacional de Río Cuarto, Argentina*
Dr Christopher Batchelor-McAuley, *Oxford University, United Kingdom*
Dr Daniel Belton, *University of Huddersfield, United Kingdom*
Mr Christopher Birch, *Hull York Medical School, United Kingdom*
Dr Paul W. Bohn, *University of Notre Dame, USA*
Miss Gülçin Bolat, *Hacettepe University, Turkey*
Professor Alan Bond, *Monash University, Australia*
Miss Rebecca Brodie, *Royal Society of Chemistry, United Kingdom*
Professor Andreas Bund, *Ilmenau University of Technology, Germany*
Mrs Paula Caldevilla, *DropSens, Spain*
Professor Aicheng Chen, *Lakehead University, Canada*
Dr Jingyuan Chen, *University of Fukui, Japan*
Miss Mehrin Chowdhury, *Durham University, United Kingdom*
Professor Taek Dong Chung, *Seoul National University, Korea*
Miss Nicola Coles, *Royal Society of Chemistry, United Kingdom*
Professor Richard Compton, *Oxford University, United Kingdom*
Mr Nicholas Contento, *University of Notre Dame, USA*
Dr Sara Dale, *University of Bath, United Kingdom*
Mr Martijn van Dijk, Metrohm *Autolab, The Netherlands*
Dr Yuehua Dou, *University of Hull, United Kingdom*
Dr Luigi Falciola, *Università degli Studi di Milano, Spain*
Dr Sarah Farley, *Royal Society of Chemistry, United Kingdom*
Professor Peter Fielden, *Lancaster University, United Kingdom*
Dr Neville Freeman, *NanoFlex Ltd, United Kingdom*
Mr Mohammed Haque, *University College London, United Kingdom*
Professor Rob Hillman, *University of Leicester, United Kingdom*
Miss Meetal Hirani, *University College London, United Kingdom*
Dr Manyi Ho, Schlumberger *Cambridge Research Centre, United Kingdom*
Dr Benjamin Horrocks, *Newcastle University, United Kingdom*
Ms Corie Horwood, *University of Calgary, Canada*
Dr Nay Ming Huang, *University of Malaya, Malaysia*
Professor Graham Hutchings, *Cardiff University, United Kingdom*
Dr Adriana Ispas, *Ilmenau University of Technology, Germany*
Dr Christopher Johnson, *Imperial College, United Kingdom*
Professor Frederic Kanoufi, *ESPCI ParisTech/CNRS UMR7195, France*
Mrs Aleksandra Karczmarczyk, *Polish Academy of Sciences, Poland*
Dr Ritu Kataky, *Durham University, United Kingdom*
Dr Nicola Kay, *NanoFlex Ltd, United Kingdom*
Professor Alexei Kornyshev, *Imperial College London, United Kingdom*

Mr Clemens Kubeil, *Ilmenau University of Technology, Germany*
Dr Hong Ngee Lim, *Universiti Putra Malaysia, Malaysia*
Professor Frank Marken, *University of Bath, United Kingdom*
Professor Andrew Mount, *The University of Edinburgh, United Kingdom*
Dr Wojciech Nogala, *Polish Academy of Sciences, United Kingdom*
Miss Nyasha Ntola, *Durham University, United Kingdom*
Dr Anthony O'Mullane, *RMIT University, Australia*
Dr Alan O'Riordan, *University College Cork, Ireland*
Dr Richard O'Rorke, *University of Leeds, United Kingdom*
Dr Dominika Ogończyk, *Polish Academy of Sciences, Poland*
Dr Andrew Osborne, *Metrohm UK, United Kingdom*
Mr Lee Partington, *Hull York Medical School, United Kingdom*
Miss Valentina Pifferi, *Università degli Studi di Milano, Italy*
Dr Krishnakumar Pillai, *ITT Dublin, Ireland*
Dr Blake Plowman, *Oxford University, United Kingdom*
Professor Martin Pumera, *Nanyang Technological University, Singapore*
Dr Neil Rees, *University of Birmingham, United Kingdom*
Miss Ilka Schmueser, *University of Edinburgh, United Kingdom*
Professor Patrik Schmuki, *University of Erlangen–Nuremberg, Germany*
Dr Monika Schoenleber, *University of Hull, United Kingdom*
Professor Wolfgang Schuhmann, *Ruhr-Universität Bochum, Germany*
Dr Baljit Singh, *ITT Dublin, Ireland*
Dr Michael Spencelayh, *Royal Society of Chemistry, United Kingdom*
Dr Yusran Sulaiman, *Universiti Putra Malaysia, Malaysia*
Ms Reshma Sultana, *NanoFlex Ltd, United Kingdom*
Mrs Ozge Surucu, *Hacettepe University, Turkey*
Dr Jonathan Terry, *University of Edinburgh, United Kingdom*
Mrs Rachel Thompson, *Royal Society of Chemistry, United Kingdom*
Dr Kristina Tschulik, *Oxford University, United Kingdom*
Dr Jay Wadhawan, *Hull University, United Kingdom*
Miss Amelie Wahl, *University College Cork, Ireland*
Dr Andy Wain, *Oxford University, United Kingdom*
Dr Alain Walcarius, *University of Lorraine, France*
Professor Anthony Walton, *University of Edinburgh, United Kingdom*
Professor Joseph Wang, *University of California, San Diego, USA*
Dr David Whitcombe, *NanoFlex Ltd, United Kingdom*
Professor David Williams, *University of Auckland, New Zealand*
Mr Kevin Wright, *University of Hull, United Kingdom*
Professor Guobao Xu, *Changchun Institute of Applied Chemistry, China*
Professor Dongping Zhan, *Xiamen University, China*

Index of contributors*

Abbas, Yawar, 93
Acevedo, Diego F., **147**
Albu, Sergiu P., **107**
Al-Hinai, Mariam N., **71**
Altomare, Marco, **107**
Amatore, Christian, **33**, 93, 219, 315, 411
Amiot, Fabien, **241**
Bae, Je Hyun, 219, **361**
Barbero, Cesar A., 93, **147**, 219, 315, 411
Barker, Robert, **391**
Barry, Seán, **283**, 377
Batchelor-McAuley, Christopher, 93, 219, 315, 411
Beebee, Charlotte, **391**
Bohn, Paul W., **57**, 93, 219, 315, 411
Bond, Alan, 93, 219, 315
Bund, Andreas, 219, 315, **339**, 411
Cao, Yongzhi, **189**
Chen, Aicheng, 93, **135**, 219
Chung, Taek Dong, 219, **361**, 411
Collignon, Manon Guille, **33**
Combellas, Catherine, **241**
Compton, Richard, 93, 219, 315, 411
Correa, Elon, **283**
Dale, Sara E. C., **249**
Dalgliesh, Robert M., **391**
Dawson, Karen, **377**, 283
Diab, Nizam, **19**
Dong, Shen, **189**
Etienne, Mathieu, **241**
Falciola, Luigi, 315
Fedala, Yasmina, **241**
Fielden, Peter, 411
Freeman, Neville J., **295**, 315
Fullarton, Claire, **391**
Glidle, Andrew, **391**
Goodacre, Royston, **283**
Grützke, Stefanie, **19**
Han, Ji-Hyung, **361**
Han, Donghyeop, **361**
Han, Lianhuan, **189**

Hassanien, Reda, **71**
Herzog, Grégoire, **241**
Hillman, A. Robert, 219, 315, **391**, 411
Hirani, Meetal, 93, 219
Horrocks, Benjamin R., **71**, 93
Horsfall, Alton B., **71**
Horwood, Corie, 219
Houlton, Andrew, **71**
Hu, Lianzhe, **175**
Hu, Zhenjiang, **189**
Hutchings, Graham, 315
Jia, Jingchun, **189**
Kanoufi, Frédéric, 93, 219, **241**, 315, 411
Kataky, Ritu, 93, 219, 315, 411
Kornyshev, Alexei A., 93, **117**, 219, 315, 411
Kubeil, Clemens, **339**, 411
Lafleur, Todd, **135**
Lee, Kiyoung, **107**
Lemaître, Frédéric, 33
Liu, Ning, **107**
M. Brown, Rachel, **391**
MacHale, John, **377**
Marken, Frank, **249**, 411
Mirabolghasemi, Hamed, **107**
Miras, Maria C., **147**
Moncada, Angelica Baena, **147**
Mount, Andrew R., 93, 219, **295**, 315, 411
Munteanu, Sorin, **241**
Najdovski, Ilija, **199**
Nebel, Michaela, **19**
Niu, Wenxin, **175**
O'Mullane, Anthony P., **199**, 219
O'Riordan, Alan, **283**, 315, 377, 411
Oleinick, Alexander, **33**
O'Rorke, Richard, 411
Partington, Lee, 411
Pearson, Andrew, **199**
Planes, Gabriel A., **147**
Plowman, Blake J., **199**, 411
Pumera, Martin, 93, **275**, 315

Quinn, Aidan J., **377**
Rodriguez, Rusbel Coneo, **147**
Roger, Jean Paul, **241**
Ryder, Karl S., **391**
Schmueser, Ilka, **295**
Schmuki, Patrik, **107**, 219
Schuhmann, Wolfgang, **19**, 93, 219, 315, 411
Schulte, Albert, **19**
Shah, Badal, **135**
Sibottier, Emilie, **241**
Skoda, Maximilian, **391**
Svir, Irina, 33
Terry, Jonathan G., **295**
Tessier, Gilles, **241**
Tian, Zhong-Qun, **189**
Tian, Zhao-Wu, **189**
Tighineanu, Alexei, **107**
Toh, Rou Jun, **275**
Tschulik, Kristina, 93, 219, 315, 411
Wahl, Amélie, 315, **377**, 411
Wain, Andy, 219, 411

Walcarius, Alain, 93, **241**, 315
Walton, Anthony J., **295**
Wang, Joseph, **9**, 93, 315
Watkins, Erik, **391**
Williams, David E., 93, 219, 315, **437**
Woodvine, Helena L., **295**
Wright, Nicholas G., **71**, 411
Xu, Guobao, **175**, 219, 315, 411
Yan, Yongda, **189**
Yuan, Yali, **175**
Yuan, Tao, **175**
Yuan, Ye, **189**
Zaino, Lawrence P., **57**
Zhan, Dongping, **189**
Zhang, Ling, **175**
Zhang, Jie, **189**
Zhao, Jianming, **175**
Zhao, Xuesen, **189**
Zhao, Jing, **57**
Zhu, Shuyun, **175**

*The page numbers in **bold** type indicate papers submitted for discussion.